ROCK MECHANICS AND THE DESIGN OF STRUCTURES IN ROCK

About the authors...

LEONARD OBERT received his B.A. and M.A. in physics from the University of California at Los Angeles and his Ph.D., also in physics, from Johns Hopkins University. He has received Department of the Interior Awards for Excellence and for Distinguished Service. Presently he is Science Advisor for Mining Research at the United States Bureau of Mines in Denver.

WILBUR I. DUVALL received his B.S. and his M.S. in physics from the University of Maryland. The Department of the Interior has honored him with its Distinguished Service Award and its Outstanding Achievement Award. He is presently Research Manager in Ground Control Science with the United States Bureau of Mines in Denver.

Both authors have been actively engaged in rock mechanics and mining research for the past twenty-eight years. They have written more than ninety articles on related subjects, and are responsible for the development of many instruments and techniques used in this field of engineering.

ROCK MECHANICS
AND THE DESIGN
OF STRUCTURES
IN ROCK

LEONARD OBERT

Science Advisor—Mining Research
U.S. Bureau of Mines, Denver, Colorado

WILBUR I. DUVALL

Supervisory Research Physicist
U.S. Bureau of Mines, Denver, Colorado

JOHN WILEY & SONS, INC.

NEW YORK · LONDON · SYDNEY

Library of Congress Catalog Card Number: 66-26753
Printed in the United States of America

PREFACE

Unlike many modern fields of engineering that have been preceded by or developed with their companion science, the creation of underground structures in rock for the extraction of minerals or for civil works has been practiced for centuries without the benefit of scientific guidance. Metals, nonmetals, and gems were mined in the pre-Christian area; as early as 400 B.C., an inclined shaft had been sunk to a depth of 368 feet. In the late Renaissance period, Georgius Agricola in his treatice on mining, *De Re Metalurgia* (1556), described rather mature methods of mining involving shafts, winzes, and levels, as well as man-, horse-, and water-powered hoists and pumps. Improvement in the state of the art continued. But it was not until the nineteenth century that engineering societies in Europe and North America were founded and engineers began to report their experiences in technical publications. At first these reports were mostly qualitative, based on visual observations, but with time quantitative information was included, such as measurement of surface subsidence and roof and floor convergence. Young and Stoek's comprehensive summary on subsidence,* which was published in 1916, lists over 100 papers prepared in the preceding 60 years that dealt with the mechanics of subsidence, mostly in relation to European coal fields.

In the first few decades of this century technical reports began to appear which treated rock as an engineering material that could be used for construction purposes, or excavations of mines or civil works. These technical papers included information on subjects such as the mechanical properties of rock, criterion of failure for rock, deep mining, and rock bursts. Laboratory studies of both photoelastic and rock models were reported as well as theoretical and empirical generalizations regarding the

* L. E. Young, and H. H. Stoek, *Subsidence Resulting from Mining*, *Bull. 91*, Engineering Experiment Station, University of Illinois, August, 1916.

v

state of stress around surface and underground excavations in rock. The larger part of the more fundamental investigations were made in the engineering laboratories of universities or other research institutions.

In the past three decades, the production of scientific information related to rock properties and the design and stability of structures in rock accelerated rapidly. In 1951 the First International Convention on Rock Pressure and Ground Support was held in Liege, Belgium, and since that date there have been frequent national and international symposia and conventions on similar subjects. The extent of interest in this subject can be appreciated by considering that in 1961 over 200 technical reports dealing with rock and structures in rock were published in various parts of the world. This contribution to our understanding of rock phenomena is the result of a collective effort on the part of mining, civil, petroleum, and geological engineers, and geologists. Out of this accumulation of knowledge there has evolved a new branch of engineering—rock mechanics.

The rapid growth of rock mechanics during the past two decades can be attributed to a number of causes: first, a part of this growth is probably a consequence of the general increase in scientific activity that has produced among other things, theory, instruments, and measuring procedures adaptable to rock mechanics investigations; second, in the field of mining as the more accessible ore bodies are depleted, there is an ever-increasing incentive to extract a higher percentage of mineral from a deposit and to mine at greater depths. Both of these factors require a more thorough understanding in the mechanics of structures in rock; third, in the field of civil works there has been a great increase in the number and magnitude of projects, such as dams, diversion tunnels, subsurface powerhouses, subways, and automobile and railroad tunnels; fourth, because it has been demonstrated that certain items such as liquid petroleum products, chemicals, foods, and medical supplies can be stored underground economically, there is an increasing interest in the design and structural stability of underground storage chambers; fifth, since the end of World War II a need for underground military installations (hardened bases) has developed; and sixth, overshadowing all of these causes there is always the prime requirement for improving the safety of all underground structures.

In preparing this book, we have attempted to bring together and present in one place the general fundamentals and pertinent information required for an understanding of rock mechanics and the design of structures in rock. More specifically this book contains (1) a brief, but what is hoped, sufficient mathematical treatment of stress, strain, elasticity, and inelastic effects to provide the reader with the necessary theoretical background for analyzing the stresses, strains, and deformations in structures; (2) a discussion of the methods and procedures for measuring the mechanical properties of rock, and a consideration of mechanisms of failure; (3) a description of the instruments and procedures for measuring stress, strain, deformation, and other related quantities, together with results from both laboratory and field investigations; and (4) a discussion of procedures based on both theoretical and empirical results for designing, analyzing, and evaluating the stability of underground structures.

This book is intended to be a general exposition and reference for the engineering profession, although the subject material is presented at a level such that it should be appropriate for college seniors or first-year graduate students who have completed prerequisite courses in calculus, mechanics, and strength of materials. In Part One the derivation of some of the simpler boundary problems has been included to illustrate mathematical processes. However, in Parts Two and Three mathematical detail has been kept to a minimum, but appropriate references are cited.

As indicated in the preceding paragraphs, rock mechanics is a relatively new branch of engineering. Although a large number of technical reports have been prepared on this subject, there still remain areas in which our knowledge is either deficient or completely lacking. Also, the literature contains some conflicting hypotheses and experimental results. We have tried to keep this controversial material to a minimum—where it has been included it has been qualified accordingly. Also, if alternate hypotheses or results exist, an attempt has been made to evaluate the merits and limitations of each.

At the end of each chapter a list of references has been included which either were specifically cited in the chapter or are of sufficient interest to be recommended for supplemental reading. However, it is not presumed or intended that this is a complete list of references to the related subject material. Also, since we both have been engaged in rock mechanics investigations for over 25 years, either as investigators or project supervisors, it is inevitable in writing on this subject to cite from personal experience, sometimes to the exclusion of equal or better source material. Where we have been guilty of this fault, it is not because we have considered our material superior—rather it is because we have followed a natural path of least effort.

We are deeply indebted to Professor Don Deere, University of Illinois, and other members of the faculty of the University of Illinois for a very thoughtful reading of the manuscript, and for many helpful comments and criticisms. Professor Howard Pincus, Department of Geology, Ohio State University, reviewed parts of the manuscript and offered many valuable suggestions, especially on geological matters for which we are appreciative. We are indebted to the Literary Executor of the late

Sir Ronald A. Fisher, F.R.S., Cambridge, to Dr. Frank Yates, F.R.S., Rothamsted, and to Messrs. Oliver & Boyd Ltd., Edinburgh, for permission to reprint tables from their book, *Statistical Tables for Biological, Agricultural, and Medical Research*. Finally, we wish to point out that this book is not an official statement or representation of the United States Bureau of Mines.

Leonard Obert

November 1966   Wilbur I. Duvall

CONTENTS

PARTIAL LIST OF SYMBOLS

A	area
ASG	apparent specific gravity
A_e	excavated area
A_p	cross-sectional area of pillar
A_t	total mined area
A_0	amplitude of electromagnetic wave
a	centrifugal acceleration
a	intercept of straight line through sample data
a, b	inner and outer radii of thick-wall cylinder
b	slope of straight line through sample data
b	width
C	stress optical constant
C_p	compressive strength of pillar rock
C_0	compressive strength (uniaxial)
c	cohesion
c	number of columns in a factorial design
c	propagation velocity of an elastic wave
c	propagation velocity of light in a vacuum
c	radius of plastic zone
c	$\frac{1}{2}$ crack length
D	deviation of a value from the average
D, d	diameter
d	displacement
E	voltage
E	Young's modulus
e	volumetric strain
F	force
F	line load

F	model fringe constant
F	ratio of two variances
F_b	rock bolt tension
F_c	compressive load at failure
F_s	safety factor
F_t	tensile load at failure
F_x, F_y, F_z	component of force
F_z	axial load on a specimen
f	distributed load
f	frequency
f	number of readings in a given range
f	stress concentration
f_b	fundamental longitudinal frequency
f_s	fundamental torsional frequency
G	modulus of rigidity
G	strain gage factor
g.	acceleration of gravity
H_o	height of opening
H_0	null hypothesis
h	depth below surface
h	height
I	intensity of a light beam
I	moment of inertia
I_1, I_2, I_3	invariants of stress, invariants of strain
i	angle of incidence
K	bulk modulus, modulus of volume expansion
K	maximum compressive stress concentration
L	confidence limit
L	gage length
L	length
M	mass
M	moment of force
M_c	applied torque at failure
m	magnification
m	ratio of applied normal stresses
N	fringe order
N	number of items in a population
n	index of refraction
n	normal
n	number of items in a sample
n	rotational speed
P	probability

P, Q	maximum and minimum principal stresses
P', Q'	secondary principal stresses in a plane
P, Q, R	principal stresses
p	applied pressure
p, q, r	parameters of a mapping function
p_c	fracture pressure
p_i	internal pressure
p_o	external pressure
p_0	pore pressure
p_s	shut-in pressure
p_z	axial pressure
Q	total sum of the deviations squared
Q_i	input quantity
Q_o	output quantity
q	distributed load
R	angle of refraction
R	extraction ratio
R	radius of curvature
R	resistance
RF	reinforcement factor
R_a	areal extraction ratio
R_r	radial body force intensity
R_0	flexural strength (outer fiber tensile strength)
R_1	reaction force
r	angle of reflection
r	number of rows
r, θ	polar coordinates
r, θ, z	cylindrical coordinates
r, θ, Φ	spherical coordinates
S	standard deviation of the sample
S^2	sample variance
S_d	deformation sensitivity
$S_E{}^2$	variance for data about straight line through data
S_h	applied horizontal stress
$S_m{}^2$	variance of the mean
S_n	resultant stress on a plane
S_{nx}, S_{ny}, S_{nz}	resultant stresses on a plane
$\bar{S}_p$	average pillar stress
S_p	pillar stress
$S_p{}^2$	pooled sample variance
S_s	strain sensitivity
$S_t{}^2$	total variance

S_v	applied vertical stress
S_x, S_y, S_z	applied normal stresses
S_0	shear strength
T	surface tension
T_{xy}, S_{xy}	applied shear stress
T_0	tensile strength
t	deviation of sample or population means in terms of sample standard deviations
t	thickness
t	time
Δt	travel time
U	borehole deformation (diametral)
u	a coded value of X
$\bar{u}$	mean of the coded data
$\bar{u}, \bar{v}, \bar{w}$	applied components of displacement at boundary
u, v, w	displacement
V	shear force
V	volume
V_c	exterior volume of a specimen
v	velocity of light
v_b	longitudinal bar velocity
v_p	longitudinal wave velocity
v_s	shear wave velocity
W	energy
W_b	rock bolt load
W_p	width of pillar
W_r	dry weight of specimen
W_s	saturated weight of specimen
W_v	saturated weight of specimen immersed in water
W_0	strain energy per unit volume
W_o	width of opening
w	width
X	value for one item in a sample
$\bar{X}$	sample mean
$\bar{\bar{X}}$	grand average
$\bar{X}, \bar{Y}, \bar{Z}$	body force intensities
$\bar{X}_n, \bar{Y}_n, \bar{Z}_n$	applied components of stress on a surface
x, y, z	Cartesian (rectangular) coordinates
Y	amplitude for standard normal distribution
Z	deviation measured in terms of standard deviation
α	angle of rupture
α	intercept for a straight line through population data

α	size of the type 1 error
β	slope of a straight line through population data
γ	shear strain
γ	unit weight
γ_w	unit weight of water
$\gamma_{xy},\ \gamma_{yz},\ \gamma_{zx}$	components of shear strain
Δ	phase difference
δ	phase constant
ϵ	strain
$\dot{\epsilon}$	strain rate
$\epsilon_p,\ \epsilon_q$	principal strains in a plane
$\epsilon_x,\ \epsilon_y,\ \epsilon_z$	component of strain
η	coefficient of viscosity
η	deflection of a beam
Θ	sum of the principal stresses
λ	Lame's constant
λ	wave length
μ	coefficient of internal friction
μ	mean of the population
$\mu\epsilon$	micro inch per inch
ν	Poisson's ratio
ν	degrees of freedom
ρ	density
ρ	radius of curvature
σ	standard deviation of the population
$\sigma_{nn},\ \sigma$	normal stress
$\sigma_x,\ \sigma_y,\ \sigma_z$	normal stress components
σ^2	population variance
$\sigma_1,\ \sigma_2,\ \sigma_3$	principal stresses
τ	shear stress
τ_{nt}	shear stress
$\tau_{xy},\ \tau_{yz},\ \tau_{zx}$	shear stress components
Φ	Airy stress function
ϕ	angle of internal friction
ω	angular frequency $= 2\pi f$

PART ONE

THEORY OF ELASTIC
AND INELASTIC BODIES

INTRODUCTION

The design and stability of underground structures in rock is the subject of this book. By an underground structure we are referring to any excavated or natural subsurface opening, or system of openings that is virtually self-supporting. A self-supporting system is one in which the structural stresses are carried on the walls, pillars, and other unexcavated parts of the opening(s) rather than on linings, packs, chocks, etc., placed within the openings. The Bureau of Mines' oil shale mine at Rifle, Colo., Fig. 1, is an example of this type of opening. This specification does not preclude the use of artificial support within the structure, but it does presuppose that these artificial devices offer only local support which will not significantly affect the stresses in the self-supported structure. Thus, this includes most mines, tunnels, shafts, and underground openings constructed for military purposes, storage of food, chemicals, liquids, liquid petroleum gases, or for other civil or industrial purposes. Rock mechanics is that branch of engineering concerned with the mechanical properties of rock and the application of this knowledge to engineering problems dealing with rock. Rock is considered to be the mineral or organic (for example, coal) matter that comprises the solid part of the earth's crust excluding soil which Terzaghi* has defined as "—sediments and other unconsolidated accumulations of solid particles produced by the mechanical or chemical disintegration of rocks—." Thus, the distinction between rock and soil is in the degree of consolidation and in the limit on the size of the particles.

When designing an underground structure in rock or evaluating the stability of an existing structure, we must determine (1) the stresses and/or deformation in the structure resulting from external or body loads, and (2) the ability of the structure to withstand these stresses or deformations. A limit on this ability is generally evaluated in terms of the stress necessary

* Terzaghi, K., *Theoretical Soil Mechanics*, John Wiley and Sons, 1943, p. 1.

to cause a structural failure, which is manifested usually by a sudden collapse of one or more structural components, although sometimes excessive deformation may also be a limiting factor. Designing a structure in rock is in a number of ways a more difficult problem than designing or evaluating the stability of surface structures made of steel, concrete, or

Fig. 1. U.S. Bureau of Mines oil shale mine. Rifle, Colorado.

other conventional construction materials. One difficulty becomes evident when it is realized that subsurface rock is under a preexisting and unknown state of stress due to both the weight of the overlying cover and possible tectonic forces. Although the subsurface state of stress may be estimated by assuming that it is due only to the weight of the superincumbent rock, stress measurements made in existing structures indicate that estimates made on this basis may be in substantial error.

The absence of specific information on the mechanical properties of in situ rock before underground access is possible creates another problem. The mechanical properties of most conventional structural materials can be obtained or produced to a given specification and a structure can be designed to utilize these materials; however, the design of underground structures is usually limited because of a lack of information about the properties and behavior of the in situ rock. Although some knowledge may be obtained from a geological study of the area or by examining and testing exploratory drill cores, the rock and the geology in an area are generally so variable that drill cores do not provide an adequate sample.

A third problem arises in relation to calculating the stresses and deformations in the various parts of a rock structure. To a first approximation, an underground structure is an opening or system of openings in a semi-infinite body of rock. If the body of rock is isotropic and homogeneous, it can be considered as a continuous medium. If, in addition, the rock is linear-elastic and the openings are of comparatively simple geometrical shape, then the stresses and deformations can be determined by mathematical methods. However, as the configuration of openings becomes more complex, or as the rock becomes more variable, intersected by dikes, faults, or systems of joints, the complete structure will become too difficult to treat mathematically and we must resort to model studies or, better still, to in situ measurement if underground access is possible.

In designing a rock structure its intended use is another factor that should be considered. For example, civil engineers dealing with the construction of an underground structure, such as a liquid-petroleum gas storage-chamber or a hardened (underground) military base, must keep in mind that these structures should be designed for a long lifetime, the safety factor should be high, and that the design should be such that the excavation can be mined in a minimum time. Also, the opportunity for in situ measurement and experimentation is either limited or not possible at all, and any modification in the design after the start of excavation is usually very costly. Considering these factors collectively, and taking into account that the cost of the principal structural material, rock, is generally inexpensive, the tendency to overdesign this type of structure is almost universal.

The design objectives in underground mining are almost the opposite of those for civil engineering projects. Generally the economics of mining are such that it is necessary to extract as large a percentage of the deposit as possible; hence the size and number of pillars, barriers, and other structural supports left within the economic limits of the ore body are kept to a minimum. Although there are a variety of methods for recovering the ore from a deposit, the common practice is to abandon local areas (stopes) as they are mined out. After abandonment the stopes may partially or completely fail after a period ranging from days to years. Thus, stopes (and sometimes complete mines) have an operating lifetime, and this lifetime is usually short compared with the intended lifetime of conventional structures. The concept of an operating lifetime together with the economic advantage of operating with a minimum margin of safety creates design problems that are unique to mining. Fortunately, the development of a mine is so slow that there is usually an opportunity for experimentation, and design modifications are effected as the results of experiment and experience warrant.

Because of these problems, the engineer cannot apply the same analytic procedures, or achieve the same preciseness in designing rock structures as is realized in the design of conventional structures. In fact, it must be accepted from the beginning that a different approach to the problem of designing rock structures should be developed—one in which the engineer realizes that only an approximate design can be made before underground access is possible, and that this design must be modified as information from underground investigation becomes available.

In the initial design, the accumulation of knowledge that comprises rock mechanics should be considered for its full value, but, in addition, practical experience based on personal or reported visual observation should be excogitated, especially if the observations were made on an equivalent structure in a similar rock type and at a comparable depth. As data pertaining to the properties of the rock and the nature of the stress field become available, the assumption used in the original design can be re-evaluated in the light of this new information and the design modified accordingly. There is no alternative approach to the design of structures in rock, although design has been expedited in a number of instances preliminary to the start of an underground project by mining experimental rooms, adits, or shafts so that geological data and a better sampling of the rock for mechanical property testing can be obtained. This expedient also makes it possible to measure the preexisting state of stress in the rock, using the methods described in Section 14.3. Obviously the problem of evaluating the stability of existing rock structures is not similarly handicapped because in this case underground access is possible; hence measurement can be made as dictated by the nature of the problem.

Rock mechanics like all engineering disciplines, should have a theoretical basis, and in this book the theory of elasticity will be used as that basis. The dependence of engineers on elastic theory and on elasticity as the most important property of materials is expressed very clearly by Freudenthal*:

The engineer usually considers elasticity of his materials to be the most important property on which to base his design for service. He realizes, however, that the actual performance of the structure might be very different from the assumed performance of the designed structure, and that both the differences and similarities in the behavior of the real and designed structures, as well as the safety of the real structure, are essentially results of the deviation of the behavior of real materials from the linear elasticity assumed in design. Hence, the interest of the engineer necessarily centers on the deviation from conditions instinctively considered 'normal,' that is, elastic, rather than on generally defined conditions of rheological behavior. This deviation from elastic conditions may be considerable, but its importance is still mainly in its relation to an elastic component.

Part One considers the analysis of stress and strain (Chapters 1 and 2), the theory of elasticity (Chapter 3), procedures for solving problems in elastic theory (Chapter 4), and the solutions to a number of problems related to rock mechanics (Chapter 5). Hence, as a starting point in considering design and stability problems it will be assumed that rock is an isotropic, homogeneous, linear-elastic material, and that the deformational response of a rock body to an applied force can be determined from elastic theory, providing the problem is not intractable. This is known to be a reasonable assumption for many rock types, providing that the pieces of rock are small, such as mechanical property specimens or small rock models, and the magnitude of the stress does not exceed that normally encountered in underground operations. However, in situ rock contains fractures, joints, bedding planes, partings which may be filled with clay or organic material, faults, or other defects of geological origin, and the departure from linear elasticity may be considerable. For rocks in this class, elastic theory has been modified as treated in Chapter 6 to accommodate materials with specified inelastic properties, although the result is not a theory of inelasticity but simply a modified elastic theory.

* Freudenthal, A. M., *The Inelastic Behavior of Engineering Materials and Structures*, John Wiley and Sons, 1950, p. 22.

CHAPTER 1

ANALYSIS OF STRESS

1.1 INTRODUCTION

Mathematical theories of elasticity, viscosity or plasticity all follow a similar pattern in that concepts of force, stress, deformation, and strain are developed first. Next, relationships between stress and strain are assumed which idealize the behavior of actual materials. Finally, equations of equilibrium or motion are established which enable the state of stress and strain to be calculated for a body subjected to a prescribed set of boundary conditions and applied forces.

Analysis of stress is a branch of statics which deals with the description of the state of stress throughout a body. In this chapter, the concepts of body forces, surface forces, stresses, normal stress, and shear stress are developed by means of physical definitions, and mathematical relations between these quantities are formulated. The principles to be developed can be applied to any body composed of a continuous distribution of matter, such as steel, rock, wood, or plastic. Later, restrictions as to the nature of the materials will be made. However, these restrictions will not be imposed until required for further mathematical development, thereby permitting the most general use of these concepts.

1.2 DEFINITIONS AND BASIC CONCEPTS

There are two types of forces which can act upon a body composed of continuous matter—body forces and surface forces. Body forces act throughout the body and are produced without physical contact with other bodies. Examples are gravitational, magnetic, and inertial forces. Surface forces act on external surfaces of a body and result from physical contact with other bodies.

The intensity of a body force is defined as the force per unit volume. Let the cartesian components of the body force intensities be denoted by X, Y, and Z respectively. If a body element of volume ΔV has the resultant forces ΔF_x, ΔF_y, and ΔF_z acting on it, the components of the body force intensities at a point within ΔV are defined as:

$$X = \lim_{\Delta V \to 0} \frac{\Delta F_x}{\Delta V}$$

$$Y = \lim_{\Delta V \to 0} \frac{\Delta F_y}{\Delta V} \qquad (1.2.1)$$

$$Z = \lim_{\Delta V \to 0} \frac{\Delta F_z}{\Delta V}$$

In discussing surface forces, a distinction between exterior and interior surface must be made. An exterior surface is any surface comprising a part of the boundary of a body, including the surface of any interior holes in the body. An interior surface is created only by an imaginary cut through the body.

The term stress means force per unit area. More specifically the stress S_n, at a point P in a plane whose normal is n, is given by the expression

$$S_n = \lim_{\Delta A \to 0} \frac{\Delta F}{\Delta A} \qquad (1.2.2)$$

where ΔF denotes the resultant force acting on the area ΔA surrounding the point P. In general the direction of ΔF or S_n will not coincide with the direction of the normal n to the area ΔA (Fig. 1.2.1).

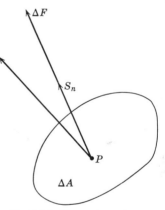

Fig. 1.2.1. Stress at a point.

The resultant force ΔF acting on an area ΔA can be resolved into cartesian components ΔF_x, ΔF_y, and ΔF_z in the x, y, and z directions respectively. The cartesian components of stress acting on the element of area ΔA are by definition

$$S_{nx} = \lim_{\Delta A \to 0} \frac{\Delta F_x}{\Delta A}$$

$$S_{ny} = \lim_{\Delta A \to 0} \frac{\Delta F_y}{\Delta A} \qquad (1.2.3)$$

$$S_{nz} = \lim_{\Delta A \to 0} \frac{\Delta F_z}{\Delta A}$$

When two subscripts are used on a symbol representing a stress, the first refers to the direction of the normal to the area, and the second to the direction of the stress. Thus S_{nx} is the stress in the x direction acting on the element of area whose normal is in the n direction. It should be noted that stress is a tensor quantity, whereas force is a vector quantity. A force is defined if both its magnitude and direction are given. To define a stress,

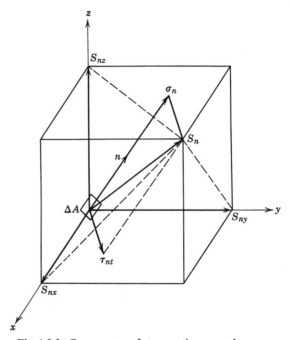

Fig. 1.2.2. Components of stress acting on a plane.

the plane on which the stress acts must be given as well as the magnitude and direction of the stress. The resultant of the three stresses, S_{nx}, S_{ny}, and S_{nz} is S_n. The relationships between these stresses and their direction cosines as determined from Fig. 1.2.2 are

$$S_n^2 = S_{nx}^2 + S_{ny}^2 + S_{nz}^2$$
$$S_{nx} = S_n \cos (S_n, x) \qquad (1.2.4)$$
$$S_{ny} = S_n \cos (S_n, y)$$
$$S_{nz} = S_n \cos (S_n, z)$$

where $\cos (S_n, x)$ is the cosine of the angle between the direction of S_n and the x axis. Similar definitions apply for $\cos (S_n, y)$ and $\cos (S_n, z)$.

The resultant force ΔF can also be resolved into two components—ΔF_n normal to the area ΔA and ΔF_t parallel to the area ΔA. The stresses acting on the area ΔA are by definition

$$\sigma_{nn} = \lim_{\Delta A \to 0} \frac{\Delta F_n}{\Delta A}$$

$$\tau_{nt} = \lim_{\Delta A \to 0} \frac{\Delta F_t}{\Delta A}$$

(1.2.5)

where t is in the direction of the intersection of the plane formed by σ_{nn} and S_n and the plane tangent to the area ΔA at the point P.

A stress perpendicular to an area is called a normal stress and is denoted by σ. A stress parallel to an area is called a shear stress and is denoted by τ. The resultant of σ_{nn} and τ_{nt} is also S_n; therefore these stresses are related by the following:

$$S_n{}^2 = \sigma_{nn}{}^2 + \tau_{nt}{}^2$$

$$\sigma_{nn} = S_n \cos(S_n, n)$$

$$\tau_{nt} = S_n \cos(S_n, t)$$

(1.2.6)

The relationships given by Eqs. 1.2.4 and 1.2.6 are shown geometrically in Fig. 1.2.2. It is important to note that these relationships apply only to stresses acting on the same plane. In order to determine the stresses acting on other planes through the point P, additional considerations are required which will be discussed later.

Suppose a point P in a body is surrounded by a small rectangular parallelepiped whose faces are parallel to a set of cartesian coordinates x, y, and z. The convention for designation of normal and shear stresses is given in Fig. 1.2.3. The normal stress acting on the plane normal to the x axis is designated σ_x.* The shear stress acting in the y direction and on a plane normal to the x axis is designated τ_{xy}. The shear stress acting in the z direction and on the plane normal to the x axis is designated τ_{xz}. Similar definitions exist for σ_y, σ_z, τ_{yx}, τ_{yz}, τ_{zx}, and τ_{zy}. All these stresses are assumed positive if directed as given in Fig. 1.2.3.

The normal stresses σ_x, σ_y, and σ_z, are assumed positive when directed outward from the surface, and represent tensile stresses. Negative normal stresses are compressive stresses and are represented by arrows into a surface element.

The shear stresses, $\tau_{xy}, \tau_{yx}, \tau_{yz}, \tau_{zy}, \tau_{zx}$, and τ_{xz}, are positive when directed in a positive cartesian direction while acting on a plane whose outward normal points in a positive direction, or, when directed in a negative

---

* The second subscript x has been omitted as the symbol σ is used only for normal stress; thus σ_x means σ_{xx}.

cartesian direction while acting on a plane whose outward normal points in a negative cartesian direction. The shear stresses are negative if their direction is the reverse of that given above.

It can be shown that of the six cartesian components of shear stresses only three are independent. Consider a rectangular parallepiped (Fig. 1.2.3), as an element of a body about a point P. As this element is reduced

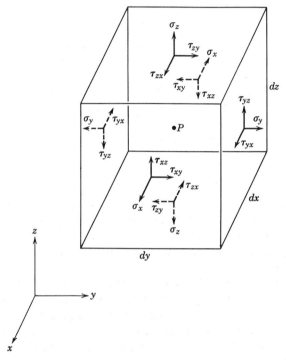

Fig. 1.2.3. Components of stress for rectangular coordinates.

in size, the variation in the nine components of stress along the faces and between parallel faces will become so small that the stresses on opposite faces can be assumed constant and equal. Also as the size of the element is reduced, body forces will approach zero faster than interior surface forces, therefore body forces can be neglected in comparison to surface forces. Let the point P be at the center of the element and consider the moments of the forces about an axis through the point P and in the x direction. Since the body is assumed in equilibrium, the sum of the moments is given by

$$\sum M_x = \frac{\tau_{yz}\, dx\, dz\, dy}{2} + \frac{\tau_{yz}\, dx\, dz\, dy}{2} - \frac{\tau_{zy}\, dx\, dy\, dz}{2} - \frac{\tau_{zy}\, dx\, dy\, dz}{2} = 0$$

Similar equations for $\Sigma\,M_y$ and $\Sigma\,M_z$ can be obtained from a consideration of axes of rotation in the y and z directions. Division of each equation by $dx\,dy\,dz$ shows that the following relations exist between the six shear components

$$\tau_{xy} = \tau_{yx}$$
$$\tau_{yz} = \tau_{zy} \qquad (1.2.7)$$
$$\tau_{zx} = \tau_{xz}$$

Equations 1.2.7 show that pairs of shear stresses are equal in magnitude and sign. However, they are opposite in angular direction as can be seen from Fig. 1.2.3.

1.3 STRESS IN A PLANE

Many problems in the theory of elasticity can be solved by considerations of stresses in only two dimensions. As mathematical difficulties are reduced in going from three to two dimensions, the developments in the

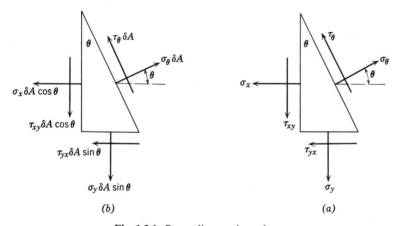

(b) (a)

Fig. 1.3.1. Stress diagram in a plane.

next two sections will be restricted to two dimensions. Later on, the concepts and procedures developed for two-dimensional analysis will be extended to three dimensions.

Figure 1.3.1a is a stress diagram for an element of a body in the xy plane. Assume that the stresses σ_x, σ_y, and τ_{xy} are known and that the stresses σ_θ and τ_θ acting on the plane whose normal makes an angle θ with respect to the x axis are desired. Let δA be the area of the plane whose normal makes an angle θ with respect to the x axis, and construct the force diagram shown in Fig. 1.3.1b. As equilibrium of the element is assumed, summation of the forces in the σ_θ direction gives

$$\sum F_{\sigma_\theta} = \sigma_\theta \delta A - \sigma_x \delta A \cos^2 \theta - \sigma_y \delta A \sin^2 \theta - 2\tau_{xy} \delta A \cos \theta \sin \theta = 0$$

Summation of forces in the τ_θ direction gives

$$\sum F_{\tau_\theta} = \tau_\theta \delta A + \tau_{xy} \delta A \sin^2 \theta - \tau_{xy} \delta A \cos^2 \theta$$
$$+ \sigma_x \delta A \cos \theta \sin \theta - \sigma_y \delta A \sin \theta \cos \theta = 0$$

Division of these equations by δA, rearrangement of terms and simplification gives

$$\sigma_\theta = \sigma_x \cos^2 \theta + \sigma_y \sin^2 \theta + 2\tau_{xy} \sin \theta \cos \theta$$
$$\tau_\theta = \tau_{xy} (\cos^2 \theta - \sin^2 \theta) - (\sigma_x - \sigma_y) \sin \theta \cos \theta \qquad (1.3.1)$$

By means of trigonometric identities, Eqs. 1.3.1 can be rewritten as

$$\sigma_\theta = \tfrac{1}{2}(\sigma_x + \sigma_y) + \tfrac{1}{2}(\sigma_x - \sigma_y) \cos 2\theta + \tau_{xy} \sin 2\theta \qquad (1.3.2)$$

$$\tau_\theta = \tau_{xy} \cos 2\theta - \tfrac{1}{2}(\sigma_x - \sigma_y) \sin 2\theta \qquad (1.3.3)$$

Equations 1.3.2 and 1.3.3 give the magnitude and sign of the stresses σ_θ and τ_θ acting over the inclined section in terms of the given stresses σ_x, σ_y, and τ_{xy} acting on sections normal to the x and y axes. These stresses are periodic in π because of the terms $\sin 2\theta$ and $\cos 2\theta$. Thus they must have maximum and minimum values or be constant. The derivative of the normal stress σ_θ with respect to θ equated to zero gives

$$\frac{d\sigma_\theta}{d\theta} = -(\sigma_x - \sigma_y) \sin 2\theta_1 + 2\tau_{xy} \cos 2\theta_1 = 0$$

where θ_1 has been substituted for θ to denote a specific angle. Solving for θ_1 gives

$$\theta_1 = \tfrac{1}{2} \tan^{-1} \frac{2\tau_{xy}}{\sigma_x - \sigma_y} \qquad (1.3.4)$$

In Eq. 1.3.4, two values of θ_1 are possible, θ_1 and $\theta_1 + 90°$. One angle gives the direction of the maximum normal stress and the other the direction of the minimum normal stress.

Substitution of Eq. 1.3.4 into 1.3.2 and simplification gives

$$P = \sigma_{\max} = \tfrac{1}{2}(\sigma_x + \sigma_y)^2 + \tfrac{1}{2}\sqrt{(\sigma_x - \sigma_y) + 4\tau_{xy}^2} \qquad (1.3.5)$$

and

$$Q = \sigma_{\min} = \tfrac{1}{2}(\sigma_x + \sigma_y) - \tfrac{1}{2}\sqrt{(\sigma_x - \sigma_y)^2 + 4\tau_{xy}^2} \qquad (1.3.6)$$

The maximum and minimum normal stresses are called principal stresses and are denoted by P and Q. The sections over which the principal stresses act are called principal planes. Substitution of Eq. 1.3.4 into 1.3.3

shows that the shear stress is identically zero. Thus no shear stresses act on planes where the normal stresses are maximum or minimum.

The derivative of the shear stress τ_θ with respect to θ, in Eq. 1.3.3 equated to zero gives

$$\frac{d\tau_\theta}{d\theta} = -2\tau_{xy} \sin 2\theta_2 - (\sigma_x - \sigma_y) \cos 2\theta_2 = 0$$

Solving for θ_2 gives

$$\theta_2 = \tfrac{1}{2} \tan^{-1} \frac{-(\sigma_x - \sigma_y)}{2\tau_{xy}} \tag{1.3.7}$$

Again, two values of θ_2 are possible, θ_2 and $\theta_2 + 90°$. Substitution of Eq. 1.3.7 into 1.3.3 gives

$$\tau_{\substack{\max \\ \min}} = \pm\tfrac{1}{2}\sqrt{(\sigma_x - \sigma_y)^2 + 4\tau_{xy}^2} \tag{1.3.8}$$

Equation 1.3.8 appears, at first glance, to contradict the results given in Eqs. 1.2.7. However, inspection of the angular sense of $\tau_{\min}$ shows that its direction must be reversed to conform to the conventional notation given in Fig. 1.2.3. This reversal of angular direction accounts for the negative sign of $\tau_{\min}$. Also, this change of sign occurs if θ is replaced by $\theta + 90°$ in Eq. 1.3.3. Thus $\tau_\theta = (-)\tau_{\theta+90°}$ and the sense of $\tau_{\theta+90°}$ must be reversed to conform to the standard notation.

Comparison of Eq. 1.3.7 with 1.3.4 shows that the angles θ_1 and θ_2 differ by 45°. The maximum normal stresses and maximum shear stresses occur on sections at 45° to each other.

Substitution of Eq. 1.3.7 into 1.3.2 gives

$$\sigma_\theta = \sigma_{\theta+90°} = \frac{\sigma_x + \sigma_y}{2} \tag{1.3.9}$$

Equation 1.3.9 shows that the normal stresses are equal on sections where the shear stresses are maximum. Thus planes of maximum shear stress are not in general free of normal stress.

A state of pure shear is said to exist if planes of maximum shear stress are free of any normal stress. Pure shear can be obtained by making $\sigma_x = -\sigma_y$. That is, any rectangular element subjected to a tensile stress in one direction and an equal compressive stress in the other direction is in a state of pure shear.

Subtracting Eq. 1.3.6 from 1.3.5, and comparing the result with Eq. 1.3.8 shows that

$$\tau_{\max} = \frac{P - Q}{2} \tag{1.3.10}$$

Thus the maximum shear stress is one-half the difference of the two principal stresses and occurs on sections 45° from the principal planes.

There may be occasions when the principal stresses are known and the normal and shear stresses on a section at an angle θ with respect to the principal plane are required. The equations for calculating these stresses can easily be derived as in the preceding. The results are

$$\sigma_\theta = \frac{P+Q}{2} + \frac{P-Q}{2}\cos 2\theta, \tag{1.3.11}$$

and

$$\tau_\theta = -\frac{P-Q}{2}\sin 2\theta \tag{1.3.12}$$

1.4 MOHR'S CIRCLE OF STRESS

A geometrical solution for stresses in any required direction is provided by Mohr's circle. Consider the stress diagram in Fig. 1.4.1a. The steps in the construction of Mohr's circle are as follows:

1. Construct a set of orthogonal axes and label the vertical axis τ and the horizontal axis σ as shown in Fig. 1.4.1b. It is necessary that the scale for these two axes be equal.

2. Plot the normal stresses σ_x and σ_y given in Fig. 1.4.1a on the normal stress axis.

3. Plot the shear stress τ_{xy} acting on the right-hand edge of the element in Fig. 1.4.1a directly below or above the point representing σ_x on the normal axis. If the shear stress is counter-clockwise relative to the center of the element, plot τ_{xy} below the normal stress axis and, if the shear stress is clockwise relative to the center of the element, plot τ_{xy} above the normal stress axis.

4. Plot the shear stress τ_{yx}, acting on the section common with σ_y, either above or below the point σ_y but on the opposite side of the normal stress axis as used in step 3.

5. Join the two shear points with a straight line. This line will intercept the normal stress axis at the point $\frac{1}{2}(\sigma_x + \sigma_y)$.

6. Draw a circle with center on the normal axis at the point $\frac{1}{2}(\sigma_x + \sigma_y)$ and diameter equal to the length of the line joining the two shear points.

From Fig. 1.4.1b, it is seen that the projection of the radius of the circle on the shear axis gives the shear stress at any angle and that the projections of the ends of the diameter on the normal stress axis give the normal stresses at any angle. The radius of the circle is the maximum shear stress and is

$$r = \tau_{\max} = \frac{1}{2}\sqrt{(\sigma_x - \sigma_y)^2 + 4\tau_{xy}^2}$$

The intersections of the circle with the normal stress axis are the principal stresses. The tangent of the angle 2θ is $2\tau_{xy}/(\sigma_x - \sigma_y)$, and 2θ is

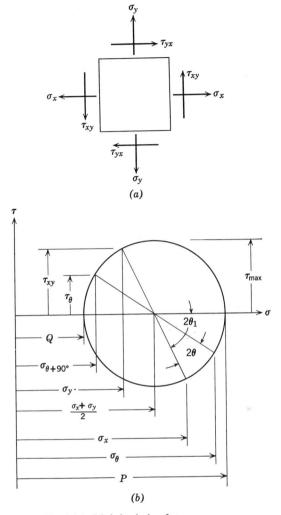

Fig. 1.4.1. Mohr's circle of stress.

twice the angle between the x axis and the direction of the principal stress given in Eq. 1.3.4. The direction of rotation of the radius from its original constructed position to where the circle intersects the normal stress axis is in the same angular sense as the direction of rotation of the element for the normal stress to become the principal stress. Notice also that the shear stress is zero when the normal stresses are maximum and minimum. Also, when the shear stress is maximum, the normal stresses are equal to one-half the sum of the original normal stresses. As the center of the circle is always at the point

$$\frac{\sigma_x + \sigma_y}{2} = \frac{\sigma_\theta + \sigma_{\theta+90°}}{2}$$

the sum of the normal stresses is an invariant with the angular rotation of the element.

1.5   STRESS IN THREE DIMENSIONS

The state of stress in a body at a point P is fully determined if the six cartesian components of stress σ_x, σ_y, σ_z, τ_{xy}, τ_{yz}, and τ_{zx} are given. From

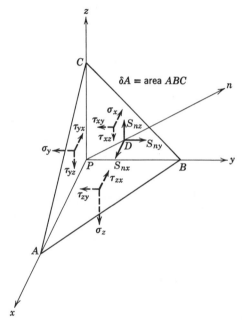

Fig. 1.5.1. Stress diagram for three dimensions.

these components of stress it is possible to calculate the normal and shear stresses on any other plane through the point P.

Consider a small tetrahedronal element, with three of its sides parallel to the three coordinate planes, cut out of a body at the point P. Let the six cartesian components of stress acting on the element before it was removed be given as shown in Fig. 1.5.1. The area of triangle ABC is δA, and the line PD extended outward is the normal n to the plane ABC. As the tetrahedron is reduced in size, body forces become small compared to surface forces and variations of surface stresses over the areas of the element become small compared to their magnitude, thus body forces and variation of surface stresses can be neglected.

The volume of a tetrahedron is

$$V = \tfrac{1}{3}(\text{altitude})(\text{base area})$$

Thus

$$\tfrac{1}{3}(\overline{PD})(\delta A) = \tfrac{1}{3}(\overline{AP})(\text{area } \Delta PCB)$$

But

$$\overline{PD} = \overline{AP} \cos(n, x)$$

Therefore

$$\text{area } \Delta PCB = \delta A \cos(n, x)$$

By similar treatments for ΔACP and ΔPAB, it can be shown that

$$\text{area } \Delta ACP = \delta A \cos(n, y)$$
$$\text{area } \Delta PAB = \delta A \cos(n, z)$$

As the tetrahedron is in equilibrium, summation of the forces in the x direction gives

$$S_{nx}\,\delta A = \sigma_x\,\delta A \cos(n, x) + \tau_{yx}\,\delta A \cos(n, y) + \tau_{zx}\,\delta A \cos(n, z)$$

Summation of the forces in the y and z directions will give two similar equations. Division of these equations by δA gives

$$S_{nx} = \sigma_x \cos(n, x) + \tau_{yx} \cos(n, y) + \tau_{zx} \cos(n, z)$$
$$S_{ny} = \sigma_y \cos(n, y) + \tau_{zy} \cos(n, z) + \tau_{xy} \cos(n, x) \qquad (1.5.1)$$
$$S_{nz} = \sigma_z \cos(n, z) + \tau_{xz} \cos(n, x) + \tau_{yz} \cos(n, y)$$

Equations 1.5.1 give only the cartesian components of stress on the inclined plane. To obtain the normal and shear stresses for the inclined plane, consider the stress diagram shown in Fig. 1.5.2. Let a new set of axes be x', y', and z' with x' in the direction of the outward drawn normal n. The normal and shear components of stress $\sigma_{x'}$ and $\tau_{x'y'}$ can be determined from the cartesian components of stress S_{nx}, S_{ny}, and S_{nz} by the cosine projection. By considering two more tetrahedrons with inclined planes having outward drawn normals parallel to the y' and z' axes, and following the above procedure, the following set of equations can be derived:

$$\sigma_{x'} = S_{nx} \cos(x', x) + S_{ny} \cos(x', y) + S_{nz} \cos(x', z)$$
$$\tau_{x'y'} = S_{nx} \cos(y', x) + S_{ny} \cos(y', y) + S_{nz} \cos(y', z)$$
$$\sigma_{y'} = S_{nx} \cos(y', x) + S_{ny} \cos(y', y) + S_{nz} \cos(y', z) \qquad (1.5.2)$$
$$\tau_{y'z'} = S_{nx} \cos(z', x) + S_{ny} \cos(z', y) + S_{nz} \cos(z', z)$$
$$\sigma_{z'} = S_{nx} \cos(z', x) + S_{ny} \cos(z', y) + S_{nz} \cos(z', z)$$
$$\tau_{z'x'} = S_{nx} \cos(x', x) + S_{ny} \cos(x', y) + S_{nz} \cos(x', z)$$

When we combine Eqs. 1.5.1 and 1.5.2, (noting that n in Eqs. 1.5.1 can be replaced by x' when using the first two equations in Eqs. 1.5.2, by y' when using the middle two equations and by z' when using the last two equations in Eqs. 1.5.2), we obtain

$$
\begin{aligned}
\sigma_{x'} =\ & \sigma_x \cos^2(x', x) + \sigma_y \cos^2(x', y) + \sigma_z \cos^2(x', z) \\
& + 2\tau_{xy} \cos(x', y) \cos(x', x) + 2\tau_{yz} \cos(x', z) \cos(x', y) \\
& + 2\tau_{zx} \cos(x', x) \cos(x', z) \\
\tau_{x'y'} =\ & \sigma_x \cos(x', x) \cos(y', x) + \sigma_y \cos(x', y) \cos(y', y) \\
& + \sigma_z \cos(x', z) \cos(y', z) \\
& + \tau_{xy}[\cos(x', y) \cos(y', x) + \cos(x', x) \cos(y', y)] \\
& + \tau_{yz}[\cos(x', z) \cos(y', y) + \cos(x', y) \cos(y', z)] \\
& + \tau_{zx}[\cos(x', x) \cos(y', z) + \cos(x', z) \cos(y', x)] \\
\sigma_{y'} =\ & \sigma_x \cos^2(y', x) + \sigma_y \cos^2(y', y) + \sigma_z \cos^2(y', z) \\
& + 2\tau_{xy} \cos(y', x) \cos(y', y) + 2\tau_{yz} \cos(y', y) \cos(y', z) \\
& + 2\tau_{zx} \cos(y', z) \cos(y', x) \\
\tau_{y'z'} =\ & \sigma_x \cos(y', x) \cos(z', x) + \sigma_y \cos(y', y) \cos(z', y) \\
& + \sigma_z \cos(y', z) \cos(z', z) \\
& + \tau_{xy}[\cos(y', y) \cos(z', x) + \cos(y', x) \cos(z', y)] \\
& + \tau_{yz}[\cos(y', z) \cos(z', y) + \cos(y', y) \cos(z', z)] \\
& + \tau_{zx}[\cos(y', x) \cos(z', z) + \cos(y', z) \cos(z', x)] \\
\sigma_{z'} =\ & \sigma_x \cos^2(z', x) + \sigma_y \cos^2(z', y) + \sigma_z \cos^2(z', z) \\
& + 2\tau_{xy} \cos(z', y) \cos(z', x) + 2\tau_{yz} \cos(z', z) \cos(z', y) \\
& + 2\tau_{zx} \cos(z', x) \cos(z', z) \\
\tau_{z'x'} =\ & \sigma_x \cos(z', x) \cos(x', x) + \sigma_y \cos(z', y) \cos(x', y) \\
& + \sigma_z \cos(z', z) \cos(x', z) \\
& + \tau_{xy}[\cos(z', y) \cos(x', x) + \cos(z', x) \cos(x', y)] \\
& + \tau_{yz}[\cos(z', z) \cos(x', y) + \cos(z', y) \cos(x', z)] \\
& + \tau_{zx}[\cos(z', x) \cos(x', z) + \cos(z', z) \cos(x', x)]
\end{aligned}
\tag{1.5.3}
$$

Equations 1.5.3 can be used to calculate the normal and shear stresses on planes with outward normals in the x', y', and z' directions if the six cartesians stress components, σ_x, σ_y, σ_z τ_{xy}, τ_{yz}, and τ_{zx} are given. Thus these equations completely describe the state of stress at the point P.

It should be noted that all of the direction cosines in Eqs. 1.5.3 are not independent. From analytical geometry, it is shown that the cosine of an

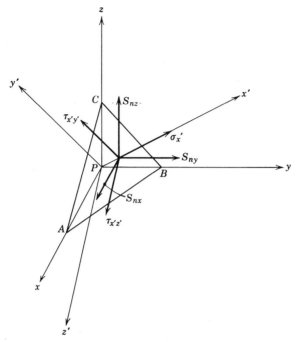

Fig. 1.5.2. Stress diagram for rotation of coordinate axes.

angle between two lines is the sum of the products of the paired direction cosines of the two lines, that is, if n and m are two lines that intersect, then

$$\cos (n, m) = \cos (n, x) \cos (m, x) + \cos (n, y) \cos (m, y)$$
$$+ \cos (n, z) \cos (m, z) \quad (1.5.4)$$

By means of this rule the relationships between the direction cosines in Eqs. 1.5.3 are

$$\cos^2 (x', x) + \cos^2 (x', y) + \cos^2 (x', z) = 1$$
$$\cos^2 (y', x) + \cos^2 (y', y) + \cos^2 (y', z) = 1$$
$$\cos^2 (z', x) + \cos^2 (z', y) + \cos^2 (z', z) = 1 \quad (1.5.6)$$

$$\cos (x', x) \cos (y', x) + \cos (x', y) \cos (y' y) + \cos (x', z) \cos (y', z) = 0$$
$$\cos (y', x) \cos (z', x) + \cos (y', y) \cos (z', y) + \cos (y', z) \cos (z', z) = 0$$
$$\cos (z', x) \cos (x', x) + \cos (z', y) \cos (x', y) + \cos (z', z) \cos (x', z) = 0$$

It is possible to show that there is one set of direction cosines which will reduce all shear stresses $\tau_{x'z'}$, and $\tau_{y'z'}$, $\tau_{z'x'}$ to zero and which will also make the normal stresses have their extreme values. The three mutually perpendicular planes where this condition exists are called the principal planes, and the normal stresses acting on these planes are the principal stresses.

The resultant normal stress acting on any inclined plane whose outward normal is n, is related to S_{nx}, S_{ny}, and S_{nz}, the cartesian components of stress acting on the inclined plane, by

$$\sigma_n = S_{nx} \cos (n, x) + S_{ny} \cos (n, y) + S_{nz} \cos (n, z) \qquad (1.5.6)$$

Substitution of Eqs. 1.5.1 into 1.5.6 gives the resultant normal stress as

$$
\begin{aligned}
\sigma_n = {} & \sigma_x \cos^2 (n, x) + \sigma_y \cos^2 (n, y) + \sigma_z \cos^2 (n, z) \\
& + 2\tau_{xy} \cos (n, x) \cos (n, y) + 2\tau_{yz} \cos (n, y) \cos (n, z) \\
& + 2\tau_{zx} \cos (n, z) \cos (n, x) \qquad (1.5.7)
\end{aligned}
$$

Equation 1.5.7 can be shown to represent a quadratic surface. Let the length of a line OP, parallel to the normal direction n and extending from the origin O of the xyz coordinate system to the point P, be equal to the reciprocal of the square root of the absolute value of the stress σ_n, that is,

$$\overline{OP} = \frac{1}{\sqrt{|\sigma_n|}}$$

The coordinates of the point P are

$$x = OP \cos (n, x) = \frac{\cos (n, x)}{\sqrt{|\sigma_n|}}$$

$$y = OP \cos (n, y) = \frac{\cos (n, y)}{\sqrt{|\sigma_n|}}$$

$$z = OP \cos (n, z) = \frac{\cos (n, z)}{\sqrt{|\sigma_n|}}$$

Substitution of these values of the direction cosines into Eq. 1.5.7 gives

$$\sigma_x x^2 + \sigma_y y^2 + \sigma_z z^2 + 2\tau_{xy} xy + 2\tau_{yz} yz + 2\tau_{zx} zx = \pm 1 \qquad (1.5.8)$$

Equation 1.5.8 represents a quadratic surface for any given state of stress at a point P. If the coordinate system is rotated, the components of stress will change but the quadratic surface is unaltered. It is always possible to find one orientation of the coordinate system for a general second-degree quadratic surface such that the coefficient of the cross product terms are zero. Let $x_i y_i z_i$ be such a coordinate system. Then, Eq. 1.5.8 becomes

$$\sigma_1 x_i^2 + \sigma_2 y_i^2 + \sigma_3 z_i^2 = \pm 1 \qquad (1.5.9)$$

This new coordinate system is the principal coordinate system, the $y_i z_i$, $z_i x_i$, and $x_i y_i$ planes are the principal planes, and the normal stresses σ_1, σ_2, and σ_3 are the principal stresses. All shear stresses are zero on these principal planes.

Referring to Fig. 1.5.2, if $\tau_{x_i y_i}$ and $\tau_{x_i z_i}$ are zero and $\sigma_{x'}$ is designated as σ_i the normal stress on one of the principal planes, then

$$\sigma_i \cos (n_i, x) = S_{ix}$$
$$\sigma_i \cos (n_i, y) = S_{iy} \qquad (1.5.10)$$
$$\sigma_i \cos (n_i, z) = S_{iz}$$

Combining Eqs. 1.5.10 with Eqs. 1.5.1 gives

$$(\sigma_x - \sigma_i) \cos (n_i, x) + \tau_{yx} \cos (n_i, y) + \tau_{zx} \cos (n_i, z) = 0$$
$$\tau_{xy} \cos (n_i, x) + (\sigma_y - \sigma_i) \cos (n_i, y) + \tau_{zy} \cos (n_i, z) = 0$$
$$\tau_{xz} \cos (n_i, x) + \tau_{yz} \cos (n_i, y) + (\sigma_z - \sigma_i) \cos (n_i, z) = 0$$
$$(1.5.11)$$

The determinant of the coefficients of the direction cosines in Eqs. 1.5.11 must equal zero if a nontrivial solution exists. Forming this determinant and expanding gives

$$\sigma_i^3 - (\sigma_x + \sigma_y + \sigma_z)\sigma_i^2 + (\sigma_x\sigma_y + \sigma_y\sigma_z + \sigma_z\sigma_x - \tau_{xy}^2 - \tau_{yz}^2 - \tau_{zx}^2)\sigma_i$$
$$- (\sigma_x\sigma_y\sigma_z - \sigma_x\tau_{yz}^2 - \sigma_y\tau_{zx}^2 - \sigma_z\tau_{xy}^2 + 2\tau_{xy}\tau_{yz}\tau_{zx}) = 0$$
$$(1.5.12)$$

The three roots of the cubic equation 1.5.12 are real and are the principal stresses σ_1, σ_2, and σ_3. The planes on which the principal stresses act are the principal planes. The direction of the normals to the principal planes can be found by means of Eqs. 1.5.11 and the relationships

$$\cos^2 (n_i, x) + \cos^2 (n_i, y) + \cos^2 (n_i, z) = 1$$

Solving these equations gives

$$\cos (n_i, x) = \frac{A_i}{\sqrt{A_i^2 + B_i^2 + C_i^2}}$$

$$\cos (n_i, y) = \frac{B_i}{\sqrt{A_i^2 + B_i^2 + C_i^2}} \qquad (1.5.13)$$

$$\cos (n_i, z) = \frac{C_i}{\sqrt{A_i^2 + B_i^2 + C_i^2}}$$

where

$$A_i = (\sigma_y - \sigma_i)(\sigma_z - \sigma_i) - \tau_{zy}\tau_{yz}$$
$$B_i = \tau_{zy}\tau_{xz} - \tau_{xy}(\sigma_z - \sigma_i)$$
$$C_i = \tau_{xy}\tau_{yz} - \tau_{xz}(\sigma_y - \sigma_i)$$

The principal stresses on the principal planes remain unaltered by changes in coordinate systems, therefore the coefficients of Eq. 1.5.12 are invariants which will remain constant under any transformation of the coordinate system. Thus the three invariants of the six cartesian components of stress are

$$I_1 = \sigma_x + \sigma_y + \sigma_z$$
$$I_2 = \sigma_x\sigma_y + \sigma_y\sigma_z + \sigma_z\sigma_x - \tau_{xy}{}^2 - \tau_{yz}{}^2 - \tau_{zx}{}^2 \qquad (1.5.14)$$
$$I_3 = \sigma_x\sigma_y\sigma_z - \sigma_x\tau_{yz}{}^2 - \sigma_y\tau_{zx}{}^2 - \sigma_z\tau_{xy}{}^2 + 2\tau_{xy}\tau_{yz}\tau_{zx}$$

Assume that the principal coordinate system is x, y, and z, and that the principal stresses in the directions x, y, and z, are σ_1, σ_2, and σ_3 respectively. The cartesian components of stress on a plane whose normal is n are obtained from Eqs. 1.5.1 as

$$S_{nx} = \sigma_1 \cos (n, x)$$
$$S_{ny} = \sigma_2 \cos (n, y) \qquad (1.5.15)$$
$$S_{nz} = \sigma_3 \cos (n, z)$$

The resultant stress on the plane whose normal is n is obtained by substitution of Eq. 1.5.15 into the first equation of Eqs. 1.2.4. Thus

$$S_n{}^2 = \sigma_1{}^2 \cos^2 (n, x) + \sigma_2{}^2 \cos^2 (n, y) + \sigma_3{}^2 \cos^2 (n, z) \quad (1.5.16)$$

The normal stress on this plane is obtained by the substitution of Eqs. 1.5.15 into Eq. 1.5.16. Thus

$$\sigma_n = \sigma_1 \cos^2 (n, x) + \sigma_2 \cos^2 (n, y) + \sigma_3 \cos^2 (n, z) \qquad (1.5.17)$$

Substitution of Eqs. 1.5.16 and 1.5.17 into the first equation of Eqs. 1.2.6 gives the shear stress on the plane whose normal is n. Thus

$$\tau_n{}^2 = \sigma_1{}^2 \cos^2 (n, x) + \sigma_2{}^2 \cos (n, y) + \sigma_3{}^2 \cos{}^2 (n, z)$$
$$- [\sigma_1 \cos^2 (n, x) + \sigma_2 \cos^2 (n, y) + \sigma_3 \cos^2 (n, z)]^2$$

Since the sum of the squares of the direction cosines is unity, the above equation can be rewritten as

$$\tau_n{}^2 = (\sigma_1 - \sigma_2)^2 \cos^2 (n, x) \cos^2 (n, y) + (\sigma_2 - \sigma_3)^2 \cos^2 (n, y) \cos^2 (n, z)$$
$$+ (\sigma_3 - \sigma_1)^2 \cos^2 (n, z) \cos^2 (n, x) \quad (1.5.18)$$

Equations 1.5.16, 1.5.17, and 1.5.18 give the resultant, normal, and shear stresses in terms of the principal stress on any plane whose normal is n. Equation 1.5.18 can be used to determine the maximum shears and the direction of the normals to the planes, where these maximum shears occur. For example, the maximum shear resulting from the principal stresses σ_1 and σ_2 must occur in a plane whose normal makes a right angle to the z axis, or σ_3 direction, thus $\cos(n, z) = 0$. Similarly, the maximum shear resulting from the principal stresses σ_2 and σ_3 must occur when $\cos(n, x) = 0$, and the maximum shear resulting from σ_3 and σ_1 must occur when $\cos(n, y) = 0$. Therefore the following three equations result from Eq. 1.5.18

$$(\tau_n)_{xy} = (\sigma_1 - \sigma_2) \cos(n, x) \cos(n, y)$$
$$(\tau_n)_{yz} = (\sigma_2 - \sigma_3) \cos(n, y) \cos(n, z)$$
$$(\tau_n)_{zx} = (\sigma_3 - \sigma_1) \cos(n, z) \cos(n, x)$$

Maximizing these three equations gives

$$(\tau_{\max})_{xy} = \frac{\sigma_1 - \sigma_2}{2}$$

$$(\tau_{\max})_{yz} = \frac{\sigma_2 - \sigma_3}{2} \qquad (1.5.19)$$

$$(\tau_{\max})_{zx} = \frac{\sigma_3 - \sigma_1}{2}$$

These maximum shear stresses occur on planes whose normals make an angle of $45°$ with the σ_1, σ_2, and σ_3 directions respectively.

It should be noted that if $\sigma_1 > \sigma_2 > \sigma_3$, then $\sigma_1 = \sigma_{\max}$ and $\sigma_3 = \sigma_{\min}$ are the maximum and minimum principal normal stresses at a point. Therefore the maximum shear stress at a point is given by $\tau_{\max} = \frac{1}{2}(\sigma_{\max} - \sigma_{\min})$.

The necessary equations for calculating the normal and shear stresses from the principal stresses for any other orientation of a cartesian coordinate system x', y', z', are obtained from Eqs. 1.5.3 by letting the shear stresses τ_{xy}, τ_{yz}, and τ_{zx} be zero. Thus

$$\sigma_{x'} = \sigma_1 \cos^2(x', x) + \sigma_2 \cos^2(x', y) + \sigma_3 \cos^2(x', z)$$
$$\sigma_{y'} = \sigma_1 \cos^2(y', x) + \sigma_2 \cos^2(y', y) + \sigma_3 \cos^2(y', z)$$
$$\sigma_{z'} = \sigma_1 \cos^2(z', x) + \sigma_2 \cos^2(z', y) + \sigma_3 \cos^2(z', z)$$
$$\tau_{x'y'} = \sigma_1 \cos(x', x) \cos(y', x) + \sigma_2 \cos(x', y) \cos(y', y)$$
$$+ \sigma_3 \cos(x', z) \cos(y', z) \qquad (1.5.20)$$
$$\tau_{y'z'} = \sigma_1 \cos(y', x) \cos(z', x) + \sigma_2 \cos(y', y) \cos(z', y)$$
$$+ \sigma_3 \cos(y', z) \cos(z', z)$$
$$\tau_{z'x'} = \sigma_1 \cos(z', x) \cos(x', x) + \sigma_2 \cos(z', y) \cos(x', y)$$
$$+ \sigma_3 \cos(z', z) \cos(x', z)$$

1.6   OCTAHEDRAL STRESSES

Let the x, y, z, coordinate axes correspond to the principal axes of stress at a given point in a body, and let σ_1, σ_2, and σ_3, be the principal stresses in the x, y, and z, directions, respectively. Assume that the given point is surrounded by an octahedron. Let the normals to the faces of the octahedron have the following direction cosines

$$\cos (n, x) = \pm \frac{1}{\sqrt{3}}$$

$$\cos (n, y) = \pm \frac{1}{\sqrt{3}} \qquad (1.6.1)$$

$$\cos (n, z) = \pm \frac{1}{\sqrt{3}}$$

The eight planes of the octahedron are established by taking all possible combinations of the plus and minus signs in Eqs. 1.6.1. The resultant, normal, and shear stresses on any face of the octahedron can be calculated from Eqs. 1.5.16, 1.5.17, 1.5.18, and 1.6.1. Thus

$$(S_n)_{\text{oct}} = \sqrt{\frac{\sigma_1{}^2 + \sigma_2{}^2 + \sigma_3{}^2}{3}}$$

$$(\sigma_n)_{\text{oct}} = \frac{\sigma_1 + \sigma_2 + \sigma_3}{3} \qquad (1.6.2)$$

$$(\tau_n)_{\text{oct}} = \tfrac{1}{3}\sqrt{(\sigma_1 - \sigma_2)^2 + (\sigma_2 - \sigma_3)^2 + (\sigma_3 - \sigma_1)^2}$$

The stresses $(\sigma_n)_{\text{oct}}$ and $(\tau_n)_{\text{oct}}$ are called the octahedral normal and shear stress, respectively. These stresses play an important part in the criterion of failure for plastic yielding.

1.7   SECONDARY PRINCIPAL STRESSES

The concept of secondary principal stresses is one that generally is not treated in books on the theory of elasticity or strength of materials. However, it is a concept that is useful in photoelasticity or experimental stress analysis.

Consider a general three-dimensional stress system as defined by Fig. 1.2.3. The six independent stress components are σ_x, σ_y, σ_z, τ_{xy}, τ_{yz}, and τ_{zx}. Secondary principal stresses are defined for a given direction i as the principal stresses resulting from the stress components which lie in a plane normal to the given direction i and are denoted by $(P_i{}', Q_i{}')$. The planes on which P' and Q' act are secondary principal planes and these planes are free of shear stress in the direction at right angles to the given direction i. The secondary principal planes are not necessarily free of shear stress in the direction i.

By rotation of the xy axes about the z axis it is possible to find one and only one orientation of the xy axes which will result in the normal stresses being maximum and minimum and shear stress being zero in the plane normal to the z axis. Thus, the secondary principal stresses for the z axis are the maximum and minimum stresses resulting from the stress components σ_x, σ_y, and τ_{xy}. From Eqs. 1.3.5 and 1.3.6, the secondary principal stresses are given by

$$(P', Q')_z = \frac{\sigma_x + \sigma_y}{2} \pm \frac{1}{2} \sqrt{(\sigma_x - \sigma_y)^2 + 4\tau_{xy}^2} \qquad (1.7.1)$$

The orientation of the new xy axes with respect to the original direction is given by

$$(\theta')_z = \tfrac{1}{2} \tan^{-1} \frac{2\tau_{xy}}{\sigma_x - \sigma_y} \qquad (1.7.2)$$

It should be noted that where there is only one set of principal stresses, there are an infinite number of secondary principal stresses. However, for any given direction i, there is only one set of secondary principal stresses.

1.8   EQUATIONS OF EQUILIBRIUM

In discussing stresses at a point P, the element of the body surrounding the point has always been assumed in equilibrium. The volume of the element has been considered small so that body forces and variation of surface forces could be neglected. These simplifications were made for ease of derivation. However, the same results can be obtained by considering variation of surface stresses. This section will derive the necessary relations between stresses acting on a small rectangular parallelepiped element of a body when variation in surface stresses are allowed to exist and body forces are considered.

Assume that at the point $P(x, y, z)$ of a body under stress that a small rectangular parallelepiped with edges dx, dy, and dz, is removed from the body and the stresses that existed before removal are restored (Fig. 1.8.1). In dealing with continuous media, the stresses in a body must be continuously varying in all directions. Thus, the stresses will be continuous functions of the coordinates and, for small elements in the body, the differences in stresses on opposite faces of the element can be represented by increments of stress. For the element of the body given in Fig. 1.8.1 to be in equilibrium, the following equilibrium conditions must be satisfied

$$\sum F_x = \sum F_y = \sum F_z = 0$$

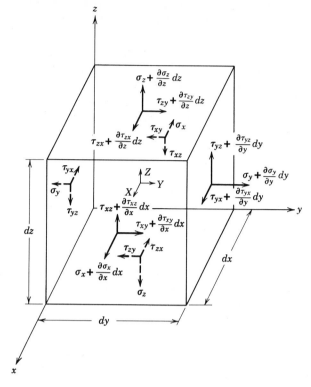

Fig. 1.8.1. Equilibrium conditions for rectangular coordinates.

Summation of the forces in the x direction gives

$$\sum F_x = \left(\sigma_x + \frac{\partial \sigma_x}{\partial x}\,dx\right) dy\,dz - \sigma_x\,dy\,dz + \left(\tau_{yx} + \frac{\partial \tau_{yx}}{\partial y}\,dy\right) dz\,dx$$

$$- \tau_{yx}\,dz\,dx + \left(\tau_{zx} + \frac{\partial \tau_{zx}}{\partial z}\,dz\right) dx\,dy$$

$$- \tau_{zx}\,dx\,dy + X\,dx\,dy\,dz = 0$$

Similar equations for the forces in the y and z direction can be written. Simplifying these equations and dividing by $dx\,dy\,dz$ gives

$$\frac{\partial \sigma_x}{\partial x} + \frac{\partial \tau_{yx}}{\partial y} + \frac{\partial \tau_{zx}}{\partial z} + X = 0$$

$$\frac{\partial \tau_{xy}}{\partial x} + \frac{\partial \sigma_y}{\partial y} + \frac{\partial \tau_{zy}}{\partial z} + Y = 0 \qquad (1.8.1)$$

$$\frac{\partial \tau_{xz}}{\partial x} + \frac{\partial \tau_{yz}}{\partial y} + \frac{\partial \sigma_z}{\partial z} + Z = 0$$

Equations 1.8.1 contain the equations of equilibrium which relate the stresses from point to point in a body. The six cartesian components of stress that vary continuously throughout the body must satisfy the above three conditions.

If the moments about the x, y, and z axes are considered in succession, the result is the same as that given in Eq. 1.2.7.

REFERENCES

1. Durelli, A. J., E. A. Phillips, and C. H. Tsao, *Introduction to the Theoretical and Experimental Analysis of Stress and Strain*, Chapter 1, McGraw-Hill Book Co., New York, 1958.
2. Frocht, M. M., *Photoelasticity*, Vol. I, Chapters 1 and 2, John Wiley and Sons, 1941.
3. Jaeger, J. C., *Elasticity, Fracture and Flow*, Chapter 1, Methuen & Co., London, 1962.
4. Lee, George H., *An Introduction to Experimental Stress Analysis*, Chapter 1, John Wiley and Sons, New York, 1950.
5. Love, A. E. H., *The Mathematical Theory of Elasticity*, Chapter 2, Dover Publications, New York, 1944.
6. Sechler, Earnest E., *Elasticity in Engineering*, Chapters 1, 2, and 3, John Wiley and Sons, New York, 1952.
7. Sokolnikoff, I. S., *Mathematical Theory of Elasticity*, Chapter 2, McGraw-Hill Book Co., New York, 1946.
8. Timoshenko, S., and J. N. Goodier, *Theory of Elasticity*, Chapters 1, 2, and 8, McGraw-Hill Book Co., New York, 1951.

CHAPTER 2

ANALYSIS OF STRAIN

2.1 INTRODUCTION

Any body composed of a continuous distribution of matter will undergo a deformation when subjected to a set of stresses, that is, the relative positions of various points in the body will be altered with respect to each other, and with respect to a fixed set of coordinate axes. The mathematical techniques for describing these deformations in a body is the subject of analysis of strain, and is basically a geometrical problem.

Deformation of a body is the change in its size and shape that occurs as a result of relative displacements between points in the body. It is important to distinguish between a displacement that results in rigid body motion and a displacement that results in deformation. In rigid body motion the size and shape of a body remain unchanged; however, the position of the body with respect to some fixed reference is altered. Rigid body motion can occur by translation or rotation.

2.2 DEFINITIONS

Two types of deformation can occur in a body, one is a change in length of a straight line and the other a change in the angle of intersection of two straight lines. The change in length per unit original length of a straight line is defined as longitudinal strain. More specifically, the longitudinal strain ϵ at a point P on a line segment ΔL is defined by the expression

$$\epsilon = \lim_{\Delta L \to 0} \frac{\delta L}{\Delta L}$$

where δL is the change in length of the given line segment. A positive longitudinal strain corresponds to an increase in length and a negative longitudinal strain corresponds to a decrease in length.

30

Shear strain, γ, is defined as the angular change in a right angle. A positive shear strain represents a decrease in the right angle and a negative shear strain represents an increase in the right angle.

The mathematical developments that follow are based upon the assumptions that (1) the quantities ϵ and γ are so small compared to unity that their squares and products are negligible, and (2) in the neighborhood of a point P in a body, the strains are homogeneous, that is, changes in

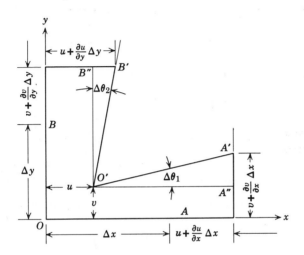

Fig. 2.2.1. Strain-displacement relation.

deformation with position in the body are linear over small areas within the body. As a result of these assumptions, it can be shown that in the neighborhood of a point in a body, all straight lines remain straight after straining and parallel lines remain parallel, although their direction may change with straining.* Also, as a result of the above two assumptions, it can be shown that the order of application of two deformations has no effect upon the final configuration of the body.* That is, the "principle of superposition" applies. The linear theory of elasticity is based upon the principle of superposition both for stress and strain.

Figure 2.2.1 shows one corner of a rectangular parallelepiped cut out of a strained body. The reference axes Ox and Oy are assumed to coincide with the sides of the element before straining. After straining, point O has moved to point O', point A has moved to A', and point B has moved to B'. The displacement of a point will be represented by three rectangular

* A. J., Durelli, E. A. Phillips, and C. H. Tsao, *Introduction to the Theoretical and Experimental Analysis of Stress and Strain*, McGraw-Hill Book Co., New York, 1958, pp. 36–43.

components, u, v, and w, parallel to the x, y, and z axes. These displacements are assumed to be continuous functions of the coordinates (x, y, z). Thus, if u is the displacement of point O in the x direction, the displacement of a neighboring point A in the x direction is $u + (\partial u/\partial x)\Delta x$. Higher derivatives and products are not required because of the two assumptions that strains are small compared to unity and that in the neighborhood of a point the strain is homogeneous.

From Fig. 2.2.1, the length of the line segment $O'A''$ is given by

$$O'A'' = \Delta x + u + \frac{\partial u}{\partial x}\Delta x - u = \Delta x + \frac{\partial u}{\partial x}\Delta x$$

and the length of the line segment $A'A''$ is

$$A'A'' = v + \frac{\partial v}{\partial x}\Delta x - v = \frac{\partial v}{\partial x}\Delta x$$

Therefore the length of the line segment $O'A'$ is given by

$$O'A' = \sqrt{\left(\Delta x + \frac{\partial u}{\partial x}\Delta x\right)^2 + \left(\frac{\partial v}{\partial x}\Delta x\right)^2} = \Delta x\sqrt{1 + 2\frac{\partial u}{\partial x} + \left(\frac{\partial u}{\partial x}\right)^2 + \left(\frac{\partial v}{\partial x}\right)^2}$$

Expanding the expression under the radical sign and neglecting second-order terms give

$$O'A' = \Delta x\left(1 + \frac{\partial u}{\partial x}\right)$$

Therefore the change in length of the line segment OA is

$$\delta x = O'A' - OA = \Delta x\left(1 + \frac{\partial u}{\partial x}\right) - \Delta x = \frac{\partial u}{\partial x}\Delta x$$

By definition, the strain

$$\epsilon_x = \lim_{\Delta x \to 0}\frac{\delta x}{\Delta x} = \frac{\partial u}{\partial x}\frac{\Delta x}{\Delta x}$$

Thus

$$\epsilon_x = \frac{\partial u}{\partial x} \tag{2.2.1}$$

By referring to Fig. 2.2.1 and remembering that angles $\Delta\theta_1$ and $\Delta\theta_2$ are small and that strains are also small compared to unity, the following equations can be written:

$$\tan \Delta\theta_1 = \Delta\theta_1 = \frac{A'A''}{O'A''} = \frac{(\partial v/\partial x)\,\Delta x}{\Delta x + (\partial u/\partial x)\,\Delta x} = \frac{\partial v}{\partial x}$$

$$\tan \Delta\theta_2 = \Delta\theta_2 = \frac{B'B''}{O'B''} = \frac{(\partial u/\partial y)\,\Delta y}{\Delta y + (\partial v/\partial y)\,\Delta y} = \frac{\partial u}{\partial y}$$

By definition, the shear strain γ_{xy}, in the right angle AOB is $\Delta\theta_1 + \Delta\theta_2$; therefore

$$\gamma_{xy} = \frac{\partial u}{\partial y} + \frac{\partial v}{\partial x} \tag{2.2.2}$$

By a similar treatment for the yz and zx planes, the six components of strain can be shown to be related to the displacements by the following six equations:

$$
\begin{aligned}
\epsilon_x &= \frac{\partial u}{\partial x} & \gamma_{xy} &= \frac{\partial u}{\partial y} + \frac{\partial v}{\partial x} \\[2mm]
\epsilon_y &= \frac{\partial v}{\partial y} & \gamma_{yz} &= \frac{\partial v}{\partial z} + \frac{\partial w}{\partial y} \\[2mm]
\epsilon_z &= \frac{\partial w}{\partial z} & \gamma_{zx} &= \frac{\partial w}{\partial x} + \frac{\partial u}{\partial z}
\end{aligned}
\tag{2.2.3}
$$

The three equations on the left are the components of normal strain and the three equations on the right are the components of shear strain in the cartesian directions. If u, v, and w, are continuous functions of the space coordinates x, y, and z, of a body, Eqs. 2.2.3 completely specify the state of strain at a point in the body.

2.3 STRAIN IN A PLANE

The problem to be considered next is that of determining the components of strain in any direction about a point P, given the cartesian components of strain at the point P. To simplify the mathematical derivations and the visualization of these various strain components, the discussion to follow will be confined to two dimensions. Many problems in the theory of elasticity can be solved by a consideration of strains in a plane.

Figure 2.3.1 shows an element of a body which is deformed by normal strains in the x and y directions and by a shear strain in the xy plane. As a result of each of these strains, any line segment Δr not in the direction of x or y has suffered a strain also. The amount of strain in a line segment Δr that makes an angle θ with the x axis can be determined from the geometrical relations shown in Fig. 2.3.1.

From Fig. 2.3.1a and the definition of strain

$$\delta x = \Delta x \epsilon_x$$

The change in length of Δr is δr and is given by

$$\delta r = \delta x \cos\theta = \Delta x \epsilon_x \cos\theta$$

The length of Δr is

$$\Delta r = \frac{\Delta x}{\cos\theta}$$

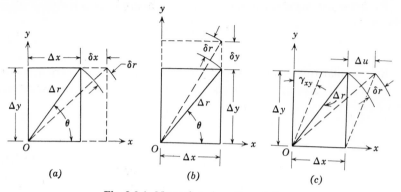

Fig. 2.3.1. Normal strains after rotation.

Therefore the strain ϵ_θ in the direction of Δr as a result of the normal strain ϵ_x is

$$\epsilon_\theta = \frac{\delta r}{\Delta r} = \epsilon_x \cos^2 \theta \qquad (2.3.1)$$

By a similar procedure using Fig. 2.3.1b

$$\delta y = \Delta y \epsilon_y$$

and

$$\delta r = \delta y \sin \theta = \epsilon_y \Delta y \sin \theta$$

And

$$\Delta r = \frac{\Delta y}{\sin \theta}$$

Therefore the strain ϵ_θ in the direction Δr as a result of the normal strain ϵ_y is

$$\epsilon_\theta = \frac{\delta r}{\Delta r} = \epsilon_y \sin^2 \theta \qquad (2.3.2)$$

From Fig. 2.3.1c the length of Δu, for small values of γ_{xy}, is

$$\Delta u = \Delta y \tan \gamma_{xy} = \Delta y \gamma_{xy}$$

The length of δr is

$$\delta r = \Delta u \cos \theta = \Delta y \gamma_{xy} \cos \theta$$

and the length of Δr is

$$\Delta r = \frac{\Delta y}{\sin \theta}$$

Thus the normal strain ϵ_θ in the direction Δr, as a result of the shear strain γ_{xy}, is

$$\epsilon_\theta = \gamma_{xy} \sin \theta \cos \theta \qquad (2.3.3)$$

By use of the principle of superposition Eqs. 2.3.1, 2.3.2., and 2.3.3 can be added together to give the total normal strain in the direction of Δr as a result of normal strains ϵ_x and ϵ_y and the shear strain γ_{xy}. Thus

$$\epsilon_\theta = \epsilon_x \cos^2 \theta + \epsilon_y \sin^2 \theta + \gamma_{xy} \sin \theta \cos \theta \qquad (2.3.4)$$

Simplifying Eq. 2.3.4 by means of trigonometric identities gives

$$\epsilon_\theta = \frac{\epsilon_x + \epsilon_y}{2} + \frac{\epsilon_x - \epsilon_y}{2} \cos 2\theta + \frac{\gamma_{xy}}{2} \sin 2\theta \qquad (2.3.5)$$

Comparison of Eq. 2.3.5 with Eq. 1.3.2 shows that the two equations differ only in the factor $\frac{1}{2}$ in the last term. This difference is explained by the fact that two shear stresses τ_{xy} and τ_{yx} are required to produce the shear strain of γ_{xy}. This difference can be eliminated by inserting the factor of $\frac{1}{2}$ on the right-hand side of Eq. 2.2.2, the defining equation for shear strain. Such definitions are used in tensor notations, but by tradition, engineering shear strain is defined by Eq. 2.2.2.

Figure 2.3.2 illustrates the geometrical relations between the cartesian components of strain for an element of a body in the xy plane and the shear strains for another element of the body that makes an angle θ with the x axis. From Fig. 2.3.2a and the definition of strain

$$\delta x = \Delta x \epsilon_x$$

also

$$r \, d\theta = \delta x \sin \theta = \epsilon_x \Delta x \sin \theta$$

The length of r is

$$r = \frac{\Delta x}{\cos \theta}$$

Therefore

$$d\theta = \epsilon_x \sin \theta \cos \theta$$

In a similar way it can be shown that

$$d\theta_1 = \epsilon_x \sin \theta \cos \theta$$

The shear strain γ_θ in the element that makes an angle θ with the x axis is the angular change in the right angle and is the algebraic sum of $d\theta$ and $d\theta_1$. Noting that both $d\theta$ and $d\theta_1$ increased the right angle, γ_θ is given by

$$\gamma_\theta = -2\epsilon_x \sin \theta \cos \theta \qquad (2.3.6)$$

From Fig. 2.3.2b and the definition of strain

$$\delta y = \Delta y \epsilon_y$$

also

$$r \, d\theta = \delta y \cos \theta$$

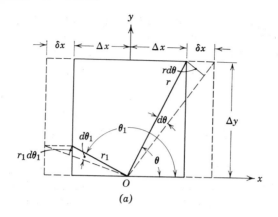

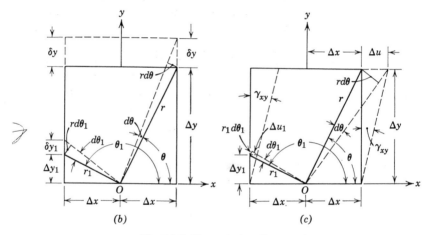

Fig. 2.3.2. Shear strains after rotation.

The length of r is

$$r = \frac{\Delta y}{\sin \theta}$$

Therefore

$$d\theta = \epsilon_y \sin \theta \cos \theta$$

In a similar way it can be shown that $d\theta_1$ is

$$d\theta_1 = \epsilon_y \sin \theta \cos \theta$$

The shear strain γ_θ is the algebraic sum of $d\theta$ and $d\theta_1$. Noting that both $d\theta$ and $d\theta_1$ decrease the right angle, we obtain

$$\gamma_\theta = 2\epsilon_y \sin \theta \cos \theta \qquad (2.3.7)$$

From Fig. 2.3.2c and noting that γ_{xy} is small

$$\Delta u = \Delta y \tan \gamma_{xy} = \Delta y \gamma_{xy}$$

and

$$\Delta y = r \sin \theta$$

Thus

$$\Delta u = r \gamma_{xy} \sin \theta$$

Also

$$r \, d\theta = \Delta u \sin \theta$$

Therefore

$$d\theta = \gamma_{xy} \sin^2 \theta$$

In a similar way it can be shown that

$$d\theta_1 = \gamma_{xy} \cos^2 \theta$$

The shear strain γ_θ is the algebraic sum of $d\theta$ and $d\theta_1$. Noting that $d\theta$ increased the right angle and $d\theta_1$ decreased the right angle, we obtain

$$\gamma_\theta = \gamma_{xy}(\cos^2 \theta - \sin^2 \theta) \tag{2.3.8}$$

By the principle of superposition, the total shear strain γ_θ in the right angle whose side makes an angle θ with the x axis, is the sum of Eqs. 2.3.6, 2.3.7, and 2.3.8; thus

$$\gamma_\theta = -2(\epsilon_x - \epsilon_y) \sin \theta \cos \theta + \gamma_{xy}(\cos^2 \theta - \sin^2 \theta) \tag{2.3.9}$$

Simplifying by means of trigonometric identities gives

$$\gamma_\theta = \gamma_{xy} \cos 2\theta - (\epsilon_x - \epsilon_y) \sin 2\theta \tag{2.3.10}$$

Comparison of Eq. 2.3.10 with 1.3.3 shows that they differ only in the factor of $\frac{1}{2}$ in the last term. This difference is explained by the fact that each normal stress results in two shear stresses, whereas each normal strain results in only one shear strain. Again this difference results because of the specific definition used for shear strain.

If a rectangular element of a body has undergone a normal strain ϵ_x and ϵ_y and a shear strain γ_{xy}, the values of the normal strain ϵ_θ and γ_θ for another rectangular element rotated an angle θ with respect to the x axis are given by Eqs. 2.3.5 and 2.3.10. These new strains are periodic in π because of the terms $\sin 2\theta$ and $\cos 2\theta$. Thus they must have maximum values or be constant. The derivative of the normal strain ϵ_θ with respect to θ equated to zero gives

$$\frac{d\epsilon_\theta}{d\theta} = -(\epsilon_x - \epsilon_y) \sin 2\theta_1 + \gamma_{xy} \cos 2\theta_1 = 0$$

Solving for θ_1 gives

$$\theta_1 = \tfrac{1}{2} \tan^{-1} \frac{\gamma_{xy}}{\epsilon_x - \epsilon_y} \tag{2.3.11}$$

Two angles are possible in Eq. 2.3.11: θ_1 and $\theta_1 + 90°$. One angle gives the direction of the maximum normal strain and the other gives the direction of the minimum normal strain. The planes in these directions are the principal planes of strain and, as will be shown later, they are also the principal planes of stress.

Substituting Eq. 2.3.11 into 2.3.5 and simplifying gives

$$\epsilon_p = \epsilon_{\theta_1} = \frac{\epsilon_x + \epsilon_y}{2} + \frac{1}{2}\sqrt{(\epsilon_x - \epsilon_y)^2 + \gamma_{xy}{}^2} \tag{2.3.12}$$

and

$$\epsilon_q = \epsilon_{\theta_1 + 90°} = \frac{\epsilon_x + \epsilon_y}{2} - \frac{1}{2}\sqrt{(\epsilon_x - \epsilon_y)^2 + \gamma_{xy}{}^2} \tag{2.3.13}$$

Equations 2.3.12 and 2.3.13 give the values of the principal strains. The maximum principal strain is denoted by ϵ_p and the minimum principal strain is denoted by ϵ_q.

Substitution of Eq. 2.3.11 into 2.3.10 shows that the shear strain is identically zero on principal planes. Thus principal planes remain normal during the application of strain.

The derivative of the shear strain (Eq. 2.3.10) with respect to θ equated to zero gives

$$\frac{d\gamma_\theta}{d\theta} = -2\gamma_{xy} \sin 2\theta_2 - 2(\epsilon_x - \epsilon_y) \cos 2\theta_2 = 0$$

Solving for θ_2 gives

$$\theta_2 = \tfrac{1}{2} \tan^{-1} - \frac{\epsilon_x - \epsilon_y}{\gamma_{xy}} \tag{2.3.14}$$

Again two values of the angle are possible, θ_2 and $\theta_2 + 90°$. Comparison of Eqs. 2.3.14 and 2.3.11 shows that θ_1 and θ_2 differ by 45°. Thus maximum normal strain and maximum shear strain occur on planes that are 45° apart. The maximum shear strain is obtained by substituting Eq. 2.3.14 into 2.3.10, giving

$$\gamma_{\substack{\max \\ \min}} = \gamma_{\theta_2} = \pm\sqrt{(\epsilon_x - \epsilon_y)^2 + \gamma_{xy}{}^2} \tag{2.3.15}$$

The value of the normal strain on the plane of maximum shear is found by substituting Eq. 2.3.14 into Eq. 2.3.5, giving

$$\epsilon_{\theta_2} = \epsilon_{\theta_2 + 90°} = \frac{\epsilon_x + \epsilon_y}{2} \tag{2.3.16}$$

Equation 2.3.16 shows that the normal strains are equal to each other on sections where the shear strains are maximum or minimum. Thus planes of maximum shear are not in general free of normal strain.

A state of pure shear strain is said to exist if the normal strains are of equal magnitude but of opposite sign. For this condition, Eq. 2.3.16 is identically zero and the planes of maximum shear strain are free of normal strains.

Subtracting Eq. 2.3.13 from 2.3.12 and comparing with Eq. 2.3.15 shows that

$$\gamma_{\max} = \epsilon_p - \epsilon_q \qquad (2.3.17)$$

Thus the maximum shear strain is equal to the difference between the principal normal strains.

Adding Eqs. 2.3.13 and 2.3.12 gives

$$\epsilon_p + \epsilon_q = \epsilon_x + \epsilon_y \qquad (2.3.18)$$

Thus the sum of the normal strains is an invariant under rotation of the coordinate axes.

There may be occasions when the principal strains are known and the normal and shear strains on a section at an angle θ with respect to the principal plane is desired. The necessary equations for calculating these strains can be derived as in the preceding and are

$$\epsilon_\theta = \frac{\epsilon_p + \epsilon_q}{2} + \frac{\epsilon_p - \epsilon_q}{2} \cos 2\theta \qquad (2.3.19)$$

and

$$\gamma_\theta = -(\epsilon_p - \epsilon_q) \sin 2\theta \qquad (2.3.20)$$

2.4 MOHR'S CIRCLE OF STRAIN

A geometrical solution for strains in any direction is provided by Mohr's circle of strain. The only difference between the circle of strain and the circle of stress lies in the fact that in the circle of stress the ordinate represents the whole shear stress, whereas in the circle of strain the ordinate represents only one-half the shear strain. The steps in the construction of Mohr's circle of strain are as follows:

1. Construct a set of orthogonal axes and label the ordinate axis $\gamma/2$ and the abscissa axis ϵ.

2. Plot the normal strains ϵ_x and ϵ_y on the normal strain axis.

3. Plot one-half the shear strain, occurring in the right angle of the first quadrant of the cartesian coordinates, directly above or below the point representing ϵ_x on the normal axis. If the shear strain produces a decrease in the right angle, plot below the normal axis, and if the shear

strain produces an increase in the right angle, plot above the normal strain axis.

4. Plot one-half the shear strain, occurring in the adjacent right angle of the element, directly above or below the point representing ϵ_y on the normal strain axis. As the shear strain in adjacent right angles of an element are always of opposite sign, this point is plotted on the opposite side of the normal strain axis as used in step 3.

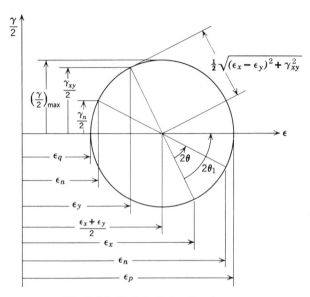

Fig. 2.4.1. Mohr's circle of strain.

5. Draw a diagonal connecting the two shear strain points. This diagonal will intersect the normal strain axis at the point $(\epsilon_x + \epsilon_y)/2$.

6. Draw a circle with center on the normal strain axis at the point $(\epsilon_x + \epsilon_y)/2$ and diameter equal to the length of the line joining the two shear strain points.

From Fig. 2.4.1 it is seen that the projection of the radius of the circle on the shear strain gives one-half the shear strain at any angle and that the projections of the ends of the diameter of the circle on the normal strain axis gives the normal strain at any angle. The radius of the circle is one-half the maximum shear strain and is $\frac{1}{2}\sqrt{(\epsilon_x - \epsilon_y)^2 + \gamma_{xy}^2}$. The intersections of the circle with the normal strain axis are the principal strains. The shear strain is zero when the normal strains are maximum and minimum. The tangent of the angle $2\theta_1$ is $\gamma_{xy}/(\epsilon_x - \epsilon_y)$, and $2\theta_1$ is twice the angle between the x axis and the direction of the principal strain. The direction of rotation of the radius from its original position to where the circle intersects the normal strain axis is in the same angular sense as the direction of rotation of the element for the normal strain to become the principal strain. When the shear strains are maximum, the normal strains are equal to one-half the sum of the principal strains, or to one-half the sum of any other normal strains. As the center of the circle is always at the point $(\epsilon_\theta + \epsilon_{\theta+90°})/2$, the sum of the normal strains is an invariant.

2.5 STRAIN ROSETTES

Normal strain at a point can be determined experimentally more easily and more accurately than shear strain. Thus methods for calculating principal strains from a series of normal strain measurements are required. The fundamental equation in the rosette method of determining principal strain is

$$\epsilon_\theta = \frac{\epsilon_x + \epsilon_y}{2} + \frac{\epsilon_x - \epsilon_y}{2}\cos 2\theta + \frac{\gamma_{xy}}{2}\sin 2\theta \qquad (2.5.1)$$

Inspection of Eq. 2.5.1 shows that ϵ_x, ϵ_y, and γ_{xy} can be determined if three values of ϵ_θ are determined in three known directions, θ_1, θ_2, and θ_3. The three values of normal strain, ϵ_1, ϵ_2, ϵ_3, with their respective angles result in three equations which can be solved simultaneously for ϵ_x, ϵ_y, and γ_{xy}. Calculations can be greatly simplified if certain specific angles are used.

Assume that a rectangular rosette is used, that is, the angles are $0°$, $45°$, and $90°$. Substitution of these values into Eq. 2.5.1 gives the following three equations:

$$\epsilon_1 = \epsilon_x$$

$$\epsilon_2 = \frac{\epsilon_x + \epsilon_y}{2} + \frac{\gamma_{xy}}{2}$$

$$\epsilon_3 = \epsilon_y$$

Solving these equations gives

$$\epsilon_x = \epsilon_1$$

$$\epsilon_y = \epsilon_3 \qquad (2.5.2)$$

$$\gamma_{xy} = 2\epsilon_2 - (\epsilon_1 + \epsilon_3)$$

To obtain the principal strains, substitute Eq. 2.5.2 into Eqs. 2.3.12 and 2.3.13, obtaining

$$\epsilon_p = \frac{\epsilon_1 + \epsilon_3}{2} + \frac{\sqrt{2}}{2}\sqrt{(\epsilon_1 - \epsilon_2)^2 + (\epsilon_2 - \epsilon_3)^2}$$

$$\qquad (2.5.3)$$

$$\epsilon_q = \frac{\epsilon_1 + \epsilon_3}{2} - \frac{\sqrt{2}}{2}\sqrt{(\epsilon_1 - \epsilon_2)^2 + (\epsilon_2 - \epsilon_3)^2}$$

The directions of the principal strains are found from Eqs. 2.3.11. thus

$$\theta_{p,q} = \tfrac{1}{2} \tan^{-1} \frac{2\epsilon_2 - (\epsilon_1 + \epsilon_3)}{\epsilon_1 - \epsilon_3} \tag{2.5.4}$$

The direction of θ_p can be determined by inspection, for θ_p must make an angle of less than 45° with the algebraic greater normal strain ϵ_x or ϵ_y.

Another commonly used method is the equiangular rosette, in which the angles are 0°, 60°, and 120°. Substitution of these angles into Eq. 2.5.1 and solving gives

$$\epsilon_x = \epsilon_1$$

$$\epsilon_y = \tfrac{1}{3}(2\epsilon_2 + 2\epsilon_3 - \epsilon_1) \tag{2.5.5}$$

$$\gamma_{xy} = \frac{2}{\sqrt{3}}(\epsilon_2 - \epsilon_3)$$

By means of Eqs. 2.5.5., 2.3.12, 2.3.13, and 2.3.11, the principal strains and their directions can be obtained as

$$\epsilon_p = \tfrac{1}{3}(\epsilon_1 + \epsilon_2 + \epsilon_3) + \frac{\sqrt{2}}{3} \sqrt{(\epsilon_1 - \epsilon_2)^2 + (\epsilon_2 - \epsilon_3)^2 + (\epsilon_3 - \epsilon_1)^2}$$

$$\epsilon_q = \tfrac{1}{3}(\epsilon_1 + \epsilon_2 + \epsilon_3) - \frac{\sqrt{2}}{3} \sqrt{(\epsilon_1 - \epsilon_2)^2 + (\epsilon_2 - \epsilon_3)^2 + (\epsilon_3 - \epsilon_1)^2} \tag{2.5.6}$$

and

$$\theta_{p,q} = \tfrac{1}{2} \tan^{-1} \frac{\sqrt{3}(\epsilon_2 - \epsilon_3)}{2\epsilon_1 - \epsilon_2 - \epsilon_3}$$

The specific direction of the principal strain can be determined by the following rule: The algebraically maximum strain ϵ_p always lies in the angle formed by the greater of the two normal strains and the major shear diagonal which is the diagonal elongated by the shear strain.

2.6 STRAIN IN THREE DIMENSIONS

The state of strain in a body at a point P is completely specified if the three normal strains ϵ_x, ϵ_y, and ϵ_z and the three shear strains γ_{xy}, γ_{yz}, and γ_{zx} are given. From these six cartesian components of strain the normal and shear strains in any other direction can be calculated. Suppose that the origin of a cartesian coordinate system is located at a point P and that the state of stress is specified by the six cartesian components of strain ϵ_x, ϵ_y, ϵ_z, γ_{xy}, γ_{yz}, and γ_{zx}. Assume that a second orthogonal coordinate system $Px'y'z'$ is located with its origin at the point P and with the x' axis in the direction of the line segment PP_1 (see Fig. 2.6.1). Let the direction cosines of Px' with respect to Px, Py, and Pz be $\cos(x', x)$, $\cos(x', y)$ and $\cos(x', z)$.

Using the principles developed in Sections 1.5 and 2.5, it can be shown that the normal strains $\epsilon_{x'}$, $\epsilon_{y'}$, and $\epsilon_{z'}$, in the line segment PP_1, PP_2, and PP_3 respectively and the shear strains $\gamma_{x'y'}$, $\gamma_{y'z'}$, and $\gamma_{z'x'}$, in the right angles P_1PP_2, P_2PP_3, and P_3PP_1 respectively are related to the six cartesian components by the following equations:

$$
\begin{aligned}
\epsilon_{x'} =\ & \epsilon_x \cos^2(x', x) + \epsilon_y \cos^2(x', y) + \epsilon_z \cos^2(x', z) \\
& + \gamma_{xy} \cos(x', x) \cos(x', y) + \gamma_{yz} \cos(x', y) \cos(x', z) \\
& + \gamma_{zx} \cos(x', z) \cos(x', x) \\
\epsilon_{y'} =\ & \epsilon_x \cos^2(y', x) + \epsilon_y \cos^2(y', y) + \epsilon_z \cos^2(y', z) \\
& + \gamma_{xy} \cos(y', x) \cos(y', y) + \gamma_{yz} \cos(y', y) \cos(y', z) \\
& + \gamma_{zx} \cos(y', z) \cos(y', x) \\
\epsilon_{z'} =\ & \epsilon_x \cos^2(z', x) + \epsilon_y \cos^2(z', y) + \epsilon_z \cos^2(z', z) \\
& + \gamma_{xy} \cos(z', x) \cos(z', y) + \gamma_{yz} \cos(z', y) \cos(z', z) \\
& + \gamma_{zx} \cos(z', z) \cos(z', x) \\
\gamma_{x'y'} =\ & 2\epsilon_x \cos(x', x) \cos(y,' x) + 2\epsilon_y \cos(x', y) \cos(y', y) \\
& + 2\epsilon_z \cos(x', z) \cos(y', z) \\
& + \gamma_{xy}[\cos(x', x) \cos(y', y) + \cos(x', y) \cos(y', x)] \\
& + \gamma_{yz}[\cos(x', y) \cos(y', z) + \cos(x', z) \cos(y', y)] \\
& + \gamma_{zx}[\cos(x', z) \cos(y', x) + \cos(x', x) \cos(y', z)] \\
\gamma_{y'z'} =\ & 2\epsilon_x \cos(y', x) \cos(z', x) + 2\epsilon_y \cos(y', y) \cos(z', y) \\
& + 2\epsilon_z \cos(y', z) \cos(z', z) \\
& + \gamma_{xy}[\cos(y', x) \cos(z', y) + \cos(y', y) \cos(z', x)] \\
& + \gamma_{yz}[\cos(y', y) \cos(z', z) + \cos(y', z) \cos(z', y)] \\
& + \gamma_{zx}[\cos(y', z) \cos(z', x) + \cos(y', x) \cos(z', z)] \\
\gamma_{z'x} =\ & 2\epsilon_x \cos(z', x) \cos(x', x) + 2\epsilon_y \cos(z', y) \cos(x', y) \\
& + 2\epsilon_z \cos(z', z) \cos(x', z) \\
& + \gamma_{xy}[\cos(z', x) \cos(x', y) + \cos(z', y) \cos(x', x)] \\
& + \gamma_{yz}[\cos(z', y) \cos(x', z) + \cos(z', z) \cos(x', y)] \\
& + \gamma_{zx}[\cos(z', z) \cos(x', x) + \cos(z', x) \cos(x', z)]
\end{aligned}
\tag{2.6.1}
$$

Equations 2.6.1 are the transformation equations for strain and are identical in form to the transformation equations for stress as given in Eqs. 1.53. To obtain one set of equations from the other it is only necessary to replace σ_x, σ_y, and σ_z by ϵ_x, ϵ_y, and ϵ_z, respectively, and τ_{xy}, τ_{yz}, and τ_{zx} by $\gamma_{xy}/2$, $\gamma_{yz}/2$, and $\gamma_{zx}/2$, respectively.

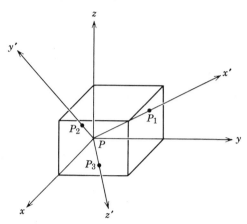

Fig. 2.6.1. Rotation of axes for rectangular coordinates.

Because the transformation laws for stress and strain are identical, the properties of these transformations are also the same. Thus there are principal strains, principal planes of strain, and principal directions. Furthermore, there is an ellipse of strain which can be derived by the same procedures used in Section 1.5, and the three invariants of strain, by analogy, are

$$I_1 = \epsilon_x + \epsilon_y + \epsilon_z$$

$$I_2 = \epsilon_x \epsilon_y + \epsilon_y \epsilon_z + \epsilon_z \epsilon_x - \frac{\gamma_{xy}^2}{4} - \frac{\gamma_{yz}^2}{4} - \frac{\gamma_{zx}^2}{4} \qquad (2.6.2)$$

$$I_3 = \epsilon_x \epsilon_y \epsilon_z - \frac{\epsilon_x \gamma_{yz}^2}{4} - \frac{\epsilon_y \gamma_{zx}^2}{4} - \frac{\epsilon_z \gamma_{xy}^2}{4} + \frac{\gamma_{xy} \gamma_{yz} \gamma_{zx}}{4}$$

The first invariant of strain is, under the assumption of infinitesimal deformation, the change in volume per unit volume or the volume dilatation e. This can be shown by considering the rectangular parallelepiped in Fig. 2.6.2. The volume dilatation e is by definition

$$e = \frac{\Delta V}{V} = \frac{V' - V}{V}$$

where V and V' are the original and final volumes of the rectangular parallelepiped. From Fig. 2.6.2

$$V = \Delta x \Delta y \Delta z$$

and

$$V' = (\Delta x + \delta x)(\Delta y + \delta y)(\Delta z + \delta z)$$

By definition of normal strain

$$\epsilon_x = \frac{\delta x}{\Delta x}, \qquad \epsilon_y = \frac{\delta y}{\Delta y} \qquad \text{and} \quad \epsilon_z = \frac{\delta z}{\Delta z}$$

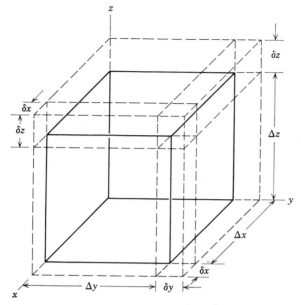

Fig. 2.6.2. Dilatation of a cube.

Therefore

$$e = (1 + \epsilon_x)(1 + \epsilon_y)(1 + \epsilon_z) - 1$$

Expanding and neglecting products of strain give

$$e = \epsilon_x + \epsilon_y + \epsilon_z \tag{2.6.3}$$

Thus the volume dilatation at any point is the sum of the three orthogonal normal strains at the point.

2.7 EQUATIONS OF COMPATIBILITY

From Eqs. 2.2.3, the six components of strain ϵ_x, ϵ_y, ϵ_z, γ_{xy}, γ_{yz}, and γ_{zx} are related to the three displacement components, u, v, and w by means of space derivatives. Thus

$$\epsilon_x = \frac{\partial u}{\partial x}, \qquad \gamma_{xy} = \frac{\partial u}{\partial y} + \frac{\partial v}{\partial x}$$

$$\epsilon_y = \frac{\partial v}{\partial y}, \qquad \gamma_{yz} = \frac{\partial v}{\partial z} + \frac{\partial w}{\partial y} \tag{2.7.1}$$

$$\epsilon_z = \frac{\partial w}{\partial z}, \qquad \gamma_{zx} = \frac{\partial w}{\partial x} + \frac{\partial u}{\partial z}$$

If the displacements are given as continuous functions of the space coordinates x, y, z, the six strain components can always be determined by means of the six equations in Eq. 2.7.1. Since there are three displacement components and six strain components, the six strain components in a body cannot be specified arbitrarily; otherwise they may not be compatible. Strains are compatible if they result in displacements that produce no separation between small elements of the body. Thus there must exist additional relations between the six strain components that must be satisfied at all times to assure compatible displacements. These additional relationships can be derived easily. Taking second partial derivatives of ϵ_x and ϵ_y with respect to y and x, respectively, gives

$$\frac{\partial^2 \epsilon_x}{\partial y^2} = \frac{\partial^3 u}{\partial y^2 \, \partial x} \quad \text{and} \quad \frac{\partial^2 \epsilon_y}{\partial x^2} = \frac{\partial^3 v}{\partial x^2 \, \partial y}$$

Taking the partial derivatives of γ_{xy} with respect to x and y gives

$$\frac{\partial^2 \gamma_{xy}}{\partial x \, \partial y} = \frac{\partial^3 u}{\partial y^2 \, \partial x} + \frac{\partial^3 v}{\partial x^2 \, \partial y}$$

Adding these two equations gives one of the compatibility equations as

$$\frac{\partial^2 \epsilon_x}{\partial y^2} + \frac{\partial^2 \epsilon_x}{\partial x^2} = \frac{\partial^2 \gamma_{xy}}{\partial x \, \partial y}$$

Taking the partial derivations of ϵ_x with respect to y and z gives

$$\frac{\partial^2 \epsilon_x}{\partial y \, \partial z} = \frac{\partial^3 u}{\partial x \, \partial y \, \partial z}$$

Taking the partial derivations of γ_{yz}, γ_{zx}, and γ_{xy}, with respect to x, y, and z respectively gives

$$\frac{\partial \gamma_{yz}}{\partial x} = \frac{\partial^2 v}{\partial x \, \partial z} + \frac{\partial^2 w}{\partial x \, \partial y}$$

$$\frac{\partial \gamma_{zx}}{\partial y} = \frac{\partial^2 u}{\partial y \, \partial z} + \frac{\partial^2 w}{\partial x \, \partial y}$$

$$\frac{\partial \gamma_{xy}}{\partial z} = \frac{\partial^2 u}{\partial y \, \partial z} + \frac{\partial^2 v}{\partial x \, \partial y}$$

Combining the last three equations as indicated below and taking partial derivatives with respect to x gives

$$\frac{\partial}{\partial x}\left(-\frac{\partial \gamma_{yz}}{\partial x} + \frac{\partial \gamma_{zx}}{\partial y} + \frac{\partial \gamma_{xy}}{\partial z} \right)$$

$$= -\frac{\partial^3 v}{\partial x^2 \, \partial z} - \frac{\partial^3 w}{\partial x^2 \, \partial y} + \frac{\partial^3 u}{\partial x \, \partial y \, \partial z} + \frac{\partial^3 w}{\partial x^2 \, \partial y} + \frac{\partial^3 u}{\partial x \, \partial y \, \partial z} + \frac{\partial^3 v}{\partial x^2 \, \partial z}$$

Therefore a second compatibility equation is

$$2\frac{\partial^2 \epsilon_x}{\partial y \, \partial z} = \frac{\partial}{\partial x}\left(-\frac{\partial \gamma_{yz}}{\partial x} + \frac{\partial \gamma_{zx}}{\partial y} + \frac{\partial \gamma_{xy}}{\partial z}\right)$$

By similar procedure, four additional equations can be derived. Thus the following six compatibility equations can be obtained:

$$\frac{\partial^2 \epsilon_x}{\partial y^2} + \frac{\partial^2 \epsilon_y}{\partial x^2} = \frac{\partial^2 \gamma_{xy}}{\partial x \, \partial y}$$

$$\frac{\partial^2 \epsilon_y}{\partial z^2} + \frac{\partial^2 \epsilon_z}{\partial y^2} = \frac{\partial^2 \gamma_{yz}}{\partial y \, \partial z}$$

$$\frac{\partial^2 \epsilon_z}{\partial x^2} + \frac{\partial^2 \epsilon_x}{\partial z^2} = \frac{\partial^2 \gamma_{zx}}{\partial z \, \partial x}$$

$$2\left(\frac{\partial^2 \epsilon_x}{\partial y \, \partial z}\right) = \frac{\partial}{\partial x}\left(-\frac{\partial \gamma_{yz}}{\partial x} + \frac{\partial \gamma_{zx}}{\partial y} + \frac{\partial \gamma_{xy}}{\partial z}\right)$$

$$2\left(\frac{\partial^2 \epsilon_y}{\partial z \, \partial x}\right) = \frac{\partial}{\partial y}\left(\frac{\partial \gamma_{yz}}{\partial x} - \frac{\partial \gamma_{zx}}{\partial y} + \frac{\partial \gamma_{xy}}{\partial z}\right)$$

$$2\left(\frac{\partial^2 \epsilon_z}{\partial x \, \partial y}\right) = \frac{\partial}{\partial z}\left(\frac{\partial \gamma_{yz}}{\partial x} + \frac{\partial \gamma_{zx}}{\partial y} - \frac{\partial \gamma_{xy}}{\partial z}\right)$$

(2.7.2)

These six compatibility equations must be satisfied if the six strain components as given by Eq. 2.7.1 and the displacement components u, v, and w are to be continuous functions of the space coordinates.

REFERENCES

1. Durelli, A. J., E. A. Phillips, and C. H. Tsao, *Introduction to the Theoretical and Experimental Analysis of Stress and Strain*, Chapter 2, McGraw-Hill Book Co., New York, 1958.
2. Frocht, M. M., *Photoelasticity*, Vol. I, Chapters 1 and 2, John Wiley and Sons, New York, 1941.
3. Jaeger, J. C., *Elasticity, Fracture and Flow*, Chapter 1, Methuen and Co., London, 1962.
4. Lee, George H, *An Introduction to Experimental Stress Analysis*, Chapter 1, John Wiley and Sons, New York, 1950.
5. Love, A. E. H., *The Mathematical Theory of Elasticity*, Chapter 1, Dover Publications, New York, 1944.
6. Sechler, Earnest E., *Elasticity in Engineering*, Chapter 4, John Wiley and Sons, New York, 1952.
7. Sokolnikoff, I. S., *Mathematical Theory of Elasticity*, Chapter 1, McGraw-Hill Book Co., New York, 1946.
8. Timoshenko, S. and J. N. Goodier, *Theory of Elasticity*, Chapters 1, 2, and 8, McGraw-Hill Book Co., New York, 1951.

CHAPTER 3

THEORY OF ELASTICITY

3.1 INTRODUCTION

Chapters 1 and 2 presented a discussion of stress and strain at a point in a body. Specifically, it was shown that stresses result from the application of surface and body forces and that the components of stress in any direction at a point can be determined if the six cartesian components of stress are given. Similarly, it was shown that the components of strain in any direction could be calculated from the six cartesian components of strain at a point. Strain was shown to be related to deformation of the body. The only restrictive assumptions necessary for the mathematical developments were: (1) that the body be composed of a continuous distribution of matter; (2) that strain be small compared to unity, and (3) that stress and deformation in the body be continuous functions of the space coordinates.

The restrictive assumptions used in Chapters 1 and 2 were quite general, for the elastic properties of the material of the body were not specified. Elastic properties of different materials can vary widely in magnitude and character. Thus before relationships between stress and strain can be developed, the type of material and its elastic properties must be specified.

The different branches of the science of mechanics are based upon certain idealized elastic properties of materials. Thus, for example, fluid mechanics deals with gases and liquids, and the theory of viscosity deals with viscous fluids. The classical theory of elasticity is restricted to solid materials processing the following idealized elastic properties.

1. *Linearity between stress and strain.* If a body is subjected to a stress, then the strain in the direction of the stress is directly proportional to the applied stress. That is, Hooke's law applies.

2. *Homogeneity*. The material of a body is uniformly distributed through-out the volume of the body and the elastic properties of the material are the same at all points in the body.

3. *Isotropy*. The elastic properties of the material are the same in all directions.

4. *Perfect elasticity*. Upon removal of deforming forces, the size and shape of a body return precisely to their original state.

No actual materials satisfy exactly all of these requirements. However, the deviations from the assumed idealized conditions for many materials are so slight that results predicted on the basis of the theory of elasticity are verified by experiment. Most structural materials and many rocks possess characteristics and elastic properties which permit the theory of elasticity to be used in practice. However, it is important to keep in mind the basic assumptions regarding the properties of solid materials when attempting to use elastic theory to solve problems in rock mechanics. The elastic properties of the rocks should be determined to ascertain if these rocks satisfy approximately the four basic assumptions given above.

3.2   STRESS-STRAIN RELATIONS

Relationships between the components of stress and those of strain have been established by experimentation; these relationships are known as Hooke's law. Assume a rectangular parallelepiped, with its sides parallel to the coordinate axes, acted upon by a normal stress σ_x uniformly distrib-uted over two opposite sides. The magnitude of the normal strain ϵ_x is given by

$$\epsilon_x = \frac{\sigma_x}{E}$$

where E is the modulus of elasticity. Extension of the body in the x direction is accompanied by a lateral contraction in both the y and z direction; thus

$$\epsilon_y = -\nu \frac{\sigma_x}{E} \quad \text{and} \quad \epsilon_z = -\nu \frac{\sigma_x}{E}$$

where ν is a constant known as Poisson's ratio. For most rocks and structural materials E is usually greater than 10^6 psi; thus stresses of the order of magnitude of 10^4 psi result in strains of less than 10^{-2} in./in. Poisson's ratio for many materials is between 0.15 and 0.35 and often is assumed equal to 0.25.

If the preceding rectangular parallelepiped is subjected to the simul-taneous action of normal stresses σ_x, σ_y, and σ_z uniformly distributed over its sides, then the resultant normal strains by the principle of superposition are

$$\epsilon_x = \frac{1}{E}\left[\sigma_x - \nu(\sigma_y + \sigma_z)\right]$$

$$\epsilon_y = \frac{1}{E}\left[\sigma_y - \nu(\sigma_z + \sigma_x)\right] \qquad (3.2.1)$$

$$\epsilon_z = \frac{1}{E}\left[\sigma_z - \nu(\sigma_x + \sigma_y)\right]$$

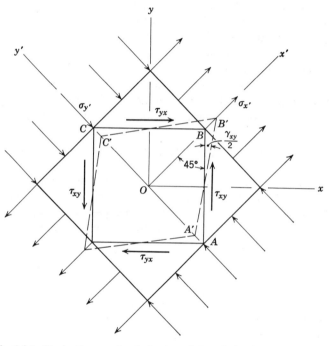

Fig. 3.2.1. Strain diagram for derivation of the relation between E and G.

Consider a rectangular element of a body with its sides parallel to x, y, z coordinates, acted upon by the shear stresses τ_{xy} and τ_{yx}. The normal stresses on the sides of another rectangular element at 45° to the original element are by Eq. 1.3.2 (see Fig. 3.2.1).

$$\sigma_{x'} = -\sigma_{y'} = \tau_{xy}$$

Since the normal stresses $\sigma_{x'}$ and $\sigma_{y'}$ are equal in magnitude but of opposite sign, the elongation of a line segment OB is equal to the shortening of a line segment OA. The right angle ABC will decrease in magnitude upon application of the shear stress τ_{xy}. This decrease can be calculated from triangle OBA. After application of the stress,

$$\frac{\overline{OA'}}{\overline{OB'}} = \tan\left(\frac{\pi}{4} - \frac{\gamma_{xy}}{2}\right) = \frac{1 + \epsilon_{y'}}{1 + \epsilon_{x'}}$$

From Eq. 3.2.1 and the values of $\sigma_{x'}$ and $\sigma_{y'}$ given above, the strains $\epsilon_{x'}$ and $\epsilon_{y'}$ are obtained as follows:

$$\epsilon_{x'} = \frac{1}{E}(\sigma_{x'} - \nu\sigma_{y'}) = \frac{(1 + \nu)\sigma_{x'}}{E} = \frac{1 + \nu}{E}\tau_{xy}$$

and

$$\epsilon_{y'} = \frac{1}{E}(\sigma_{y'} - \nu\sigma_{x'}) = \frac{-(1 + \nu)}{E}\sigma_{x'} = \frac{-(1 + \nu)}{E}\tau_{xy}$$

For small $\gamma_{xy}/2$ expansion of $\tan(\pi/4 - \gamma_{xy}/2)$ gives

$$\tan\left(\frac{\pi}{4} - \frac{\gamma_{xy}}{2}\right) = \frac{\tan(\pi/4) - \tan(\gamma_{xy}/2)}{1 + \tan(\pi/4)\tan(\gamma_{xy}/2)} = \frac{1 - (\gamma_{xy}/2)}{1 + (\gamma_{xy}/2)}$$

From these four equations, the following relationship can be obtained:

$$\frac{1 - (\gamma_{xy}/2)}{1 + (\gamma_{xy}/2)} = \frac{1 - [(1 + \nu)/E]\tau_{xy}}{1 + [(1 + \nu)/E]\tau_{xy}}$$

Simplification of this equation gives

$$\gamma_{xy} = \frac{2(1 + \nu)}{E}\tau_{xy} = \frac{\tau_{xy}}{G} \tag{3.2.2}$$

where

$$G = \frac{E}{2(1 + \nu)} \tag{3.2.3}$$

Consideration of the shear stresses τ_{yz} and τ_{zx} results in two more equations similar to Eq. 3.2.2. Thus the relationship between the components of shear stress and shear strain are

$$\gamma_{xy} = \frac{\tau_{xy}}{G}, \quad \gamma_{yz} = \frac{\tau_{yz}}{G}, \quad \gamma_{zx} = \frac{\tau_{zx}}{G} \tag{3.2.4}$$

The quantity G is the modulus of rigidity and relates shear stress to shear strain. The six equations in Eqs. 3.2.1 and 3.2.4 relate the components of stress to the components of strain and are known as Hooke's law equations for an isotropic solid.

One of the invariants of stress is (see Eq. 1.5.14)

$$\Theta = \sigma_x + \sigma_y + \sigma_z \tag{3.2.5}$$

and one of the invariants of strain is (see Eqs. 2.6.2 or 2.6.3)

$$e = \epsilon_x + \epsilon_y + \epsilon_z \tag{3.2.6}$$

where e is the dilatation.

Addition of Eqs. 3.2.1 and the use of Eqs. 3.2.5 and 3.2.6 gives

$$e = \frac{1 - 2\nu}{E} \Theta \tag{3.2.7}$$

For a hydrostatic stress field p,

$$\sigma_x = \sigma_y = \sigma_z = p$$

Therefore Eq. 3.2.7 becomes

$$e = \frac{3(1 - 2\nu)}{E} p = \frac{p}{K} \tag{3.2.8}$$

where

$$K = \frac{E}{3(1 - 2\nu)} \tag{3.2.9}$$

The quantity $E/3(1 - 2\nu)$ is the modulus of volume expansion or bulk modulus and is denoted by K.

The six stress-strain relations, Eqs. 3.2.1 and 3.2.4, can be rewritten to express stresses in terms of strains. Thus

$$\sigma_x = \frac{\nu E}{(1 + \nu)(1 - 2\nu)} e + \frac{E}{1 + \nu} \epsilon_x$$

$$\sigma_y = \frac{\nu E}{(1 + \nu)(1 - 2\nu)} e + \frac{E}{1 + \nu} \epsilon_y$$

$$\sigma_z = \frac{\nu E}{(1 + \nu)(1 - 2\nu)} e + \frac{E}{1 + \nu} \epsilon_z \tag{3.2.10}$$

$$\tau_{xy} = \frac{E}{2(1 + \nu)} \gamma_{xy}, \quad \tau_{yz} = \frac{E}{2(1 + \nu)} \gamma_{yz}, \quad \tau_{zx} = \frac{E}{2(1 + \nu)} \gamma_{zx}$$

The quantity $\nu E/(1 + \nu)(1 - 2\nu)$ is known as Lame's constant and is denoted by λ. Equations 3.2.10 can be rewritten, using λ and G, as

$$\sigma_x = \lambda e + 2G\epsilon_x, \quad \tau_{xy} = G\gamma_{xy}$$
$$\sigma_y = \lambda e + 2G\epsilon_y, \quad \tau_{yx} = G\gamma_{yz}$$
$$\sigma_z = \lambda e + 2G\epsilon_z, \quad \tau_{zx} = G\gamma_{zx} \tag{3.2.11}$$

The stress-strain relations in Eqs. 3.2.1 and 3.2.4 are sometimes written as

$$\epsilon_x = \frac{1}{E}[\sigma_x(1 + \nu) - \nu\Theta], \qquad \gamma_{xy} = \frac{2(1 + \nu)}{E}\tau_{xy}$$

$$\epsilon_y = \frac{1}{E}[\sigma_y(1 + \nu) - \nu\Theta], \qquad \gamma_{yz} = \frac{2(1 + \nu)}{E}\tau_{yz} \qquad (3.2.12)$$

$$\epsilon_z = \frac{1}{E}[\sigma_z(1 + \nu) - \nu\Theta], \qquad \gamma_{zx} = \frac{2(1 + \nu)}{E}\tau_{zx}$$

It should be noted that for an isotropic material there are only two independent elastic constants. If any two are known, the others can be calculated. The necessary equations for calculating any elastic constant from any two elastic constants are summarized in Table 3.2.1.

Table 3.2.1. Equations Relating the Elastic Constants of an Isotropic Body

$$E = 2G(1 + \nu) = 3K(1 - 2\nu) = \frac{\lambda(1 + \nu)(1 - 2\nu)}{\nu} = \frac{9KG}{3K + G} = \frac{G(3\lambda + 2G)}{\lambda + G}$$

$$= \frac{9K(K + \lambda)}{3K - \lambda}$$

$$G = \frac{E}{2(1 + \nu)} = \frac{3K(1 - 2\nu)}{2(1 + \nu)} = \frac{\lambda(1 - 2\nu)}{2\nu} = \frac{3EK}{9K - E} = \frac{3(K - \lambda)}{2}$$

$$= \frac{\sqrt{(E + \lambda)^2 + 8\lambda^2} + E - 3\lambda}{4}$$

$$K = \frac{E}{3(1 - 2\nu)} = \frac{2G(1 + \nu)}{3(1 - 2\nu)} = \frac{\lambda(1 + \nu)}{3\nu} = \frac{EG}{3(3G - E)} = \frac{3\lambda + 2G}{3}$$

$$= \frac{\sqrt{(E + \lambda)^2 + 8\lambda^2} + E + 3\lambda}{6}$$

$$\lambda = \frac{\nu E}{(1 + \nu)(1 - 2\nu)} = \frac{2G\nu}{1 - 2\nu} = \frac{3K\nu}{1 + \nu} = \frac{G(2G - E)}{E - 3G} = \frac{3K - 2G}{3}$$

$$= \frac{3K(3K - E)}{9K - E}$$

$$\nu = \frac{E - 2G}{2G} = \frac{3K - E}{6K} = \frac{\lambda}{2(\lambda + G)} = \frac{\lambda}{3K - \lambda} = \frac{3K - 2G}{2(3K + G)}$$

$$= \frac{\sqrt{(E + \lambda)^2 + 8\lambda^2} - (E + \lambda)}{4\lambda}$$

3.3 BASIC DIFFERENTIAL EQUATIONS IN ELASTIC THEORY

The problem in the theory of elasticity is to determine in every direction at every point within an elastic body the six stress components (σ_x, σ_y, σ_z, τ_{xy}, τ_{yz}, τ_{zx}), the six strain components (ϵ_x, ϵ_y, ϵ_z, γ_{xy}, γ_{yz}, γ_{zx}), and the three components of displacements (u, v, w), given the elastic constants of the body, the size and shape of the body, and the boundary conditions. The boundary conditions may be prescribed as applied loads, as applied displacements, or both. Body forces if present are usually prescribed as forces per unit volume or unit mass. As there are 15 unknowns, six stresses, six strains, and three displacements, theoretically there will have to be 15 independent partial differential equations in order to determine the 15 unknowns. If the 15 unknowns are determined as functions of the space coordinates, then the transformation equations for stresses and strains at a point (Eqs. 1.5.3 and 2.6.1) can be used to compute the stresses and strains in any direction at every point in the body.

The necessary and sufficient conditions which the components of stress, strain, and displacement must satisfy in order to obtain a solution of an elasticity problem are the following:

1. The stress-strain relations (Eqs. 3.2.1 and 3.2.4, or 3.2.11).
2. The strain-displacement relations (Eq. 2.2.3).
3. The equilibrium conditions (Eq. 1.8.1).
4. The compatibility conditions (Eq. 2.7.2).
5. The prescribed boundary conditions at the exterior surfaces of the body.

The first four conditions must be satisfied at every point both inside the body and on its surfaces. The fifth condition is concerned only with the exterior surfaces of the body and therefore is a boundary condition.

The necessary equations for specifying the boundary conditions can be obtained from Eqs. 1.5.1. If $\bar{X}_n$, $\bar{Y}_n$, $\bar{Z}_n$ are the applied components of stress on the surface of the body, then the stress components at points on the boundary are given by

$$\sigma_x \cos(n, x) + \tau_{xy} \cos(n, y) + \tau_{zx} \cos(n, z) = \bar{X}_n$$

$$\tau_{xy} \cos(n, x) + \sigma_y \cos(n, y) + \tau_{zy} \cos(n, z) = \bar{Y}_n \qquad (3.3.1)$$

$$\tau_{zx} \cos(n, x) + \tau_{yz} \cos(n, y) + \sigma_z \cos(n, z) = \bar{Z}_n$$

If known displacements are prescribed on the boundary of the body, the necessary boundary conditions are

$$u = \bar{u}$$
$$v = \bar{v} \qquad (3.3.2)$$
$$w = \bar{w}$$

where $\bar{u}$, $\bar{v}$, and $\bar{w}$ are the applied displacements. Boundary conditions can be prescribed by either Eqs. 3.3.1 or 3.3.2.

There are six stress-strain relations (Eqs. 3.2.1 and 3.2.4), six strain-displacement relations (Eqs. 2.2.3), and three equilibrium equations (Eqs. 1.8.1), for a total of 15 equations. Theoretically these 15 partial differential equations are necessary and sufficient to determine the 15 unknowns. The boundary conditions are used to evaluate the constants of integration that arise, and the compatibility equations are used to assure that the displacements are continuous functions of the space variables.

Some simplification can be obtained by combining the different sets of equations. For example, the six stress-strain relations can be combined with the six strain-displacement relations to eliminate the six strain components. The resulting equations are

$$\frac{\partial u}{\partial x} = \frac{1}{E}[\sigma_x - \nu(\sigma_y + \sigma_z)], \qquad \frac{\partial u}{\partial y} + \frac{\partial v}{\partial x} = \frac{1}{G}\tau_{xy},$$

$$\frac{\partial v}{\partial y} = \frac{1}{E}[\sigma_y - \nu(\sigma_z + \sigma_x)], \qquad \frac{\partial v}{\partial z} + \frac{\partial w}{\partial y} = \frac{1}{G}\tau_{yz}, \qquad (3.3.3)$$

$$\frac{\partial w}{\partial z} = \frac{1}{E}[\sigma_z - \nu(\sigma_x + \sigma_y)], \qquad \frac{\partial w}{\partial x} + \frac{\partial u}{\partial z} = \frac{1}{G}\tau_{zx}$$

or

$$\sigma_x = \lambda\left(\frac{\partial u}{\partial x} + \frac{\partial v}{\partial y} + \frac{\partial w}{\partial z}\right) + 2G\frac{\partial u}{\partial x}, \qquad \tau_{xy} = G\left(\frac{\partial u}{\partial y} + \frac{\partial v}{\partial x}\right)$$

$$\sigma_y = \lambda\left(\frac{\partial u}{\partial x} + \frac{\partial v}{\partial y} + \frac{\partial w}{\partial z}\right) + 2G\frac{\partial v}{\partial y}, \qquad \tau_{yz} = G\left(\frac{\partial v}{\partial z} + \frac{\partial w}{\partial y}\right) \quad (3.3.4)$$

$$\sigma_z = \lambda\left(\frac{\partial u}{\partial x} + \frac{\partial v}{\partial y} + \frac{\partial w}{\partial z}\right) + 2G\frac{\partial w}{\partial z}, \qquad \tau_{zx} = G\left(\frac{\partial w}{\partial x} + \frac{\partial u}{\partial z}\right)$$

The three equilibrium equations are

$$\frac{\partial \sigma_x}{\partial x} + \frac{\partial \tau_{yx}}{\partial y} + \frac{\partial \tau_{zx}}{\partial z} + X = 0$$

$$\frac{\partial \tau_{xy}}{\partial x} + \frac{\partial \sigma_y}{\partial y} + \frac{\partial \tau_{zy}}{\partial z} + Y = 0 \qquad (3.3.5)$$

$$\frac{\partial \tau_{xz}}{\partial x} + \frac{\partial \tau_{yz}}{\partial y} + \frac{\partial \sigma_z}{\partial z} + Z = 0$$

The six stress-displacement equations and three equilibrium equations form a set of nine equations with nine unknowns (six stresses and three displacements). This system of nine equations can be solved with the requirement that the stress components and displacement components satisfy the boundary conditions. The strain components can be calculated from the displacement components by means of the strain-displacement relations. Since the strains are calculated from the displacements, the compatibility equations will be satisfied automatically.

The compatibility Eqs. 2.7.2 can be expressed in terms of stresses by use of the stress-strain relations in Eq. 3.2.12 and the equilibrium equations. The resulting equations are

$$\nabla^2 \sigma_x + \frac{1}{1+\nu}\frac{\partial^2 \Theta}{\partial x^2} = \frac{-\nu}{1-\nu}\left(\frac{\partial X}{\partial x} + \frac{\partial Y}{\partial y} + \frac{\partial Z}{\partial z}\right) - 2\frac{\partial X}{\partial x}$$

$$\nabla^2 \sigma_y + \frac{1}{1+\nu}\frac{\partial^2 \Theta}{\partial y^2} = \frac{-\nu}{1-\nu}\left(\frac{\partial X}{\partial x} + \frac{\partial Y}{\partial y} + \frac{\partial Z}{\partial z}\right) - 2\frac{\partial Y}{\partial y}$$

$$\nabla^2 \sigma_z + \frac{1}{1+\nu}\frac{\partial^2 \Theta}{\partial z^2} = \frac{-\nu}{1-\nu}\left(\frac{\partial X}{\partial x} + \frac{\partial Y}{\partial y} + \frac{\partial Z}{\partial z}\right) - 2\frac{\partial Z}{\partial z} \qquad (3.3.6)$$

$$\nabla^2 \tau_{yz} + \frac{1}{1+\nu}\frac{\partial^2 \Theta}{\partial y\,\partial z} = -\left(\frac{\partial Y}{\partial z} + \frac{\partial Z}{\partial y}\right)$$

$$\nabla^2 \tau_{zx} + \frac{1}{1+\nu}\frac{\partial^2 \Theta}{\partial z\,\partial x} = -\left(\frac{\partial Z}{\partial x} + \frac{\partial X}{\partial z}\right)$$

$$\nabla^2 \tau_{xy} + \frac{1}{1+\nu}\frac{\partial^2 \Theta}{\partial x\,\partial y} = -\left(\frac{\partial X}{\partial y} + \frac{\partial Y}{\partial x}\right)$$

where

$$\nabla^2 = \frac{\partial^2}{\partial x^2} + \frac{\partial^2}{\partial y^2} + \frac{\partial^2}{\partial z^2} \qquad (3.3.7)$$

If the body forces are constant or zero, then Eqs. 3.3.6 become

$$(1+\nu)\nabla^2 \sigma_x + \frac{\partial^2 \Theta}{\partial x^2} = 0, \qquad (1+\nu)\nabla^2 \tau_{yz} + \frac{\partial^2 \Theta}{\partial y\,\partial z} = 0$$

$$(1+\nu)\nabla^2 \sigma_y + \frac{\partial^2 \Theta}{\partial y^2} = 0, \qquad (1+\nu)\nabla^2 \tau_{zx} + \frac{\partial^2 \Theta}{\partial z\,\partial x} = 0 \qquad (3.3.8)$$

$$(1+\nu)\nabla^2 \sigma_z + \frac{\partial^2 \Theta}{\partial z^2} = 0, \qquad (1+\nu)\nabla^2 \tau_{xy} + \frac{\partial^2 \Theta}{\partial x\,\partial y} = 0$$

Equations 3.3.6 with the equilibrium conditions constitute a set of nine equations which can be solved for the six stress components with the requirement that the stress components satisfy the boundary conditions. The excess of equations is not a real one as the equilibrium conditions were used in the derivation of Eqs. 3.3.6. By means of the stress-strain and/or strain-displacement equations, the six strains and three displacements can be determined from the stresses.

The equilibrium equations can be expressed in terms of displacements by use of the stress-displacement equations (3.3.3). The resulting equations are

$$\nabla^2 u + \frac{1}{1-2\nu} \frac{\partial}{\partial x}\left(\frac{\partial u}{\partial x} + \frac{\partial v}{\partial y} + \frac{\partial w}{\partial z}\right) + \frac{X}{G} = 0$$

$$\nabla^2 v + \frac{1}{1-2\nu} \frac{\partial}{\partial y}\left(\frac{\partial u}{\partial x} + \frac{\partial v}{\partial y} + \frac{\partial w}{\partial z}\right) + \frac{Y}{G} = 0 \qquad (3.3.9)$$

$$\nabla^2 w + \frac{1}{1-2\nu} \frac{\partial}{\partial z}\left(\frac{\partial u}{\partial x} + \frac{\partial v}{\partial y} + \frac{\partial w}{\partial z}\right) + \frac{Z}{G} = 0$$

Substitution of Eqs. 3.3.4 into Eqs. 3.3.1 gives the stress-boundary conditions in terms of displacements as

$$\lambda e \cos(n, x) + G\left[2\frac{\partial u}{\partial x}\cos(n, x) + \left(\frac{\partial u}{\partial y} + \frac{\partial v}{\partial x}\right)\cos(n, y)\right.$$

$$\left. + \left(\frac{\partial u}{\partial z} + \frac{\partial w}{\partial x}\right)\cos(n, z)\right] = \bar{X}_n$$

$$\lambda e \cos(n, y) + G\left[2\frac{\partial v}{\partial y}\cos(n, y) + \left(\frac{\partial v}{\partial z} + \frac{\partial w}{\partial y}\right)\cos(n, z)\right. \qquad (3.3.10)$$

$$\left. + \left(\frac{\partial v}{\partial x} + \frac{\partial u}{\partial y}\right)\cos(n, x)\right] = \bar{Y}_n$$

$$\lambda e \cos(n, z) + G\left[2\frac{\partial w}{\partial z}\cos(n, z) + \left(\frac{\partial w}{\partial x} + \frac{\partial u}{\partial z}\right)\cos(n, x)\right.$$

$$\left. + \left(\frac{\partial w}{\partial y} + \frac{\partial v}{\partial z}\right)\cos(n, y) = \bar{Z}_n\right.$$

Solutions of the stress-displacement equations (3.3.3) with the requirement that the displacements satisfy the boundary conditions as given by Eq. 3.3.2 will determine the three components of displacement. The strain-displacement equations are used to calculate the components of strain and the stress components are obtained from the stress-strain relations.

3.4 PLANE STRESS EQUATIONS

Problems in the theory of elasticity are considerably simplified if all stresses are parallel to one plane. There are many problems in which the stress distribution is essentially plane. The typical example for plane stress is that of a thin plate subjected to forces parallel to the plane of the plate and uniformly distributed over the thickness of the plate (see Fig. 3.4.1).

Plane stress conditions are obtained when σ_z, τ_{xz}, and τ_{yz} and all variations of stress with respect to z are zero. Thus for plane stress the stress-strain relations as given by Eqs. 3.2.1 and 3.2.4 reduce to the following:

Fig. 3.4.1. Plane stress conditions in thin plate.

$$\epsilon_x = \frac{1}{E}(\sigma_x - \nu\sigma_y)$$

$$\epsilon_y = \frac{1}{E}(\sigma_y - \nu\sigma_x)$$

$$\epsilon_z = -\frac{\nu}{E}(\sigma_x + \sigma_y) \tag{3.4.1}$$

$$\gamma_{xy} = \frac{\tau_{xy}}{G}$$

Also the equilibrium conditions in Eqs. 1.8.1, the strain-displacement relations in Eqs. 2.2.3, and the compatibility conditions in Eqs. 2.7.2, are, respectively

$$\frac{\partial\sigma_x}{\partial x} + \frac{\partial\tau_{xy}}{\partial y} + X = 0, \qquad \frac{\partial\tau_{xy}}{\partial x} + \frac{\partial\sigma_y}{\partial y} + Y = 0 \tag{3.4.2}$$

$$\epsilon_x = \frac{\partial u}{\partial x}, \qquad \epsilon_y = \frac{\partial v}{\partial y}, \qquad \gamma_{xy} = \frac{\partial u}{\partial y} + \frac{\partial v}{\partial x} \tag{3.4.3}$$

$$\frac{\partial^2\epsilon_x}{\partial y^2} + \frac{\partial^2\epsilon_y}{\partial x^2} = \frac{\partial^2\gamma_{xy}}{\partial x\,\partial y} \tag{3.4.4}$$

Substitution of Eqs. 3.4.1 into Eq. 3.4.4 gives

$$\frac{\partial^2}{\partial y^2}(\sigma_x - \nu\sigma_y) + \frac{\partial^2}{\partial x^2}(\sigma_y - \nu\sigma_x) = 2(1 + \nu)\frac{\partial^2\tau_{xy}}{\partial x\,\partial y}$$

Differentiation of the first equilibrium equation with respect to x, and the second one with respect to y, and addition of these results gives

$$2\frac{\partial^2\tau_{xy}}{\partial x\,\partial y} = -\frac{\partial^2\sigma_x}{\partial x^2} - \frac{\partial^2\sigma_y}{\partial y^2} - \frac{\partial X}{\partial x} - \frac{\partial Y}{\partial y}$$

Substitution of this equation into the preceding one gives

$$\left(\frac{\partial^2}{\partial x^2} + \frac{\partial^2}{\partial y^2}\right)(\sigma_x + \sigma_y) = -(1 + \nu)\left(\frac{\partial X}{\partial x} + \frac{\partial Y}{\partial y}\right) \qquad (3.4.5)$$

If body forces are constant or zero, Eq. 3.27 becomes

$$\left(\frac{\partial^2}{\partial x^2} + \frac{\partial^2}{\partial y^2}\right)(\sigma_x + \sigma_y) = 0 \qquad (3.4.6)$$

There are three unknown stress components. The equations necessary to obtain a solution are the two equilibrium equations (3.4.2) and the compatibility equation (either 3.4.5 or 3.4.6), together with the boundary conditions.

3.5 PLANE STRAIN EQUATIONS

Simplification of elasticity problems also occurs when a state of plane strain or plane deformation exists. If the displacements of all points of a deformed body are in planes normal to the length of a body, a state of plane strain exists. It does not complicate the problem if a uniform extension in the direction of the axis of the body is superimposed upon the plane deformation. A good example of plane strain is a long horizontal tunnel at depth in a rock mass.

For plane strain to exist, γ_{xz} and γ_{yz} must be zero throughout the body and the variation of ϵ_z with respect to z must be zero. Thus ϵ_z is zero or a constant. If ϵ_z is set equal to zero in the last equation of 3.2.1, the result is

$$\sigma_z = \nu(\sigma_x + \sigma_y) \qquad (3.5.1)$$

Thus the stress-strain relations for plane strain with $\epsilon_z = 0$ become

$$\epsilon_x = \frac{1}{E}[(1 - \nu^2)\sigma_x - \nu(1 + \nu)\sigma_y]$$

$$\epsilon_y = \frac{1}{E}[(1 - \nu^2)\sigma_y - \nu(1 + \nu)\sigma_x] \qquad (3.5.2)$$

$$\gamma_{xy} = \frac{\tau_{xy}}{G} = \frac{2(1 + \nu)}{E}\tau_{xy}$$

The equilibrium conditions, the strain-displacement equations, and the compatibility conditions are the same for plane strain as for plane stress. Substitution of Eqs. 3.5.2 into Eq. 3.4.4 gives

$$(1 - \nu)\left[\frac{\partial^2\sigma_x}{\partial y^2} + \frac{\partial^2\sigma_y}{\partial x^2}\right] - \nu\left[\frac{\partial^2\sigma_y}{\partial y^2} + \frac{\partial^2\sigma_x}{\partial x^2}\right] = 2\frac{\partial^2\tau_{xy}}{\partial x\,\partial y}$$

Differentiating the equilibrium equations as before, adding the results, and substituting into the above equation gives

$$\left(\frac{\partial^2}{\partial x^2} + \frac{\partial^2}{\partial y^2}\right)(\sigma_x + \sigma_y) = -\frac{1}{1-\nu}\left(\frac{\partial X}{\partial x} + \frac{\partial Y}{\partial y}\right) \qquad (3.5.3)$$

If the body forces are constant or zero, then Eq. 3.5.3 becomes

$$\left(\frac{\partial^2}{\partial x^2} + \frac{\partial^2}{\partial y^2}\right)(\sigma_x + \sigma_y) = 0 \qquad (3.5.4)$$

Thus, in the case of constant body forces, the compatibility equations in terms of stress is the same for both plane strain and plane stress. Also it should be noted that neither the compatibility equation nor the equilibrium equations contain the elastic constants of the body. Hence the stress distribution is the same for all isotropic materials provided the state of stress or strain is two-dimensional.

The equilibrium conditions and the compatibility condition together with the boundary condition provide the necessary equations for obtaining solutions to plane strain problems.

3.6 AIRY STRESS FUNCTION

It has been shown that two-dimensional problems in elastic theory reduce to solving the two differential equations of equilibrium and the compatibility differential equation. Constants of integration are evaluated by means of the boundary conditions. The usual method of solving these equations when body forces are constant or zero is to introduce a new function known as the Airy stress function.

Let the body forces be zero; then the Airy stress function Φ is defined as

$$\sigma_x = \frac{\partial^2 \Phi}{\partial y^2}, \quad \sigma_y = \frac{\partial^2 \Phi}{\partial x^2}, \quad \tau_{xy} = -\frac{\partial^2 \Phi}{\partial x\,\partial y} \qquad (3.6.1)$$

The two-dimensional equilibrium equations for zero body forces are

$$\frac{\partial \sigma_x}{\partial x} + \frac{\partial \tau_{xy}}{\partial y} = 0$$

$$\frac{\partial \sigma_y}{\partial y} + \frac{\partial \tau_{xy}}{\partial x} = 0 \qquad (3.6.2)$$

Substitution of Eq. 3.6.1 into 3.6.2 shows that Airy's stress function satisfies the equilibrium conditions. The two-dimensional compatibility condition is

$$\left(\frac{\partial^2}{\partial x^2} + \frac{\partial^2}{\partial y^2}\right)(\sigma_x + \sigma_y) = 0 \qquad (3.6.3)$$

Substitution of Eq. 3.6.1 into 3.6.3 gives

$$\left(\frac{\partial^2}{\partial x^2} + \frac{\partial}{\partial y^2}\right)\left(\frac{\partial^2\Phi}{\partial x^2} + \frac{\partial^2\Phi}{\partial y^2}\right) = \frac{\partial^4\Phi}{\partial x^4} + 2\frac{\partial^4\Phi}{\partial x^2\,\partial y^2} + \frac{\partial^4\Phi}{\partial y^4} = 0 \quad (3.6.4)$$

Any function Φ that satisfies Eq. 3.6.4 also satisfies the compatibility conditions and the equilibrium conditions. Thus, two-dimensional problems involving zero body forces reduce to solving a fourth-degree, biharmonic equation as given by Eq. 3.6.4.

3.7 SAINT VENANT'S PRINCIPLE

The solution of groups of simultaneous partial differential equations is a difficult task. Direct methods for obtaining solutions are not always available; therefore, we usually recourse to other methods. One such method is the educated guess of the final solution. The assumed solution is then checked against the basic equations and boundary conditions. If all conditions are not satisfied, revisions in the assumed solution are made and the process repeated until a solution is found that satisfies all conditions of the problem.

It has been found that the boundary conditions met in practice frequently cannot be precisely specified mathematically. Solutions to such problems sometimes can be found for other boundary conditions. In such cases, a principle proposed by Saint Venant is used. Briefly, this principle states: If a system of forces acting on one portion of the boundary is replaced by a statically equivalent system of forces acting on the same portion of the boundary, then the stresses, strains, and nonrigid body displacements in the parts of the body sufficiently far moved from this portion of the boundary remain approximately the same. This principle has been verified theoretically and experimentally so often that it is now accepted as a fundamental law in the theory of elasticity.

3.8 POLAR COORDINATES

In general the mathematical difficulties of problems in the theory of elasticity are reduced if the boundaries of a body can be made to coincide with the coordinate system used. Thus, for example, problems involving spheres or spherical openings in an infinite medium can be solved easier if spherical coordinates are used with the center of the coordinate system at the center of the sphere.

General expressions for the stress-strain relations, the strain-displacement relations, and the equilibrium conditions can be derived for any general curvalinear coordinate system (Sokolnikoff, 1956). These derivations are beyond the scope of this chapter and will not be given. Instead the basic equations for plane polar coordinates will be derived to illustrate the principles involved.

The relationships between polar coordinates and rectangular coordinates are from Fig. 3.8.1a

$$x = r \cos \theta$$
$$y = r \sin \theta$$

(3.8.1)

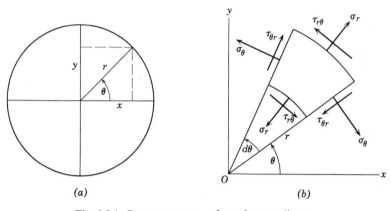

(a)   *(b)*

Fig. 3.8.1. Stress components for polar coordinates.

The positive normal and shear-stress components for polar coordinates are shown in Fig. 3.8.1b. By summing moments of forces, we can show that $\tau_{r\theta} = \tau_{\theta r}$ at a point.

The displacement of a point in the radial, or r direction, is designated as u and the displacement of a point in the tangential, or θ direction, is designated as v. The strain-displacement relations can be determined from Fig. 3.8.2. In Fig. 3.8.2a the line segment ab becomes $a'b'$ after deformation. Thus the radial strain is

$$\epsilon_r = \frac{a'b' - ab}{ab} = \frac{u + \frac{\partial u}{\partial r} dr - u}{dr}$$

Thus

$$\epsilon_r = \frac{\partial u}{\partial r}$$

(3.8.2)

As a result of a radial displacement u, the line segment ac becomes $a'c''$. The length of the line segment ac is $r\, d\theta$, and the length of the line segment $a'c''$ is $(r + u)\, d\theta$. Therefore the tangential strain resulting from the radial

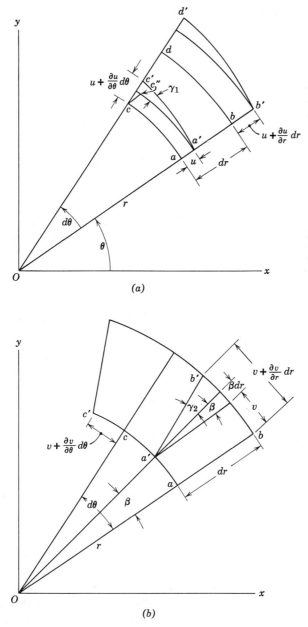

Fig. 3.8.2. Strain components in polar coordinates.

displacement u is

$$\epsilon_{\theta_1} = \frac{(r + u)\, d\theta - r\, d\theta}{r\, d\theta} = \frac{u}{r}$$

There is also a tangential strain as a result of tangential displacements. From Fig. 3.8.2b, the line segment ac becomes $a'c'$ after tangential deformation. The length of ac is $r\, d\theta$. Thus the tangential strain is

$$\epsilon_{\theta_2} = \frac{v + (\partial v/\partial \theta)\, d\theta - v}{r\, d\theta} = \frac{1}{r}\frac{\partial v}{\partial \theta}$$

The total tangential strain ϵ_θ is the sum of ϵ_{θ_1} and ϵ_{θ_2}, thus

$$\epsilon_\theta = \frac{u}{r} + \frac{1}{r}\frac{\partial v}{\partial \theta} \tag{3.8.3}$$

The shear strain $\gamma_{r\theta}$ can be calculated from the change in the right angle cab. From Fig. 3.8.2a and the fact that changes in angles are small

$$\tan \gamma_1 = \gamma_1 = \frac{\left(\dfrac{\partial u}{\partial \theta}\right) d\theta}{r\, d\theta} = \frac{1}{r}\frac{\partial u}{\partial \theta}$$

From Fig. 3.8.2b and by remembering that changes in angles are small, we see that

$$\tan \beta = \beta = \frac{v}{r}$$

and

$$\tan \gamma_2 = \gamma_2 = \frac{v + (\partial v/\partial r)\, dr - v - \beta\, dr}{dr}$$

Combining these two equations gives

$$\gamma_2 = \frac{(\partial v/\partial r)\, dr - (v/r)\, dr}{dr} = \frac{\partial v}{\partial r} - \frac{v}{r}$$

Therefore the total shear strain $(\gamma_1 + \gamma_2)$ is

$$\gamma_{r\theta} = \frac{1}{r}\frac{\partial u}{\partial \theta} + \frac{\partial v}{\partial r} - \frac{v}{r} \tag{3.8.4}$$

The equilibrium equation can be derived by a consideration of the normal and shear stresses acting on an element of a body cut out by two radii and two circular arcs, Fig. 3.8.3. The position of a point at the center of this element is given by r and θ. The stresses acting at this point are σ_r, σ_θ, and $\tau_{r\theta}$. However, the stresses acting on the faces of the element about the point will be different because of variation of stress with position. Thus all the normal and shear stresses acting on the element are designated with a subscript to associate the stress with the sides of the element. The

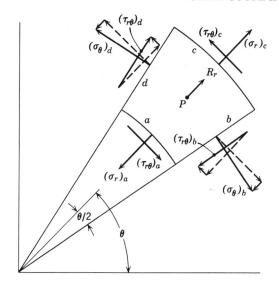

Fig. 3.8.3. Equilibrium conditions for polar coordinates.

length of line a is $r_a\, d\theta$, the length of line c is $r_c\, d\theta$, and the length of lines b and d are $r_c - r_a = dr$. Also the radial body force is R_r.

Summation of forces in a radial direction through the center of the element gives

$$\sum F_r = (\sigma_r)_c r_c\, d\theta - (\sigma_r)_a r_a\, d\theta + (\tau_{r\theta})_d \cos\frac{d\theta}{2}\, dr - (\tau_{r\theta})_b \cos\frac{d\theta}{2}\, dr$$

$$- (\sigma_\theta)_b \sin\frac{d\theta}{2}\, dr - (\sigma_\theta)_d \sin\frac{d\theta}{2}\, dr + R_r r\, d\theta\, dr = 0$$

Summation of forces in a tangential direction normal to the radial direction through the center of the element gives

$$\sum F_\theta = (\sigma_\theta)_d \cos\frac{d\theta}{2}\, dr - (\sigma_\theta)_b \cos\frac{d\theta}{2}\, dr + (\tau_{r\theta})_d \sin\frac{d\theta}{2}\, dr$$

$$+ (\tau_{r\theta})_c r_c\, d\theta - (\tau_{r\theta})_a r_a\, d\theta + (\tau_{r\theta})_b \sin\frac{d\theta}{2}\, dr = 0$$

Since $d\theta/2$ is small, $\sin(d\theta/2) = d\theta/2$ and $\cos(d\theta/2) = 1$. Dividing the preceding equations by $dr\, d\theta$ gives

$$\frac{(\sigma_r r)_c - (\sigma_r r)_a}{dr} - \frac{(\sigma_\theta)_b + (\sigma_\theta)_d}{2} + \frac{(\tau_{r\theta})_d - (\tau_{r\theta})_b}{d\theta} + R_r r = 0$$

and

$$(3.8.5)$$

$$\frac{(\sigma_\theta)_d - (\sigma_\theta)_b}{d\theta} + \frac{(\tau_{r\theta} r)_c - (\tau_{r\theta} r)_a}{dr} + \frac{(\tau_{r\theta})_d + (\tau_{r\theta})_b}{2} = 0$$

As the dimensions of the element are made smaller and smaller, the following relations will hold in the limit:

$$\frac{(\sigma_r r)_c - (\sigma_r r)_a}{dr} \to \frac{\partial(\sigma_r r)}{\partial r} = r\frac{\partial \sigma_r}{\partial r} + \sigma_r$$

$$(\sigma_\theta)_b \to (\sigma_\theta)_a \to \sigma_\theta$$

$$\frac{(\tau_{r\theta})_d - (\tau_{r\theta})_b}{d\theta} \to \frac{\partial \tau_{r\theta}}{\partial \theta}$$

$$\frac{(\sigma_\theta)_d - (\sigma_\theta)_b}{d\theta} \to \frac{\partial \sigma_\theta}{\partial \theta}$$

$$\frac{(\tau_{r\theta}r)_c - (\tau_{r\theta}r)_a}{dr} \to \frac{\partial(\tau_{r\theta}r)}{\partial r} = r\frac{\partial \tau_{r\theta}}{\partial r} + \tau_{r\theta}$$

$$(\tau_{r\theta})_d \to (\tau_{r\theta})_b \to \tau_{r\theta}$$

Substitution of these relations into Eqs. 3.8.5 gives the equilibrium equations in polar coordinates as

$$\frac{\partial \sigma_r}{\partial r} + \frac{\sigma_r - \sigma_\theta}{r} + \frac{1}{r}\frac{\partial \tau_{r\theta}}{\partial \theta} + R_r = 0$$
$$\frac{1}{r}\frac{\partial \sigma_\theta}{\partial \theta} + \frac{\partial \tau_{r\theta}}{\partial r} + \frac{2\tau_{r\theta}}{r} = 0$$

$$(3.8.6)$$

The compatibility equations in polar coordinates can be derived from the three strain-displacement equations

$$\epsilon_r = \frac{\partial u}{\partial r}$$

$$\epsilon_\theta = \frac{u}{r} + \frac{1}{r}\frac{\partial v}{\partial \theta}$$

$$(3.8.7)$$

$$\gamma_{r\theta} = \frac{1}{r}\frac{\partial u}{\partial \theta} + \frac{\partial v}{\partial r} - \frac{v}{r}$$

The second partial derivative of ϵ_r with respect to θ gives

$$\frac{\partial^2 \epsilon_r}{\partial \theta^2} = \frac{\partial^3 u}{\partial \theta^2 \partial r}$$

$$(3.8.8)$$

The partial derivative of $r\epsilon_\theta$ with respect to r gives

$$\frac{\partial r\epsilon_\theta}{\partial r} = \frac{\partial u}{\partial r} + \frac{\partial^2 v}{\partial r\, \partial \theta} = \epsilon_r + \frac{\partial^2 v}{\partial r\, \partial \theta}$$

and the partial derivative of this equation with respect to r gives

$$\frac{\partial^2 r\epsilon_\theta}{\partial r^2} = \frac{\partial \epsilon_r}{\partial r} + \frac{\partial^3 v}{\partial r^2 \partial \theta} \tag{3.8.9}$$

The partial derivative of $r\gamma_{r\theta}$ with respect to r gives

$$\frac{\partial r\gamma_{r\theta}}{\partial r} = \frac{\partial^2 u}{\partial r \partial \theta} + r\frac{\partial^2 v}{\partial r^2}$$

and the partial derivative of this equation with respect to θ gives

$$\frac{\partial^2 r\gamma_{r\theta}}{\partial \theta \, \partial r} = \frac{\partial^3 u}{\partial r \partial \theta^2} + r\frac{\partial^3 v}{\partial \theta \, \partial r^2} \tag{3.8.10}$$

Substitution of Eqs. 3.8.8 and 3.8.9 into 3.8.10 gives

$$\frac{\partial^2 r\gamma_{r\theta}}{\partial \theta \, \partial r} = \frac{\partial^2 \epsilon_r}{\partial \theta^2} + r\frac{\partial^2 r\epsilon_\theta}{\partial r^2} - r\frac{\partial \epsilon_r}{\partial r} \tag{3.8.11}$$

Equation 3.8.11 is the compatibility equation which must be satisfied if the strains are to be compatible with the displacement such that the displacements are continuous functions of the space coordinates. This equation can be expressed in terms of stresses only by making use of the stress-strain relations and the equilibrium conditions in polar coordinates.

The stress-strain relations in plane polar coordinates are

$$\epsilon_r = \frac{1}{E}(\sigma_r - \nu\sigma_\theta)$$

$$\epsilon_\theta = \frac{1}{E}(\sigma_\theta - \nu\sigma_r) \tag{3.8.12}$$

$$\gamma_{r\theta} = \frac{2(1 + \nu)}{E}\tau_{r\theta}$$

Substitution of Eqs. 3.8.12 into 3.8.11 gives

$$2(1 + \nu)\frac{\partial^2 r\tau_{r\theta}}{\partial \theta \, \partial r} = \frac{\partial^2(\sigma_r - \nu\sigma_\theta)}{\partial \theta^2} + r\frac{\partial^2 r(\sigma_\theta - \nu\sigma_r)}{\partial r^2} - r\frac{\partial(\sigma_r - \nu\sigma_\theta)}{\partial r}$$

Expanding the term on the left-hand side and the middle term on the right-hand side of the equation, and then collecting these terms gives

$$2(1 + \nu)\left(\frac{\partial \tau_{r\theta}}{\partial \theta} + r\frac{\partial^2 \tau_{r\theta}}{\partial r \partial \theta}\right) = \frac{\partial^2 \sigma_r}{\partial \theta^2} + 2r\frac{\partial \sigma_\theta}{\partial r} + r^2\frac{\partial^2 \sigma_\theta}{\partial r^2} - r\frac{\partial \sigma_r}{\partial r}$$
$$- \nu\left(\frac{\partial^2 \sigma_\theta}{\partial \theta^2} + 2r\frac{\partial \sigma_r}{\partial r} + r^2\frac{\partial^2 \sigma_r}{\partial r^2} - r\frac{\partial \sigma_\theta}{\partial r}\right) \tag{3.8.13}$$

If the body force is zero, the partial derivative of the first equilibrium equation with respect to r is

$$\frac{\partial^2 \sigma_r}{\partial r^2} - \frac{1}{r^2}\frac{\partial \tau_{r\theta}}{\partial \theta} + \frac{1}{r}\frac{\partial^2 \tau_{r\theta}}{\partial r \partial \theta} - \frac{(\sigma_r - \sigma_\theta)}{r^2} + \frac{1}{r}\frac{\partial(\sigma_r - \sigma_\theta)}{\partial r} = 0 \quad (3.8.14)$$

The partial derivative of the second equilibrium equation with respect to θ, multiplied by $1/r$ is

$$\frac{1}{r^2}\frac{\partial^2 \sigma_\theta}{\partial \theta^2} + \frac{1}{r}\frac{\partial^2 \tau_{r\theta}}{\partial r \partial \theta} + \frac{2}{r^2}\frac{\partial \tau_{r\theta}}{\partial \theta} = 0 \quad (3.8.15)$$

Multiplying the first equilibrium equation by $1/r$ and solving for $(\sigma_r - \sigma_\theta)/r^2$ gives

$$\frac{\sigma_r - \sigma_\theta}{r^2} = -\frac{1}{r}\frac{\partial \sigma_r}{\partial r} - \frac{1}{r^2}\frac{\partial \tau_{r\theta}}{\partial \theta} \quad (3.8.16)$$

Substitution of Eq. 3.8.16 into 3.8.14 gives

$$\frac{\partial^2 \sigma_r}{\partial r^2} + \frac{1}{r}\frac{\partial^2 \tau_{r\theta}}{\partial r \partial \theta} + \frac{1}{r}\frac{\partial \sigma_r}{\partial r} + \frac{1}{r}\frac{\partial(\sigma_r - \sigma_\theta)}{\partial r} = 0 \quad (3.8.17)$$

Addition of Eqs. 3.8.17 and 3.8.15 gives

$$\frac{2}{r^2}\frac{\partial \tau_{r\theta}}{\partial \theta} + \frac{2}{r}\frac{\partial^2 \tau_{r\theta}}{\partial r \partial \theta} + \frac{\partial^2 \sigma_r}{\partial r^2} + \frac{1}{r^2}\frac{\partial^2 \sigma_\theta}{\partial \theta^2} + \frac{1}{r}\frac{\partial \sigma_r}{\partial r} + \frac{1}{r}\frac{\partial(\sigma_r - \sigma_\theta)}{\partial r} = 0$$

This equation can be written as

$$2\left(\frac{\partial \tau_{r\theta}}{\partial r} + r\frac{\partial^2 \tau_{r\theta}}{\partial r \partial \theta}\right) = -r^2\frac{\partial^2 \sigma_r}{\partial r^2} - \frac{\partial^2 \sigma_\theta}{\partial \theta^2} - 2r\frac{\partial \sigma_r}{\partial r} + r\frac{\partial \sigma_\theta}{\partial r} \quad (3.8.18)$$

Substitution of Eq. 3.8.18 into Eq. 3.8.13 gives

$$-r^2\frac{\partial^2 \sigma_r}{\partial r^2} - \frac{\partial^2 \sigma_\theta}{\partial \theta^2} - 2r\frac{\partial \sigma_r}{\partial r} + r\frac{\partial \sigma_\theta}{\partial r} = \frac{\partial^2 \sigma_r}{\partial \theta^2} + 2r\frac{\partial \sigma_\theta}{\partial r} + r^2\frac{\partial^2 \sigma_\theta}{\partial r^2} - r\frac{\partial \sigma_r}{\partial r}$$

Simplification and division by r^2 gives

$$\frac{\partial^2(\sigma_r + \sigma_\theta)}{\partial r^2} + \frac{1}{r^2}\frac{\partial^2(\sigma_r + \sigma_\theta)}{\partial \theta^2} + \frac{1}{r}\frac{\partial(\sigma_r + \sigma_\theta)}{\partial r} = 0$$

or as

$$\left(\frac{\partial^2}{\partial r^2} + \frac{1}{r^2}\frac{\partial^2}{\partial \theta^2} + \frac{1}{r}\frac{\partial}{\partial r}\right)(\sigma_r + \sigma_\theta) = 0 \quad (3.8.19)$$

Equation 3.8.19 is the compatibility equation in polar coordinates and corresponds to Eq. 3.4.6 for cartesian coordinates. This equation plus the equilibrium equations and the boundary conditions provide the necessary equations to obtain solutions to two-dimensional problems in polar coordinates for zero body forces.

An Airy stress function can be found for polar coordinates just as was possible for the two-dimensional cartesian case. The stress components for the Airy stress function are given by

$$\sigma_r = \frac{1}{r}\frac{\partial \Phi}{\partial r} + \frac{1}{r^2}\frac{\partial^2 \Phi}{\partial \theta^2}$$

$$\sigma_\theta = \frac{\partial^2 \Phi}{\partial r^2} \tag{3.8.20}$$

$$\tau_{r\theta} = \frac{1}{r^2}\frac{\partial \Phi}{\partial \theta} - \frac{1}{r}\frac{\partial^2 \Phi}{\partial r\,\partial \theta}$$

By direct substitution, it can be checked that these stress functions satisfy the equilibrium conditions. Substitution of Eqs. 3.8.20 into 3.8.19 gives

$$\left(\frac{\partial^2}{\partial r^2} + \frac{1}{r^2}\frac{\partial^2}{\partial \theta^2} + \frac{1}{r}\frac{\partial}{\partial r}\right)\left(\frac{\partial^2 \Phi}{\partial r^2} + \frac{1}{r^2}\frac{\partial^2 \Phi}{\partial \theta^2} + \frac{1}{r}\frac{\partial \Phi}{\partial r}\right) = 0 \tag{3.8.21}$$

This equation in polar coordinates corresponds to Eq. 3.6.4 in cartesian coordinates. Problems in the theory of elasticity are solved in polar coordinates by finding a stress function Φ that satisfies Eq. 3.8.21 and whose corresponding stress components, as given by Eqs. 3.8.20, satisfy the boundary conditions.

3.9 STRAIN ENERGY

The application of a force to a surface of a body produces deformation in the body and, as a result, the surface on which the force acts moves as the force is applied. Thus a certain amount of work is done on the body. For perfect elastic materials this work is converted to potential energy of strain.

Consider a uniform bar loaded in simple tension by a normal stress acting on the ends of the bar as shown in Fig. 3.9.1a. The force on the ends of the bar is $\sigma_x\,dy\,dz$; after application of this force, the end of the bar is extended $\epsilon_x\,dx$. The relation between these two quantities during this application of the load is as shown in Fig. 3.9.1b. The work, dW, done during application of the load is the area of the triangle OAB, thus

$$dW = \tfrac{1}{2}\sigma_x\epsilon_x\,dx\,dy\,dz$$

A consideration of the other components of stress and strain gives similar equations. Thus the total work per unit volume, W_0, resulting from

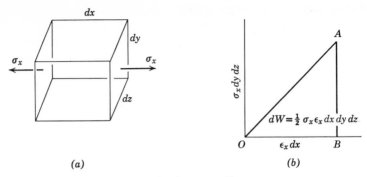

Fig. 3.9.1. Strain energy diagram.

the simultaneous application of the six stress components σ_x, σ_y, σ_z, τ_{xy}, τ_{yz}, and τ_{zx} is

$$W_0 = \tfrac{1}{2}(\sigma_x \epsilon_x + \sigma_y \epsilon_y + \sigma_z \epsilon_z + \tau_{xy}\gamma_{xy} + \tau_{yz}\gamma_{yz} + \tau_{zx}\gamma_{zx}) \quad (3.9.1)$$

The quantity W_0 is the strain energy per unit volume and can be expressed in terms of stress components or strain components only by means of Hooke's law equations. Thus using Eqs. 3.2.1 and 3.2.4

$$W_0 = \frac{1}{2E}(\sigma_x^{\,2} + \sigma_y^{\,2} + \sigma_z^{\,2}) - \frac{\nu}{E}(\sigma_x\sigma_y + \sigma_y\sigma_z + \sigma_z\sigma_x)$$
$$+ \frac{1}{2G}(\tau_{xy}^{\,2} + \tau_{yz}^{\,2} + \tau_{zx}^{\,2}) \quad (3.9.2)$$

Or, using Eq. 3.2.11

$$W_0 = \tfrac{1}{2}\lambda e^2 + G(\epsilon_x^{\,2} + \epsilon_y^{\,2} + \epsilon_z^{\,2}) + \tfrac{1}{2}G(\gamma_{xy}^{\,2} + \gamma_{yz}^{\,2} + \gamma_{zx}^{\,2}) \quad (3.9.3)$$

In Eq. 3.9.3 it is easy to show that the derivative of W_0 with respect to any one strain component is the corresponding stress component. Thus, for example,

$$\frac{\partial W_0}{\partial \epsilon_x} = \lambda e + 2G\epsilon_x = \sigma_x \quad (3.9.4)$$

Similar equations for the other components of stress can be obtained.

The strain energy per unit volume can be shown to be composed of two parts, one due to a change in volume and the other due to distortion. Let the mean strain e_m be defined as

$$e_m = \frac{\epsilon_x + \epsilon_y + \epsilon_z}{3} = \frac{e}{3} \quad (3.9.5)$$

Let the deviations of the normal strains from the mean strain be e_x, e_y, and e_z. Then

$$e_x = \epsilon_x - e_m = \frac{2\epsilon_x - \epsilon_y - \epsilon_z}{3}$$

$$e_y = \epsilon_y - e_m = \frac{2\epsilon_y - \epsilon_z - \epsilon_x}{3} \qquad (3.9.6)$$

$$e_z = \epsilon_z - e_m = \frac{2\epsilon_z - \epsilon_x - \epsilon_y}{3}$$

With some algebra and the relations in Eqs. 3.9.5 and 3.9.6, the following relationships can be derived:

$$e_x + e_y + e_z = \epsilon_x + \epsilon_y + \epsilon_z - 3e_m = 0 \qquad (3.9.7)$$

$$e_x{}^2 + e_y{}^2 + e_z{}^2 = -2(e_x e_y + e_y e_z + e_z e_x) \qquad (3.9.8)$$

$$e_x{}^2 + e_y{}^2 + e_z{}^2 = (\epsilon_x - \epsilon_y)^2 + (\epsilon_y - \epsilon_z)^2 + (\epsilon_z - \epsilon_x)^2 \qquad (3.9.9)$$

Substitution of Eqs. 3.9.6 into the strain-energy function as given by Eq. 3.9.3 gives

$$W_0 = \tfrac{1}{2}\lambda(3e_m)^2 + G[(e_x + e_m)^2 + (e_y + e_m)^2 + (e_z + e_m)^2] + \tfrac{1}{2}G(\gamma_{xy}{}^2 + \gamma_{yz}{}^2 + \gamma_{zx}{}^2)$$

Expansion of this equation, using Eq. 3.9.7 and the relationships in Table 3.2.1, gives the strain-energy function as

$$W_0 = \frac{K}{2}(3e_m)^2 + G(e_x{}^2 + e_y{}^2 + e_z{}^2) + \frac{G}{2}(\gamma_{xy}{}^2 + \gamma_{yz}{}^2 + \gamma_{zx}{}^2) \qquad (3.9.10)$$

Using Eq. 3.9.9, the strain energy function in terms of the original strains can be written as

$$W_0 = \frac{Ke^2}{2} + G[(\epsilon_x - \epsilon_y)^2 + (\epsilon_y - \epsilon_z)^2 + (\epsilon_z - \epsilon_x)^2]$$

$$+ \frac{G}{2}(\gamma_{xy}{}^2 + \gamma_{yz}{}^2 + \gamma_{zx}{}^2) \qquad (3.9.11)$$

The term $Ke^2/2$ in either Eq. 3.9.10 or 3.9.11 is the strain energy resulting from a change in volume. The strain energy resulting from distortion is the sum of the last two terms in either Eq. 3.9.10 or 3.9.11.

The same result can be obtained by starting with stresses instead of strains. Also principal stresses or principal strains could have been used. By using principal stresses, the mean stress is

$$\sigma_m = \frac{\sigma_1 + \sigma_2 + \sigma_3}{3} \qquad (3.9.12)$$

and the principal deviation stresses are

$$s_1 = \sigma_1 - \sigma_m$$
$$s_2 = \sigma_2 - \sigma_m \qquad (3.9.13)$$
$$s_3 = \sigma_3 - \sigma_m$$

The strain-energy function for principal stresses is, from Eq. 3.9.2

$$W_0 = \frac{1}{2E}(\sigma_1^2 + \sigma_2^2 + \sigma_3^2) - \frac{\nu}{E}(\sigma_1\sigma_2 + \sigma_2\sigma_3 + \sigma_3\sigma_1) \quad (3.9.14)$$

After some algebra and making use of Eqs. 3.9.12, 3.9.13, and the relations in Table 3.2.1, Eq. 3.9.14 can be written as

$$W_0 = \frac{\sigma_m^2}{2K} + \frac{s_1^2 + s_2^2 + s_3^2}{4G} \qquad (3.9.15)$$

Using Eq. 3.9.13, the strain-energy function can also be written as

$$W_0 = \frac{1 - 2\nu}{6E}(\sigma_1 + \sigma_2 + \sigma_3)^2 + \frac{(\sigma_1 - \sigma_2)^2 + (\sigma_2 - \sigma_3)^2 + (\sigma_3 - \sigma_1)^2}{12G}$$
$$(3.9.16)$$

The first terms on the right-hand side of Eq. 3.9.15 or 3.9.16 represent the strain energy due to change in volume, and the second terms represent the strain energy due to distortion.

The similarity of the terms in Eq. 3.9.16 and those in Eq. 1.6.2 for the octahedral normal and shear stresses should be noted.

3.10   SUMMARY OF SOME IMPORTANT FORMULAE

The strain-displacement and equilibrium equations are summarized for rectangular, cylindrical, and spherical coordinates. The displacement components are u, v, and w in the x, y, and z or r, θ, and z or the r, θ, and ϕ directions respectively.

Rectangular Coordinates (x, y, z) Strain-Displacement Equations

$$\epsilon_x = \frac{\partial u}{\partial x} \qquad \gamma_{xy} = \frac{\partial u}{\partial y} + \frac{\partial v}{\partial x}$$

$$\epsilon_y = \frac{\partial v}{\partial y} \qquad \gamma_{yz} = \frac{\partial v}{\partial z} + \frac{\partial w}{\partial y} \qquad (3.10.1)$$

$$\epsilon_z = \frac{\partial w}{\partial z} \qquad \gamma_{zx} = \frac{\partial w}{\partial x} + \frac{\partial u}{\partial z}$$

Equilibrium Conditions for No Body Forces

$$\frac{\partial \sigma_x}{\partial x} + \frac{\partial \tau_{xy}}{\partial y} + \frac{\partial \tau_{zx}}{\partial z} = 0$$

$$\frac{\partial \tau_{xy}}{\partial x} + \frac{\partial \sigma_y}{\partial y} + \frac{\partial \tau_{yz}}{\partial z} = 0 \qquad (3.10.2)$$

$$\frac{\partial \tau_{zx}}{\partial x} + \frac{\partial \tau_{yz}}{\partial y} + \frac{\partial \sigma_z}{\partial z} = 0$$

Cylindrical Coordinates (r, θ, z) Strain-Displacement Equations

$$\epsilon_r = \frac{\partial u}{\partial r} \qquad\qquad \gamma_{r\theta} = \frac{1}{r}\frac{\partial u}{\partial \theta} + \frac{\partial v}{\partial r} - \frac{v}{r}$$

$$\epsilon_\theta = \frac{u}{r} + \frac{1}{r}\frac{\partial v}{\partial \theta} \qquad \gamma_{\theta z} = \frac{\partial v}{\partial z} + \frac{1}{r}\frac{\partial w}{\partial \theta} \qquad (3.10.3)$$

$$\epsilon_z = \frac{\partial w}{\partial z} \qquad\qquad \gamma_{zr} = \frac{\partial w}{\partial r} + \frac{\partial u}{\partial z}$$

Equilibrium Conditions for No Body Forces

$$\frac{\partial \sigma_r}{\partial r} + \frac{1}{r}\frac{\partial \tau_{r\theta}}{\partial \theta} + \frac{\partial \tau_{rz}}{\partial z} + \frac{\sigma_r - \sigma_\theta}{r} = 0$$

$$\frac{\partial \tau_{r\theta}}{\partial r} + \frac{1}{r}\frac{\partial \sigma_\theta}{\partial \theta} + \frac{\partial \tau_{\theta z}}{\partial z} + \frac{2}{r}\tau_{r\theta} = 0 \qquad (3.10.4)$$

$$\frac{\partial \tau_{rz}}{\partial r} + \frac{1}{r}\frac{\partial \tau_{\theta z}}{\partial \theta} + \frac{\partial \sigma_z}{\partial z} + \frac{1}{r}\tau_{rz} = 0$$

Spherical Coordinates (r, θ, ϕ) Strain-Displacement Equations

$$\epsilon_r = \frac{\partial u}{\partial r} \qquad\qquad \gamma_{r\theta} = \frac{1}{r}\frac{\partial u}{\partial \theta} + \frac{\partial v}{\partial r} - \frac{v}{r}$$

$$\epsilon_\theta = \frac{u}{r} + \frac{1}{r}\frac{\partial v}{\partial \theta} \qquad\qquad \gamma_{\theta\phi} = \frac{1}{r \sin \theta}\frac{\partial v}{\partial \phi} + \frac{1}{r}\frac{\partial w}{\partial \theta} - \frac{w \cot \theta}{r}$$

$$\epsilon_\phi = \frac{u}{r} + \frac{v}{r}\cot \theta + \frac{1}{r \sin \theta}\frac{\partial w}{\partial \phi} \qquad \gamma_{\phi r} = \frac{\partial w}{\partial r} - \frac{w}{r} + \frac{1}{r \sin \theta}\frac{\partial u}{\partial \phi}$$

$$(3.10.5)$$

Equilibrium Conditions for No Body Forces

$$\frac{\partial \sigma_r}{\partial r} + \frac{1}{r}\frac{\partial \tau_{r\theta}}{\partial \theta} + \frac{1}{r \sin \theta}\frac{\partial \tau_{r\phi}}{\partial \phi} + \frac{2\sigma_r - \sigma_\theta - \sigma_\phi + \tau_{r\theta} \cot \theta}{r} = 0$$

$$\frac{\partial \tau_{r\theta}}{\partial r} + \frac{1}{r}\frac{\partial \sigma_\theta}{\partial \theta} + \frac{1}{r \sin \theta}\frac{\partial \tau_{\theta\phi}}{\partial \phi} + \frac{3\tau_{r\theta} + (\sigma_\theta - \sigma_\phi) \cot \theta}{r} = 0 \quad (3.10.6)$$

$$\frac{\partial \tau_{r\phi}}{\partial r} + \frac{1}{r}\frac{\partial \tau_{\theta\phi}}{\partial \theta} + \frac{1}{r \sin \theta}\frac{\partial \sigma_\phi}{\partial \phi} + \frac{3\tau_{r\phi} + 2\tau_{\theta\phi} \cot \theta}{r} = 0$$

REFERENCES

1. Durelli, A. J., E. A. Phillips, and C. H. Tsao, *Introduction to the Theoretical and Experimental Analysis of Stress and Strain*, Chapter 4, McGraw-Hill Book Co., New York, 1958.
2. Frocht, M. M., *Photoelasticity*, Vol. I, Chapters 1 and 2, John Wiley and Sons, New York, 1941.
3. Jaeger, J. C., *Elasticity, Fracture and Flow*, Chapter 1, Methuen and Co., London, 1962.
4. Lee, George H., *An Introduction to Experimental Stress Analysis*, Chapter 1, John Wiley and Sons, New York, 1950.
5. Love, A. E. H., *The Mathematical Theory of Elasticity*, Chapters 3 and 4, Dover Publications, New York, 1944.
6. Sechler, Earnest E., *Elasticity in Engineering*, Chapter 5, John Wiley and Sons, New York, 1952.
7. Sokolnikoff, I. S., *Mathematical Theory of Elasticity*, Chapter 3 and Chapter 4, Section 48, McGraw-Hill Book Co., New York, 1946.
8. Timoshenko, S., and J. N. Goodier, *Theory of Elasticity*, Chapters 1, 2, and 8, McGraw-Hill Book Co., New York, 1951.

CHAPTER 4

SOLUTION OF PROBLEMS
IN POLAR COORDINATES

4.1 INTRODUCTION

The differential equations developed in Chapter 3 relate the stress, strain, or deformation in an elastic body to the space coordinates. Integration of these differential equations, with the requirement that compatibility and boundary conditions are satisfied, will provide solutions in the form of algebraic equations which describe the stress, strain, or deformation for all points in the given body.

To illustrate some of the methods used in obtaining solutions for these differential equations, a number of simpler problems in polar coordinates will be solved. However, it is not the purpose of this chapter to develop the methods for solving these equations, but to demonstrate how established methods can be used to solve some of the problems that relate to structural rock mechanics.

4.2 SOLUTIONS FOR AIRY'S STRESS FUNCTION IN POLAR COORDINATES

In Section 3.8 the compatibility equation in polar coordinates was derived from the stress-displacement equations. By means of the polar equilibrium equations, for the condition of no body forces, the compatibility equation in terms of stresses was developed. The stress components in terms of an Airy Stress function Φ were defined as

$$\sigma_r = \frac{1}{r}\frac{\partial \Phi}{\partial r} + \frac{1}{r^2}\frac{\partial^2 \Phi}{\partial \theta^2}$$

$$\sigma_\theta = \frac{\partial^2 \Phi}{\partial r^2}$$

$$\tau_{r\theta} = \frac{1}{r^2}\frac{\partial \Phi}{\partial \theta} - \frac{1}{r}\frac{\partial^2 \Phi}{\partial r\, \partial \theta}$$

(4.2.1)

From these equations and the compatibility equations in terms of stresses, a fourth-order biharmonic partial differential equation in terms of the function Φ was derived. Thus

$$\nabla^2 \cdot \nabla^2 \Phi = \left(\frac{\partial^2}{\partial r^2} + \frac{1}{r}\frac{\partial}{\partial r} + \frac{1}{r^2}\frac{\partial^2}{\partial \theta^2}\right)\left(\frac{\partial^2 \Phi}{\partial r^2} + \frac{1}{r}\frac{\partial \Phi}{\partial r} + \frac{1}{r^2}\frac{\partial^2 \Phi}{\partial \theta^2}\right) = 0 \quad (4.2.2)$$

Solutions to this equation can be obtained by the method of separation of variables. Assume that

$$\Phi(r, \theta) = R(r) \cdot \Psi(\theta) \quad (4.2.3)$$

where $R(r)$ is a function of r only and $\Psi(\theta)$ is a function of θ only. Substitution of Eq. 4.2.3 into Eq. 4.2.2 gives

$$\Psi\left[\frac{d^4R}{dr^4} + \frac{2}{r}\frac{d^3R}{dr^3} - \frac{1}{r^2}\frac{d^2R}{dr^2} + \frac{1}{r^3}\frac{dR}{dr}\right]$$
$$+ \frac{d^2\Psi}{d\theta^2}\left[\frac{2}{r^2}\frac{d^2R}{dr^2} - \frac{2}{r^3}\frac{dR}{dr} + \frac{4R}{r^4}\right] + \frac{R}{r^4}\frac{d^4\Psi}{d\theta^4} = 0 \quad (4.2.4)$$

Total derivative symbols are used as R is a function of r only and Ψ is a function of θ only.

If Ψ is a constant, then the derivatives of Ψ with respect to θ are zero and the following equation results from Eq. 4.2.4

$$\frac{d^4R}{dr^4} + \frac{2}{r}\frac{d^3R}{dr^3} - \frac{1}{r^2}\frac{d^2R}{dr^2} + \frac{1}{r^3}\frac{dR}{dr} = 0 \quad (4.2.5)$$

Equation 4.2.5 can be rewritten as

$$\frac{1}{r}\frac{d}{dr}\left\{r\frac{d}{dr}\left[\frac{1}{r}\frac{d}{dr}\left(r\frac{dR}{dr}\right)\right]\right\} = 0 \quad (4.2.6)$$

Successive integration of Eq. 4.2.6 gives

$$R(r) = A_0 r^2 + B_0 r^2 \log r + C_0 \log r + D_0 \quad (4.2.7)$$

Equation 4.2.7 is a general solution of Eq. 4.2.5 as it contains the necessary four arbitrary constants. When Ψ is a constant, or when the stress function is independent of the angle variable θ, Eq. 4.2.7 is a solution of Eq. 4.2.2.

If

$$\Psi = A_0 \theta \quad (4.2.8)$$

then

$$\frac{d^2\Psi}{d\theta^2} = \frac{d^4\Psi}{d\theta^4} = 0$$

and Eq. 4.2.4 becomes

$$A_0\theta\left[\frac{d^4R}{dr^4} + \frac{2}{r}\frac{d^3R}{dr^3} - \frac{1}{r^2}\frac{d^2R}{dr^2} + \frac{1}{r^3}\frac{dR}{dr}\right] = 0 \quad (4.2.9)$$

For Eq. 4.2.9 to be satisfied for all values of θ, the expression in the bracket must equal zero. The expression in the bracket set equal to zero is Eq. 4.2.5 whose solution is Eq. 4.2.7. Therefore a solution to Eq. 4.2.2, when $\Psi = A\theta$, is

$$\Phi(r\theta) = E_0\theta r^2 + F_0\theta r^2 \log r + G_0\theta \log r + H_0\theta \qquad (4.2.10)$$

Let

$$\Psi = \begin{Bmatrix} a \sin \theta \\ b \cos \theta \end{Bmatrix}$$

then

$$\frac{d^2\Psi}{d\theta^2} = -\begin{Bmatrix} a \sin \theta \\ b \cos \theta \end{Bmatrix} \qquad (4.2.11)$$

and

$$\frac{d^4\Psi}{d\theta^4} = \begin{Bmatrix} a \sin \theta \\ b \cos \theta \end{Bmatrix}$$

Substitution of Eqs. 4.2.11 into Eq. 4.2.4 gives

$$\left[\frac{d^4R}{dr^4} + \frac{2}{r}\frac{d^3R}{dr^3} - \frac{3}{r^2}\frac{d^2R}{dr^2} + \frac{3}{r^3}\frac{dR}{dr} - \frac{3R}{r^4}\right]\begin{Bmatrix} a \sin \theta \\ b \cos \theta \end{Bmatrix} = 0 \quad (4.2.12)$$

As R is independent of θ, the expression in the bracket in Eq. 4.2.12 must be equal to zero. This expression, set equal to zero, can be rewritten as

$$\frac{d}{dr}\left\{\frac{1}{r}\frac{d}{dr}\left[r\frac{d}{dr}\left(\frac{1}{r}\frac{d}{dr}\{rR\}\right)\right]\right\} = 0 \qquad (4.2.13)$$

Successive integration of Eq. 4.2.13 gives

$$R = Ar + Br^{-1} + Cr^3 + Dr \log r \qquad (4.2.14)$$

Thus when $\Psi = \begin{Bmatrix} a \sin \theta \\ b \cos \theta \end{Bmatrix}$, another solution to Eq. 4.2.2 is

$$\Phi(r\theta) = (A_1r + B_1r^{-1} + C_1r^3 + D_1r \log r) \sin \theta$$
$$+ (E_1r + F_1r^{-1} + G_1r^3 + H_1r \log r) \cos \theta \qquad (4.2.15)$$

Let $\Psi = \theta \sin \theta$, or $\Psi = \theta \cos \theta$
Then

$$\frac{d^2\Psi}{d\theta^2} = 2 \cos \theta - \theta \sin \theta$$

or

$$\frac{d^2\Psi}{d\theta^2} = -2 \sin \theta - \theta \cos \theta$$

and   (4.2.16)

$$\frac{d^4\Psi}{d\theta^4} = -4 \cos \theta + \theta \sin \theta$$

or

$$\frac{d^4\Psi}{d\theta^4} = 4 \sin \theta + \theta \cos \theta$$

Substitution of Eqs. 4.2.16 into 4.2.4 gives

$$\theta \begin{Bmatrix} \sin \theta \\ \cos \theta \end{Bmatrix} \left[\frac{d^4R}{dr^4} + \frac{2}{r} \frac{d^3R}{dr^3} - \frac{3}{r^2} \frac{d^2R}{dr^2} + \frac{3}{r^3} \frac{dR}{dr} - \frac{3R}{r^4} \right]$$

$$+ \begin{Bmatrix} 4 \cos \theta \\ -4 \sin \theta \end{Bmatrix} \left[\frac{1}{r^2} \frac{d^2R}{dr^2} - \frac{1}{r^3} \frac{dR}{dr} + \frac{R}{r^4} \right] = 0 \quad (4.2.17)$$

For Eq. 4.2.17 to hold for all values of θ both the bracketed expressions must be zero; therefore

$$\frac{1}{r^2} \frac{d^2R}{dr^2} - \frac{1}{r^3} \frac{dR}{dr} + \frac{R}{r^4} = 0$$

and

$$(4.218)$$

$$\frac{d^4R}{dr^4} + \frac{2}{r} \frac{d^3R}{dr^3} - \frac{3}{r^2} \frac{d^2R}{dr^2} + \frac{3}{r^3} \frac{dR}{dr} - \frac{3R}{r^4} = 0$$

The last equation in Eqs. 4.2.18 is identical to Eq. 4.2.12, therefore its solution is also Eq. 4.2.14. The first equation of Eqs. 4.2.18 can be rewritten as

$$\frac{1}{r^2} \frac{d}{dr} \left[r \frac{d}{dr} \left(\frac{R}{r} \right) \right] = 0 \quad (4.2.19)$$

Successive integration of Eq. 4.2.19 gives

$$R = J_1 r + K_1 r \log r \quad (4.2.20)$$

The two terms in Eq. 4.2.20 are the same as the first and last terms of Eq. 4.2.14, thus these two terms are the only solutions that will satisfy both equations of 4.2.18 simultaneously. Therefore when $\Psi = \theta \begin{pmatrix} \sin \theta \\ \cos \theta \end{pmatrix}$, a solution to Eq. 4.2.2 is

$$\Phi(r, \theta) = (J_1 r + K_1 r \log r)(\theta \sin \theta) + (L_1 r + M_1 r \log r)(\theta \cos \theta) \quad (4.2.21)$$

Assume that

$$\Psi = \begin{pmatrix} \sin n\theta \\ \cos n\theta \end{pmatrix}$$

then

$$\frac{d^2\Psi}{d\theta^2} = -n^2 \begin{pmatrix} \sin n\theta \\ \cos n\theta \end{pmatrix} \quad (4.2.22)$$

and

$$\frac{d^4\Psi}{d\theta^4} = n^4 \begin{pmatrix} \sin n\theta \\ \cos n\theta \end{pmatrix}$$

Substitution of Eqs. 4.2.22 into Eq. 4.2.4 gives after dividing through by $\begin{pmatrix} \sin n\theta \\ \cos n\theta \end{pmatrix}$

$$\frac{d^4R}{dr^4} + \frac{2}{r}\frac{d^3R}{dr^3} - \frac{1+2n^2}{r^2}\frac{d^2R}{dr^2} + \frac{1+2n^2}{r^3}\frac{dR}{dr} + \frac{n^4-4n^2}{r^4}R = 0 \quad (4.2.23)$$

Equation 4.2.23 is an ordinary differential equation with variable coefficients which can be changed to an ordinary differential equation with constant coefficients by letting

$$t = \log r \quad \text{or} \quad r = e^t$$

and

$$\left(\frac{dt}{dr} = \frac{1}{r}\right) \quad (4.2.24)$$

Using the chain rule for differentiation with respect to the second variable gives

$$\frac{dR}{dr} = \frac{dR}{dt}\frac{dt}{dr} = \frac{1}{r}\frac{dR}{dt}$$

$$\frac{d^2R}{dr^2} = \frac{d}{dr}\left(\frac{1}{r}\frac{dR}{dt}\right) = -\frac{1}{r^2}\frac{dR}{dt} + \frac{1}{r}\frac{d^2R}{dt^2}\frac{dt}{dr} = \frac{1}{r^2}\left(\frac{d^2R}{dt^2} - \frac{dR}{dt}\right)$$

$$\frac{d^3R}{dr^3} = \frac{d}{dr}\left[\frac{1}{r^2}\left(\frac{d^2R}{dt^2} - \frac{dR}{dt}\right)\right] = -\frac{2}{r^3}\left(\frac{d^2R}{dt^2} - \frac{dR}{dt}\right) + \frac{1}{r^2}\left(\frac{d^3R}{dt^3}\frac{dt}{dr} - \frac{d^2R}{dt^2}\frac{dt}{dr}\right)$$

$$= \frac{1}{r^3}\left[\frac{d^3R}{dt^3} - 3\frac{d^2R}{dt^2} + 2\frac{dR}{dt}\right] \quad (4.2.25)$$

$$\frac{d^4R}{dr^4} = \frac{d}{dr}\left[\frac{1}{r^3}\left(\frac{d^3R}{dt^3} - 3\frac{d^2R}{dt^2} + 2\frac{dR}{dt}\right)\right]$$

$$= -\frac{3}{r^4}\left[\frac{d^3R}{dt^3} - 3\frac{d^2R}{dt^2} + 2\frac{dR}{dt}\right] + \frac{1}{r^3}\left[\frac{d^4R}{dt^4}\frac{dt}{dr} - 3\frac{d^3R}{dt^3}\frac{dt}{dr} + 2\frac{d^2R}{dt^2}\frac{dt}{dr}\right]$$

$$= \frac{1}{r^4}\left[\frac{d^4R}{dt^4} - 6\frac{d^3R}{dt^3} + 11\frac{d^2R}{dt^2} - 6\frac{dR}{dt}\right]$$

Substitution of Eqs. 4.2.25 in 4.2.23 gives

$$\frac{d^4R}{dt^4} - 4\frac{d^3R}{dt^3} + 2(2-n^2)\frac{d^2R}{dt^2} + 4n^2\frac{dR}{dt} + (n^4-4n^2)R = 0 \quad (4.2.26)$$

Equation 4.2.26 is an ordinary differential equation with constant coefficients which may be solved by letting $R(t) = e^{pt}$, which when substituted into Eq. 4.2.26, gives

$$p^4 - 4p^3 + 2(2 - n^2)p^2 + 4n^2p + n^4 - 4n^2 = 0 \qquad (4.2.27)$$

The roots of Eq. 4.2.27 if $n > 1$, are n, $-n$, $(2 - n)$, and $(2 + n)$, therefore

$$R(t) = ae^{nt} + be^{-nt} + ce^{(2-n)t} + de^{(2+n)t}$$

and

$$R(r) = Ar^n + Br^{-n} + Cr^{(2-n)} + Dr^{(2+n)}$$

Thus when $\Psi = \begin{pmatrix} \sin n\theta \\ \cos n\theta \end{pmatrix}$ and $n > 1$, a solution to Eq. 4.2.2 is

$$\begin{aligned} \Phi(r, \theta) = {}& \sin n\theta[A_n r^n + B_n r^{-n} + C_n r^{2-n} + D_n r^{2+n}] \\ & + \cos n\theta[E_n r^n + F_n r^{-n} + G_n r^{2-n} + H_n r^{2+n}] \end{aligned} \qquad (4.2.28)$$

Various possible solutions to Eq. 4.2.2 are given by Eqs. 4.2.7, 4.2.10, 4.2.15, 4.2.21, and 4.2.28. The sum of all of these solutions is also a solution, thus a general expression for the solutions to Eq. 4.2.2 is*

$$\begin{aligned} \Phi(r, \theta) = {}& A_0 r^2 + B_0 r^2 \log r + C_0 \log r + D_0 \\ & + (E_0 r^2 + F_0 r^2 \log r + G_0 \log r + H_0)\theta \\ & + (A_1 r + B_1 r^{-1} + C_1 r^3 + D_1 r \log r) \sin \theta \\ & + (E_1 r + F_1 r^{-1} + G_1 r^3 + H_1 r \log r) \cos \theta \\ & + (J_1 r + K_1 r \log r)\theta \sin \theta + (L_1 r + M_1 r \log r)\theta \cos \theta \\ & + \sum_{n=2}^{\infty} [A_n r^n + B_n r^{-n} + C_n r^{2-n} + D_n r^{2+n}] \sin n\theta \\ & + \sum_{n=2}^{\infty} [E_n r^n + F_n r^{-n} + G_n r^{2-n} + H_n r^{2+n}] \cos n\theta \end{aligned} \qquad (4.2.29)$$

That Eq. 4.2.29 is a solution of Eq. 4.2.2 can be shown by direct substitution. Various terms of this solution can be used to obtain solutions to specific problems in polar coordinates for the case of no body forces. The boundary conditions for each specific problem are used to evaluate the unknown constants A, B, C, etc. It is not necessary to start with all the terms of the general expression since any of the constants can be assumed equal to zero. However, a sufficient number of terms must be used to satisfy the boundary conditions of the problem under study.

* An expression very similar to this one was given originally by J. H. Michell, "On the Determination of Stress in the Elastic Solid with Applications to the Theory of Plates," *London Math. Society*, **31**, pp. 100–124 (April, 1899). Also see Timoshenko, S. and J. N. Goodier, *Theory of Elasticity*, McGraw-Hill Book Co., 1951, p. 116.

4.3 CONCENTRATED LOAD ON BOUNDARY OF SEMI-INFINITE PLATE

Assume a semi-infinite plate of thickness t acted upon by a concentrated line load $-F$, normal to the upper surface of the plate and uniformly distributed over the thickness of the plate. Let the origin of the coordinate system be at the point of application of the load as indicated in Fig. 4.3.1a.

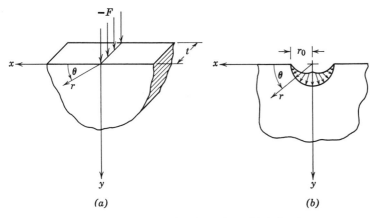

Fig. 4.3.1. Line load on boundary of infinite plate.

The boundary conditions of the problem can be specified as

$$\sigma_\theta = \tau_{r\theta} = 0 \quad \text{at} \quad \theta = 0 \text{ or } \pi, \quad (r > 0)$$

$$\sigma_r = \sigma_\theta = \tau_{r\theta} \to 0 \quad \text{as} \quad r \to \infty$$

$$\left.\begin{array}{l} \displaystyle\int_0^\pi S_{ry} tr \, d\theta = -F \\[2mm] \displaystyle\int_0^\pi S_{rx} tr \, d\theta = 0 \end{array}\right\} \quad (r > 0) \tag{4.3.1}$$

where S_{ry} and S_{rx} are the vertical and horizontal components of stress acting on any cylindrical surface of radius r and center at the origin as indicated in Fig. 4.3.1b.

Assume a stress function of the form

$$\Phi = Cr\theta \cos \theta \tag{4.3.2}$$

As this function is part of the general solution, it must satisfy the Airy stress function Eq. 4.2.2. Thus the stress components are given by Eq. 4.2.1

as

$$\sigma_r = \frac{1}{r}\frac{\partial \Phi}{\partial r} + \frac{1}{r^2}\frac{\partial^2 \Phi}{\partial \theta^2} = \frac{C\theta \cos \theta}{r} + \frac{C}{r}(-2 \sin \theta - \theta \cos \theta) = \frac{-2C \sin \theta}{r}$$

$$\sigma_\theta = \frac{\partial^2 \Phi}{\partial r^2} = 0 \tag{4.3.3}$$

$$\tau_{r\theta} = \frac{1}{r^2}\frac{\partial \Phi}{\partial \theta} - \frac{1}{r}\frac{\partial^2 \Phi}{\partial r \partial \theta} = \frac{C}{r}(\cos \theta - \theta \sin \theta) - \frac{C}{r}(\cos \theta - \theta \sin \theta) = 0$$

The only nonzero stress component is σ_r. When θ equals zero or π, σ_r is zero and as r approaches large values, σ_r approaches zero. Thus the first two boundary conditions are satisfied.

From Fig. 4.3.1*b* and Eq. 4.3.3

$$S_{ry} = \sigma_r \sin \theta = \frac{-2C \sin^2 \theta}{r}$$

$$S_{rx} = \sigma_r \cos \theta = \frac{-2C \sin \theta \cos \theta}{r} \tag{4.3.4}$$

Substitution of Eqs. 4.3.4 into the last two boundary conditions gives

$$-2Ct\int_0^\pi \sin^2 \theta \, d\theta = -Ct\pi = -F$$

$$-2Ct\int_0^\pi \sin \theta \cos \theta \, d\theta = 0 \tag{4.3.5}$$

From the first equation of Eqs. 4.3.5

$$C = \frac{F}{t\pi} \tag{4.3.6}$$

Substitution of Eq. 4.3.6 into Eq. 4.3.3 gives the stress distribution for a concentrated load on the edge of a semi-infinite plate as

$$\sigma_r = -\frac{2F \sin \theta}{\pi t r}$$

$$\sigma_\theta = 0 \tag{4.3.7}$$

$$\tau_{r\theta} = 0$$

The discontinuity in the stress at $r = 0$ can be removed by replacing the concentrated force by its statical equivalent acting on a small cylindrical surface of radius a. If the stress distribution on this surface is given by

$$(\sigma_r)_{r=a} = -\frac{2F \sin \theta}{\pi t a}$$

the vertical component is F and the horizontal component is zero. Such a stress distribution is the statical equivalent of F as shown in Fig. 4.3.1*b*.

The x and y components of stress can be obtained by application of Eqs. 1.3.1. Thus from Fig. 4.3.1a and Eqs. 3.8.1

$$\sigma_x = \sigma_r \cos^2 \theta = -\frac{2F \cos^2 \theta \sin \theta}{\pi t r} = -\frac{2F x^2 y}{\pi t (x^2 + y^2)^2}$$

$$\sigma_y = \sigma_r \sin^2 \theta = -\frac{2F \sin^3 \theta}{\pi t r} = -\frac{2F y^3}{\pi t (x^2 + y^2)^2} \qquad (4.3.8)$$

$$\tau_{xy} = \sigma_r \sin \theta \cos \theta = -\frac{2F \sin^2 \theta \cos \theta}{\pi t r} = -\frac{2F x y^2}{\pi t (x^2 + y^2)^2}$$

The stress distributions represented by Eqs. 4.3.8 are shown plotted in Fig. 4.3.2a. Distance along the x axis is given as units of $y = $ constant. The stresses are given in units of $-2F/\pi t y$. Thus the functions shown plotted are those on the right-hand side of the following equations:

$$-\frac{\sigma_x}{2F/\pi t y} = \frac{(x/y)^2}{[(x/y)^2 + 1]^2}$$

$$-\frac{\sigma_y}{2F/\pi t y} = \frac{1}{[(x/y)^2 + 1]^2}$$

$$-\frac{\tau_{xy}}{2F/\pi t y} = \frac{x/y}{[(x/y)^2 + 1]^2}$$

Figure 4.3.2b gives the stress distribution along the y axis, or along any radius. The distance along the radius is given in terms of the radius of the small cylindrical surface where the stress distribution is assumed to be given by

$$(\sigma_r)_{r=a} = -\frac{2F \sin \theta}{\pi t a}$$

The stress is given in terms of the applied stress on the small cylindrical surface. Thus the function plotted is

$$\frac{\sigma_r}{(\sigma_r)_{r=a}} = \frac{a}{r}$$

The displacements resulting from the application of a concentrated load on the edge of a plate will be obtained next.

Combining Eqs. 3.8.7 and 3.8.12 gives

$$\frac{\partial u}{\partial r} = \frac{1}{E} (\sigma_r - \nu \sigma_\theta)$$

$$\frac{u}{r} + \frac{1}{r} \frac{\partial v}{\partial \theta} = \frac{1}{E} (\sigma_\theta - \nu \sigma_r) \qquad (4.3.9)$$

$$\frac{1}{r} \frac{\partial u}{\partial \theta} + \frac{\partial v}{\partial r} - \frac{v}{r} = \frac{2(1 + \nu)}{E} \tau_{r\theta}$$

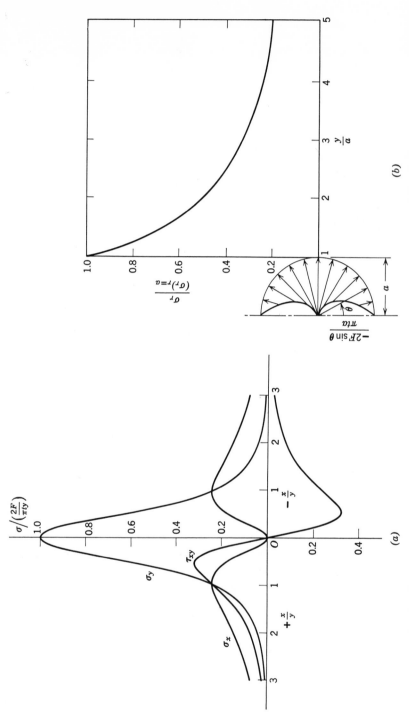

Fig. 4.3.2(b). Stress distribution along y axis for line load on edge of infinite plate.

Fig. 4.3.2(a). Stress distribution along a line $y =$ constant for line load on boundary of infinite plate.

84

Substitution of Eq. 4.3.7 into 4.3.9 gives

$$\frac{\partial u}{\partial r} = - \frac{2F}{\pi Et} \frac{\sin \theta}{r}$$

$$\frac{u}{r} + \frac{1}{r} \frac{\partial v}{\partial \theta} = \frac{2vF}{\pi Et} \frac{\sin \theta}{r} \tag{4.3.10}$$

$$\frac{1}{r} \frac{\partial u}{\partial \theta} + \frac{\partial v}{\partial r} - \frac{v}{r} = 0$$

Integration of the first equation in Eqs. 4.3.10 gives

$$u = - \frac{2F \sin \theta}{\pi Et} \log r + f(\theta) \tag{4.3.11}$$

when $f(\theta)$ is a function of θ only.

Substitution of Eq. 4.3.11 into the second equation of Eqs. 4.3.10 gives, after simplification,

$$\frac{\partial v}{\partial \theta} = \left[\frac{2vF}{\pi Et} + \frac{2F \log r}{\pi Et} \right] \sin \theta - f(\theta) \tag{4.3.12}$$

Integration of Eq. 4.3.12 gives

$$v = - \left[\frac{2F}{\pi Et} (\log r + v) \right] \cos \theta - \int f(\theta) \, d\theta + g(r) \tag{4.3.13}$$

where $g(r)$ is a function of r only.

The partial derivative of Eq. 4.3.11 with respect to θ and of Eq. 4.3.13 with respect to r are respectively

$$\frac{\partial u}{\partial \theta} = - \frac{2F}{\pi Et} \cos \theta \log r + \frac{df(\theta)}{d\theta}$$

and

$$\frac{\partial v}{\partial r} = - \frac{2F \cos \theta}{\pi Etr} + \frac{dg(r)}{dr} \tag{4.3.14}$$

By use of Eqs. 4.3.13 and 4.3.14, the last equation of Eqs. 4.3.10 becomes

$$\left[\int f(\theta) \, d\theta + \frac{df(\theta)}{d\theta} - \frac{2F(1 - v) \cos \theta}{\pi Et} \right] + \left[r \frac{dg(r)}{dr} - g(r) \right] = 0$$

As r and θ are independent, each part of the above equation must equal some constant k or be identically zero therefore

$$\int f(\theta) \, d\theta + \frac{df(\theta)}{d\theta} - \frac{2F(1 - v) \cos \theta}{\pi Et} = k$$

$$r \frac{dg(r)}{dr} - g(r) = k \tag{4.3.15}$$

A solution to the first equation of Eq. 4.3.15 is

$$f(\theta) = \frac{F(1 - \nu)\theta \cos \theta}{\pi Et} + A \sin \theta + B \cos \theta \qquad (4.3.16)$$

and a solution to the second equation of Eqs. 4.3.15 is

$$g(r) = Cr - k \qquad (4.3.17)$$

where A, B, and C are arbitrary constants.

Substitution of Eqs. 4.3.16 and 4.3.17 into Eq. 4.3.11 and 4.3.13 gives

$$u = - \frac{2F \sin \theta \log r}{\pi Et} + \frac{F(1 - \nu)\theta \cos \theta}{\pi Et} + A \sin \theta + B \cos \theta$$

$$v = - \frac{2F}{\pi Et} \log r \cos \theta - \frac{2\nu F}{\pi Et} \cos \theta - \frac{F(1 - \nu)}{\pi Et} \cos \theta \qquad (4.3.18)$$

$$- \frac{F(1 - \nu)\theta \sin \theta}{\pi Et} + A \cos \theta - B \sin \theta + Cr$$

From the symmetry of the problem, $v = 0$ when $\theta = \pi/2$ for all values of r, thus

$$v = - \frac{F(1 - \nu)}{2Et} - B + Cr = 0$$

or

$$C = 0 \quad \text{and} \quad B = - \frac{F(1 - \nu)}{2Et}$$

To evaluate the constant A, let the displacement resulting from $A \sin \theta$ be u', and that from $A \cos \theta$ be v'. The vector sum of u' and v' is always in the y direction and is independent of distance or direction from the origin. Therefore, A must represent a rigid body motion and can be set equal to zero. Substitution of the values of A, B, and C into Eq. 4.3.18 gives

$$u = - \frac{2F}{\pi tE} \sin \theta \log r + \frac{F(1 - \nu)}{\pi tE} \left(\theta - \frac{\pi}{2} \right) \cos \theta$$

$$(4.3.19)$$

$$v = - \frac{2F}{\pi tE} \cos \theta \log r - \frac{F(1 + \nu)}{\pi tE} \cos \theta - \frac{F(1 - \nu)}{\pi tE} \left(\theta - \frac{\pi}{2} \right) \sin \theta$$

Equations 4.3.19 give the radial and tangential displacements for all points in the plate except the origin. On the x axis, or free boundary, the horizontal displacement is constant and always directed toward the origin. For example

$$u_{\theta=0} = - \frac{F(1 - \nu)}{2tE} \quad \text{and} \quad u_{\theta=\pi} = - \frac{F(1 - \nu)}{2tE}$$

The vertical displacements on the free boundary are

$$v_{\theta=0} = -\frac{2F}{\pi tE}\log r - \frac{F(1+\nu)}{\pi tE}$$

$$v_{\theta=\pi} = \frac{2F}{\pi tE}\log r + \frac{F(1+\nu)}{\pi tE}$$

When $r = e^{-1+\nu/2}$, the vertical displacements on the free boundary are zero. For distances along the x axis less than $e^{-1+\nu/2}$, the vertical displacements are downward, and for distances greater than $e^{-1+\nu/2}$, the vertical displacements are upward.

4.4 DISTRIBUTED LOAD ON BOUNDARY OF SEMI-INFINITE PLATE

Assume a semi-infinite plate of thickness t acted upon by a stress f normal to the upper edge of the plate and uniformly distributed over an area of length $2a$ and width t, the thickness of the plate, as shown in Fig. 4.4.1a.

The results of the previous section can be used to solve the problem of the distributed load. The stress components for a small concentrated load $ft\,dx$ acting at point x are from Eqs. 4.3.8

$$d\sigma_x = -\frac{2ft\,dx}{\pi t} \cdot \frac{\cos^2\theta\sin\theta}{r}$$

$$d\sigma_y = -\frac{2ft\,dx}{\pi t} \cdot \frac{\sin^3\theta}{r}$$

$$d\tau_{xy} = -\frac{2ft\,dx}{\pi t} \cdot \frac{\sin^2\theta\cos\theta}{r}$$

From Fig. 4.4.1a $rd\theta = dx\sin\theta$, thus the above equations become

$$d\sigma_x = -\frac{2f}{\pi}\cos^2\theta\,d\theta$$

$$d\sigma_y = -\frac{2f}{\pi}\sin^2\theta\,d\theta \qquad (4.4.1)$$

$$d\tau_{xy} = -\frac{2f}{\pi}\sin\theta\cos\theta\,d\theta$$

Integration of Eqs. 4.4.1 from θ_1 to θ_2 will give the stress components

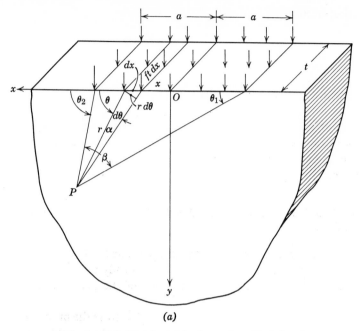

Fig. 4.4.1(a). Distributed load on edge of infinite plate.

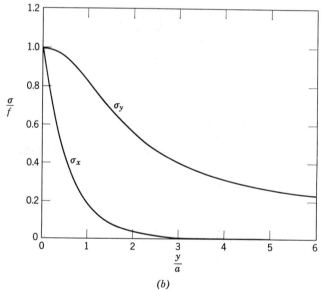

Fig. 4.4.1(b). Stress distributions along y axis for distributed load on edge of infinity plate.

resulting from the total load $F = 2fta$. Thus

$$\sigma_x = -\frac{2f}{\pi}\int_{\theta_1}^{\theta_2}\cos^2\theta\,d\theta = -\frac{f}{2\pi}[2(\theta_2 - \theta_1) + (\sin 2\theta_2 - \sin 2\theta_1)]$$

$$\sigma_y = -\frac{2f}{\pi}\int_{\theta_1}^{\theta_2}\sin^2\theta\,d\theta = -\frac{f}{2\pi}[2(\theta_2 - \theta_1) - (\sin 2\theta_2 - \sin 2\theta_1)]$$

$$\tau_{xy} = -\frac{f}{\pi}\int_{\theta_1}^{\theta_2}\sin 2\theta\,d\theta = \frac{f}{2\pi}[\cos 2\theta_2 - \cos 2\theta_1] \qquad (4.4.2)$$

Equations 4.4.2 can be transformed by means of trigometric identities to

$$\sigma_x = -\frac{f}{\pi}[\theta_2 - \theta_1 + \sin(\theta_2 - \theta_1)\cos(\theta_2 + \theta_1)]$$

$$\sigma_y = -\frac{f}{\pi}[\theta_2 - \theta_1 - \sin(\theta_2 - \theta_1)\cos(\theta_2 + \theta_1)]$$

$$\tau_{xy} = -\frac{f}{\pi}[\sin(\theta_2 + \theta_1)\sin(\theta_2 - \theta_1)]$$

If β is the included angle at point P in Fig. 4.4.1a, then $\beta = \theta_2 - \theta_1$ and when P lies on the y axis, $\theta_2 + \theta_1 = \pi$. Thus for points on the y axis, the above equations become

$$\sigma_x = -\frac{f}{\pi}[\beta - \sin\beta]$$

$$\sigma_y = -\frac{f}{\pi}[\beta + \sin\beta] \qquad (4.4.3)$$

$$\tau_{xy} = 0$$

Equations 4.4.3 give the stress distribution along the y axis. This distribution of stress is shown in Fig. 4.4.1b. Comparison of the curves in Fig. 4.4.1b with the one given in Fig. 4.3.2b shows that there is very little difference in the two stress distributions for large values of y/a.

4.5 UNIFORM PRESSURE ON INSIDE AND OUTSIDE OF A THICK WALL CYLINDER

Consider a cylinder, as shown in Fig. 4.5.1, whose inner and outer radii are a and b, and whose inner and outer surfaces are acted upon by uniform pressure $-p_i$ and $-p_o$, respectively. The boundary conditions of the problem can be stated as follows:

$$\begin{aligned}\sigma_r &= -p_i \quad \text{and} \quad \tau_{r\theta} = 0 \quad \text{at} \quad r = a\\ \sigma_r &= -p_o \quad \text{and} \quad \tau_{r\theta} = 0 \quad \text{at} \quad r = b\end{aligned} \qquad (4.5.1)$$

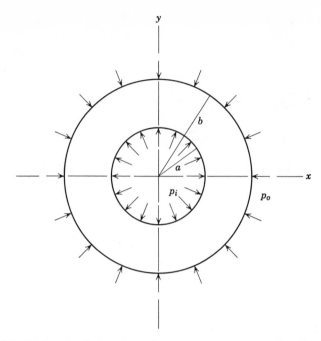

Fig. 4.5.1. Thick wall cylinder with uniform pressures on inside and outside surfaces.

As the boundary stresses are independent of the angle variable θ, a stress function independent of the angle variable is suggested. From Eq. 4.2.29, there are three terms which are independent of the angle. For a trial test select the first and third terms, thus the stress function is

$$\Phi = Ar^2 + C \log r \qquad (4.5.2)$$

As Eq. 4.5.2 satisfies Eq. 4.2.2, the stress components are given by

$$\sigma_r = \frac{1}{r}\frac{\partial \Phi}{\partial r} + \frac{1}{r^2}\frac{\partial^2 \Phi}{\partial \theta^2} = 2A + \frac{C}{r^2}$$

$$\sigma_\theta = \frac{\partial^2 \Phi}{\partial r^2} = 2A - \frac{C}{r^2} \qquad (4.5.3)$$

$$\tau_{r\theta} = \frac{1}{r^2}\frac{\partial \Phi}{\partial \theta} - \frac{1}{r}\frac{\partial^2 \Phi}{\partial r \, \partial \theta} = 0$$

When $r = a$

$$(\sigma_r)_{r=a} = 2A + \frac{C}{a^2} = -p_i \qquad \text{and} \quad \tau_{r\theta} = 0$$

When $r = b$ $\qquad\qquad\qquad\qquad\qquad\qquad\qquad\qquad (4.5.4)$

$$(\sigma_r)_{r=b} = 2A + \frac{C}{b^2} = -p_o \qquad \text{and} \quad \tau_{r\theta} = 0$$

Thus the boundary conditions are satisfied if the constants A and C are chosen to satisfy Eqs. 4.5.4. Solving Eqs. 4.5.4 for A and C gives

$$A = \frac{a^2 p_i - b^2 p_o}{2(b^2 - a^2)}$$

$$C = \frac{a^2 b^2 (p_o - p_i)}{b^2 - a^2} \tag{4.5.5}$$

Substitution of Eqs. 4.5.5 into 4.5.3 gives

$$\sigma_r = \frac{a^2 p_i - b^2 p_o}{b^2 - a^2} + \frac{1}{r^2} \cdot \frac{a^2 b^2 (p_o - p_i)}{b^2 - a^2}$$

$$\sigma_\theta = \frac{a^2 p_i - b^2 p_o}{b^2 - a^2} - \frac{1}{r^2} \cdot \frac{a^2 b^2 (p_o - p_i)}{b^2 - a^2} \tag{4.5.6}$$

$$\tau_{r\theta} = 0$$

As $\tau_{r\theta}$ is identically zero, the polar components of stress σ_r and σ_θ are the principal stresses. The sum of the stress components is a constant independent of r and θ, thus

$$\sigma_r + \sigma_\theta = \frac{2(a^2 p_i - b^2 p_o)}{b^2 - a^2} \tag{4.5.7}$$

For plane stress conditions, from Eqs. 3.4.1

$$\epsilon_z = -\frac{\nu}{E}(\sigma_r + \sigma_\theta)$$

Therefore the cylinder undergoes a uniform extension or contraction in the axial direction, and cross sections of the cylinder remain plane after application of the load.

A problem of special interest in the rock mechanics field is that of a thick wall cylinder subjected to external pressure only. This problem corresponds to the problem of a tunnel or shaft lining in a rock formation having a hydrostatic stress field. Setting $p_i = 0$ in Eq. 4.5.6 gives

$$\sigma_r = -\frac{b^2 p_o}{b^2 - a^2}\left(1 - \frac{a^2}{r^2}\right)$$

$$\sigma_\theta = -\frac{b^2 p_o}{b^2 - a^2}\left(1 + \frac{a^2}{r^2}\right) \tag{4.5.8}$$

$$\tau_{r\theta} = 0$$

When $r = a$, $\sigma_r = 0$ and

$$\sigma_\theta = -\frac{2b^2 p_o}{b^2 - a^2}$$

When $r = b$, $\sigma_r = -p_o$ and

$$\sigma_\theta = - \frac{p_o(b^2 + a^2)}{b^2 - a^2}$$

The tangential stresses at the boundaries and in the interior exceed the applied normal stress. The tangential and radial stresses are reduced for a given p_o as the thickness of the wall is increased. The stress distribution throughout the wall is shown graphically in Fig. 4.5.2a for a cylinder with a ratio of $a/b = 0.8$.

The stresses are plotted in terms of σ/p_o, or what is commonly known as stress concentration. Distance through the wall of the cylinder is given in units of the inner radius a.

The sum of the principal stresses is

$$\sigma_r + \sigma_\theta = \frac{-2p_o b^2}{b^2 - a^2} = (\sigma_\theta)_{max} \tag{4.5.9}$$

This sum is also the maximum value of the tangential stress and occurs at the inner boundary of the cylinder. Figure 4.5.2b is a plot of $(\sigma_\theta/p_o)_{max}$ as a function of the ratio b/a. Thus this graph gives the maximum stress concentration as a function of the ratio of the outer radii to the inner radii. From this graph it can be seen that as b/a approaches one, σ_θ/p_o approaches infinity, and as b/a approaches infinity, that σ_θ/p_o approaches two.

Next, the displacements for the thick wall cylinder will be obtained. Because of the symmetry of the problem there can be no tangential displacements, thus only a radial displacement exists. From Eqs. 4.3.9 and 4.5.6, the radial stress-displacement equation for plane stress conditions is

$$\frac{\partial u}{\partial r} = \frac{1}{E(b^2 - a^2)} \left[(a^2 p_i - b^2 p_o)(1 - \nu) + \frac{a^2 b^2 (p_o - p_i)}{r^2} (1 + \nu) \right] \tag{4.5.10}$$

Integration of Eq. 4.5.10 gives

$$u = \frac{1}{E(b^2 - a^2)} \left[(a^2 p_i - b^2 p_o)(1 - \nu)r - \frac{a^2 b^2 (p_o - p_i)(1 + \nu)}{r} \right] \tag{4.5.11}$$

where the constant of integration which is normally a function of θ only is zero. When $p_i = 0$, Eq. 4.5.11 becomes

$$u = - \frac{b^2 p_o}{E(b^2 - a^2)} \left[(1 - \nu)r + \frac{a^2}{r} (1 + \nu) \right] \tag{4.5.12}$$

When $r = a$, Eq. 4.5.12 becomes

$$u = - \frac{2ab^2 p_o}{E(b^2 - a^2)} \tag{4.5.13}$$

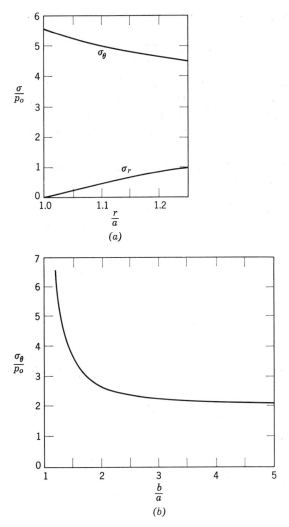

Fig. 4.5.2(a). Stress distribution in thick wall cylinder. (*b*) Maximum stress concentration versus ratio *b/a* for thick wall cylinder.

Equation 4.5.13 gives the radial displacement for any point on the inner surface of a thick wall cylinder. If a gage is inserted inside a cylinder to measure the change in its diameter while a uniform pressure is applied to the outside of the cylinder, the observed change in diameter will be just twice the value given by Eq. 4.5.13.

It should be noted in Eq. 4.5.13 that the change in radius per unit radius is a function of the inner and outer radii of the thick wall cylinder. Thus

$$\frac{u}{a} = - \frac{2b^2}{b^2 - a^2} \cdot \frac{p_o}{E} \qquad (4.5.14)$$

The first factor on the right-hand side of Eq. 4.5.14, is the maximum stress concentration factor for the tangential stress, see Eq. 4.5.9. This factor is also the sensitivity factor for the rate of change of u with applied pressure. Thus, Fig. 4.5.2b can be used with either Eq. 4.5.14 or 4.5.9.

4.6 CONCENTRATED DIAMETRAL LOADS ON A CIRCULAR DISK

Consider a circular disk of thickness t, and radius R subjected to concentrated line loads of $-F$, uniformly distributed over the thickness of the disk and acting along a diameter AB as shown in Fig. 4.6.1. Let the origin of the xy coordinate system be at the center of the disk.

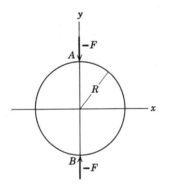

The solution to this problem can be derived from the results given in Section 4.3. The stress distribution for a concentrated line load acting on the boundary of a semi-infinite plate is entirely radial and is given by (see Eq. 4.3.7)

$$\sigma_r = - \frac{2F \sin \theta}{\pi t r} \qquad (4.6.1)$$

$$\sigma_\theta = \tau_{r\theta} = 0$$

Fig. 4.6.1. Concentrated diametral loads on circular disk.

Consider a circle of diameter D tangent to the surface of the semi-infinite plate at the origin O_1. From Fig. 4.6.2a the following relationship can be written

$$r_1 = D \sin \theta_1 \qquad (4.6.2)$$

Equation 4.6.2 gives the values of r_1 for points on the circumference of the circle of diameter D. Substitution of Eq. 4.6.2 into Eq. 4.6.1 gives the radial stress on the circumference of the tangent circle as

$$\sigma_r = - \frac{2F}{\pi t D} \qquad (4.6.3)$$

Thus the radial stress is constant on the circumference of this tangent circle. Therefore if a circular disk of diameter D and thickness t is subjected to a concentrated force $-F$ at the point O_1 and a constant boundary stress of $-(2F/\pi tD)$ directed toward the point O_1, the stress distribution in the disk is given by

$$\sigma_{r_1} = -\frac{2F \sin \theta_1}{\pi t r_1} \qquad (4.6.4)$$

where r_1 is the radial distance from the point O_1 and θ_1 is the angle measured in the direction as shown in Fig. 4.6.2a.

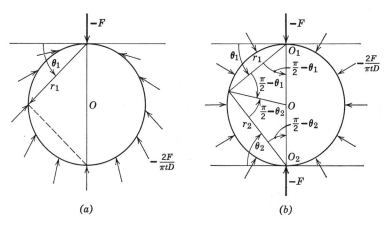

(a)   (b)

Fig. 4.6.2. Boundary stresses required to balance concentrated load.

Consider next that the circular disk is acted upon by a concentrated force $-F$ at the point O_2 diametrically opposite to point O_1 and a constant boundary stress of $-(2F/\pi tD)$ directed at the point O_2. The stress distribution for this set of forces is

$$\sigma_{r_2} = -\frac{2F \sin \theta_2}{\pi t r_2} \qquad (4.6.5)$$

where r_2 is the radial distance from point O_2 and θ_2 is the angle measured in the direction as shown in Fig. 4.6.2b.

On the circular boundary the radial lines r_1 and r_2 are everywhere perpendicular to each other, and therefore the stresses σ_{r_1} and σ_{r_2} are perpendicular to each other on the circular boundary also. From Fig. 4.6.2b, the following relationship can be written:

$$\left(\frac{\pi}{2} - \theta_1\right) + \left(\frac{\pi}{2} - \theta_2\right) = \frac{\pi}{2}$$

or   (4.6.6)

$$\theta_1 + \theta_2 = \frac{\pi}{2}$$

The normal boundary stress resulting from σ_{r_1} and σ_{r_2} is

$$\sigma_r = \sigma_{r_1} \cos^2\left(\frac{\pi}{2} - \theta_1\right) + \sigma_{r_2} \cos^2\left(\frac{\pi}{2} - \theta_2\right)$$

Expanding the cos terms and replacing σ_{r_1} and σ_{r_2} with $-(2F/\pi tD)$, their value on the boundary, gives

$$\sigma_r = -\frac{2F}{\pi tD} [\sin^2 \theta_1 + \sin^2 \theta_2]$$

Substitution of Eq. 4.6.6 into the above equation gives

$$\sigma_r = -\frac{2F}{\pi tD} [\sin^2 \theta_1 + \cos^2 \theta_2] = -\frac{2F}{\pi tD} \qquad (4.6.7)$$

Thus the normal boundary stress is constant and equal to $-(2F/\pi tD)$.

Assume now that an additional normal boundary stress of $2F/\pi tD$ is applied to the circular disk, then the resultant normal boundary stress is zero everywhere except at the point O_1 and O_2 where the concentrated loads of $-F$ are acting. The stress distribution for a normal boundary stress of $2F/\pi tD$ can be obtained from Eq. 4.5.8 by letting $a \to 0$ and replacing $-p_0$ with $2F/\pi tD$, thus

$$\sigma_{r_3} = \frac{2F}{\pi tD}$$

$$\sigma_{\theta_3} = \frac{2F}{\pi tD} \qquad (4.6.8)$$

Thus, the three systems of stresses acting on the circular disk produce the following stress distributions

$$\sigma_{r_1} = -\frac{2F \sin \theta_1}{\pi tr_1}, \quad \sigma_{\theta_1} = \tau_{r_1\theta_1} = 0$$

$$\sigma_{r_2} = -\frac{2F \sin \theta_2}{\pi tr_2}, \quad \sigma_{\theta_2} = \tau_{r_2\theta_2} = 0 \qquad (4.6.9)$$

$$\sigma_{r_3} = \frac{2F}{\pi tD}, \quad \sigma_{\theta_3} = \frac{2F}{\pi tD}, \quad \tau_{r_3\theta_3} = 0$$

The x and y components of these stresses are obtained from Fig. 4.6.3 as

$$\sigma_x = \sigma_{r_1} \cos^2 \theta_1 + \sigma_{r_2} \cos^2 \theta_2 + \sigma_{r_3} \sin^2 \alpha + \sigma_{\theta_3} \cos^2 \alpha$$

$$\sigma_y = \sigma_{r_2} \sin^2 \theta_1 + \sigma_{r_2} \sin^2 \theta_2 + \sigma_{r_3} \cos^2 \alpha + \sigma_{\theta_3} \sin^2 \alpha \qquad (4.6.10)$$

$$\tau_{xy} = -\sigma_{r_1} \sin \theta_1 \cos \theta_1 + \sigma_{r_2} \sin \theta_2 \cos \theta_2$$

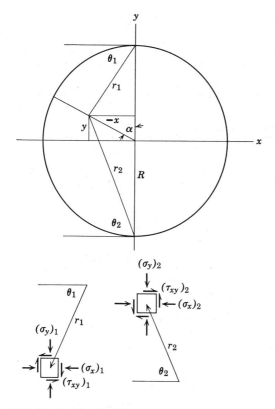

Fig. 4.6.3. Method for obtaining x and y components of stress.

Substitution of Eqs. 4.6.9 into 4.6.10 gives

$$\sigma_x = -\frac{2F}{\pi t}\left[\frac{\cos^2\theta_1\sin\theta_1}{r_1} + \frac{\cos^2\theta_2\sin\theta_2}{r_2} - \frac{1}{D}\right]$$

$$\sigma_y = -\frac{2F}{\pi t}\left[\frac{\sin^3\theta_1}{r_1} + \frac{\sin^3\theta_2}{r_2} - \frac{1}{D}\right] \qquad (4.6.11)$$

$$\tau_{xy} = -\frac{2F}{\pi t}\left[-\frac{\sin^2\theta_1\cos\theta_1}{r_1} + \frac{\sin^2\theta_2\cos\theta_2}{r_2}\right]$$

From Fig. 4.6.3

$$\sin\theta_1 = \frac{R-y}{r_1} \qquad\qquad \cos\theta_1 = \frac{-x}{r_1}$$

$$\sin\theta_2 = \frac{R+y}{r_2} \qquad\qquad \cos\theta_2 = \frac{-x}{r_2} \qquad (4.6.12)$$

$$r_1^2 = x^2 + (R-y)^2 \qquad r_2^2 = x^2 + (R+y)^2$$

Substitution of Eqs. 4.6.12 into 4.6.11 gives

$$\sigma_x = -\frac{2F}{\pi t}\left[\frac{(R-y)x^2}{\{x^2+(R-y)^2\}^2} + \frac{(R+y)x^2}{\{x^2+(R+y)^2\}^2} - \frac{1}{D}\right]$$

$$\sigma_y = -\frac{2F}{\pi t}\left[\frac{(R-y)^3}{\{x^2+(R-y)^2\}^2} + \frac{(R+y)^3}{\{x^2+(R+y)^2\}^2} - \frac{1}{D}\right] \quad (4.6.13)$$

$$\tau_{xy} = \frac{2F}{\pi t}\left[\frac{(-x)(R-y)^2}{\{x^2+(R-y)^2\}^2} - \frac{(-x)(R+y)^2}{\{x^2+(R+y)^2\}^2}\right]$$

notice that the minus sign on x should be dropped when τ_{xy} is calculated in the positive x direction. For points on the x axis, $y = 0$, therefore Eqs. 4.6.13 reduce to

$$\sigma_x = \frac{F}{\pi t R}\left[\frac{R^2 - x^2}{R^2 + x^2}\right]^2$$

$$\sigma_y = -\frac{F}{\pi t R}\left[\frac{(3R^2 + x^2)(R^2 - x^2)}{(R^2 + x^2)}\right] \quad (4.6.14)$$

$$\tau_{xy} = 0$$

Thus on the x axis, σ_x is always positive or tensile, and σ_y always negative or compressive for an applied compressive load.

For points on the y axis, $x = 0$, and Eqs. 4.6.13 reduce to

$$\sigma_x = \frac{F}{\pi t R}$$

$$\sigma_y = -\frac{F}{\pi t R}\left[\frac{3R^2 + y^2}{R^2 - y^2}\right] \quad (4.6.15)$$

$$\tau_{xy} = 0$$

Thus on the y axis, σ_x is a constant positive or tensile stress, and σ_y is always negative or compressive for an applied compressive load. Because σ_x is a constant tensile stress on the y axis, a short cylinder placed under a diametrical load makes an effective tensile strength test for rock samples (Section 11.6).

4.7   INFINITE PLATE WITH A CIRCULAR HOLE

The problem of a hole in an infinite plate is of special interest in the rock mechanics field because it corresponds to the problem of a long horizontal tunnel at depth in a uniform rock formation. Only a few cross-sectional shapes for the opening can be analyzed theoretically; however, by means of photoelastic methods of stress analysis, openings of any given cross-sectional shape can be studied.

Consider an infinite plate of thickness t with a circular hole of radius a located at the origin as illustrated in Fig. 4.7.1. Let the applied stress in the x direction be S_x and in the y direction be S_y. At a large distance from the hole, the polar components of stress will be those resulting from the applied stress only. Thus from Eqs. 1.3.2 and 1.3.3, the boundary conditions at $R = \infty$ are

$$(\sigma_r)_{r=\infty} = \tfrac{1}{2}(S_x + S_y) + \tfrac{1}{2}(S_x - S_y)\cos 2\theta$$
$$(\tau_{r\theta})_{r=\infty} = -\tfrac{1}{2}(S_x - S_y)\sin 2\theta \tag{4.7.1}$$

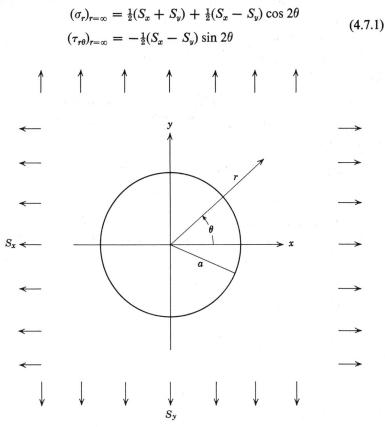

Fig. 4.7.1. Circular hole in infinite plate.

The boundary conditions at $r = a$ are

$$(\sigma_r)_{r=a} = (\tau_{r\theta})_{r=a} = 0 \tag{4.7.2}$$

To solve the problem of the circular hole in an infinite plate, assume a stress function of the form

$$\Phi = A \log r + Br^2 + (Cr^2 + Dr^4 + Er^{-2} + F)\cos 2\theta \tag{4.7.3}$$

As Eq. 4.7.3 is part of the general solution of Eq. 4.2.2 the stress components are given by Eqs. 4.2.1. Thus

$$\sigma_r = \frac{1}{r}\frac{\partial \Phi}{\partial r} + \frac{1}{r^2}\frac{\partial^2 \Phi}{\partial \theta^2} = \frac{A}{r^2} + 2B + (-2C - 6Er^{-4} - 4Fr^{-2})\cos 2\theta$$

$$\sigma_\theta = \frac{\partial^2 \Phi}{\partial r^2} = -\frac{A}{r^2} + 2B + (2C + 12Dr^2 + 6Er^{-4})\cos 2\theta \qquad (4.7.4)$$

$$\tau_{r\theta} = \frac{1}{r^2}\frac{\partial \Phi}{\partial \theta} - \frac{1}{r}\frac{\partial^2 \Phi}{\partial r\,\partial\theta} = (2C + 6Dr^2 - 6Er^{-4} - 2Fr^{-2})\sin 2\theta$$

When $r = \infty$

$$(\sigma_r)_{r=\infty} = \tfrac{1}{2}(S_x + S_y) + \tfrac{1}{2}(S_x - S_y)\cos 2\theta = 2B - 2C\cos 2\theta$$
$$(\tau_{r\theta})_{r=\infty} = -\tfrac{1}{2}(S_x - S_y)\sin 2\theta = [2C + 6D(\infty)^2]\sin 2\theta$$

When $r = a$

$$(\sigma_r)_{r=a} = 0 = \frac{A}{a^2} + 2B + (-2C - 6Ea^{-4} - 4Fa^{-2})\cos 2\theta$$

$$(\tau_{r\theta})_{r=a} = 0 = (2C + 6Da^2 - 6Ea^{-4} - 2Fa^{-2})\sin 2\theta$$

As the above equations must be true for all values of θ and as the stress cannot become infinite as $r \to \infty$, the following equations must hold:

$$D = 0$$
$$2B = \tfrac{1}{2}(S_x + S_y)$$
$$-2C = \tfrac{1}{2}(S_x - S_y)$$
$$\frac{A}{a^2} + 2B = 0 \qquad (4.7.5)$$
$$-2C - 6Ea^{-4} - 4Fa^{-2} = 0$$
$$2C - 6Ea^{-4} - 2Fa^{-2} = 0$$

Solving the equations in Eqs. 4.7.5 gives

$$A = -\frac{a^2}{2}(S_x + S_y)$$
$$2B = \tfrac{1}{2}(S_x + S_y)$$
$$2C = -\tfrac{1}{2}(S_x' - S_y)$$
$$D = 0 \qquad (4.7.6)$$
$$E = -\tfrac{1}{4}(S_x - S_y)a^4$$
$$F = \tfrac{1}{2}(S_x - S_y)a^2$$

Substitution of these equations (4.7.6) into those of Eq. 4.7.4 gives*

$$\sigma_r = \tfrac{1}{2}(S_x + S_y)\left(1 - \frac{a^2}{r^2}\right) + \tfrac{1}{2}(S_x - S_y)\left(1 + \frac{3a^4}{r^4} - \frac{4a^2}{r^2}\right) \cos 2\theta$$

$$\sigma_\theta = \tfrac{1}{2}(S_x + S_y)\left(1 + \frac{a^2}{r^2}\right) - \tfrac{1}{2}(S_x - S_y)\left(1 + \frac{3a^4}{r^4}\right) \cos 2\theta \qquad (4.7.7)$$

$$\tau_{r\theta} = -\tfrac{1}{2}(S_x - S_y)\left(1 - \frac{3a^4}{r^4} + \frac{2a^2}{r^2}\right) \sin 2\theta$$

Equations 4.7.7 are the stress components in an infinite plate containing a circular hole when the applied-stress field at infinity is $\sigma_x = S_x$ and $\sigma_y = S_y$.

Consider first the case when $S_x = S_y = -p$; Eqs. 4.7.7 become

$$\sigma_r = -p\left(1 - \frac{a^2}{r^2}\right)$$

$$\sigma_\theta = -p\left(1 + \frac{a^2}{r^2}\right) \qquad (4.7.8)$$

$$\tau_{r\theta} = 0$$

These equations are identical to those for the thick wall cylinder with pressure on the outside only and the outer radii equal to infinity. See Eqs. 4.5.8 and divide numerator and denominator by b^2 and let b approach infinity.

The maximum tangential stress occurs at the boundary of the circular opening and is equal to two times the applied stress if $S_x = S_y$. The radial stress is zero at the boundaries, thus the maximum shear stress at the boundary is equal to the applied stress and occurs on planes at 45° to the boundary.

When $S_x = 0$, Eqs. 4.7.7 become

$$\sigma_r = \frac{S_y}{2}\left(1 - \frac{a^2}{r^2}\right) - \frac{S_y}{2}\left(1 + \frac{3a^4}{r^4} - \frac{4a^4}{r^2}\right) \cos 2\theta$$

$$\sigma_\theta = \frac{S_y}{2}\left(1 + \frac{a^2}{r^2}\right) + \frac{S_y}{2}\left(1 + \frac{3a^4}{r^4}\right) \cos 2\theta \qquad (4.7.9)$$

$$\tau_{r\theta} = \frac{S_y}{2}\left(1 - \frac{3a^4}{r^4} + \frac{2a^2}{r^2}\right) \sin 2\theta$$

For $S_x = 0$, the maximum tangential stress is three times the applied stress and occurs at the boundary on the x axis, that is, where $\theta = 0$ or π.

* This solution was first obtained by G. Kirsch; see Z. Ver. dent. Inq., **42**, 1898.

When $\theta = \pi/2$ or $3\pi/2$, the tangential stress at the boundary of the opening is equal to the applied stress but is of opposite sign.

The stress distribution along axes of symmetry and around the boundary of the circular opening are shown graphically for the condition $S_x = 0$ in Figs. 4.7.2a and 4.7.2b.

The displacements for the circular opening in an infinitely wide plate are obtained by integrating the stress displacement equations. From Eqs. 3.8.7. and 3.8.12, the stress displacement relations for plane stresses are

$$\frac{\partial u}{\partial r} = \frac{1}{E}(\sigma_r - \nu\sigma_\theta)$$

$$\frac{u}{r} + \frac{1}{r}\frac{\partial v}{\partial \theta} = \frac{1}{E}(\sigma_\theta - \nu\sigma_r) \qquad (4.7.10)$$

$$\frac{1}{r}\frac{\partial u}{\partial \theta} + \frac{\partial v}{\partial r} - \frac{v}{r} = \frac{2(1+\nu)}{E}\tau_{r\theta}$$

Combining Eqs. 4.7.10 and 4.7.7 gives

$$\frac{\partial u}{\partial r} = \frac{1}{E}\left[\left(\frac{S_x + S_y}{2}\right)\left(1 - \frac{a^2}{r^2}\right) + \left(\frac{S_x - S_y}{2}\right)\left(1 + \frac{3a^4}{r^4} - \frac{4a^2}{r^2}\right)\cos 2\theta\right]$$
$$- \frac{\nu}{E}\left[\left(\frac{S_x + S_y}{2}\right)\left(1 + \frac{a^2}{r^2}\right) - \left(\frac{S_x - S_y}{2}\right)\left(1 + \frac{3a^4}{r^4}\right)\cos 2\theta\right]$$

$$(4.7.11)$$

$$\frac{u}{r} + \frac{1}{r}\frac{\partial v}{\partial \theta} = \frac{1}{E}\left[\left(\frac{S_x + S_y}{2}\right)\left(1 + \frac{a^2}{r^2}\right) - \left(\frac{S_x - S_y}{2}\right)\left(1 + \frac{3a^4}{r^4}\right)\cos 2\theta\right]$$
$$- \frac{\nu}{E}\left[\left(\frac{S_x + S_y}{2}\right)\left(1 - \frac{a^2}{r^2}\right) + \left(\frac{S_x - S_y}{2}\right)\left(1 + \frac{3a^4}{r^4} - \frac{4a^2}{r^2}\right)\cos 2\theta\right]$$

$$\frac{1}{r}\frac{\partial u}{\partial \theta} + \frac{\partial v}{\partial r} - \frac{v}{r} = - \frac{2(1+\nu)}{E}\left(\frac{S_x - S_y}{2}\right)\left(1 - \frac{3a^4}{r^4} + \frac{2a^2}{r^2}\right)\sin 2\theta$$

Integration of the first equation of Eqs. 4.7.11 gives

$$u = \frac{1}{E}\left[\left(\frac{S_x + S_y}{2}\right)\left(r + \frac{a^2}{r}\right) + \left(\frac{S_x - S_y}{2}\right)\left(r - \frac{a^4}{r^3} + \frac{4a^2}{r}\right)\cos 2\theta\right]$$
$$- \frac{\nu}{E}\left[\left(\frac{S_x + S_y}{2}\right)\left(r - \frac{a^2}{r}\right) - \left(\frac{S_x - S_y}{2}\right)\left(r - \frac{a^4}{r^3}\right)\cos 2\theta\right] + g_1(\theta)$$

$$(4.7.12)$$

where $g_1(\theta)$ is a function of θ only.

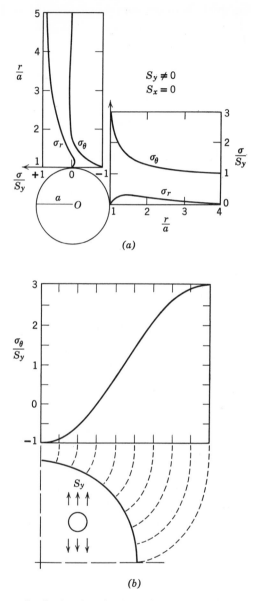

(a)

(b)

Fig. 4.7.2. Stress distribution for circular hole in a plate for plane uniaxial stress conditions.

Combining Eq. 4.7.12 with the second equation of Eqs. 4.7.11 gives

$$\frac{\partial v}{\partial \theta} = \frac{1}{E}\left[-2\left(\frac{S_x - S_y}{2}\right)\left(r + \frac{2a^2}{r} + \frac{a^4}{r^3}\right)\cos 2\theta\right]$$

$$- \frac{\nu}{E}\left[2\left(\frac{S_x - S_y}{2}\right)\left(r - \frac{2a^2}{r} + \frac{a^4}{r^3}\right)\cos 2\theta\right] - g_1(\theta)$$

Integration of the above equation gives

$$v = \frac{1}{E}\left[-\left(\frac{S_x - S_y}{2}\right)\left(r + \frac{2a^2}{r} + \frac{a^4}{r^3}\right)\sin 2\theta\right]$$

$$- \frac{\nu}{E}\left[\left(\frac{S_x - S_y}{2}\right)\left(r - \frac{2a^2}{r} + \frac{a^4}{r^3}\right)\sin 2\theta\right] - \int g_1(\theta)\,d\theta + g_2(r) \quad (4.7.13)$$

where $g_2(r)$ is a function of r only.

Differentiation of Eq. 4.7.12 with respect to θ gives

$$\frac{\partial u}{\partial \theta} = \frac{1}{E}\left[-2\left(\frac{S_x - S_y}{2}\right)\left(r + \frac{4a^2}{r} - \frac{a^4}{r^3}\right)\sin 2\theta\right]$$

$$+ \frac{\nu}{E}\left[2\left(\frac{S_x - S_y}{2}\right)\left(r - \frac{a^4}{r^3}\right)\sin 2\theta\right] + \frac{dg_1(\theta)}{d\theta} \quad (4.7.14)$$

Differentiation of Eq. 4.7.13 with respect to r gives

$$\frac{\partial v}{\partial r} = \frac{1}{E}\left[-\left(\frac{S_x - S_y}{2}\right)\left(1 - \frac{2a^2}{r^2} - \frac{3a^4}{r^4}\right)\sin 2\theta\right]$$

$$- \frac{\nu}{E}\left[\left(\frac{S_x - S_y}{2}\right)\left(1 + \frac{2a^2}{r^2} - \frac{3a^4}{r^4}\right)\sin 2\theta\right] + \frac{dg_2(r)}{dr} \quad (4.7.15)$$

The third equation of Eqs. 4.7.11, together with Eqs. 4.7.13, 4.7.14, and 4.7.15, reduce after some algebra to

$$\left[\frac{dg_1(\theta)}{d\theta} + \int g_1(\theta)\,d\theta\right] + \left[r\frac{dg_2(r)}{dr} - g_2(r)\right] = 0 \quad (4.7.16)$$

As $g_1(\theta)$ is a function of θ only and $g_2(r)$ is a function of r only, each part of Eq. 4.7.10 must equal some constant k, thus

$$r\frac{dg_2(r)}{dr} - g_2(r) = k$$

and   (4.7.17)

$$\frac{dg_1(\theta)}{d\theta} + \int g_1(\theta)\,d\theta = k$$

Solutions to these ordinary differential equations are

$$g_2(r) = Cr - k$$
$$g_1(\theta) = A \sin \theta + B \cos \theta \tag{4.7.18}$$

Substitution of Eqs. 4.7.18 into 4.7.12 and 4.7.13 gives the displacements as

$$u = \frac{1}{E}\left[\left(\frac{S_x + S_y}{2}\right)\left(r + \frac{a^2}{r}\right) + \left(\frac{S_x - S_y}{2}\right)\left(r - \frac{a^4}{r^3} + \frac{4a^2}{r}\right)\cos 2\theta\right]$$
$$- \frac{\nu}{E}\left[\left(\frac{S_x + S_y}{2}\right)\left(r - \frac{a^2}{r}\right) - \left(\frac{S_x - S_y}{2}\right)\left(r - \frac{a^4}{r^3}\right)\cos 2\theta\right]$$
$$+ A \sin \theta + B \cos \theta$$

$$\tag{4.7.19}$$

and

$$v = \frac{\nu}{E}\left[-\left(\frac{S_x - S_y}{2}\right)\left(r + \frac{2a^2}{r} + \frac{a^4}{r^3}\right)\sin 2\theta\right]$$
$$- \frac{\nu}{E}\left[\left(\frac{S_x - S_y}{2}\right)\left(r - \frac{2a^2}{r} + \frac{a^4}{r^3}\right)\sin 2\theta\right]$$
$$+ A \cos \theta - B \sin \theta + Cr$$

The boundary conditions for the displacements are $v = 0$ when $\theta = 0$ or $\pi/2$ for all values of r. Thus

$$v_{\theta=0} = 0 = A + Cr$$
$$v_{\theta=\pi/2} = 0 = -B + Cr$$

Solving these equations for A, B, and C gives

$$A = B = C = 0$$

Therefore, the displacements are given by

$$u = \frac{1}{E}\left[\left(\frac{S_x + S_y}{2}\right)\left(r + \frac{a^2}{r}\right) + \left(\frac{S_x - S_y}{2}\right)\left(r - \frac{a^4}{r^3} + \frac{4a^2}{r}\right)\cos 2\theta\right]$$
$$- \frac{\nu}{E}\left[\left(\frac{S_x + S_y}{2}\right)\left(r - \frac{a^2}{r}\right) - \left(\frac{S_x - S_y}{2}\right)\left(r - \frac{a^4}{r^3}\right)\cos 2\theta\right]$$

and $$\tag{4.7.20}$$

$$v = \frac{1}{E}\left[-\left(\frac{S_x - S_y}{2}\right)\left(r + \frac{2a^2}{r} + \frac{a^4}{r^3}\right)\sin 2\theta\right]$$
$$- \frac{\nu}{E}\left[\left(\frac{S_x - S_y}{2}\right)\left(r - \frac{2a^2}{r} + \frac{a^4}{r^3}\right)\sin 2\theta\right]$$

When $r = a$, these equations become

$$u = \frac{1}{E} [(S_x + S_y)a + 2(S_x - S_y)a \cos 2\theta]$$

$$v = -\frac{1}{E} [2(S_x - S_y)a \sin 2\theta]$$

(4.7.21)

When $S_x = S_y = -p$, Eqs. 4.7.21 become

$$u = -\frac{2pa}{E} \quad \text{and} \quad v = 0 \tag{4.7.22}$$

These displacements should be compared to those obtained for the thick wall cylinder when $b \to \infty$ and $p_i = 0$.

When $S_x = 0$, Eqs. 4.7.20 become

$$u = \frac{S_y}{E} [a - 2a \cos 2\theta]$$

$$v = \frac{S_y}{E} [2a \sin 2\theta]$$

(4.7.23)

Equations 4.7.23 give the displacement at the boundary of the circular hole in an infinite plate when $S_y \neq 0$ and $S_x = 0$. The radial displacement is maximum when $\theta = \pi/2$, or $3\pi/2$ and has the value of $3aS_y/E$. When

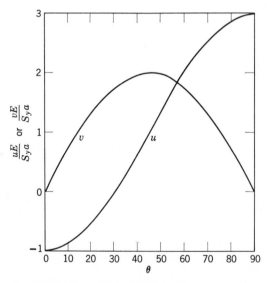

Fig. 4.7.3. Displacements, u and v, on the boundary of circular hole in a plate under a condition of plane uniaxial stress.

$\theta = 0$, or π, the radial displacement is $-(aS_y)/E$. The displacements on the boundary of the circular opening as a function of angle are shown graphically in Fig. 4.7.3.

The displacements that occur when a circular hole is subjected to a two-dimensional stress field under the condition of plain strain can be determined by integrating the stress-displacement equations for plain strain. The stress-displacement relations in polar coordinates are from Eqs. 3.5.2 and 3.8.7.

$$\frac{\partial u}{\partial r} = \frac{1}{E}[(1 - v^2)\sigma_r - v(1 + v)\sigma_\theta]$$

$$\frac{u}{r} + \frac{1}{r}\frac{\partial v}{\partial \theta} = \frac{1}{E}[(1 - v^2)\sigma_\theta - v(1 + v)\sigma_r] \qquad (4.7.24)$$

$$\frac{1}{r}\frac{\partial u}{\partial \theta} + \frac{\partial v}{\partial r} - \frac{v}{r} = \frac{2(1 - v)}{E}\tau_{r\theta}$$

Combining Eqs. 4.7.24 with Eqs. 4.7.7 gives

$$\frac{\partial u}{\partial r} = \frac{1 - v^2}{E}\left[\left(\frac{S_x + S_y}{2}\right)\left(1 - \frac{a^2}{r^2}\right)\right.$$
$$\left. + \left(\frac{S_x - S_y}{2}\right)\left(1 + \frac{3a^4}{r^4} - \frac{4a^2}{r^2}\right)\cos 2\theta\right]$$
$$- \frac{v(1 + v)}{E}\left[\left(\frac{S_x + S_y}{2}\right)\left(1 + \frac{a^2}{r^2}\right)\right.$$
$$\left. - \left(\frac{S_x - S_y}{2}\right)\left(1 + \frac{3a^4}{r^4}\right)\cos 2\theta\right]$$

$$\frac{u}{r} + \frac{1}{r}\frac{\partial v}{\partial \theta} = \frac{1 - v^2}{E}\left[\left(\frac{S_x + S_y}{2}\right)\left(1 + \frac{a^2}{r^2}\right)\right. \qquad (4.7.25)$$
$$\left. - \left(\frac{S_x - S_y}{2}\right)\left(1 + \frac{3a^4}{r^4}\right)\cos 2\theta\right]$$
$$- \frac{v(1 + v)}{E}\left[\left(\frac{S_x + S_y}{2}\right)\left(1 - \frac{a^2}{r^2}\right)\right.$$
$$\left. + \left(\frac{S_x - S_y}{2}\right)\left(1 + \frac{3a^4}{r^4} - \frac{4a^2}{r^2}\right)\cos 2\theta\right]$$

$$\frac{1}{r}\frac{\partial u}{\partial \theta} + \frac{\partial v}{\partial r} - \frac{v}{r} = -\frac{2(1 - v)}{E}\left(\frac{S_x - S_y}{2}\right)\left(1 - \frac{3a^4}{r^4} + \frac{2a^2}{r^2}\right)\sin 2\theta$$

Integration of Eqs. 4.7.25 gives

$$
\begin{aligned}
u = \frac{1 - v^2}{E}\Bigg[& \left(\frac{S_x + S_y}{2}\right)\left(r + \frac{a^2}{r}\right) \\
& + \left(\frac{S_x - S_y}{2}\right)\left(r - \frac{a^4}{r^3} + \frac{4a^2}{r}\right)\cos 2\theta \Bigg] \\
- \frac{v(1 + v)}{E}\Bigg[& \left(\frac{S_x + S_y}{2}\right)\left(r - \frac{a^2}{r}\right) \\
& - \left(\frac{S_x - S_y}{2}\right)\left(r - \frac{a^4}{r^3}\right)\cos 2\theta \Bigg]
\end{aligned}
$$

(4.7.26)

$$
\begin{aligned}
v = \frac{1 - v^2}{E}\Bigg[& -\left(\frac{S_x - S_y}{2}\right)\left(r + \frac{2a^2}{r} + \frac{a^4}{r^3}\right)\sin 2\theta \Bigg] \\
- \frac{v(1 + v)}{E}\Bigg[& \left(\frac{S_x - S_y}{2}\right)\left(r - \frac{2a^2}{r} + \frac{a^4}{r^3}\right)\sin 2\theta \Bigg]
\end{aligned}
$$

When $r = a$

$$
u = \frac{1 - v^2}{E}\,[a(S_x + S_y) + 2a(S_x - S_y)\cos 2\theta]
$$

$$
v = -\,\frac{1 - v^2}{E}\,[2a(S_x - S_y)\sin 2\theta]
$$

(4.7.27)

Equations 4.7.27 differ from Eqs. 4.7.21 only in the factor $1 - v^2$. Thus the displacements at the boundary of the circular hole, for the condition of plane strain, are approximately 94% of the displacement for the condition of plane stress when Poisson's ratio is 0.25.

4.8   THICK WALL CYLINDER IN TRIAXIAL STRESS FIELD

The determination of the stress distribution in a thick wall cylinder subjected to a triaxial stress field is a three-dimensional problem that can be solved by two-dimensional methods. Figure 4.8.1 gives the coordinate system and the nomenclature. The normal stresses in the r, θ, and z directions are σ_r, σ_θ, and σ_z; the normal strains in these directions are ϵ_r, ϵ_θ, and ϵ_z; and the displacements are u, v, and w, respectively. The shear stresses for the $r\theta$, θz, and zr planes are $\tau_{r\theta}$, $\tau_{\theta z}$, and τ_{zr}; the shear strains are $\gamma_{r\theta}$, $\gamma_{\theta z}$, and γ_{zr}, respectively. The assumed boundary conditions are an internal pressure $-p_i$ at $r = a$, an external pressure $-p_o$ at $r = b$, and an axial pressure $-p_z$ at $z = 0$ and $z = 1$.

The problem of the thick wall cylinder subjected to only an internal and external pressure was solved for the case of plane stress in Section 4.5. Thus from Eqs. 4.5.6, the stress distribution for the two-dimensional problem is

$$\sigma_r = \frac{a^2 b^2 (p_o - p_i)}{(b^2 - a^2) r^2} + \frac{p_i a^2 - p_o b^2}{b^2 - a^2}$$

$$\sigma_\theta = -\frac{a^2 b^2 (p_o - p_i)}{(b^2 - a^2) r^2} + \frac{p_i a^2 - p_o b^2}{b^2 - a^2} \qquad (4.8.1)$$

$$\tau_{r\theta} = 0$$

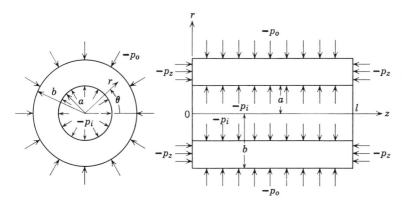

Fig. 4.8.1. Notation and coordinate system for thick wall cylinder under triaxial load.

The sum $\sigma_r + \sigma_\theta$ is independent of r and θ. Thus as a result of σ_r and σ_θ, there is a uniform strain produced in the axial or z direction. As the strain in the z direction is constant, plane sections normal to the axis of the cylinder remain plane during application of $-p_i$ and $-p_o$. Therefore, it is possible to load the thick wall cylinder with a uniformly distributed stress $-p_z$ to produce a total uniform strain ϵ_z in the axial direction without introducing any distortion in sections normal to the axis of the cylinder. Thus $\gamma_{\theta z}$ and γ_{zr} are zero and, as $\tau_{r\theta}$ is zero from Eqs. 4.8.1, then $\gamma_{r\theta}$ is zero also. Therefore, a solution for the thick wall cylinder for the case of plane strain, with ϵ_z = constant, will be the desired solution.

For two-dimensional problems involving no body forces, the stress distribution is the same for the case of plane stress as it is for the case of plane strain (see Section 3.5). However, the strains and displacements for the two cases are different because the stress-strain relations are different.

The three-dimensional stress-strain relations in cylindrical coordinates are

$$\epsilon_r = \frac{1}{E}\left[\sigma_r - \nu(\sigma_\theta + \sigma_z)\right]$$

$$\epsilon_\theta = \frac{1}{E}\left[\sigma_\theta - \nu(\sigma_z + \sigma_r)\right] \tag{4.8.2}$$

$$\epsilon_z = \frac{1}{E}\left[\sigma_z - \nu(\sigma_r + \sigma_\theta)\right]$$

As ϵ_z is assumed constant, the third equation of Eqs. 4.8.2 is solved for σ_z and the resulting equation substituted into the first two equations of Eqs. 4.8.2 to give

$$\sigma_z = -p_z = E\epsilon_z + \nu(\sigma_r + \sigma_\theta)$$

$$\epsilon_r = \frac{1}{E}\left[(1 - \nu^2)\sigma_r - \nu(1 + \nu)\sigma_\theta\right] - \nu\epsilon_z \tag{4.8.3}$$

$$\epsilon_\theta = \frac{1}{E}\left[(1 - \nu^2)\sigma_\theta - \nu(1 + \nu)\sigma_r\right] - \nu\epsilon_z$$

The necessary strain-displacement relations in cylindrical coordinates are from Eqs. 3.10.3

$$\epsilon_r = \frac{\partial u}{\partial r}$$

$$\epsilon_\theta = \frac{u}{r} + \frac{1}{r}\frac{\partial v}{\partial \theta} \tag{4.8.4}$$

$$\epsilon_z = \frac{\partial w}{\partial z}$$

Substitution of Eqs. 4.8.1 into Eqs. 4.8.3 gives

$$\epsilon_z = -\frac{p_z}{E} - \frac{2\nu}{E}\left[\frac{p_i a^2 - p_o b^2}{b^2 - a^2}\right]$$

$$\epsilon_r = \frac{1 + \nu}{E}\left[\frac{p_i a^2 - p_o b^2}{b^2 - a^2}(1 - 2\nu) + \frac{a^2 b^2(p_o - p_i)}{(b^2 - a^2)r^2}\right] - \nu\epsilon_z \tag{4.8.5}$$

$$\epsilon_\theta = \frac{1 + \nu}{E}\left[\frac{p_i a^2 - p_o b^2}{b^2 - a^2}(1 - 2\nu) - \frac{a^2 b^2(p_o - p_i)}{(b^2 - a^2)r^2}\right] - \nu\epsilon_z$$

Combining Eqs. 4.8.5 and 4.8.4 gives

$$\frac{\partial u}{\partial r} = \frac{1 + \nu}{E}\left[\frac{p_i a^2 - p_o b^2}{b^2 - a^2}(1 - 2\nu) + \frac{a^2 b^2(p_o - p_i)}{(b^2 - a^2)r^2}\right] - \nu\epsilon_z$$

$$\frac{u}{r} + \frac{1}{r}\frac{\partial v}{\partial \theta} = \frac{1 + \nu}{E}\left[\frac{p_i a^2 - p_o b^2}{b^2 - a^2}(1 - 2\nu) - \frac{a^2 b^2(p_o - p_i)}{(b^2 - a^2)r^2}\right] - \nu\epsilon_z \tag{4.8.6}$$

$$\frac{\partial w}{\partial z} = -\frac{p_z}{E} - \frac{2\nu}{E}\frac{p_i a^2 - p_o b^2}{b^2 - a^2}$$

Integration of the first equation in Eqs. 4.8.6 gives

$$u = \frac{1+v}{E}\left[\frac{p_i a^2 - p_o b^2}{b^2 - a^2}(1 - 2v)r - \frac{a^2 b^2 (p_o - p_i)}{(b^2 - a^2)r}\right] - v\epsilon_z r \quad (4.8.7)$$

Because of symmetry, the arbitrary function of θ and z that should appear in Eq. 4.8.7 is zero. If Eq. 4.8.7 is substituted into the second equation of Eq. 4.8.6, the result is $\partial v/\partial\theta = 0$. Therefore, v can only be a rigid body displacement and is taken as zero. Integration of the third equation of Eqs. 4.8.6 gives

$$w = -\frac{1}{E}\left[p_z + 2v\frac{p_i a^2 - p_o b^2}{b^2 - a^2}\right]z \quad (4.8.8)$$

where again the arbitrary function from integration is set equal to zero as it can only represent rigid body motion. Thus the stresses, strains, and displacements for a thick wall cylinder subjected to an internal pressure $-p_i$, an external pressure $-p_o$, and an axial pressure $-p_z$ are given by Eqs. 4.8.1, 4.8.5, 4.8.7, and 4.8.8.

When thick wall cylinders are used in triaxial testing, only the axial pressure $-p_z$ and external pressure $-p_o$ are applied. Gages are used to measure the axial strain or deformation and the internal deformation of the cylindrical hole. If the above equations are evaluated at $r = a$ for no internal pressure, the results are

$$\sigma_r = 0$$

$$\sigma_\theta = -p_o\frac{2b^2}{b^2 - a^2} \quad (4.8.9)$$

$$\sigma_z = -p_z = E\epsilon_z - \frac{2vp_o b^2}{b^2 - a^2}$$

$$\epsilon_r = \frac{2v(1+v)p_o b^2}{E(b^2 - a^2)} - v\epsilon_z = \frac{2vp_o b^2}{E(b^2 - a^2)} + \frac{v}{E}p_z$$

$$\epsilon_\theta = -\frac{2(1-v^2)p_o b^2}{E(b^2 - a^2)} - v\epsilon_z = -\frac{2p_o b^2}{E(b^2 - a^2)} + \frac{vp_z}{E} \quad (4.8.10)$$

$$\epsilon_z = -\frac{p_z}{E} + \frac{2vp_o b^2}{E(b^2 - a^2)}$$

$$u = -\frac{2(1-v^2)p_o b^2 a}{E(b^2 - a^2)} - v\epsilon_z a = -\frac{2p_o b^2 a}{E(b^2 - a^2)} + \frac{vap_z}{E}$$

$$v = 0 \quad (4.8.11)$$

$$w = -\epsilon_z Z = -\frac{1}{E}\left[p_z - 2v\frac{p_o b^2}{b^2 - a^2}\right]Z$$

REFERENCES

1. Durelli, A. J., E. A. Phillips, and C. H. Tsao, *Introduction to the Theoretical and Experimental Analysis of Stress and Strain*, Chapter 6, McGraw-Hill Book Co., New York, 1958.
2. Frocht, M. M., *Photoelasticity*, Vol. 2, Chapters 1, 2, 4, and 5, John Wiley and Sons, New York, 1948.
3. Jaeger, J. C., *Elasticity, Fracture and Flow*, Methuen and Co., London, 1962.
4. Kirsch, G., *Z. Ver. dent. Inq.*, Vol. 42, 1898.
5. Love, A. E. H., *The Mathematical Theory of Elasticity*, Dover Publications, New York, 1944.
6. Michell, J. H., "On the Determination of Stress in an Elastic Solid with Applications to the Theory of Plates," *London Math. Society*, **31**, pp. 100–124 (April 1899).
7. Timoshenko, S. and J. N. Goodier, *Theory of Elasticity*, Chapter 4, McGraw-Hill Book Co., New York, 1951.

CHAPTER 5

STRESS DISTRIBUTION IN SIMPLE STRUCTURES

5.1 INTRODUCTION

Elastic theory provides the solution to a large number of problems that have a direct application in structural rock mechanics. In this chapter a number of these problems are considered in which the treatment is restricted to a statement of the problem and the boundary conditions. Moreover, only the part of the results that is required in the analysis of rock structure problems is given. Reference to original sources will be made so that the interested reader can obtain the details of the mathematical solutions.

As pointed out in Chapter 3, two-dimensional problems involving constant body forces have the same stress distribution for plane-strain or plane-stress conditions. This is true because the two equilibrium equations (3.4.2) and the compatibility equation (3.4.6) in terms of stress are the same for plane-strain or plane-stress conditions. These three equations do not contain the elastic constants and involve only the three unknown stresses σ_x, σ_y, and τ_{xy}. Thus their solution, together with the boundary conditions, must result in stress distributions that are independent of the elastic constants and the conditions of plane strain or plane stress. However, if strains or displacements are required, these quantities will involve the elastic constants (because the stress-strain relations contain the elastic constants) and they will depend upon the conditions of plane strain or plane stress (because the stress-strain relations are different for plane-strain and plane-stress conditions).

Because stress distributions are the same for plane-strain and plane-stress conditions when body forces are constant, mathematical solutions for the stress distributions around holes in infinitely wide plates can be used as the solutions for the stress distributions around a tunnel, that is, if the applied stress field does not vary along the length of the tunnel. For example, a circular hole in an infinitely wide plate can be used to represent a long circular tunnel far removed from other openings or boundaries, provided the applied stress field for the plate is the same as for the tunnel. The effect of the variable body force of gravity can be neglected when the height of the opening is small compared to the depth below the surface.[7]

It has been found that the maximum stress around holes in wide plates always occur on the inner boundaries of the holes. In the discussions that follow, only the stress distributions on the inner boundaries and along major axes will be presented. Generally the stress at other points in the plate are not required in considering structural problems.

Besides problems involving single holes in infinitely wide plates, those involving multiple holes in plates will be presented. Also a few problems involving single three-dimensional openings will be discussed. In this chapter only idealized problems will be considered; the application of these results to specific underground structural-stability problems will be dealt with in Chapter 16.

Problems involving the flexure of beams and plates are presented in this chapter. In analyzing the roof structure for underground openings in flat thinly laminated rock formations, we take the point of view that the rock lamanae overlying the opening are plates clamped at their edges and loaded only by their own weight or by additional loading from the beds above. Reference is made to this theory in Chapters 16 and 17 in relation to design of openings and in Chapter 20 with relation to the support of laminated roof with rock bolts. Also, the flexure of beams of rock is an experimental technique used for determining the elastic constant E and the tensile strength of rock samples. Thus the problem of determining the stresses, strains, and deflections of clamped plates and beams plays an important role in structural rock mechanics.

5.2 SINGLE HOLES IN PLATES—GREENSPAN METHOD

In this section the effect of a small hole on the stresses in a uniformly loaded plate is presented. The plate is assumed to be in a state of generalized plane stress, that is, the state of stress at points remote from the hole is given by the constant normal stresses $\sigma_x = S_x$, $\sigma_y = S_y$, and the constant shearing stress by $\tau_{xy} = T_{xy}$. Greenspan[3] obtained an exact solution for the stress distribution in uniformly loaded plate containing a small hole whose boundary can be expressed in parametric form by

$$x = p \cos \beta + r \cos 3\beta, \quad y = q \sin \beta - r \sin 3\beta \qquad (5.2.1)$$

where p, q, and r are parameters, and β is an angle.

Equations 5.2.1 represent a closed curve having symmetry about both the x and y axes and, for certain values of p, q, and r, the curve is simple, that is, it does not cross itself. By adjustment of the values of p, q, and r, a variety of simple closed curves is obtained, including the circles, ellipses, approximate ovaloids, and approximate rectangles with rounded corners. Let W_0 and H_0 be the dimensions of the opening in the x and y directions, respectively, that is, the width and height. Thus W_0 is the difference in the values of x when $\beta = 0°$ and $180°$, or $W_0 = 2(p + r)$, and H_0 is the difference in the values of y when $\beta = 90°$ and $270°$, or $H_0 = 2(q + r)$.

The complete solution for the stress at all points in the plate is given by Greenspan.[3] However only the equation for calculating the tangential stresses at the boundary of the hole is discussed here. This equation is

$$[(p^2 + 6rq)\sin^2 \beta + (q^2 + 6rp)\cos^2 \beta - 6r(p + q)\cos^2 2\beta + 9r^2]\sigma_t$$

$$= (S_x + S_y)(p^2 \sin^2 \beta + q^2 \cos^2 \beta - 9r^2)$$

$$- T_{xy}(p + q)^2 \frac{p + q + 6r}{p + q + 2r} \sin 2\beta$$

$$- \frac{(p^2 - q^2)(S_x + S_y) - (p + q)^2(S_x - S_y)}{p + q - 2r}$$

$$\times [(p - 3r)\sin^2 \beta - (q - 3r)\cos^2 \beta] \qquad (5.2.2)$$

Equation 5.2.2 can be used to calculate the tangential stress for any applied stress field and for any shape of opening that can be represented by the parametric equations (5.2.1). However, it is more instructive to consider three simple applied stress field designated as

Case 1 $S_x \neq 0$ $S_y = T_{xy} = 0$

Case 2 $S_y \neq 0$ $S_x = T_{xy} = 0$

Case 3 $T_{xy} \neq 0$ $S_x = S_y = 0$

From these three cases any general applied stress field can be obtained by the principle of superposition. For calculation purposes, Eq. 5.2.2 can be rewritten in simpler form by making use of the trigonometric identity

$$\cos^2 \beta = 1 - \sin^2 \beta$$

and by writing three equations corresponding to the three applied-stress field cases. Thus

Case 1 $S_x \neq 0$ $S_y = T_{xy} = 0$

$$\frac{\sigma_t}{S_x} = \frac{D \sin^2 \beta + E}{A \sin^2 \beta + B \cos^2 2\beta + C} \tag{5.2.3}$$

Case 2 $S_y \neq 0$ $S_x = T_{xy} = 0$

$$\frac{\sigma_t}{S_y} = \frac{F \sin^2 \beta + G}{A \sin^2 \beta + B \cos 2\beta + C} \tag{5.2.4}$$

Case 3 $T_{xy} \neq 0$ $S_x = S_y = 0$

$$\frac{\sigma_t}{T_{xy}} = \frac{H \sin 2\beta}{A \sin^2 \beta + B \cos^2 2\beta + C} \tag{5.2.5}$$

where

$$A = (p - q)(p + q - 6r)$$

$$B = -6r(p + q)$$

$$C = q^2 + 6rp + 9r^2$$

$$D = (p + q)\left[(p - q) + \frac{2q(p + q - 6r)}{p + q - 2r}\right]$$

$$E = (q - 3r)\left[(q + 3r) - \frac{2q(p + q)}{p + q - 2r}\right] \tag{5.2.6}$$

$$F = (p + q)\left[(p - q) - \frac{2p(p + q - 6r)}{p + q - 2r}\right]$$

$$G = (q - 3r)\left[(q + 3r) + \frac{2p(p + q)}{p + q - 2r}\right]$$

$$H = (p + q)^2 \frac{p + q + 6r}{p + q + 2r}$$

Values of the above quantities for various values of p, q, and r are given in Table 5.2.1. By means of the constants in Table 5.2.1 and Eqs. 5.2.3, 5.2.4, and 5.2.5, the tangential boundary stresses were calculated for a $2:1$ and a $4:1$ elliptical and approximately ovaloidal opening, and a square opening with rounded corners' with its side and its diagonal parallel to the x axis. The results are shown graphically in Figs. 5.2.1 to 5.2.6, inclusive. These figures also show one quadrant of the opening that is generated by the parametric Eqs. 5.2.1 when the values of p, q, and r are those given in Table 5.2.1.

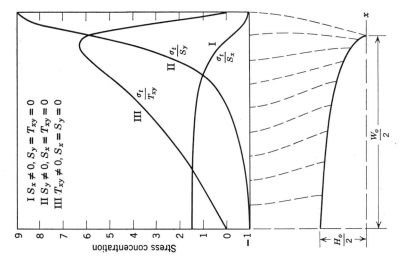

Fig. 5.2.2. Boundary stresses for elliptical opening $W_0/H_0 = 4$.

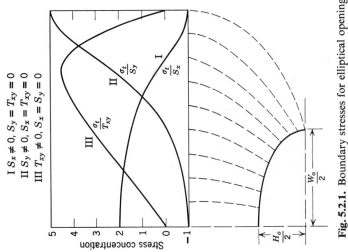

Fig. 5.2.1. Boundary stresses for elliptical opening $W_0/H_0 = 2$.

117

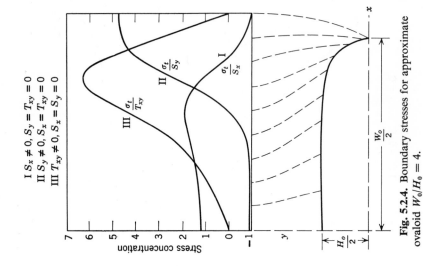

Fig. 5.2.4. Boundary stresses for approximate ovaloid $W_0/H_0 = 4$.

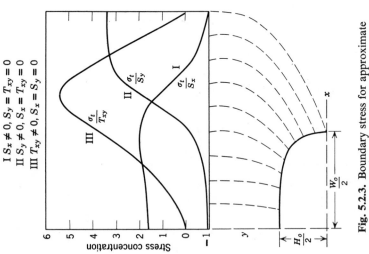

Fig. 5.2.3. Boundary stress for approximate ovaloid $W_0/H_0 = 2$.

118

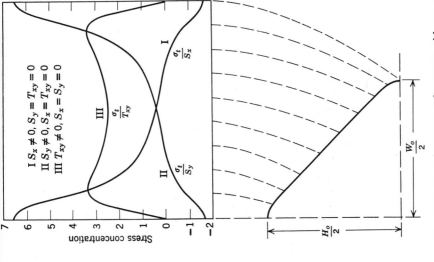

Fig. 5.2.6. Boundary stresses for square with rounded corners—diagonals parallel to axis, $W_0/H_0 = 1$.

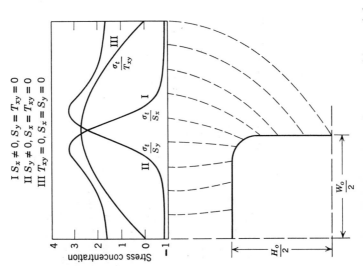

Fig. 5.2.5. Boundary stresses for square with rounded corners—sides parallel to axis, $W_0/H_0 = 1$.

119

Table 5.2.1. Constants for Calculating Boundary Stress by Greenspan's Method

Type of Opening	p	q	r	A	B	C	D	E	F	G	H
2:1 Ellipse	2	1	0	3	0	1	9	−1	−9	5	9
4:1 Ellipse	4	1	0	15	0	1	25	−1	−25	9	25
2:1 Ovaloid	2.1	1.1	−0.10	3.800	1.920	0.040	11.069	−1.779	−11.821	6.654	8.878
4:1 Ovaloid	4.19	1.19	−0.19	19.560	6.133	−3.036	30.634	−2.821	−34.895	14.867	24.545
Square with rounded corners (sides parallel to co-ordinate axes)	1	1	−0.14	0	1.680	0.336	4.982	−1.667	−4.982	3.314	2.696
Square with rounded corners (diagonals parallel to co-ordinate axes)	1	1	0.14	0	−1.680	2.016	2.698	−0.525	−2.698	2.173	4.984

Some generalizations can be made regarding stress concentration on the boundary of single openings in plates or long tunnels from a study of the curves presented in Figs. 5.2.1 to 5.2.6. First, for an applied normal stress parallel to one axis of the opening, the stress concentration at the end of the axis parallel to the applied stress is approximately minus one. That is, the boundary stress is of opposite sign and has about the same magnitude as the applied normal stress. Second, for an applied normal stress parallel to the minor axis of the opening, the stress concentration on the boundary at the end of the major axis increases as the ratio of the major to minor axis of the opening increases. Third, for an applied normal stress parallel to the minor axis of the opening, the maximum boundary stress concentration at the end of major axis of the opening increases as the radius of curvature of the boundary at the end of the major axis decreases. Thus the maximum stress concentrations for an ellipse are always greater than those for an ovaloid having the same major to minor axis ratio. Fourth, for an applied normal stress parallel to the major axis of the opening, the maximum boundary stress concentration at the end of the minor axis of the opening decreases for a given shape of opening as the major to minor axis ratio increases. Fifth, for an applied normal stress parallel to one axis of the opening, the maximum stress concentration is not necessarily at the end of the other axis of the opening if the smallest radius of curvature for the boundary of the opening occurs at a location other than at the end of the major and minor axes.

For an uniaxial applied-stress field, the maximum stress concentration for an elliptical opening having any ratio of major to minor axis can be obtained from Eq. 5.2.3, or 5.2.4, by setting $\beta = \pi/2$ in Eq. 5.2.3, or $\beta = 0$ in Eq. 5.2.4, and letting $r = 0$. Thus

$$\frac{\sigma_t}{S_x} = 1 + 2\frac{q}{p} = 1 + 2\frac{H_0}{W_0}$$

or

$$\frac{\sigma_t}{S_y} = 1 + 2\frac{p}{q} = 1 + 2\frac{W_0}{H_0} \tag{5.2.7}$$

where

$$2p = W_0 \quad \text{and} \quad 2q = H_0$$

For an applied biaxial-stress field that is normal and parallel to the major axis of an opening, the boundary stress distribution is obtained by adding Eqs. 5.2.3 and 5.2.4, thus

$$\sigma_t = \frac{(DS_x + FS_y)\sin^2\beta + ES_x + GS_y}{A\sin^2\beta + B\cos 2\beta + C} \tag{5.2.8}$$

For an elliptical opening, the parameter r is zero. Thus substitution of Eqs. 5.2.6, with $r = 0$, into Eq. 5.2.8 gives the boundary stress for an elliptical opening in a biaxial stress field as

$$\sigma_t = \frac{(p + q)^2(S_x - S_y) \sin^2 \beta - q^2(S_x - S_y) + 2S_y pq}{(p^2 - q^2) \sin^2 \beta + q^2} \quad (5.2.9)$$

If the boundary stress is to be constant for all values of β, then the partial derivative of σ_t with respect to β must be zero. Performing this operation and simplifying gives

$$\frac{S_x}{S_y} = \frac{p}{q} = \frac{W_0}{H_0} \quad (5.2.10)$$

Let the applied stress field be given by

$$\frac{S_x}{S_y} = m \quad (5.2.11)$$

Then the ratio of the x axis of the ellipse to the y axis is

$$\frac{W_0}{H_0} = m \quad (5.2.12)$$

and the boundary stress for the elliptical opening is constant and is given by

$$\sigma_t = (m + 1)S_y = \left(\frac{W_0}{H_0} + 1\right)S_y \quad (5.2.13)$$

5.3   SINGLE HOLES IN PLATES—COMPLEX VARIABLE METHOD

The parametric equations (5.2.1) of Section 5.2 give an exact representation of the boundary of a circle or an ellipse. However the accuracy with which these equations map the boundary of other shaped openings is rather limited. To map the boundary of rectangular openings with rounded corners for different length to width ratios and different radii of curvature at the corners, as well as ovaloids with different lengths to width ratios, requires a mapping function with more terms than is included in Eqs. 5.2.1.

A complex variable mapping function that has been used with a great deal of success is

$$Z = A\zeta + \frac{B}{\zeta} + \frac{C}{\zeta^3} + \frac{D}{\zeta^5} + \frac{E}{\zeta^7} + \cdots \quad (5.3.1)$$

where A, B, C, etc., are constants and the two complex variables are defined as

$$Z = x + iy \quad (5.3.2)$$

and

$$\zeta = e^{\alpha + i\beta} = e^{\alpha}(\cos \beta + i \sin \beta) \tag{5.3.3}$$

Expansion of Eq. 5.3.1 by means of Eqs. 5.3.2 and 5.3.3 gives

$$
\begin{aligned}
x + iy = {} & Ae^{\alpha}(\cos \beta + i \sin \beta) + Be^{-\alpha}(\cos \beta - i \sin \beta) \\
& + Ce^{-3\alpha}(\cos 3\beta - i \sin 3\beta) + De^{-5\alpha}(\cos 5\beta - i \sin 5\beta) \\
& + Ee^{-7\alpha}(\cos 7\beta - i \sin 7\beta) + \cdots
\end{aligned} \tag{5.3.4}
$$

Equation 5.3.1 or 5.3.4 is a mapping function which continuously transforms points in the $\zeta(\alpha, \beta)$-plane into points in the $Z(x, y)$-plane. Holding α constant and varying β defines a circle of radius e^{α} in the ζ-plane. This circle transforms into a closed curve in the Z-plane. Holding β constant and varying α defines a straight line in the ζ-plane which transforms into a curve that is orthogonal to the closed curves in the Z-plane obtained by keeping α constant and varying β. The unit circle in the ζ-plane, obtained by setting $\alpha = 0$, and varying β, is used to represent the boundary of the opening in the Z-plane. For $\alpha = 0$, Eq. 5.3.4 becomes, after equating real and imaginary parts,

$$
\begin{aligned}
x &= (A + B) \cos \beta + C \cos 3\beta + D \cos 5\beta + E \cos 7\beta + \cdots \\
y &= (A - B) \sin \beta - C \sin 3\beta - D \sin 5\beta - E \cos 7\beta - \cdots
\end{aligned} \tag{5.3.5}
$$

Comparison of Eqs. 5.3.5 with Eqs. 5.2.1 shows that the only difference between the two sets of equations is in the number of terms involved. Thus Eqs. 5.3.5 is a general set of parametric equations from which Eqs. 5.2.1 can be obtained by retaining only the first two terms.

Two potential functions are required to obtain the stresses in plane problems in elastic theory when the complex variable method is used.[1] These functions, in general form, are

$$
\begin{aligned}
\phi(Z) &= \sum_{n=1}^{k} A_n Z^n + \sum_{n=1}^{\infty} \frac{a_n}{Z^n} \\
\psi(Z) &= \sum_{n=1}^{k} B_n Z^n + \sum_{n=1}^{\infty} \frac{b_n}{Z^n}
\end{aligned} \tag{5.3.6}
$$

where A_n, B_n, a_n, and b_n are constants. The stress condition at infinity determines the values of A_n and B_n and the stress condition at the boundary of the opening determines the values of a_n and b_n. Therefore, A_n and B_n reflect the type of loading and a_n and b_n reflect the type of opening in the plate.

The stresses are obtained from the two potential functions by means of the following two equations:[1]

$$
\begin{aligned}
\sigma_x + \sigma_y &= 2[\phi'(Z) + \bar{\phi}'(\bar{Z})] \\
\sigma_y - \sigma_x + 2i\tau_{xy} &= 2[\bar{Z}\phi''(Z) + \psi'(Z)]
\end{aligned} \tag{5.3.7}
$$

where the prime stands for differentiation with respect to the variable in the parentheses, the bar over a function stands for the conjugate of that function and i is the square root of minus one.

Brock[1] used the above method to study the stress distribution on the boundary of square holes with rounded corners as a function of the radius of curvature of the corner. His results are summarized in Fig. 5.3.1 which shows the boundary stresses for an uniaxial applied stress field.

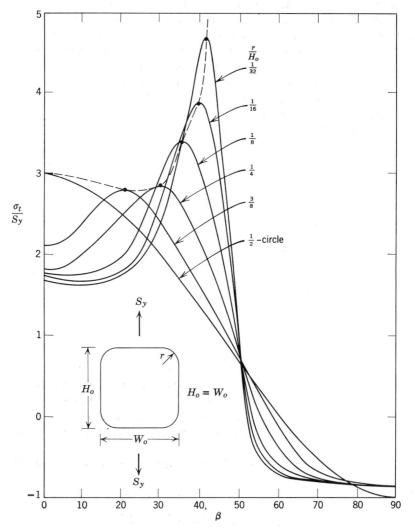

Fig. 5.3.1. Boundary stress for squares with rounded corners. (After Brock.[1])

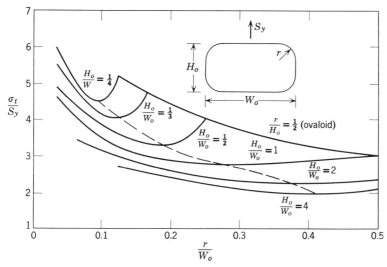

Fig. 5.3.2. Maximum stress concentrations for rectangular openings with rounded corners uniaxial stress field S_y.[4]

The most interesting result illustrated in the curves in Fig. 5.3.1 is that by proper selection of the radius of curvature at the corner of the square hole, the maximum stress concentration can be made less than that for a circle. Also, as the radius of curvature decreases below one-fourth, the maximum stress concentration at the corner increases rapidly.

Heller, Brock, and Bart[4] extended the above study to include rectangles with rounded corners where both the height-to-width ratio, H_0/W_0, and the radius of curvature-to-width ratio, r/W_0, are variables. Their results, for a uniaxial applied stress field are summarized in Fig. 5.3.2, which gives the maximum stress concentration as a function of r/W_0 and H_0/W_0. This set of curves shows that the maximum stress concentration on the boundary of a rectangular opening with rounded corners can be made smaller than that for an ovaloid of the same H_0/W_0 ratio by a suitable choice of the radius of curvature for the corner of the opening.

From Fig. 5.3.2 and Eq. 5.2.6, a comparison of the maximum boundary-stress concentrations for elliptical, ovaloidal, and rectangular openings can be made. The results are shown in Fig. 5.3.3 where maximum stress concentration is shown for different width-to-height ratios. For the rectangular opening, the radius of curvature for the smallest maximum stress concentration has been used. The curves in Fig. 5.3.3 show that a rectangle with rounded corners has the smallest maximum stress concentration for openings with a width-to-height ratio greater than 0.8 when the applied stress field is uniaxial in the height direction.

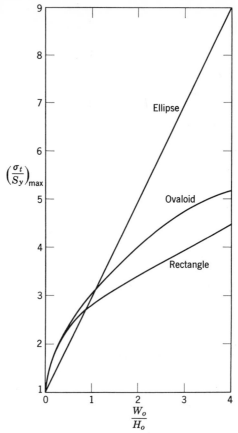

Fig. 5.3.3. Maximum stress concentration for different shape openings uniaxial stress field S_y.

The complex variable method was used to calculate the stress distribution along the major and minor axes of ellipses having a major to minor axis ratio of 2:1 and 4:1, both for an applied stress parallel to the major axis and parallel to the minor axis. Also the boundary stresses were calculated under the same two conditions. The results from these calculations are given in tabular form to facilitate the calculation of stresses for various biaxial-stress fields. As the principal of superposition applies in linear elastic theory, it is only necessary to add the stress distribution for an applied S_x to that for an applied S_y in the proper ratio to obtain the distribution for an applied biaxial-stress field of $S_x = mS_y$. The boundary stresses are given in Table 5.3.1; major and minor axial stresses are given in Table 5.3.2 for a 2:1 ellipse, and in Table 5.3.3 for a 4:1 ellipse.

Table 5.3.1. Boundary Stresses for Elliptical Openings Having Major to Minor Axis Ratios of 2:1 and 4:1

2:1 Ellipse

Coordinates		$S_y = T_{xy} = 0$	$S_x = T_{xy} = 0$	$S_x = S_y = 0$
x	y	σ_t/S_x	σ_t/S_y	σ_t/T_{xy}
2.000	0.000	−1.000	5.000	0.000
1.992	0.087	−0.911	4.821	−1.527
1.970	0.174	−0.668	4.336	−2.823
1.932	0.259	−0.331	3.662	−3.748
1.879	0.342	0.039	2.922	−4.282
1.813	0.423	0.395	2.209	−4.488
1.732	0.500	0.714	1.571	−4.454
1.638	0.574	0.987	1.026	−4.256
1.532	0.643	1.214	0.572	−3.958
1.414	0.707	1.400	0.200	−3.600
1.286	0.766	1.551	−0.102	−3.211
1.147	0.819	1.672	−0.345	−2.807
1.000	0.866	1.769	−0.539	−2.398
0.845	0.906	1.845	−0.691	−1.990
0.684	0.940	1.904	−0.808	−1.586
0.518	0.966	1.947	−0.894	−1.185
0.347	0.985	1.977	−0.954	−0.787
0.174	0.996	1.995	−0.989	−0.393
0.000	1.000	2.000	−1.000	0.000

4:1 Ellipse

Coordinates		$S_y = T_{xy} = 0$	$S_x = T_{xy} = 0$	$S_x = S_y = 0$
x	y	σ_t/S_x	σ_t/S_y	σ_t/T_{xy}
4.000	0.000	−1.000	9.000	0.000
3.985	0.087	−0.727	7.910	−3.897
3.939	0.174	−0.170	5.677	−5.887
3.864	0.259	0.337	3.654	−6.236
3.759	0.342	0.699	2.205	−5.833
3.625	0.423	0.942	1.233	−5.206
3.464	0.500	1.105	0.579	−4.558
3.277	0.574	1.217	0.131	−3.958
3.064	0.643	1.296	−0.185	−3.421
2.828	0.707	1.353	−0.412	−2.941
2.571	0.766	1.395	−0.579	−2.512
2.294	0.819	1.426	−0.703	−2.123
2.000	0.866	1.449	−0.796	−1.767
1.690	0.906	1.466	−0.866	−1.438
1.368	0.940	1.479	−0.918	−1.128
1.035	0.966	1.489	−0.955	−0.834
0.694	0.985	1.495	−0.981	−0.550
0.349	0.996	1.499	−0.995	−0.273
0.000	1.000	1.500	−1.000	0.000

127

Table 5.3.2. Stress Distribution along Major and Minor Axes for Elliptical Openings Having a Major to Minor Axis Ratio of 2:1

Stresses along Maj or Axis

Distance	$S_x \neq 0$ $S_y = T_{xy} = 0$		$S_y \neq 0$ $S_x = T_{xy} = 0$		$T_{xy} \neq 0$ $S_x = S_y = 0$
x/a	σ_y/S_x	σ_x/S_x	σ_y/S_y	σ_x/S_y	τ_{xy}/T_{xy}
1.000	−1.000	0.000	5.000	0.000	0.000
1.108	−0.235	0.030	2.675	0.734	1.746
1.228	−0.054	0.235	1.928	0.711	1.730
1.356	0.001	0.400	1.596	0.601	1.610
1.489	0.020	0.522	1.419	0.498	1.532
1.625	0.026	0.611	1.313	0.414	1.447
1.764	0.026	0.678	1.243	0.349	1.378
1.904	0.026	0.730	1.196	0.298	1.324
2.046	0.023	0.769	1.161	0.254	1.279
2.189	0.022	0.801	1.135	0.220	1.243
2.333	0.019	0.827	1.126	0.192	1.214
2.696	0.015	0.873	1.081	0.143	1.158
3.062	0.012	0.902	1.060	0.110	1.122

Stress along Minor Axis

y/b	σ_y/S_x	σ_x/S_x	σ_y/S_y	σ_x/S_y	τ_{xy}/T_{xy}
1.000	0.000	2.000	0.000	−1.000	0.000
1.383	0.132	1.620	−0.016	−0.487	0.585
1.743	0.179	1.403	0.059	−0.222	0.945
2.088	0.188	1.273	0.163	−0.085	1.089
2.422	0.185	1.194	0.271	−0.013	1.161
2.750	0.167	1.141	0.359	0.026	1.188
3.073	0.151	1.106	0.440	0.045	1.193
3.392	0.136	1.083	0.508	0.055	1.189
3.708	0.127	1.061	0.562	0.062	1.178
4.022	0.110	1.053	0.617	0.057	1.166
4.334	0.099	1.044	0.659	0.056	1.154
5.107	0.077	1.029	0.740	0.048	1.126
5.875	0.061	1.020	0.795	0.041	1.103

Table 5.3.3. Stress Distribution along Major and Minor Axes for Elliptical Openings Having a Major to Minor Axis Ratio of 4:1

Stresses along Major Axis

Distance	$S_x \neq 0$ $S_y = T_{xy} = 0$		$S_y \neq 0$ $S_x = T_{xy} = 0$		$T_{xy} \neq 0$ $S_x = S_y = 0$
x/a	σ_y/S_x	σ_x/S_x	σ_y/S_y	σ_x/S_y	τ_{xy}/T_{xy}
1.000	−1.000	0.000	9.000	0.000	0.000
1.063	−0.106	0.154	3.327	1.482	2.429
1.143	−0.008	0.420	2.210	1.143	2.135
1.234	0.012	0.581	1.773	0.859	1.870
1.333	0.014	0.683	1.548	0.663	1.678
1.438	0.014	0.750	1.414	0.528	1.542
1.546	0.013	0.799	1.325	0.430	1.443
1.656	0.012	0.833	1.263	0.357	1.368
1.769	0.010	0.860	1.219	0.301	1.312
1.884	0.009	0.881	1.184	0.258	1.266
2.000	0.008	0.897	1.158	0.223	1.231
2.295	0.006	0.925	1.113	0.162	1.168
2.594	0.005	0.943	1.083	0.123	1.128

Stresses along Minor Axis

y/b	σ_y/S_x	σ_x/S_x	σ_y/S_y	σ_x/S_y	τ_{xy}/T_{xy}
1.000	0.000	1.500	0.000	−1.000	0.000
1.750	0.056	1.336	−0.009	−0.559	0.497
2.429	0.083	1.229	0.041	−0.291	0.792
3.062	0.094	1.160	0.120	−0.132	0.961
3.668	0.094	1.114	0.207	−0.040	1.045
4.250	0.090	1.084	0.291	0.013	1.104
4.818	0.084	1.063	0.369	0.043	1.127
5.375	0.078	1.047	0.438	0.058	1.136
5.923	0.071	1.038	0.499	0.067	1.138
6.464	0.065	1.030	0.551	0.069	1.134
7.000	0.059	1.024	0.597	0.070	1.129
8.321	0.047	1.015	0.688	0.063	1.110
9.625	0.038	1.010	0.752	0.056	1.093

5.4 MULTIPLE HOLES IN PLATES

The stress distribution around an infinite row of equal size circular holes equally spaced in an infinitely wide plate, subjected to either a uniform stress normal or parallel to the line of holes, has been studied theoretically by Howland[5]. This problem corresponds to a row of circular tunnels of the same size equally spaced in a horizontal direction at given depth underground where the principal directions of the stress field before mining the openings are parallel and normal to the axes of the tunnels

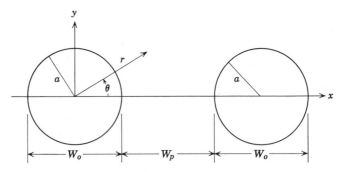

Fig. 5.4.1. Row of equal size and equally spaced circular holes in infinite plate applied stress field S_x and S_y at infinity.

with one principal direction in the vertical direction. Figure 5.4.1 gives the coordinate system and notation used in the solution of this problem.

Figure 5.4.2 shows the stress concentration around the openings (curve A), along the horizontal center line of the pillars (curve B) and along the vertical center line of the pillars (curve C) for a ratio of opening-to-pillar width of unity when the applied-stress field is normal to a line through the centers of the openings. Comparison of the results given by curve A in Fig. 5.4.2 with those for a single circular hold in an infinitely wide plate (Fig. 4.7.2) shows that the maximum stress concentration has increased from 3 to 3.26 owing to the presence of the adjacent holes. This increase is comparatively small considering the fact that the average stress on the line of centers of the holes is twice that for a single hole in an infinitely wide plate. Thus it can be inferred that the average stress in the pillar increases more rapidly than the maximum stress, and that the average stress in the pillar will approach the maximum stress as the ratio W_0/W_p increases.

An explanation for the small increase in maximum stress concentration compared to the large increase in average pillar stress can be obtained by

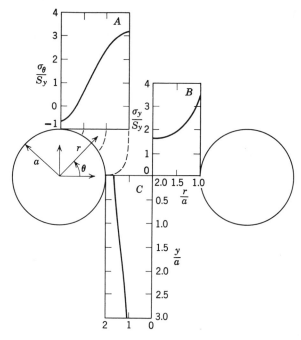

Fig. 5.4.2. Stress concentration for row of circular holes applied stress, S_y, normal to line of centers $W_0/W_b = 1$.

calculating the increase in the applied stress field at the edge of an opening resulting from the presence of two adjacent openings. For a single circular opening, the stress concentration at $r = 3a$ is 1.074 and at $r = 5a$ is 1.022 (see Eq. 4.7.9). Thus the applied stress field at the edge of any one opening is 9.6% larger because of the presence of two adjacent openings. At the center of the opening, $r = 4a$ for the adjacent openings, and the stress field is increased by only 1.074. The presence of the other openings in the row can have only a minor effect; therefore the increase in the maximum stress concentration should be greater than 7.4% and not be greater than 9.6%. From the theoretical results given in Fig. 5.4.2, the increase in the maximum stress concentration for a row of circular holes is only 8.7% more than the maximum stress concentration for a single circular hole.

Figure 5.4.3 shows the stress concentrations around the openings (curve A), above the openings (curve B), along the vertical axis of the pillar (curve C), and along the horizontal center line of the pillar (curve D) for a ratio of opening-to-pillar width of unity and an applied stress field S_x parallel to the center line of the infinite row of circular openings. Comparison of the results given by curve A in Fig. 5.4.3 with those for a

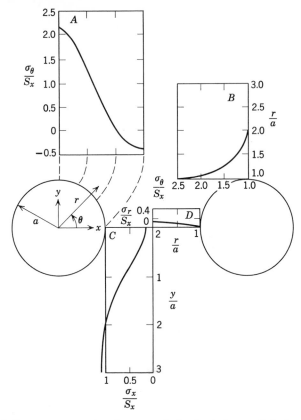

Fig. 5.4.3 Stress concentration for row of circular holes applied stress field, S_x, parallel to line of centers $W_0/W_q = 1$.

single circular hole in an infinite plate shows that the maximum stress concentration has decreased from 3.0 to 2.16 as a result of the shielding effects of the adjacent holes in the row. This shielding effect is even more pronounced on the horizontal stress σ_x along the vertical axis through the pillar (curve C) and on the horizontal stress σ_r along the horizontal center line (curve D), where the maximum radial stress σ_r never exceeds $0.14S_x$. From curve C in Fig. 5.4.2 and curves B and C in Fig. 5.4.3, we can conclude that the zone of influence in which the stress distribution is altered appreciably from the applied stress field extends only one diameter on either side of the center line of the openings.

Stress concentrations for biaxial-stress fields can be obtained by the principle of superposition. Thus, for example, the boundary-stress concentrations for an applied stress field of $S_x = S_y$ are obtained by adding the stress concentrations given by curve A in Figs. 5.4.2 and 5.4.3; the boundary-stress concentrations for an applied stress field of $S_x = \frac{1}{3}S_y$ are obtained by adding one-third of the values given in curve A of Fig. 5.4.3 to those given in curve A of Fig. 5.4.2. The resulting boundary-stress concentrations are shown in Fig. 5.4.4.

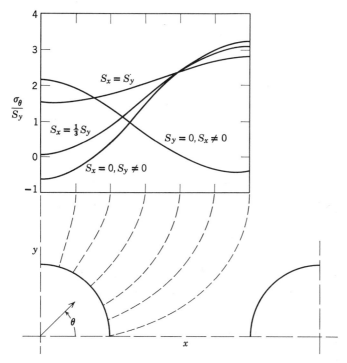

Fig. 5.4.4. Boundary stress concentration for row of circular holes in infinite plate, applied biaxial stress field, $W_0/W_p = 1$.

The solution to the problem of two equal circular holes in a plate subjected to a biaxial stress field has been given by Ling.[6] Figure 5.4.5 illustrates the problem and defines the important variables. This problem corresponds to two horizontal circular tunnels. Ling found that maximum stress concentrations occur at two points on the boundary of the openings. For a uniform biaxial stress field ($S_x = S_y \neq 0$), or for an uniaxial stress field normal to the center line of the holes ($S_y \neq 0$, $S_x = 0$), these two points are at $\theta = 0$ or $\theta = \pi$. For an applied stress field parallel to the line of center ($S_x \neq 0$, $S_y = 0$), the maximum boundary-stress concentrations occur almost at $\theta = \pm\pi/2$. The variation of the maximum

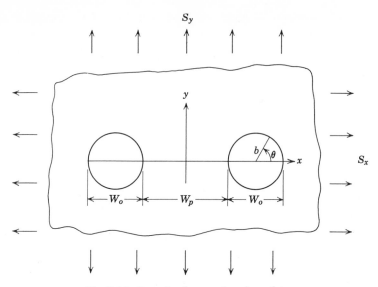

Fig. 5.4.5. Two circular openings in a plate.

stress concentrations as a function of the ratio W_p/W_0 is given in Table 5.4.1.

Comparison of the data given in Table 5.4.1 at a ratio of $W_p/W_0 = 1$ with the data given in Figs. 5.4.2, 5.4.3, and 5.4.4 shows that the stress concentration resulting from two holes in a row is considerably less than that resulting from an infinite row of holes. It should also be noted that the stress in the pillar side of the opening increases toward infinity as the pillar width decreases. Thus small ratios of W_p/W_0 should be avoided in underground structures.

Table 5.4.1. Maximum Stress Concentrations for Two Circular Holes in a Plate. (After Ling[6])

	$S_x = S_y \neq 0$		$S_y \neq 0, S_x = 0$		$S_x \neq 0, S_y = 0$
W_p/W_0	σ_t/S_y		σ_t/S_y		σ_t/S_y
	$\theta = 0$	$\theta = \pi$	$\theta = 0$	$\theta = \pi$	$\theta = \pm\pi/2$
0	2.894	0.000	3.869	0.000	2.569
0.5	2.255	2.887	3.151	3.264	2.623
1.0	2.158	2.411	3.066	3.020	2.703
2.0	2.080	2.155	3.020	2.992	2.825
4.0	2.033	2.049	3.004	2.997	2.927
7.0	2.014	2.018	3.001	2.999	2.970
10.0	2.000	2.000	3.000	3.000	3.000

5.5 THREE-DIMENSIONAL OPENINGS IN INFINITE MEDIUM

The determination of the stress distribution around three dimensional openings in an infinite medium under the action of a triaxial stress field requires the solution to the five sets of three-dimensional elasticity equations as outlined in Section 3.3. As the boundary conditions are one of the sets of equations involved, it is necessary that the boundary of the three-dimensional opening be expressible as a mathematical function of the coordinate system. Consequently, the number of three-dimensional problems that can and have been solved is rather limited. If the three-dimensional compatibility equations are expressed in terms of stresses, the resulting equations (3.3.6 or 3.3.8) contain the elastic constants. Therefore the stress distribution around three-dimensional openings in an infinite medium will depend upon the elastic constants of the medium. This result should be compared with that for the two-dimensional case where the stress distribution around openings in infinite plates is independent of the elastic constants of the medium.

Theoretical solutions for the stress distributions about spherical, spheroidal, and ellipsoidal cavities have been obtained (Southwell,[14] and Sadowsky and Sternberg[11,12]). Some of these solutions have been given in textbooks (Timoshenko and Goodier,[18] Neuber[8]), and an excellent summary of these stress distributions has been given by Terzaghi and Richart.[15]

The solution for the stress distribution about a spherical cavity in an infinite medium acted upon by a unidirectional stress field is given by Neuber,[8]

$$
\begin{aligned}
\sigma_r = \frac{S_z}{14 - 10\nu}\Bigg[&\left(10\nu - 14 + \frac{(50 - 10\nu)a^3}{r^3} - \frac{36a^5}{r^5}\right)\sin^2\theta \\
&+ \left(14 - 10\nu - \frac{(38 - 10\nu)a^3}{r^3} + \frac{24a^5}{r^5}\right)\Bigg]
\end{aligned}
$$

$$
\begin{aligned}
\sigma_\theta = \frac{S_z}{14 - 10\nu}\Bigg[&\left(14 - 10\nu - \frac{(5 - 10\nu)a^3}{r^3} + \frac{21a^5}{r^5}\right)\sin^2\theta \\
&+ \left(\frac{(9 - 15\nu)a^3}{r^3} - \frac{12a^5}{r^5}\right)\Bigg]
\end{aligned} \quad (5.5.1)
$$

$$
\begin{aligned}
\sigma_\phi = \frac{S_z}{14 - 10\nu}\Bigg[&\left(\frac{(30\nu - 15)a^3}{r^3} + \frac{15a^5}{r^5}\right)\sin^2\theta \\
&+ \left(\frac{(9 - 15\nu)a^3}{r^3} - \frac{12a^5}{r^5}\right)\Bigg]
\end{aligned}
$$

$$
\tau_{r\theta} = \frac{S_z}{14 - 10\nu}\left[10\nu - 14 - \frac{(10 + 10\nu)a^3}{r^3} + \frac{24a^5}{r^5}\right]\sin\theta\cos\theta
$$

$$
\tau_{\theta\phi} = \tau_{\phi r} = 0
$$

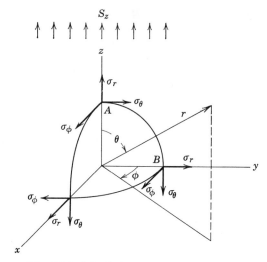

Fig. 5.5.1. Spherical coordinates and stresses.

where r, θ, and ϕ are the spherical coordinates as given in Fig. 5.5.1 and S_z is the uniformly distributed applied stress field in the z direction at a large distance from the cavity. As Eq. 5.5.1 contains Poisson's ratio, the stress distribution around a spherical cavity depends upon the material of the medium in which the cavity is located. To show how sensitive the stress distribution is to Poisson's ratio the values of σ_θ at point A, the top center

Table 5.5.2. Effect of Poisson's Ratio on
Tangential Stress at Points A and B

	A	B
ν	σ_θ/S_z	σ_θ/S_z
0	-0.214	1.929
0.1	-0.346	1.961
0.2	-0.500	2.000
0.3	-0.682	2.045
0.4	-0.900	2.100
0.5	-1.167	2.167

$$\text{For } A \quad \frac{\sigma_\theta}{S_z} = -\frac{3 + 15\nu}{14 - 10\nu}$$

$$\text{For } B \quad \frac{\sigma_\theta}{S_z} = \frac{27 - 15\nu}{14 - 10\nu}$$

of the sphere, and at point B, a typical point on the circumference of the sphere at $Z = 0$, have been calculated and are given in Table 5.5.2.

The distribution of the normal stresses along the y and z axes for the case of a unidirectional applied stress and $\nu = 0.2$ are shown in Fig. 5.5.2. Comparison of these curves with those given in Fig. 4.7.2a shows

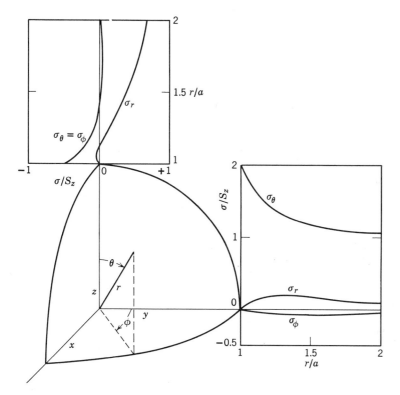

Fig. 5.5.2. Stress distribution around spherical cavity-uniaxial applied stress, S_z.

that the maximum stress concentration for the spherical cavity is less than that for a circular tunnel for a unidirectional applied-stress field. For the spherical cavity the maximum stress concentration is 2 and occurs at the boundary of the cavity at the equatorial plane. The corresponding stress concentration for a circular tunnel is 3. At the top of the spherical cavity, the tangential stress concentration is -0.5 whereas the corresponding stress concentration for a cylindrical tunnel is -1.0.

The results from Eqs. 5.5.1 for the uniaxial stress field may be superposed to obtain the stress distributions for applied biaxial or triaxial-stress fields. The most interesting triaxial-stress field is that corresponding to a hydrostatic stress condition or $S_x = S_y = S_z = -S$. For a hydrostatic-stress field of $-S$ Eqs. 5.5.1 reduces to

$$\sigma_r = -S\left(1 - \frac{a^3}{r^3}\right)$$

$$\sigma_t = \sigma_\theta = \sigma_\phi = -S\left(1 + \frac{a^3}{2r^3}\right)$$

(5.5.2)

Equations 5.5.2 do not contain Poisson's ratio; thus the stress distribution around a spherical cavity in an applied hydrostatic-stress field is not dependent upon the elastic constants of the medium. Comparison of Eqs. 5.5.2 with Eqs. 4.7.8 shows that, for a hydrostatic-stress field, larger stress concentrations occur on the boundary of a cylindrical tunnel than on the boundary of a spherical cavity. Also the zone of influence for a cylindrical tunnel is larger than that for a spherical cavity. The zone of influence is that region around an opening in which the major stress redistribution occurs when an opening is created in a uniform stress field.

Terzaghi and Richart[15] made numerical calculations of the stress distributions around oblate spheroidal cavities (minor axis vertical) using the theoretical solution of Edwards.[2] From these calculations they were able to show that the stress concentrations around spheroidal cavities are less than the stress concentrations around tunnels of the same cross-sectional shape. Also, as the ratio of the height to the diameter of the opening decreased, the maximum-stress concentration on the boundary of the opening at its equator increased but the height and width of the zone of large stress concentration decreased. This result implies that the zone of influence around spheroidal cavities is more nearly proportional to the shorter dimension of the opening rather than the larger dimension of the opening.

Sadowsky and Sternberg[11,12] determined theoretically the stress distribution about an ellipsoidal cavity in an infinite medium under an arbitrary plane-stress field perpendicular to the axis of revolution of the cavity and about a triaxial ellipsoidal cavity in an infinite medium under a triaxial applied-stress field whose principal stresses at large distances from the cavity are parallel to the axes of the cavity (see Fig. 5.5.3). The triaxial ellipsoidal cavity includes both the oblate and prolate spheroidal cavities. Thus if $b/a = 1$ and $c/b < 1$, the oblate spheroidal cavity is obtained whose minor axis is parallel to the y axis; if $c/b = 1$ and $b/a < 1$, the prolate spheroidal cavity is obtained whose major axis is parallel to the x axis (see Fig. 5.5.3).

Sadowsky and Sternberg calculated the maximum boundary-stress concentrations that occur at points A and B on the boundary of the

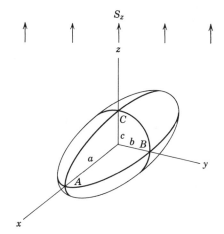

Fig. 5.5.3. Ellipsoidal cavity in uniaxial stress field.

triaxial ellipsoidal cavity for various ratios of c/b and b/a. Their results are partially reproduced in Fig. 5.5.4. Comparison of the curves in Fig. 5.5.4a with those in Fig. 5.5.4c shows that when b/a is zero (which corresponds to a long tunnel), that the maximum stress concentration at point A is always slightly less than the maximum stress concentration at point B. Note that in Fig. 5.5.4c, when $b/a = 0$, the stress concentrations at point B are those corresponding to the maximum stress concentrations for a long tunnel of elliptical cross sections as given by Eqs. 5.2.6. The results given in Fig. 5.5.4 show that the maximum-stress concentration around triaxial ellipsoidal cavities are always less than the maximum-stress concentration around tunnels having the same cross-sectional shape. Thus, in general it can be assumed that for engineering purposes the maximum stress concentrations for single isolated openings can be estimated safely by determining the maximum stress concentration around holes in plates where the cross-sectional shape of the hole is the same as a cross section of the three-dimensional opening. As two-dimensional problems are easier to solve than three-dimensional problems (both theoretically or experimental), the above result represents considerable saving in time and effort when analyzing the stability of three-dimensional openings.

Figure 5.5.5 shows the stress distribution along the x axis for a triaxial ellipsoidal cavity whose shape ratios are

$$\frac{b}{a} = \frac{c}{b} = \frac{1}{3} \tag{5.5.3}$$

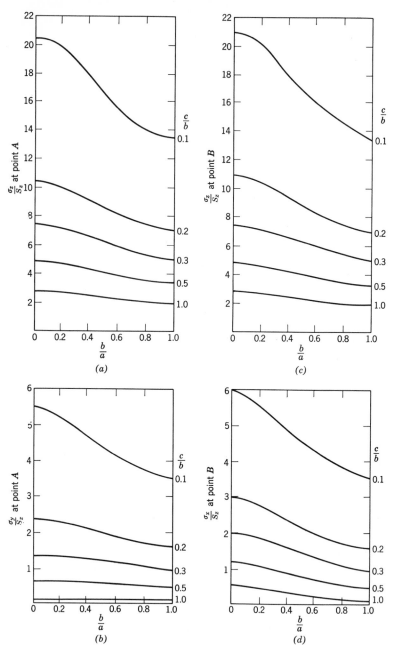

Fig. 5.5.4. Maximum boundary stresses for ellipsoidal cavity in uniaxial stress field and $v = 0.3$.[12]

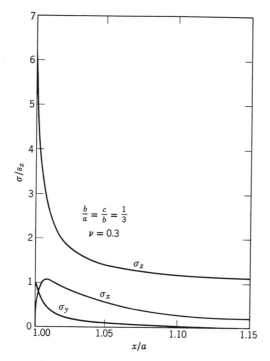

Fig. 5.5.5. Stress distribution along x axis for ellipsoidal cavity. Unidirection stress field S_z, $\nu = 0.3$.[12]

and the applied stress is S_z and Poisson's ratio is 0.3. Note that the zone of high stress concentration is restricted to a very narrow region at the tip of the cavity. From Eq. 5.5.3, the ratio of a to c is 9. The zone of high stress concentration extends to only $x/a = 1.10$ which is a distance in the medium of about one radius of the shortest axis of the opening. Thus the zone of influence about three-dimensional openings is more nearly proportional to the radius of the shortest dimension of the opening than to the longest dimension of the opening.

5.6 FLEXURE OF BEAMS

For engineering purposes the flexure of beams can be described by an approximate theory based upon the following assumptions.

1. A beam is a straight structural element whose length is long compared to its other dimensions ($L/t > 8$; see Fig. 5.6.1 for definition of symbols and coordinate system for a beam).

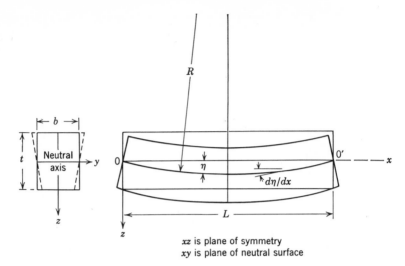

xz is plane of symmetry
xy is plane of neutral surface

Fig. 5.6.1. Nomenclature for simple beams.

2. The beam is composed of a homogeneous, isotropic, and perfect elastic material.

3. The beam has a uniform cross section and a longitudinal plane of symmetry.

4. Loads are applied normally to the longitudinal axis of the beam and in the plane symmetry.

5. Plane sections in an unloaded beam remain plane during flexure.

6. The deflection, η, and the slope, $d\eta/dx$, of the neutral surface are so small that the radius of curvature equation is approximated by

$$\frac{1}{R} = -\frac{d^2\eta/dx^2}{\left[1 + \left(\dfrac{d\eta}{dx}\right)^2\right]^{3/2}} = -\frac{d^2\eta}{dx^2} \qquad (5.6.1)$$

When a beam is acted upon by a system of concentrated forces as illustrated in Fig. 5.6.2, a system of shear forces and bending moments are created on cross sections of the beam and reaction forces are generated at the supports. Assume that the beam is cut into at the cross section n, m. For the beam to remain in equilibrium and in the same position, a system of forces must act on the section n, m.

Any system of parallel forces can be replaced by a single force and a couple. Thus the system of external forces to the left of n, m can be replaced by a shearing force V acting in the plane of n, m and by a couple M as illustrated in Fig. 5.6.2. The magnitude of the shearing force, which is the algebraic sum of the external forces to the left of n, m, is given by

$$V = R_1 - F_1 \tag{5.6.2}$$

The magnitude of the couple or bending moment which is equal to the algebraic sum of the moments of the external forces to the left of n, m with respect to the centroid of this section, is given by

$$M = R_1 x - F_1(x - c_1) \tag{5.6.3}$$

The stresses which are distributed over the cross section n, m and which represent the action of the right portion of the beam on the left portion of the beam must then be such as to balance the bending moment M and the shearing force V. Figure 5.6.2 shows how the shear force and bending moment change with distance along the beam. The reactions at the supports are given by

$$R_1 = F_1 \frac{L - C_1}{L} + F_2 \frac{L - C_2}{L}$$

$$R_2 = F_1 \frac{C_1}{L} + F_2 \frac{C_2}{L} \tag{5.6.4}$$

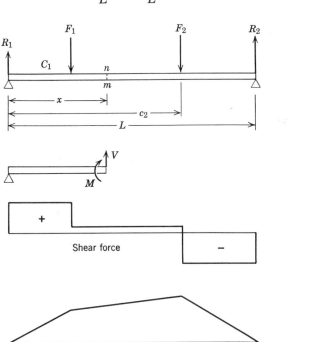

Fig. 5.6.2. Simple beam with concentrated loads and bending moment diagrams.

If a beam is acted upon by a uniformly distributed load q, as shown in Fig. 5.6.3, the reactions at the supports are

$$R_1 = R_2 = \frac{qL}{2} \qquad (5.6.5)$$

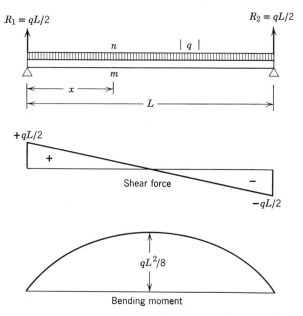

Fig. 5.6.3. Simple beam with distributed load and its shear force and bending moment diagrams.

The shear force and bending moment at some section n, m are

$$V = R_1 - qx = q\left(\frac{L}{2} - x\right) \qquad (5.6.6)$$

$$M = R_1 x - (qx)\left(\frac{x}{2}\right) = \frac{qx}{2}(L - x) \qquad (5.6.7)$$

Figure 5.6.3 gives the diagram for the shear force and bending moment in a beam acted upon by a uniformly distributed load.

If two adjacent cross sections of a beam are separated by a distance dx, as shown in Fig. 5.6.4, and if no external forces act between these two cross sections, then the shear forces on the two cross sections must be equal. However, the bending moments are not equal and the increase in the bending moment, in going from the left-hand to the right-hand cross section, is equal to the moment of the couple formed by the two equal but opposite shear forces V. Thus

$$dM = V\,dx \qquad (5.6.8)$$

For a distributed load, the shear forces on two adjacent cross sections are not equal and the change in the shear force is

$$dV = -q\,dx \qquad (5.6.9)$$

Equations 5.6.8 and 5.6.9 can be combined to give

$$\frac{d^2M}{dx^2} = -q \qquad (5.6.10)$$

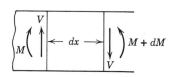 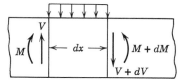

Fig. 5.6.4. Relation between shear force and bending moment.

When a beam is bent, as shown in Fig. 5.6.5, the top of the beam is in compression, the bottom of the beam is in tension, and the neutral plane is unstrained. The normal strain, ϵ_x, produced at any point p is the ratio of $\Delta s/s$. Thus from similar triangles in Fig. 5.6.5

$$\epsilon_x = \frac{Z}{R} \qquad (5.6.11)$$

where Z is the distance from the neutral axis ab and R is the radius of curvature of the neutral plane. The strain in the y direction is

$$\epsilon_y = -\nu\epsilon_x = -\nu\frac{Z}{R} \qquad (5.6.12)$$

and from Hooke's law the stress in the x direction is

$$\sigma_x = \frac{EZ}{R} \qquad (5.6.13)$$

The force acting on an elementary unit of area dA at a distance Z from the neutral axis of the cross section is

$$dF = \sigma_x\,dA = \frac{EZ}{R}\,dA \qquad (5.6.14)$$

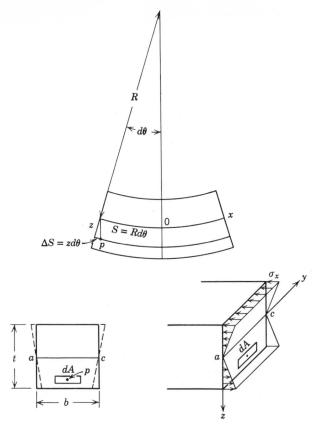

Fig. 5.6.5. Relations between strain, radius of curvature, stress, and bending moment.

The sum of all the forces on the cross section must equal a pure couple; therefore, the integral of dF must be zero, or

$$\frac{E}{R} \int Z \, dA = 0 \qquad (5.6.15)$$

Thus the neutral axis coincides with the centroid of the section as shown in Fig. 5.6.5. The moment of the force about the neutral axis is the sum of $Z \, dF$; therefore, the total moment is from Eq. 5.6.14

$$M = \frac{E}{R} \int Z^2 \, dA = \frac{E}{R} I_y \qquad (5.6.16)$$

where I_y is the moment of inertia of the cross section about the neutral axis ac. The quantity EI_y is known as the flexural rigidity of the beam.

Equations 5.6.1 and 5.6.16 can be combined to give

$$EI_y \frac{d^2\eta}{dx^2} = -M \qquad (5.6.17)$$

Equation 5.6.17 is the differential equation that defines the deflection curve for a beam acted upon by concentrated loads. Equations 5.6.10 and 5.6.17 can be combined to give

$$EI_y \frac{d^4\eta}{dx^4} = q \qquad (5.6.18)$$

Equation 5.6.18 is the differential equation that defines the deflection curve for a beam acted upon by distributed loads.

Integration of either Eq. 5.6.17 or 5.6.18 is straightforward and the arbitrary constants are evaluated by the end conditions of the beam. For a supported beam, the deflection is zero at the support. Both the deflection η and the slope $d\eta/dx$ of the neutral plane are zero at a clamped or built-in end of the beam.

The average shear stress on any section of a beam is the shear force V divided by the area A of the section. This shear stress is uniformly distributed in the y direction of the section but is not uniformly distributed in the z direction. The maximum shear stress occurs at the neutral axis of the section and for a rectangular cross section is given by[10]

$$(\tau_{xz})_{\max} = \frac{3}{2} \frac{V}{A} \qquad (5.6.19)$$

and for a circular cross section is given by

$$(\tau_{xz})_{\max} = \frac{4}{3} \frac{V}{A} \qquad (5.6.20)$$

The normal stress σ_x in a beam can be obtained by combining Eq. 5.6.13 with 5.6.16 to give

$$\sigma_x = \frac{MZ}{I_y} \qquad (5.6.21)$$

It should be noted that the stresses and strains in a beam can be determined from the shear force and bending moment. Thus complete integration of Eq. 5.6.17 or 5.6.18 is necessary only if the deflection curve is desired. Also the maximum numerical value of σ_x, for any cross section, occurs at the top and bottom surfaces of the beam and is known as the outer fiber bending-stress.

Distributed Load on Beam with Built-in Ends

Consider a beam of rectangular cross section with built-in ends acted upon by a distributed load q per unit length as shown in Fig. 5.6.6. Since this problem has many direct applications in structural rock mechanics, its solution will be discussed in detail. The boundary conditions at both ends of the beam are

$$\eta = 0 \qquad \text{at} \quad x = 0 \quad \text{or} \quad L$$

$$\frac{d\eta}{dx} = 0 \qquad \text{at} \quad x = 0 \quad \text{or} \quad L \qquad\qquad (5.6.22)$$

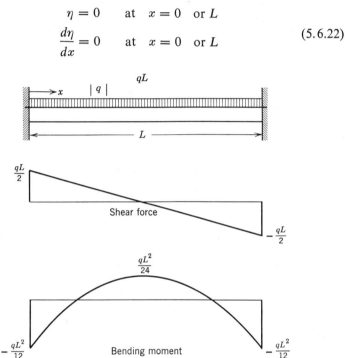

Fig. 5.6.6. Uniformly distributed load on beam with built-in ends.

Integration of Eq. 5.6.18 four times gives

$$EI_y \frac{d^3\eta}{dx^3} = qx + A$$

$$EI_y \frac{d^2\eta}{dx^2} = \frac{qx^2}{2} + Ax + B$$

$$EI_y \frac{d\eta}{dx} = \frac{qx^3}{6} + \frac{Ax^2}{2} + Bx + C \qquad\qquad (5.6.23)$$

$$EI_y\eta = \frac{qx^4}{24} + \frac{Ax^3}{6} + \frac{Bx^2}{2} + Cx + D$$

The boundary conditions are used to evaluate the constants of integration, giving

$$A = -\tfrac{1}{2}qL$$
$$B = \tfrac{1}{12}qL^2 \qquad (5.6.24)$$
$$C = D = 0$$

Thus the deflection, blending moment, and shear force are given by

$$\eta = \frac{qx^2}{24EI_y}(L - x)^2$$

$$M = -\frac{q}{12}(6x^2 - 6Lx + L^2) \qquad (5.6.25)$$

$$V = q\left(\frac{L}{2} - x\right)$$

The deflection, bending moment, and shear force evaluated at the end of the beam and at the center of the beam are

At end	At center
$\eta = 0$	$\eta = \dfrac{qL^4}{384EI_y}$
$M = -\dfrac{qL^2}{12}$	$M = \dfrac{qL^2}{24}$
$V = \dfrac{qL}{2}$	$V = 0$

$$(5.6.26)$$

For a rectangular cross section of width b, and thickness t, the moment of inertia of the cross section is

$$I_y = \tfrac{1}{12}bt^3 \qquad (5.6.27)$$

If the distributed load is that resulting only from the weight of the beam, then the load per unit length is

$$q = \gamma bt \qquad (5.6.28)$$

where γ is the unit weight. Therefore the maximum deflection at the center of the beam, the maximum shear stress at the end of the beam, and the maximum normal stress at the end of the beam are given by

$$(\eta)_{\text{max}} = \frac{\gamma L^4}{32Et^2}$$

$$(\tau_{xz})_{\text{max}} = \frac{3\gamma L}{4} \qquad (5.6.29)$$

$$(\sigma_x)_{\text{max}} = \frac{\gamma L^2}{2t}$$

Concentrated Load at Center of Beam with Supported Ends

Consider a beam with supported ends acted upon by a concentrated load P at the center of the beam as illustrated in Fig. 5.6.7. The reactions on the supports are

$$R_1 = R_2 = \frac{F}{2} \qquad (5.6.30)$$

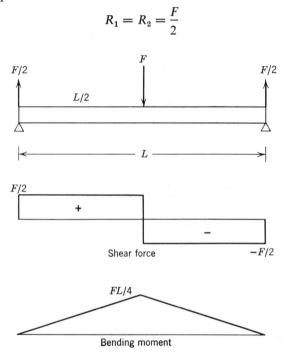

Fig. 5.6.7. Concentrated load at center of beam with supported ends.

and the two expressions for the bending moments are

$$M = R_1 x = \frac{Fx}{2} \qquad\qquad x < \frac{L}{2}$$

$$M = R_1 x - F(x - L) = \frac{F}{2}(L - x) \qquad x > \frac{L}{2} \qquad (5.6.31)$$

From Eqs. 5.6.16 and 5.6.31, the differential equations for the deflection curve are

$$EI_y \frac{d^2\eta}{dx^2} = -\frac{Fx}{2} \qquad\qquad x < \frac{L}{2}$$

$$EI_y \frac{d^2\eta}{dx^2} = -\frac{F}{2}(L - x) \qquad x > \frac{L}{2} \qquad (5.6.32)$$

Integration of these equations twice gives

$$EI_y\eta = -\frac{Fx^3}{12} + Ax + C \qquad x < \frac{L}{2}$$

$$EI_y\eta = -\frac{F}{2}\left(\frac{Lx^2}{2} - \frac{x^3}{6}\right) + Bx + D \qquad x > \frac{L}{2} \tag{5.6.33}$$

The boundary conditions are

$$\eta = 0 \qquad \text{at} \quad x = 0 \quad \text{or} \quad x = L \tag{5.6.34}$$

Also the slope $d\eta/dx$ and the deflection η must have the same value at $x = L/2$ regardless of which equation is used. Evaluation of the constants according to these boundary conditions gives

$$A = \frac{FL^2}{16}$$

$$B = \frac{3FL^2}{16} \tag{5.6.35}$$

$$C = 0$$

$$D = -\frac{FL^3}{48}$$

From Eqs. 5.6.35 and 5.6.33, the deflection equations are

$$EI_y\eta = \frac{Fx}{48}(3L^2 - 4x^2) \qquad x < \frac{L}{2}$$

$$EI_y\eta = \frac{F}{48}(-L^3 + 9L^2x - 12Lx^2 + 4x^3) \qquad x > \frac{L}{2} \tag{5.6.36}$$

The maximum bending moment, deflection, and normal stress occur at the center of the beam and are given by

$$(M)_{\text{max}} = \frac{FL}{4}$$

$$(\eta)_{\text{max}} = \frac{FL^3}{48EI_y}$$

$$(\sigma_x)_{\text{max}} = \frac{FLZ}{4I_y} \tag{5.6.37}$$

For a beam of rectangular cross section of width b and thickness t, the maximum deflection and normal stress are

$$(\eta)_{\max} = \frac{FL^3}{4Ebt^3}$$

$$(\sigma_x)_{\max} = \frac{3FL}{2bt^2}$$

(5.6.38)

For a beam of circular cross section of radius r, the moment of inertia of the section about the diameter is

$$I_y = \frac{\pi r^4}{4}$$

(5.6.39)

and the maximum deflection and normal stress are

$$(\eta)_{\max} = \frac{FL^3}{12\pi Er^4}$$

$$(\sigma_x)_{\max} = \frac{FL}{\pi r^3}$$

(5.6.40)

Two Equal Concentrated Loads at One-Third Points of Beam with Supported Ends

Consider a beam supported at both ends and acted upon by two equal concentrated loads $F/2$ as shown in Fig. 5.6.8. This problem is of interest because the two-point loading method is used as an experimental technique for determining the flexural strength of materials. The two-point loading method is often preferred over the center loading method because a uniform bending moment results between the two loads.

The reactions at the supports are

$$R_1 = R_2 = \frac{F}{2}$$

(5.6.41)

and the bending moment equations are

$$M = R_1 x = \frac{Fx}{2} \qquad\qquad x < \frac{L}{3}$$

$$M = R_1 x - \frac{F}{2}\left(x - \frac{L}{3}\right) = \frac{FL}{6} \qquad\qquad \frac{L}{3} < x < \frac{2L}{3}$$

(5.6.42)

$$M = R_1 x - \frac{F}{2}\left(x - \frac{L}{3}\right) - \frac{F}{2}\left(x - \frac{2L}{3}\right) = \frac{F}{2}(L - x) \qquad \frac{2L}{3} < x$$

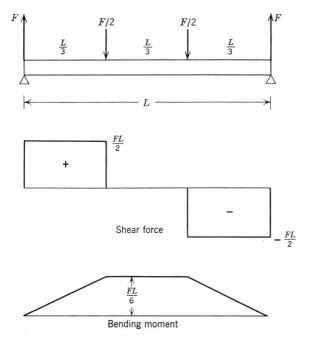

Fig. 5.6.8. Two equal concentrated loads at one-third points of beam with supported ends.

From Eq. 5.6.21 and the second equation of 5.6.42, the maximum normal stress is

$$(\sigma_x)_{\text{max}} = \frac{MZ}{I_y} = \frac{FLZ}{6I_y} \tag{5.6.43}$$

and is constant from $x = L/3$ to $x = 2L/3$. For a beam of rectangular cross section of width b and thickness t, $I_y = \frac{1}{12}bt^3$, and the maximum stress is

$$(\sigma_x)_{\text{max}} = \frac{2FL}{bt^2} \tag{5.6.44}$$

For a beam of circular cross section of radius r, $I = \frac{1}{4}\pi r^4$, and the maximum normal stress is

$$(\sigma_x)_{\text{max}} = \frac{2FL}{3\pi r^3} \tag{5.6.45}$$

Gravity-Loaded Multiple Beams with Built-in Ends

The problem of the gravity-loaded multiple beam with built-in ends can be used to study the stability of roof layers for underground openings

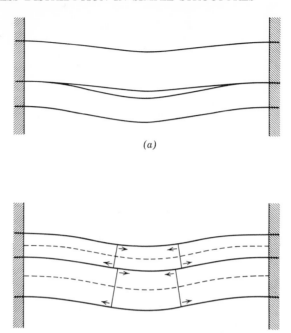

(a)

(b)

Fig. 5.6.9. Multiple beams with built-in ends.

in thinly bedded rock formations. To simplify the discussion initially, the number of beams will be restricted to two. Consider two beams of equal length L with their ends clamped together as shown in Fig. 5.6.9. There are two cases to consider depending upon the relative values of the ratios of the load per unit length to the flexural rigidity for the two beams. If this ratio for the top beam is less than that for the lower beam, then each beam acts independently and Eqs. 5.6.29 can be used to calculate the maximum deflection and stress in each beam. If this ratio for the top beam is greater than that for the lower beam, then the top beam will load the lower beam and conversely the lower beam will partially support the top beam. Thus a method for calculating this additional loading and/or support needs to be found.

Figure 5.6.9b illustrates the problem. Note that each beam has its own neutral plane, and cross sections of each beam rotate about their own neutral axis, thus causing slippage on the plane between the two beams.

In the derivation to follow, the following assumptions are made.

1. The coefficient of friction between the two beams is zero.
2. The deflection of the two beams are equal for all values of x.

3. The upper beam loads the lower beam with a uniform load per unit length of beam.

4. The lower beam supports the upper beam with an equal load per unit length.

5. Both beams are the same length and width.

The deflection of a uniformly loaded beam with built-in ends is given by Eqs. 5.6.25. For two beams under the conditions assumed above, the following equation can be written:

$$\eta_1 = \eta_2 = \frac{(q_1 + \Delta q)x^2}{24E_1I_1}(L - x)^2 = \frac{(q_2 - \Delta q)x^2}{24E_2I_2}(L - x)^2 \quad (5.6.46)$$

where the subscripts 1 and 2 stand for the lower beam and the upper beam respectively, and Δq is the added load or support which has been added to and subtracted from the lower and upper beam respectively. Equation 5.6.46 must hold for all values of x, therefore

$$\frac{q_1 + \Delta q}{E_1I_1} = \frac{q_2 - \Delta q}{E_2I_2} \quad (5.6.47)$$

Solving Eq. 5.6.47 for Δq gives

$$\Delta q = \frac{q_2E_1I_1 - q_1E_2I_2}{E_1I_1 + E_2I_2} \quad (5.6.48)$$

Substitution of Eq. 5.6.48 into the deflection equation for either the lower or upper beam (Eq. 5.6.46) gives

$$\eta_1 = \eta_2 = \frac{(q_1 + q_2)x^2}{24(E_1I_1 + E_2I_2)}(L - x)^2 \quad (5.6.49)$$

or

$$\eta_1 = \eta_2 = \frac{\dfrac{q_1 + q_2}{2}}{\dfrac{24(E_1I_1 + E_2I_2)}{2}}x^2(L - x)^2 \quad (5.6.50)$$

Thus when the upper beam loads the lower beam, the lower beam in turn supports the upper beam and each beam deflects as if its load per unit length and flexural rigidity were equal respectively to the average load per unit length and the average flexural rigidity of the two beams. For gravity loading of rectangular beams of width b and thickness t_1 and t_2, the loads per unit length and moments of inertia of the cross section for each beam are

$$q_1 = \gamma_1 b t_1$$
$$q_2 = \gamma_2 b t_2$$
$$I_1 = \tfrac{1}{12} b t_1^3$$
$$I_2 = \tfrac{1}{12} b t_2^3$$

(5.6.51)

Thus the deflection equation (5.6.49) can be written as

$$\eta_1 = \eta_2 = \frac{\gamma_1 t_1 + \gamma_2 t_2}{2(E_1 t_1^3 + E_2 t_2^3)} x^2 (L - x)^2$$

(5.6.52)

or as

$$\eta_1 = \eta_2 = \frac{\bar{\gamma}}{2\overline{Et^2}} x^2 (L - x)^2$$

(5.6.53)

where

$$\bar{\gamma} = \frac{\gamma_1 t_1 + \gamma_2 t_2}{t_1 + t_2}$$

$$\overline{Et^2} = \frac{t_1 E_1 t_1^2 + t_2 E_2 t_2^2}{t_1 + t_2}$$

(5.6.54)

The quantity $\bar{\gamma}$ is the weighted average unit weight of the two beams, weighted by the thickness of each beam. Also $\overline{Et^2}$ is the weighted average value of Et^2 for both beams, weighted by the thickness of the beams.

This procedure can be extended to any number of beams. The basic requirement is that each beam rest upon the beam below. The quantities $\bar{\gamma}$ and $\overline{Et^2}$ for k beams are given by

$$\bar{\gamma} = \frac{\displaystyle\sum_i^k \gamma_i t_i}{\displaystyle\sum_i^k t_i}$$

$$\overline{Et^2} = \frac{\displaystyle\sum_i^k t_i E_i t_i^2}{\displaystyle\sum_i^k t_i}$$

(5.6.55)

The deflection, bending moment, and shear (Eqs. 5.6.25) for a single beam can be used for multiple beams resting on each other if the average quantities $\bar{q}$ and $\overline{EI_y}$ for all the beams are used in place of q and EI_y. For gravity-loaded rectangular beams, the maximum deflection, shear stress, and normal stress (Eqs. 5.6.29) for a single beam can be used for multiple beams if the weighted average quantities $\bar{\gamma}$ and $\overline{Et^2}$ are used in place of γ and Et^2. It should be noted that even though the deflection is the same for all beams in a multiple beam arrangement, the maximum normal stress for any one beam depends upon the thickness of that beam (see the last equation of 5.6.29).

5.7   FLEXURE OF RECTANGULAR PLATES

For engineering purposes the flexure of plates is described by an approximate theory based on the following assumptions.

1. A plate is a straight, flat structural element whose width is at least more than four times the thickness and whose length is equal to or greater than its width.

2. The plate is composed of a homogeneous, isotropic, and perfect elastic material.

3. The maximum deflection of the plate is less than one-half its thickness.

4. All loads and reactions are applied normal to the plane of the plate.

5. When the plate deflects the central plane (halfway between top and bottom surfaces) remains unstressed while all other points are stressed biaxially in the plane of the plate.

6. Vertical straight lines before flexure remain straight after flexure but become inclined to the vertical; thus normal stresses in the plane of the plate are proportional to the distance from the central plane.

If a rectangular plate, whose length is long compared to its width, is uniformly loaded, it may be assumed that near the center, where the maximum deflection and stresses occur, the deflection surface is nearly cylindrical. For such a plate, an analysis of the deflection and stresses can be made by considering a strip of unit width and length a through the shorter lateral dimension of the plate as illustrated in Fig. 5.7.1a. The normal strains in the plane of the strip for a cylindrical deflection curve are, from Fig. 5.7.1b,

$$\epsilon_x = \frac{z}{R}$$

and
$$\epsilon_y = 0$$

(5.7.1)

The strain ϵ_y is zero near the center of the plate because deformation in the y direction, which is the long dimension of the plate, is prevented except near the ends. Thus the state of stress in the plane of the plate may be assumed as defined by Eqs. 3.4.1 with $\epsilon_y = 0$. Therefore

$$\sigma_x = \frac{E\epsilon_x}{1 - \nu^2}$$

(5.7.2)

$$\sigma_y = \nu\sigma_x$$

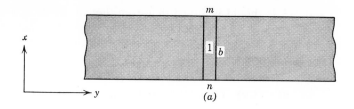

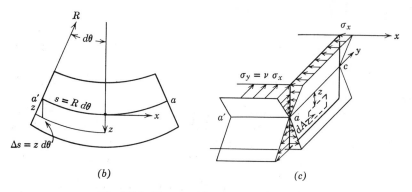

Fig. 5.7.1. Flexure of long rectangular plate.

From Eqs. 5.7.2 and 5.7.1, the normal stresses in the plane of the plate are

$$\sigma_x = \frac{EZ}{(1 - \nu^2)R}$$

and

$$\sigma_y = \frac{\nu EZ}{(1 - \nu^2)R}$$

(5.7.3)

where R is the radius of curvature of the deflection curve and is approximated by

$$\frac{1}{R} = -\frac{d^2\eta}{dx^2}$$

(5.7.4)

From Fig. 5.7.1c the moment of the force, about the neutral axis ac of the strip through the plate, is given by

$$M = \int \sigma_x Z \, dA = \frac{E}{(1 - \nu^2)R} \int Z^2 \, dA$$

or

$$M = \frac{EI_y}{(1 - \nu^2)R} = \frac{Et^3}{12(1 - \nu^2)R}$$

(5.7.5)

where I_y is the moment of inertia of the rectangular cross-sectional area of the strip through the plate about the neutral axis ab. The quantity $EI_y/(1 - \nu^2)$ or $Et^3/12(1 - \nu^2)$ is the flexural rigidity of a plate. Comparison of Eqs. 5.7.5 with Eq. 5.6.16 shows that the flexural rigidity for a strip through a plate is slightly larger than that for a beam of equivalent cross section. If Poisson's ratio is less than 0.3, the difference between the two flexural rigidities is less than 10%. Thus for engineering purposes in rock mechanics, the deflection and stresses in a plate, whose length is long compared to its width, can be approximated from simple beam theory.

Combining Eqs. 5.7.4 and 5.7.5 gives the differential equation for the deflection curve of the strip through the plate as

$$\frac{Et^3}{(1 - \nu^2)} \frac{d^2\eta}{dx^2} = -M \tag{5.7.6}$$

If Eqs. 5.7.3 are combined with Eqs. 5.7.5, the normal stress σ_x is given by

$$\sigma_x = \frac{12ZM}{t^3} \tag{5.7.7}$$

And the maximum normal stress will occur when $Z = \pm t/2$, or

$$(\sigma_x)_{\max} = \pm \frac{6M}{t^2} \tag{5.7.8}$$

Uniformly Distributed Load on Rectangular Plate with Built-in Edges

Consider a thin rectangular plate of length b, width a, and thickness t with built-in edges on all sides and supporting a uniformly distributed load. The maximum deflection occurs at the center of the plate and is given by[16]

$$(\eta)_{\max} = \alpha \frac{qa^4}{Et^3} \tag{5.7.9}$$

and the maximum bending moment occurs at the middle of the longer sides and is given by

$$M_{\max} = \beta qa^2 \tag{5.7.10}$$

where q is the load per unit area of the plate and α and β have the values, for a Poisson's ratio equal to 0.3, given in Table 5.7.1.

Table 5.7.1. Coefficients for Uniformly Loaded Rectangular Plates[17]

b/a	1.00	1.25	1.50	1.75	2.00	∞
α	0.0138	0.0199	0.0240	0.0264	0.0277	0.0284
β	0.0513	0.0665	0.0757	0.0806	0.0829	0.0833

By means of Eqs. 5.7.8 and 5.7.5, the maximum stress at the middle of the longer side is

$$(\sigma_x)_{\text{max}} = \frac{6\beta q a^2}{t^2} \tag{5.7.11}$$

For a gravity-loaded plate

$$q = \gamma t \tag{5.7.12}$$

Thus the maximum deflection and stress are given by

$$(\eta)_{\text{max}} = \frac{\alpha \gamma a^4}{E t^2}$$

and   (5.7.13)

$$(\sigma_x)_{\text{max}} = \frac{6\beta \gamma a^2}{t}$$

For a gravity-loaded plate with built-in edges whose ratio of length to width is 2 or less, Eqs. 5.7.13 with Table 5.7.1 should be used to calculate the maximum deflection and stress. However, if the ratio of b/a is greater than 2, the simple beam equations (5.6.29) may be used without introducing appreciable error.

REFERENCES

1. Brock, Joseph S., *Analytical Determination of the Stresses Around Square Holes With Rounded Corners*, David Taylor Model Basin, Report 1149.
2. Edwards, R. H., "Stress Concentrations Around Spheroidal Inclusions and Cavities," *J. Appl. Mech.*, pp. 19–30 (March 1951).
3. Greenspan, Martin, "Effect of a Small Hole on the Stresses in a Uniformly Loaded Plate," *Quarterly Appl. Math.*, **2**, pp. 60–71 (1944).
4. Heller, S. R. Jr., J. S. Brock, and R. Bart, "The Stresses Around a Rectangular Opening with Rounded Corners in a Uniformly Loaded Plate," *Trans. of the 3rd U.S. Congress on Appl. Mech.*, AIME (1958).
5. Howland, R. C. J., "Stresses in a Plate Containing an Infinite Row of Holes," *Proc. Camb. Phil. Soc.*, **30**, pp. 471–491 (1934).
6. Ling Chih-Bing, "On the Stresses in a Plate Containing Two Circular Holes," *J. Appl. Phys.*, **19**, pp. 77–82 (January 1948).
7. Mindlin, Raymond D., *Stress Distribution Around a Tunnel*, American Society of Civil Engineers, pp. 619–642, April 1939.
8. Neuber, H., *Theory of Notch Stresses*, Edwards Brothers, Ann Arbor, Michigan, 1946.
9. Prescott, John, *Applied Elasticity*, Dover Publications, New York, 1946.
10. Roark, Raymond J., *Formulas for Stress and Strain*, McGraw-Hill Book Co., 1943.
11. Sadowsky, M. A. and E. Sternberg, "Stress Concentration around an Ellipsoidal Cavity in an Infinite Body under Arbitrary Plane Stress Perpendicular to the Axis of Revolution of the Cavity," *J. Appl. Mech., Trans. ASME*, **69**, pp. A-191–A-201 (1947).

12. Sadowsky, M. A. and E. Sternberg, "Stress Concentration Around a Triaxial Ellipsoidal Cavity," *J. Appl. Mech.*, pp. 149–157 (June 1949).
13. Savin, G. N., *Stress Concentration Around Holes*, Pergamon Press, New York, 1961.
14. Southwell, R. V. and H. J. Gough, "On the Concentration of Stress in the Neighbourhood of a Small Spherical Flaw," *Phil. Mag.*, pp. 71–86 (1926).
15. Terzaghi, Karl and F. E. Richart, Jr., "Stresses in Rock Around Cavities," *Geotechnique—The International Journal of Soil Mechanics*, **III**, No. 2, pp. 57–90 (June 1952).
16. Timoshenko, S., *Strength of Materials*, Parts I and II, D. Van Nostrand Co., New York, 1947.
17. Timoshenko, S., *Theory of Plates and Shells*, McGraw-Hill Book Co., New York, p. 228, 1940.
18. Timenshenko, S. and J. M. Goodier, *Theory of Elasticity*, McGraw-Hill Book Co., New York, 1951.

CHAPTER 6

THEORIES OF INELASTIC
BEHAVIOR OF SOLIDS

6.1 INTRODUCTION

Many materials that are virtually linearly elastic at low stress levels, at room temperature, or short periods of loading deviate from an elastic behavior at high stress levels, elevated temperatures or under prolonged loading. These deviations can appear as: an incomplete instantaneous recovery of strain upon removal of stress, a change in strain with time under constant load, a change in stress with time under constant deformation, a variation of mechanical properties with applied stress, or a variation of mechanical properties with temperature, direction, and position in the body. The behavior of real materials beyond the elastic range is termed inelastic.[2] Thus the implication is that the theory of elasticity can be used to make the first approximation of the stresses, strains, and deformations in a structure under a given loading condition, and that various inelastic theories can be used to estimate the deviation from elastic behavior that may occur and to estimate the ultimate loads that a given structure can support before excessive deformation, fracture, or disintegration occurs. Inelastic theories are therefore a necessary counterpart to the theory of elasticity for a complete understanding of the mechanical behavior of structures made from real materials and for an evaluation of the stability and safety of such structures under various loading or temperature conditions.

Many inelastic theories have been developed mainly to describe the mechanical behavior of ductile, plastic, and viscous solids at high stress levels where creep and permanent deformation occur. Even though most inelastic theories were developed to describe the mechanical behavior of such engineering materials as metals, plastics, rubber, synthetic fibers, paints and tars, these theories under certain conditions apply to other materials such as rocks. There are too many inelastic theories to discuss in the space of one chapter. Therefore, only those inelastic theories that have some direct application to structural rock mechanics will be considered here. For more advanced treatment of the various inelastic theories the reader should consult the informative books listed in the references.

Most ductile materials have a fairly long plastic range before fracture occurs, whereas brittle materials such as most rocks have a very limited plastic range, if any, for uniaxial tension or compression. However, for triaxial loading conditions, most brittle materials tend to show a plastic range where permanent deformation occurs.

Theories of plasticity and viscosity usually require a redefinition of strain to account for the fact that the plastic or inelastic strains are not infinitesimal as in elastic theory. However, the inelastic and plastic range for brittle materials is relatively short, and the assumption that the inelastic strains are of the same order of magnitude as the elastic strains is justified. Therefore infinitesimal plastic and inelastic strains will be assumed.

Experimental evidence shows that plastic yielding, creep, and flow do not occur under hydrostatic compressive stress conditions, that is, when $\sigma_1 = \sigma_2 = \sigma_3 = -p$; also that deformation occurs without appreciable dilatation during yielding, creep, or flow. These experimental facts indicate that plastic and viscous flows are associated with shear stresses or distortional strain energy.

6.2 IDEALIZED INELASTIC MATERIALS

Before any mathematical theory can be developed which relates stress, strain, deformation, rate of change of stress, rate of change of strain, or rate of change of deformation, it is necessary to assume an idealized relationship between two or more of the above variables. For example, the theory of elasticity was developed by assuming that stress is directly proportional to strain. Various mathematical and mechanical models have been proposed to describe the mechanical behavior of real materials. A number of these models are described below.

The Perfectly Elastic or Hookean Solid

The uniaxial relation between stress σ and strain ϵ for a perfectly elastic solid is

$$\sigma = E\epsilon \qquad (6.2.1)$$

where E is Young's modulus and is a property of the solid. A mechanical model for the Hookean solid is simply a spring as shown in Fig. 6.2.1a.

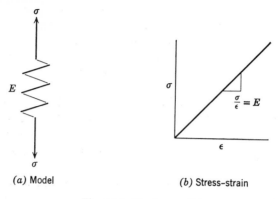

(a) Model   (b) Stress–strain

Fig. 6.2.1. Hookean solid.

The graphical representation in the stress-strain plane for a Hookean solid is a straight line that passes through the origin and has a slope of E as shown in Fig. 6.2.1b.

The Perfectly Viscous or Newtonian Material

An ideal fluid can not withstand any shear stress without deforming permanently. The perfectly viscous fluid or Newtonian material can develop shear stresses which are directly proportional to the rate of change of shear strains. Thus the mathematical model relating shear stress τ to the shear strain γ is given by

$$\tau = \eta\dot{\gamma} \tag{6.2.2}$$

where η, the coefficient of viscosity, is a constant of the material and the dot over γ stands for the partial derivative with respect to time.

Consider a tensile rod composed of a Newtonian material that is also incompressible. Let σ_1 be the normal stress in the direction of the length of the rod and $\dot{\epsilon}_1$ be the rate of change of the normal strain. Since the material is incompressible, the sum of the strains must be zero; therefore the lateral strains ϵ_2 and ϵ_3 are given by

$$\epsilon_2 = \epsilon_3 = -\tfrac{1}{2}\epsilon_1 \tag{6.2.3}$$

where the value of ν has been taken as $\tfrac{1}{2}$. The maximum shear stress is given by

$$\tau = \frac{\sigma_1 - 0}{2} = \frac{\sigma_1}{2} \tag{6.2.4}$$

and the maximum shear strain is given by

$$\gamma = \epsilon_1 - \epsilon_2 = \tfrac{3}{2}\epsilon_1 \tag{6.2.5}$$

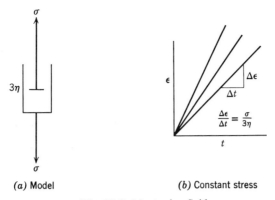

(a) Model   (b) Constant stress

Fig. 6.2.2. Newtonian fluid.

From Eqs. 6.2.2, 6.2.4, and 6.2.5, the normal stress is related to the normal strain by

$$\sigma = 3\eta\dot{\epsilon} \tag{6.2.6}$$

where the subscript 1 has been dropped so that the notation will correspond to that used in Eq. 6.2.1.

The mechanical model for a Newtonian material is a dashpot as shown in Fig. 6.2.2a. If $\epsilon = 0$ when $t = 0$, integration of Eq. 6.2.6 gives

$$\epsilon = \frac{\sigma t}{3\eta} \tag{6.2.7}$$

Therefore the graphical representation in the strain-time plane is a straight line that passes through the origin and has a slope of $\sigma/3\eta$ as shown in Fig. 6.2.2b.

The Perfectly Plastic Material

The perfectly plastic material will not deform if the applied stress is less than σ_0 and will deform permanently without limit if the applied stress

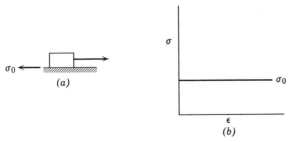

Fig. 6.2.3. Perfect plastic solid. (a) Model. (b) Stress-strain.

is equal to σ_0. Also the perfectly plastic material will not support a stress greater than σ_0. The appropriate mechanical model for such a material is a frictional contact as shown in Fig. 6.2.3a. In the stress-strain plane, the graphical representation of the perfectly plastic material is a straight line parallel to the strain axis and intersecting the stress axis at σ_0, as shown in Fig. 6.2.3b.

The Perfectly Elastoplastic or St. Venant Material

The St. Venant material is perfectly elastic for stresses less than σ_0 and is perfectly plastic for stresses equal to σ_0. The appropriate mechanical

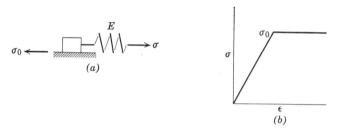

Fig. 6.2.4. St. Venant solid. (a) Model. (b) Stress-strain.

model is a spring in series with a frictional contact as shown in Fig. 6.2.4a, and the graphical representation in the stress-strain plane is illustrated in Fig. 6.2.4b.

A theory of plasticity for static equilibrium conditions based on a St. Venant material is discussed in Section 6.4; the stress distributions around a cylindrical and spherical cavity in a St. Venant material under hydrostatic loading conditions are determined in Sections 6.5 and 6.6, respectively.

The Viscoelastic or Maxwell Material

The Maxwell material is composed of an elastic element in series with a viscous element as illustrated in Fig. 6.2.5a. The same stress σ must act on the elastic element that acts on the viscous element. The total strain ϵ is the sum of the elastic strain ϵ' and viscous strain ϵ''. Thus from Eqs. 6.2.1 and 6.2.6

$$\dot{\epsilon} = \dot{\epsilon}' + \dot{\epsilon}'' = \frac{\dot{\sigma}}{E} + \frac{\sigma}{3\eta} \tag{6.2.8}$$

A similar relation can be written for shear stresses and strain since the total shear strain γ must be the sum of the elastic shear strain γ' and the viscous shear strain γ''. For the elastic element

$$\tau = G\gamma \qquad (6.2.9)$$

Therefore, from Eqs. 6.2.9 and 6.2.2,

$$\dot{\gamma} = \dot{\gamma}' + \dot{\gamma}'' = \frac{\dot{\tau}}{G} + \frac{\tau}{\eta} \qquad (6.2.10)$$

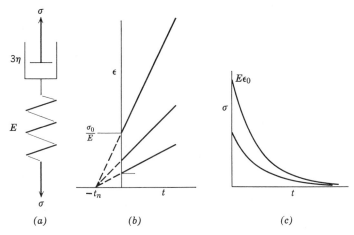

Fig. 6.2.5. Maxwell solid. (*a*) Model. (*b*) Constant stress. (*c*) Constant strain.

Assume that at time zero a constant stress σ_0 is suddenly applied. For a constant stress $\dot{\sigma}$ is zero; therefore, from Eq. 6.2.8,

$$\dot{\epsilon} = \frac{\sigma_0}{3\eta}$$

and integration gives

$$\epsilon = \frac{\sigma_0}{E} + \frac{\sigma_0 t}{3\eta} \qquad (6.2.11)$$

where the constant of integration has been made equal to the elastic strain ϵ_0 that occurs when the stress is applied at $t = 0$. Figure 6.2.5*b* shows plots of Eq. 6.2.11 for several values of σ_0. All the straight lines emanate from a single point on the time axis given by $t_\eta = -3\eta/E$.

Assume that at time $t = 0$, a sudden strain of $\epsilon = \epsilon_0$ is applied and held constant. If ϵ is constant, $\dot{\epsilon}$ is zero and Eq. 6.2.8 becomes

$$\frac{\dot{\sigma}}{E} = -\frac{\sigma}{3\eta} \qquad \text{or} \qquad \frac{d\sigma}{\sigma} = -\frac{E\,dt}{3\eta}$$

and integration gives

$$\sigma = E\epsilon_0 e^{-Et/3\eta} \qquad (6.2.12)$$

where $E\epsilon_0 = \sigma_1$ is the stress at time zero necessary to produce the strain ϵ_0. For this case the stress relaxes from its initial value of σ_1. The time required for the stress to relax to σ_1/e is the Maxwell relaxation time t_n given by

$$t_n = \frac{3\eta}{E}$$

Note that the point from which all the curves emanate in Fig. 6.2.5b is the negative relaxation time. Plots of Eq. 6.2.12 are given in Fig. 6.2.5c.

By making use of the shear Eq. 6.2.10, the constant shear-stress and constant shear-strain relations can be obtained as above. The resulting equations are

$$\gamma = \frac{\tau}{G} + \frac{\tau t}{\eta}$$

and

$$\tau = \tau_1 e^{-Gt/\eta}$$

(6.2.13)

The relaxation time for the shear stress is $t_t = \eta/G$, which for an incompressible material is equal to t_n as $E = 3G$ when $\nu = \frac{1}{2}$.

The Firmo-Viscous or Kelvin or Voigt Material

A firmo-viscous material can be represented by a spring and dashpot in parallel as shown in Fig. 6.2.6a. For this model, the strain in the elastic element must equal the strain in the viscous element; the total stress σ is the sum of the elastic stress σ' and the viscous stress σ''. Thus from Eqs. 6.2.1 and 6.2.6, the stress-strain relation is

$$\sigma = \sigma' + \sigma'' = E\epsilon + 3\eta\dot{\epsilon}$$

(6.2.14)

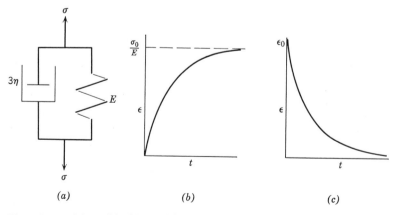

(a)   (b)   (c)

Fig. 6.2.6. Kelvin solid. (a) Model. (b) Constant stress. (c) Removal of stress.

Assume that at $t = 0$, when $\epsilon = 0$, that a constant stress σ_0 is applied to the system. Integration of Eq. 6.2.14 gives

$$\epsilon = \frac{\sigma_0}{E}(1 - e^{-Et/3\eta}) \tag{6.2.15}$$

Thus the elastic strain σ_0/E which would be obtained instantaneously for an elastic material is approached exponentially for a firmo-viscous material.

If the system is stretched until the strain is ϵ_0 and then the stress is removed, Eq. 6.2.14 becomes

$$\frac{d\epsilon}{\epsilon} = -\frac{E}{3\eta} dt$$

which on integration gives

$$\epsilon = \epsilon_0 e^{-Et/3\eta} \tag{6.2.16}$$

Thus the strain relaxes under zero stress and the time $3\eta/E$ for the strain to fall to ϵ_0/e is the relaxation time.

Complex Models

Real materials usually have a more complex behavior than that described by any of the simple linear models. However, by coupling together two or more basic linear elements, a wide variety of properties can be approximated. One of the more useful complex models is a Maxwell unit in series with a Kelvin unit as shown in Fig. 6.2.7a. This is generally known as Burger's model.

If at $t = 0$ a constant stress σ_0 is suddenly applied, the strain in the Maxwell unit is given by Eq. 6.2.11 and the strain in the Kelvin unit is given by Eq. 6.2.15. Since the total strain is the sum of these two, the resulting strain is given by

$$\epsilon = \frac{\sigma_0}{E_m} + \frac{\sigma_0 t}{3\eta_m} + \frac{\sigma_0}{E_k}(1 - e^{-E_k t/3\eta_k}) \tag{6.2.17}$$

where the subscripts m and k stand for the Maxwell and Kelvin constants, respectively.

Equation 6.2.17 comes close to representing the behavior of some rocks when subjected to suddenly applied loads. The strain is made up of three parts: (1) σ_0/E_m, which is the instantaneous elastic strain, (2) $(\sigma_0/E_k)(1 - e^{-E_k t/3\eta_k})$, which is the exponential recoverable strain, and (3) $\sigma_0 t/3\eta_m$, which is the irrecoverable strain resulting from the steady-state strain rate. Plots of Eq. 6.2.17 are shown in Fig. 6.2.7b. Note that the steady-state strain-rate lines all radiate from a common point on the

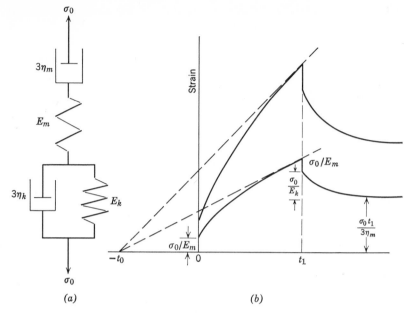

Fig. 6.2.7. Burger's solid. (*a*) Model. (*b*) Constant stress.

time axis at a value of t given by

$$t_0 = -3\eta_m \left(\frac{1}{E_m} + \frac{1}{E_k} \right) \qquad (6.2.18)$$

If after some time t_1, the load σ_0 is removed, there is an instantaneous recovery of the elastic strain σ_0/E_m, followed by an exponential recovery of strain in the Kelvin unit that decreases the strain asymptotically to the permanent deformation $(\sigma_0 t_1)/(3\eta)$. This permanent deformation remains in the Maxwell unit.

Experimental studies on rocks under uniaxial and triaxial loading conditions have shown that by a suitable adjustment of the four constants, η_m, η_k, E_m, and E_k, Eq. 6.2.17 can be made to fit experimental creep data on rocks. Further discussion of this type of inelastic behavior of rocks is given in Chapter 9.

6.3   CRITERIA FOR PLASTIC YIELDING

An essential part of any theory of inelastic behavior of solid materials is a reliable and mathematically usable criterion for specifying the maximum state of stress that can exist in a body before yielding will occur. Many different criteria of failure by fracture, flow, or yielding have been proposed. A detailed discussion of some of these criteria is given in Chapter 9. For the purposes of this chapter only criteria for plastic yielding will be considered. These criteria are based upon the experimental evidence that solid materials under hydrostatic stress conditions do not show any tendency to yield or creep. Thus yielding must be associated with distortional stresses.

The state of stress at a point in a body where plastic yielding is about to begin is independent of any coordinate system; therefore criteria for yielding must be expressible in terms of principal stresses σ_1, σ_2, and σ_3. A function of the principal stresses

$$f(\sigma_1, \sigma_2, \sigma_3) = 0 \qquad (6.3.1)$$

defines a surface within a stressed body. This surface divides the body into zones such that plastic yielding, or inelastic behavior, exists on one side of the surface and elastic behavior exists on the other side.

From Eqs. 3.9.16 and 1.6.2, the distortional energy per unit volume or the octahedral shearing stresses are dependent upon the quantity

$$(\sigma_1 - \sigma_2)^2 + (\sigma_2 - \sigma_3)^2 + (\sigma_3 - \sigma_1)^2$$

Various investigators have used the above expression as a basis for specifying a criterion of failure, for example, von Mises,[10] Hencky,[3] Nadai,[7] Huber,[5] etc. One generally accepted expression for specifying the state of stress at plastic yielding is

$$\sqrt{(\sigma_1 - \sigma_2)^2 + (\sigma_2 - \sigma_3)^2 + (\sigma_3 - \sigma_1)^2} = \sqrt{2}\sigma_0 \qquad (6.3.2)$$

where σ_0 is the yield stress in simple tension or compression.

Tresca[9] proposed a criteria for plastic yielding based upon the maximum shear stress. If $\sigma_1 > \sigma_2 > \sigma_3$, the maximum shear stress is $(\sigma_1 - \sigma_3)/2$; thus Tresca's criterion for plastic yielding is

$$\frac{\sigma_1 - \sigma_3}{2} = k \qquad (6.3.3)$$

where k is some constant.

The above two criteria for plastic yielding can be made to agree for simple tension or compression by setting $k = \sigma_0/2$. However, they will not agree for other loading conditions. For pure shear ($\sigma_1 = -\sigma_3$ and $\sigma_2 = 0$), Eq. 6.3.2 predicts plastic yielding at a value of stress given by $(\sigma_1)_m = \sigma_0/\sqrt{3}$, whereas Eq. 6.3.3 predicts plastic yielding at $(\sigma_1)_t = \sigma_0/2$. Thus for pure shear, the two criteria differ by the ratio $2/\sqrt{3}$.

For two-dimensional problems in plane strain, the stress normal to the plane is $\sigma_2 = \nu(\sigma_1 + \sigma_3)$. In the plastic zones, ν is $\frac{1}{2}$, as plastic yielding takes place at constant volume. Thus the stress normal to the plane is given by

$$\sigma_2 = \frac{\sigma_1 + \sigma_3}{2} \tag{6.3.4}$$

Equations 6.3.4 and 6.3.2 combine to give

$$\frac{\sigma_1 - \sigma_3}{2} = \frac{\sigma_0}{\sqrt{3}} \tag{6.3.5}$$

However, if Tresca's criterion were used, $(\sigma_1 - \sigma_3)/2$ would be set equal to $\sigma_0/2$. Therefore the two criteria differ for plane-strain conditions by the ratio of $2/\sqrt{3}$.

Insufficient experimental data exist to determine which of the above criteria of plastic yielding (Eq. 6.3.2 or 6.3.3) are best suited to account for plastic flow in rocks. Therefore, for plane-strain problems Tresca's criteria will be used because of its simplicity.

6.4 THEORY OF PLASTICITY

The theory of plasticity to be developed is restricted to equilibrium conditions within the stressed body. Thus the theory can determine only the state of stress, both in the plastic and elastic zones, after all deformation has ceased to change with time. Because of the assumption of equilibrium, the equilibrium equations developed in Chapter 1 must be satisfied in both zones. In the plastic zone and at the boundary between the plastic and elastic zones, the condition of plasticity as given by Eq. 6.3.3 must hold.

Because of the assumption of infinitesimal plastic strains, the strain-displacement equations as given in Chapter 2 must be satisfied both in the plastic and elastic zones. In the elastic zone, the stress-strain relations for perfectly elastic material must be satisfied and the compatibility equations in terms of stresses must also be satisfied.

To distinguish between stresses, strains, and deformations in the plastic zone from those in the elastic zone, a single prime will be used for elastic stresses, strains, and deformations; a double prime will be used for plastic stresses, strains, and deformations.

Two relatively simple problems in the theory of plasticity will be discussed. The lack of sufficient experimental data on rock structures to check the basic assumptions and the calculated results does not justify detail problem solving at this time. However, the two problems that are solved are very instructive because they illustrate the fundamental principle that plastic flow tends to relieve the high-stress concentrations that would ordinarily develop in perfectly elastic materials.

6.5 CIRCULAR TUNNEL IN AN ELASTOPLASTIC ROCK

Consider an elastoplastic rock which is under a hydrostatic stress condition and in which a long circular tunnel is mined (see Fig. 6.5.1). This problem is similar to that of an infinite plate with a circular hole as discussed in Section 4.7. The difference now is that the material of the

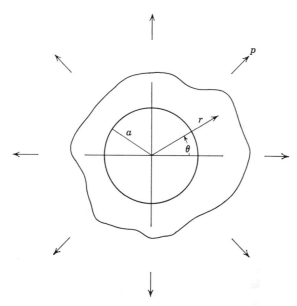

Fig. 6.5.1. Circular tunnel in hydrostatic stress field.

plate is assumed to be elastoplastic, that yielding will occur when

$$\sigma_\theta' - \sigma_r' = 2k$$

and that in the plastic region

$$\sigma_\theta'' - \sigma_r'' = 2k \qquad (6.5.1)$$

As a uniform state of stress is assumed the stresses are symmetric with respect to the coordinate θ; therefore the equilibrium conditions in polar coordinates, Eq. 3.8.6, become for the plastic zone

$$\frac{\partial \sigma_r''}{\partial r} = \frac{\sigma_\theta'' - \sigma_r''}{r} \qquad (6.5.2)$$

In the elastic zone, the equilibrium and compatibility equations for polar coordinates (Eqs. 3.8.6 and 3.8.19) are

$$\frac{\partial \sigma_r'}{\partial r} = \frac{\sigma_\theta' - \sigma_r'}{r} \tag{6.5.3}$$

$$\left(\frac{\partial^2}{\partial r^2} + \frac{1}{r}\frac{\partial}{\partial r} \right)(\sigma_r' + \sigma_\theta') = 0 \tag{6.5.4}$$

In addition to the above equations, the following boundary conditions must be satisfied:

$$\begin{aligned}
(\sigma_r'')_{r=a} = 0 && (\sigma_r'')_{r=c} = (\sigma_r')_{r=c} \\
(\sigma_\theta')_{r=\infty} = (\sigma_r')_{r=\infty} = p && (\sigma_\theta'')_{r=c} = (\sigma_\theta')_{r=c}
\end{aligned} \tag{6.5.5}$$

where a is the radius of the cylindrical hole, c is the radius of the boundary between the plastic and elastic zones, and p is the applied hydrostatic stress.

Combining Eqs. 6.5.1 and 6.5.2 gives

$$\frac{\partial \sigma_r''}{\partial r} = \frac{2k}{r}$$

which can be integrated directly and evaluated at $r = a$ to give

$$\sigma_r'' = 2k \ln \frac{r}{a} \tag{6.5.6}$$

Combining Eqs. 6.5.6 and 6.5.1 gives

$$\sigma_\theta'' = 2k \left(1 + \ln \frac{r}{a} \right) \tag{6.5.7}$$

Stresses that satisfy Eqs. 6.5.3 and 6.5.4 are from Eq. 4.7.8 of the form

$$\sigma_r = A + \frac{B}{r^2}$$
$$\sigma_\theta = A - \frac{B}{r^2} \tag{6.5.8}$$

Equations 6.5.8 evaluated at $r = \infty$ give $A = p$, and evaluated at $r = c$ give

$$\sigma_r')_{r=c} = p + \frac{B}{c^2} = 2k \ln \frac{c}{a}$$
$$(\sigma_\theta')_{r=c} = p - \frac{B}{c^2} = 2k \left(1 + \ln \frac{c}{a} \right) \tag{6.5.9}$$

Solving Eqs. 6.5.9 for B and c gives

$$B = -ka^2e^{(p-k)/k} \qquad (6.5.10)$$

$$c = ae^{(p-k)/2k} \qquad (6.5.11)$$

Thus the boundary between the plastic and elastic zones is determined by the value of the applied stress, the yielding condition, and the original value of the hole radius. Assume that the yielding shear stress is some fraction of the applied stress such that

$$k = hp \qquad (6.5.12)$$

and that the plate is subject to plane-strain conditions. Then the stresses normal to the plate are $\sigma_z = \nu(\sigma_r + \sigma_\theta)$ in the elastic zone and $\sigma_z'' = (\sigma_r'' + \sigma_\theta'')/2$ in the plastic zone. Therefore, as originally shown by Fara and Wright,[1] the complete solution for the stresses around the cylindrical tunnel are given by

$$\sigma_r'' = 2hp \ln \frac{r}{a}$$

$$\sigma_\theta'' = 2hp\left(1 + \ln \frac{r}{a}\right) \qquad a \le r \le c \qquad (6.5.13)$$

$$\sigma_z'' = hp\left(1 + 2 \ln \frac{r}{a}\right)$$

$$\sigma_r' = p\left[1 - \frac{ha^2}{r^2} e^{(1-h)/h}\right]$$

$$\sigma_\theta' = p\left[1 + \frac{ha^2}{r^2} e^{(1-h)/h}\right] \qquad c \le r \qquad (6.5.14)$$

$$\sigma_z' = 2\nu p$$

where

$$c = ae^{(1-h)/2h} \qquad (6.5.15)$$

Note that when $h = 1$

$$\sigma_r' = p\left(1 - \frac{a^2}{r^2}\right)$$

$$\sigma_\theta' = p\left(1 + \frac{a^2}{r^2}\right)$$

and

$$c = a$$

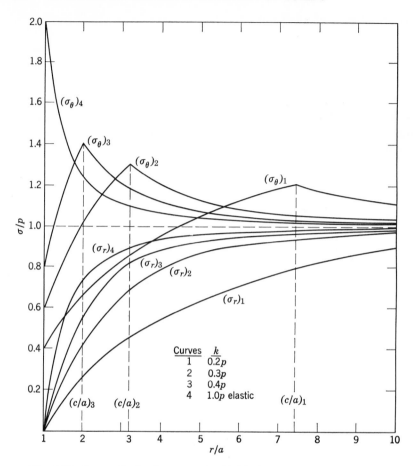

Fig. 6.5.2. Stress distribution around cylindrical tunnel in elastoplastic rock.

which is just the elastic solution. The shear stress at $r = a$ for $h = 1$ is

$$\frac{\sigma_\theta' - \sigma_r'}{2} = p$$

which is the condition for the start of plastic yielding.

Graphs of the plastic-elastic solution are shown in Fig. 6.5.2 together with the perfectly elastic solution.

The stresses near the boundary of cylindrical openings are considerably less for an elastoplastic material than for a perfectly elastic material. However, the tangential stresses outside the plastic-elastic boundary are larger than the tangential stresses at the same radial distance for the perfectly elastic material. As a consequence, the zone of influence, or that region where the presence of the opening has altered the original stress field, surrounding a cylindrical opening is much larger for an elastoplastic material than for a perfectly elastic material.

Figure 6.5.3 shows a plot of the radial and tangential stresses at the plastic-elastic boundary. For comparison purposes, the stress distribution for the perfectly elastic material is shown also. In addition a plot of Eq. 6.5.15 is shown which gives the radius of the plastic-elastic boundary as a function of k/p.

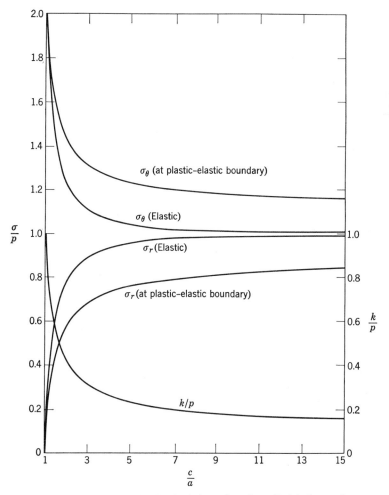

Fig. 6.5.3. Stress at plastic-elastic boundary for cylindrical tunnel.

6.6 SPHERICAL CAVITY IN A HYDROSTATIC STRESS FIELD

Consider a spherical cavity in an elastoplastic rock where the stress field, before the cavity is mined, is hydrostatic (see Fig. 6.6.1). As the stresses are symmetric with respect to the spherical coordinates θ and Φ,

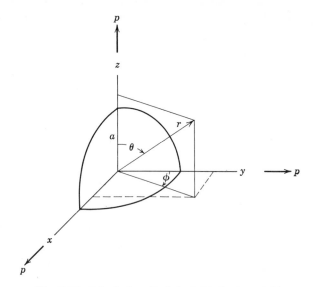

Fig. 6.6.1. Spherical cavity in hydrostatic stress field.

the equilibrium conditions in spherical coordnates reduce to

$$\frac{\partial \sigma_r}{\partial r} = \frac{2(\sigma_t - \sigma_r)}{r} \tag{6.6.1}$$

where $\sigma_t = \sigma_\theta = \sigma_\Phi$ is the tangential stress and σ_r is the radial stress. Since the displacements in the θ and Φ directions are zero and the displacement u in the r direction will not depend upon θ and Φ, the strain-displacement relations for spherical coordinates reduce to

$$\epsilon_r = \frac{\partial u}{\partial r}$$

$$\epsilon_t = \frac{u}{r} \tag{6.6.2}$$

$$\gamma_{rt} = 0$$

The stress-strain relations for spherical symmetry are

$$\epsilon_r = \frac{1}{E}(\sigma_r - 2\nu\sigma_t)$$

$$\epsilon_t = \frac{1}{E}[(1 - \nu)\sigma_t - \nu\sigma_r] \qquad (6.6.3)$$

$$\gamma_{rt} = \frac{\tau_{rt}}{G}$$

The compatibility equation in terms of stresses for spherical symmetry can be derived from the above three equations to give

$$\left[\frac{1}{r^2}\frac{\partial}{\partial r}\left(r^2\frac{\partial}{\partial r}\right)\right](\sigma_r + 2\sigma_t) = 0 \qquad (6.6.4)$$

Therefore, in the plastic zone, the equations to be satisfied are the equilibrium conditions and Tresca's yielding condition or

$$\frac{\partial \sigma_r''}{\partial r} = \frac{2(\sigma_t'' - \sigma_r'')}{r} \qquad (6.6.5)$$

$$k = \frac{\sigma_t'' - \sigma_r''}{2} \qquad (6.6.6)$$

In the elastic zone, the equations to be satisfied are the equilibrium conditions and the compatibility equations in terms of stresses or

$$\frac{\partial \sigma_r'}{\partial r} = \frac{2(\sigma_t' - \sigma_r')}{r} \qquad (6.6.7)$$

$$\left[\frac{1}{r^2}\frac{\partial}{\partial r}\left(r^2\frac{\partial}{\partial r}\right)\right](\sigma_r' + 2\sigma_t') = 0 \qquad (6.6.8)$$

In addition, the following boundary conditions must be satisfied:

$$(\sigma_r'')_{r=a} = 0 \qquad\qquad (\sigma_r'')_{r=c} = (\sigma_r')_{r=c}$$
$$(\sigma_r')_{r=\infty} = (\sigma_t')_{r=\infty} = p \qquad (\sigma_t'')_{r=c} = (\sigma_t')_{r=c} \qquad (6.6.9)$$

where c is the radius of the boundary between the plastic and elastic zones and p is the applied hydrostatic pressure.

Combining Eqs. 6.5.5 and 6.6.6 gives

$$\frac{\partial \sigma_r}{\partial r} = \frac{4k}{r}$$

This equation can be integrated directly and evaluated at $r = a$ to give

$$\sigma_r'' = 4k \ln \frac{r}{a} \tag{6.6.10}$$

combining Eqs. 6.6.10 and 6.6.6 gives

$$\sigma_t'' = 2k\left(1 + 2\ln\frac{r}{a}\right) \tag{6.6.11}$$

Stresses that satisfy Eqs. 6.6.7 and 6.6.8 are of the form

$$\sigma_r = A + \frac{2B}{r^3}$$
$$\sigma_t = A - \frac{B}{r^3} \tag{6.6.12}$$

The constants A and B in Eqs. 6.6.12 can be evaluated by means of the boundary conditions (Eqs. 6.6.9). Thus at $r = \infty$

$$(\sigma_t')_{r=\infty} = (\sigma_r')_{r=\infty} = A = p$$

and at $r = c$

$$(\sigma_r')_{r=c} = p + \frac{2B}{c^3} = 4k \ln \frac{c}{a}$$
$$(\sigma_t')_{r=c} = p - \frac{B}{c^3} = 2k\left(1 + 2\ln\frac{c}{a}\right) \tag{6.6.13}$$

Solving Eqs. 6.6.13 for B and c gives

$$c = ae^{(3p-4k)/12k}$$
$$B = -\tfrac{2}{3}ka^3 e^{(3p-4k)/4k} \tag{6.6.14}$$

Let $k = hp$; then the complete solution for the stresses in the plastic and elastic zones are given by

$$\sigma_r'' = 4hp \ln \frac{r}{a}$$
$$\sigma_t'' = 2hp\left(1 + 2\ln\frac{r}{a}\right) \qquad a \le r \le c \tag{6.6.15}$$

$$\sigma_r' = p - \frac{4hpa^3}{3r^3} e^{(3-4h)/4h}$$
$$\sigma_t' = p + \frac{2hpa^3}{3r^3} e^{(3-4h)/4h} \qquad c \le r \tag{6.6.16}$$

$$c = ae^{(3-4h)/12h} \tag{6.6.17}$$

When $h = \tfrac{3}{4}$, $c = a$, and the elastic stresses are those for a spherical cavity in an elastic medium subjected to a hydrostatic stress field, thus

$$\sigma_r = p\left(1 - \frac{a^3}{r^3}\right)$$
$$\sigma_t = p\left(1 + \frac{1}{2}\frac{a^3}{r^3}\right)$$

(6.6.18)

Therefore initiation of plastic yielding at the boundary of the spherical cavity begins when $p = \frac{4}{3}k$.

Graphs of the plastic-elastic solution are shown in Fig. 6.6.2 together with the perfectly elastic solution.

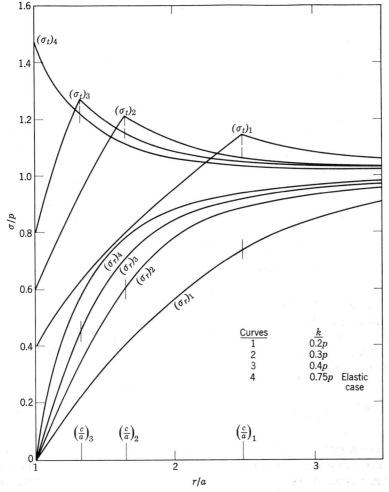

Fig. 6.6.2. Stress distributions around spherical cavity in elastoplastic rock.

The stresses near the boundary of a spherical cavity are considerably less for an elastoplastic material than they are for a perfectly elastic material. On the other hand, the tangential stresses outside the plastic-elastic boundary are larger than the tangential stresses at the same radial distance for a perfectly elastic material. Thus, the zone of influence, or that region where the presence of the opening has altered the applied stress field, is much larger for an elastoplastic material than for a perfectly elastic material.

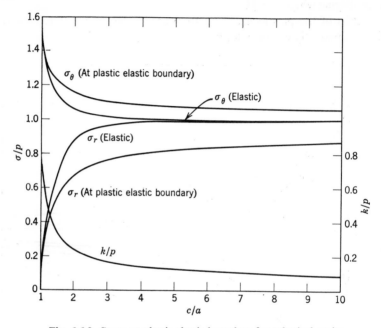

Fig. 6.6.3. Stress at plastic-elastic boundary for spherical cavity.

Figure 6.6.3 shows a plot of the radial and tangential stresses at the plastic-elastic boundary. For comparison purposes, the stress distribution for a spherical cavity in a perfectly elastic material is also shown. In addition, a plot of Eq. 6.6.17 is shown which gives the radius of the plastic-elastic boundary as a function of k/p.

6.7 THEORY OF A PURELY VISCOUS INCOMPRESSIBLE SUBSTANCE

For a perfectly viscous material, which develops a shear stress proportional to the rate of change of shear strain, the relation between the uniaxial normal stress and the rate of change of normal strain was obtained in

Section 6.2 as

$$\sigma_x = 3\eta\dot{\epsilon}_x \qquad (6.7.1)$$

For an incompressible substance $\nu = \frac{1}{2}$, and the resulting rate of change of normal strain in directions at right angles to $\dot{\epsilon}_x$ is given by

$$\dot{\epsilon}_y = \dot{\epsilon}_z = -\tfrac{1}{2}\dot{\epsilon}_y = -\tfrac{1}{2}\left(\frac{\sigma_x}{3\eta}\right) \qquad (6.7.2)$$

By applying the same logic as used in Chapter 3 to develop the Hooke's law equations for an isotropic elastic solid, we can write similar equations for an isotropic purely viscous incompressible material. For example,

$$\dot{\epsilon}_x = \frac{1}{3\eta}\left(\sigma_x - \frac{\sigma_y + \sigma_z}{2}\right) \qquad (6.7.3)$$

Inside the parentheses of Eq. 6.7.3 add and subtract $\sigma_x/2$ to obtain

$$\dot{\epsilon}_x = \frac{1}{2\eta}(\sigma_x - \sigma_m) \qquad (6.7.4)$$

where

$$\sigma_m = \frac{\sigma_x + \sigma_y + \sigma_z}{3} \qquad (6.7.5)$$

Rewriting Eq. 6.7.4 and two similar equations for $\dot{\epsilon}_y$ and $\dot{\epsilon}_z$ gives

$$\sigma_x = 2\eta\dot{\epsilon}_x + \sigma_m$$
$$\sigma_y = 2\eta\dot{\epsilon}_y + \sigma_m \qquad (6.7.6)$$
$$\sigma_z = 2\eta\dot{\epsilon}_z + \sigma_m$$

The relations between shear stress and rate of change of shear strain are by the original assumption simply

$$\tau_{xy} = \eta\gamma_{xy}$$
$$\tau_{yz} = \eta\gamma_{yz} \qquad (6.7.7)$$
$$\tau_{zx} = \eta\gamma_{zx}$$

The strains in Eqs. 6.7.3 and 6.7.7 are related to the three displacements u, v, and w by

$$\dot{\epsilon}_x = \frac{\partial\dot{u}}{\partial x} \qquad \gamma_{xy} = \frac{\partial\dot{v}}{\partial x} + \frac{\partial\dot{u}}{\partial y}$$

$$\dot{\epsilon}_y = \frac{\partial\dot{v}}{\partial y} \qquad \gamma_{yz} = \frac{\partial\dot{w}}{\partial y} + \frac{\partial\dot{v}}{\partial z} \qquad (6.7.8)$$

$$\dot{\epsilon}_z = \frac{\partial\dot{w}}{\partial z} \qquad \gamma_{zx} = \frac{\partial\dot{u}}{\partial z} + \frac{\partial\dot{w}}{\partial x}$$

where $\dot{u}$, $\dot{v}$, and $\dot{w}$ are the velocity components in x, y, and z directions.

Addition of the three normal components of rate of change of strain in Eqs. 6.7.8 gives

$$e = \epsilon_x + \epsilon_y + \epsilon_z = 0 \qquad (6.7.9)$$

as the material is assumed incompressible.

Equation 6.7.9 is the equation of continuity which expresses the fact that the mass of substance within any imaginary volume remains constant.

The three conditions of equilibrium from Chapter 1 must be satisfied also for the purely viscous substance; therefore, for no body forces,

$$\frac{\partial \sigma_x}{\partial x} + \frac{\partial \tau_{yx}}{\partial y} + \frac{\partial \tau_{zx}}{\partial z} = 0$$

$$\frac{\partial \tau_{xy}}{\partial x} + \frac{\partial \sigma_y}{\partial y} + \frac{\partial \tau_{zy}}{\partial z} = 0 \qquad (6.7.10)$$

$$\frac{\partial \tau_{xz}}{\partial x} + \frac{\partial \tau_{yz}}{\partial y} + \frac{\partial \sigma_z}{\partial z} = 0$$

Substitution of Eqs. 6.7.6 into the equilibrium conditions, Eqs. 6.7.10, and using the equation of continuity (6.7.9) results in the following three equations:

$$\eta \nabla^2 \dot{u} + \frac{\partial}{\partial x} \sigma_m = 0$$

$$\eta \nabla^2 \dot{v} + \frac{\partial}{\partial y} \sigma_m = 0 \qquad (6.7.11)$$

$$\eta \nabla^2 \dot{w} + \frac{\partial}{\partial z} \sigma_m = 0$$

Comparison of Eqs. 6.7.11 with Eqs. 3.3.9 shows that the two sets of equations can be made equivalent by assuming an incompressible solid. That is, in the relation $E = 2(1 + \nu)G$, let $\nu = \frac{1}{2}$, then $E = 3G$. If these results are substituted in Eqs. 3.3.9, the resulting equations are of the form

$$G \nabla^2 u + \frac{\partial}{\partial x} \sigma_m = 0 \qquad (6.7.12)$$

If the displacement components are replaced by the velocity components and G by η, the elastic equations become the viscosity equations. Therefore, it is possible to obtain solutions to problems involving purely viscous incompressible materials directly from solutions to problems in the theory of elasticity.

REFERENCES

1. Fara, H. D. and F. D. Wright, "Plastic and Elastic Stresses Around a Circular Shaft in a Hydrostatic Stress Field," *Trans. AIME*, **226**, pp. 319–320 (Sept. 1963).
2. Freudenthal, Alfred M., *The Inelastic Behavior of Engineering Materials and Structures*, John Wiley and Sons, New York, 1950.
3. Hencky, H., "Zur Theorie plastischer Deformationen und die hierdurch im Material herorgerufenen Nach-Spannung," *Z. ang. Math. Mech.*, **4** (1924).
4. Hill, R., *The Mathematical Theory of Plasticity*, The Clarendon Press, Oxford, 1960.
5. Huber, M., Czasopismo Techniczne, Livów (1904).*
6. Jaeger, J. C., *Elasticity, Fracture and Flow*, John Wiley and Sons, New York, 1962.
7. Nadai, A. L., *J. Appl. Phys.*, **8**, p. 205 (1937).
8. Nadai, A. L., *Theory of Flow and Fracture of Solids*, Vols. 1 and 2, McGraw-Hill Book Co., New York, 1950 and 1963.
9. Tresca, H., *Comptes Rendus*, Acad. Sci. Paris, **59**, pp. 754 and 64 (1864).
10. von Mises, R., *Mechanik der festen Körper in plastisch deformablen Zustand.* Göttingen Nachrichten, math.–phys. klasse 1913 (1913).

* Title unknown.

PART TWO

EXPERIMENTAL
ROCK MECHANICS

INTRODUCTION

In Chapters 1, 2, and 3 relationships were developed between stress, strain, and deformation for an ideally elastic, isotropic, and homogeneous continuous medium. Chapters 4 and 5 showed how these relationships could be used to determine the state of stress, strain, or deformation in bodies of different shapes subjected to various applied loads and sets of boundary conditions. Only those body shapes that were expressible by a simple mathematical equation were amenable to solution. In chapter 6, the assumptions regarding the body material were modified to include ideally elastoplastic, ideally viscous, or ideally viscoelastic materials. In structures or structural elements made of *real* materials, the ideal property requirement is never satisfied as no real material is strictly linear-elastic, elastoplastic, purely viscous, or viscoelastic, although some metals and plastics under certain load conditions exhibit characteristics approximating ideal materials. Also small pieces of rock, especially igneous rock, when subjected to a moderate applied stress may be nearly linear-elastic. Some large bodies or rock may approximate an elastic material to the degree that engineering requirements are satisfied; however, generally in situ rock contains fractures, joints, faults, and inhomogenities to the degree that significant deviations from ideal conditions occur.

In investigating rock models in the laboratory, it is usually possible to load a model in a specified direction and by a given amount; thereby satisfying the requirement that the magnitude and direction of the applied stress must be known. However, the magnitude and direction of the stress in the rock surrounding an underground opening depends on the stress field, that is, the state of stress in the rock before the opening was created. As a first approximation a gravitational stress field might be assumed, that is, a stress-field due to the weight of the overlying cover (see Section 14.9). However, to completely specify a gravitational stress field, it must be assumed that the rock is linear-elastic, isotropic, and homogeneous; that the lateral constraint is complete; that there are no stresses of tectonic origin such as those accompanying folding, shrinkage, or other distortions of the earth's crust. In some instances, these tectonic stresses, together with the effects caused by inelasticity, inhomogeniety, and anisotropy are known to affect the gravitational stress field significantly.

The requirement for a shape that can be expressed mathematically imposes a further limitation on the purely theoretical process. Simple shapes such as a circular or elliptical hole, or row of circular holes in a plate can be specified mathematically and the resulting boundary value problems for the stress, strain, or deformation have been solved (Chapters 4 and 5). More complicated shapes, such as those characteristic of openings in underground mining usually cannot be specified mathematically; but even if this were possible, the resulting boundary value problem would probably be intractable.

Because of the limitations imposed by mathematical complexity, and because of the error introduced by assuming ideal rock properties and a known stress-field, theoretical methods in themselves generally will not provide a satisfactory answer to specific rock structure problems. To some degree theoretical results may be supplemented or modified by the experience gained from day-to-day observations in underground workings. In various instances this procedure has led to erroneous conclusions, and at best lacks a quantitative aspect. The alternative to the theoretical approach is to employ empirical methods, that is, to measure rock properties and structural stresses and strains in laboratory models or in full-scale underground workings. These empirical results form the foundation for experimental rock mechanics. When this information is used to supplement or modify theory, the product is a rational basis for analyzing rock structure problems.

The experimental aspect of rock mechanics is the subject of Part Two. The first two chapters of Part Two, namely, Chapter 7, *Dimensional Analysis*, and Chapter 8, *Statistical Treatment of Data*, are not inherently a part of rock mechanics. They are, however, so fundamental to the experimental analysis of structures and the proper treatment of empirical results that Part Two is introduced with a limited exposition of these subjects. Also Chapter 9, *Instrumentation*, is included in Part Two because the class of instruments for rock mechanics is so specialized that it is not treated in other books.

CHAPTER 7

DIMENSIONAL ANALYSIS

7.1 INTRODUCTION

One of the research tools available to the design engineer is that branch of applied mathematics known as dimensional analysis. Usually a preliminary dimensional analysis of any experimental investigation discloses functional relationships between the measurable parameters involved that simplify the problem and indicate the direction to be followed in the design of the experimental program. All similitude and model studies should be based upon a dimensional analysis so that the results obtained can be applied to the prototype with confidence.

The fundamental dimensions of physical quantities in mechanics are usually taken as mass, length, and time, and are denoted by M, L, and T. The dimensions of other physical quantities follow immediately from their definitions. For example, volume has the dimensions of L^3; velocity has the dimensions of LT^{-1}; acceleration has the dimensions of LT^{-2}; and force, defined as the product of mass and acceleration by Newton's law, has the dimensions of MLT^{-2}. Thus mass, length, and time have been taken as primary quantities, and secondary quantities have been expressed in terms of the primary quantities. There are no hard and fast rules as to which measurable quantities should be considered the primary ones. In engineering mechanics, the primary quantities are often chosen as force, length, and time, or in some static problems, simply as force and length.

The dimensions of various physical quantities encountered in mechanics are summarized in Table 7.1.1, assuming that the primary quantities are either mass, length, and time or force, length, and time. Some physical quantities are dimensionless, for example, strain, Poisson's ratio, and angles. If a quantity is dimensionless, this is indicated by the symbol 1 rather than 0 as is often done. The use of 1 rather than 0 for quantities

191

Table 7.1.1. Physical Quantities and Their Dimensions

Quantity	Symbol	Dimensions for M, L, T	Dimensions for F, L, T
Length	l	L	L
Area	A	L^2	L^2
Volume	V	L^3	L^3
Time	t	T	T
Mass	m	M	$FL^{-1}T^2$
Velocity	v	LT^{-1}	LT^{-1}
Acceleration	a	LT^{-2}	LT^{-2}
Force, Load	F	MLT^{-2}	F
Mass density	ρ	ML^{-3}	$FL^{-4}T^2$
Specific weight or unit weight	γ or ρg	$ML^{-2}T^{-2}$	FL^{-3}
Angle	θ, ϕ	1	1
Angular velocity	ω	T^{-1}	T^{-1}
Angular acceleration	d	T^{-2}	T^{-2}
Pressure or stress	p, σ, τ	$ML^{-1}T^{-2}$	FL^{-2}
Work or energy	T, W	ML^2T^{-2}	FL
Momentum	mv	MLT^{-1}	FT
Power	P	ML^2T^{-3}	FLT^{-1}
Moment of force	M	ML^2T^{-2}	FL
Moment of inertia of an area	I	L^4	L^4
Moment of inertia of a mass	I	ML^2	FLT^2
Modulus of elasticity	E	$ML^{-1}T^{-2}$	FL^{-2}
Strain	ϵ, γ	1	1
Poisson's ratio	ν	1	1
Modulus of rigidity	G	$ML^{-1}T^{-2}$	FL^{-2}
Bulk modulus	K	$ML^{-1}T^{-2}$	FL^{-2}

having no dimensions permits an algebraic handling of units in functional relations. For example, the definition of strain is the change in length per unit length or $\epsilon = \Delta L/L$. If ΔL is measured in feet and L is measured in feet, the units divide out giving 1, not 0.

The most important applications of dimensional analysis in engineering are: (1) converting equations or data from one system of units to another, (2) developing relationships among variables, (3) systematizing the collection of data and reducing the number of variables that must be studied in any experimental program, and (4) establishing the principles of model design and assisting in the interpretation of the test data.

7.2 DIMENSIONAL HOMOGENEITY

The mathematical basis for dimensional analysis is based on the following two axioms. First, absolute numerical equality of quantities exists only when the quantities have the same dimensions. Second, the ratio of the magnitudes of two like quantities is independent of the units used in their measurement, provided the same units are used for both quantities.

These two axioms can be used to classify equations on a dimensional basis as either homogeneous, restricted homogeneous, or nonhomogeneous. If all the terms in a given equation reduce to the same dimensions and it does not contain dimensional constants,* the equation is said to be homogeneous. If all the terms in a given equation reduce to the same

Table 7.2.1. Examples of Homogeneous, Nonhomogeneous and Restricted Homogeneous Equations

No.	Equation	Type
1	$s = \frac{1}{2}gt^2$	Homogeneous
2	$s = 16.1t^2$	Restricted homogeneous
3	$v = at$	Homogeneous
4	$s + v = \frac{1}{2}gt^2 + at$	Nonhomogeneous
5	$W = mg$	Homogeneous
6	$W = 32.2m$	Restricted homogeneous
7	$v = \sqrt{\dfrac{E}{p}}$	Homogeneous
8	$T = \frac{1}{2}mv^2$	Homogeneous
9	$\epsilon_x = \dfrac{1}{E}[\sigma_x - \nu(\sigma_y + \sigma_z)]$	Homogeneous
10	$W_0 = \frac{1}{2}\epsilon_x\sigma_x$	Homogeneous

dimensions and it contains one or more dimensional constants, the equation is said to be restricted homogeneous. If an equation contains two or more terms on one side, the basic dimensions of which are not the same and it contains dimensionally equivalent terms on the other side, the equation is said to be nonhomogeneous. Thus a homogeneous equation is valid in all consistent systems of units, and a restricted homogeneous equation is valid in only one consistent system of units. A nonhomogeneous equation may be valid but it is of little use. Table 7.2.1 lists examples of these defined types of equations.

7.3 DIMENSIONLESS PRODUCTS

Given a set of n variables $(x_1, x_2, \ldots, x_i, \ldots, x_n)$, we can form an infinite number of products of powers of these variables as indicated:

$$u_i = x_1^{k_1}x_2^{k_2} \ldots x_i^{k_i} \ldots x_n^{k_n} \qquad (7.3.1)$$

* A dimensional constant is a proportionality constant that contains concealed dimensions. For example, in the equation for the static deflection of a spring, $d = kF$, the spring constant k has the dimensions of $M^{-1}T^2$.

The exponents k_i may have any positive or negative integral or fractional value including zero. The dimensions of these products of powers of variables may be found by replacing the symbols x_i with the symbols of its dimensions and raising the symbols to the power k_i. For example, if the variable x_i has the dimension $M^{a_i}L^{b_i}T^{c_i}$, the dimension of $x_i^{k_i}$ is $M^{a_ik_i}L^{b_ik_i}T^{c_ik_i}$. Thus the general expression for the dimensions of u_i in Eq. 7.3.1 is

$$(M^{a_1k_1}L^{b_1k_1}T^{c_1k_1})(M^{a_2k_2}L^{b_2k_2}T^{c_2k_2}) \cdots (M^{a_ik_i}L^{b_ik_i}T^{c_ik_i}) \cdots (M^{a_nk_n}L^{b_nk_n}T^{c_nk_n})$$

or

$$M^{a_1k_1+a_2k_2+\cdots a_ik_i\cdots a_nk_n}L^{b_1k_1+b_2k_2+\cdots b_ik_i\cdots b_nk_n}T^{c_1k_1+c_2k_2\cdots c_ik_i\cdots c_nk_n}$$

A dimensionless product of powers is one whose exponents of the fundamental units M, L, and T all vanish and which is designated by π and referred to as a π-term. The product of powers of the variables will be dimensionless if and only if the exponents of M, L, and T in the above expression satisfy the following equations:

$$a_1k_1 + a_2k_2 + \cdots a_ik_i + \cdots a_nk_n = 0$$
$$b_1k_1 + b_2k_2 + \cdots b_ik_i + \cdots b_nk_n = 0 \qquad (7.3.2)$$
$$c_1k_1 + c_2k_2 + \cdots c_ik_i + \cdots c_nk_n = 0$$

If the number of variables x_i, and consequently the number of exponents k_i exceeds the number of dimensions, there will be an infinite number of combinations of the k_i's which satisfy Eqs. 7.3.2. Thus the number of dimensionless products of powers of the variables is also infinite. As there is an infinite number of dimensionless products of powers or π-terms, all of these π-terms cannot be independent and interrelations between them will exist. This fact suggests the need for a method of restricting the number of π-terms to be considered.

A *complete set* of dimensionless products of powers of the variables has the following properties:

1. Any dimensionless product of powers of the given variables that is not in the set can be expressed as a product of powers of the π-terms in the set.

2. No member of the set can be expressed as a product of powers of the remaining members of the set.

We can form a matrix (which is a rectangular array of numbers) of the coefficients a_i, b_i, and c_i in Eqs. 7.3.2. Such an array of numbers is known as the dimensional matrix and is given in Table 7.3.1. If the number of rows equals the number of columns, the matrix is a square matrix of order n. If the number of rows is m and the number of columns n, the order of the matrix is $(m \times n)$ for $m \neq n$. A determinant is a square array of

Table 7.3.1. Dimensional Matrix

	x_1	x_2	$\cdots$	x_i	$\cdots$	x_n
M	a_1	a_2		a_i		a_n
L	b_1	b_2		b_i		b_n
T	c_1	c_2		c_i		c_n

numbers and its order is n if there are n rows and n columns. For a matrix of order $m \times n$, where $m > n$, we can form many determinants of order n, $n - 1$, $n - 2$, etc. The rank of a matrix is r if it contains at least one nonzero determinant of order r and if all determinants of order greater than r are zero.

From algebraic theory it can be shown (Dickson[2]) that Eqs. 7.3.2 have exactly $n - r$ linearly independent solutions where r is the rank of the dimensional matrix given in Table 7.3.1 and n is the number of variables; furthermore, it can be shown that any solution $(k_1, k_2, \ldots, k_i, \ldots, k_n)$ is a linear combination of these $(n - r)$ linearly independent solutions. The similarity between this statement and the definition of a complete set is obvious. Thus the following important theorem in dimensional analysis known as Buckingham's Pi theorem has been established: that the number of dimensionless products in a complete set is equal to the total number of variables minus the rank of their dimensional matrix. Another theorem fundamental to dimensional analysis and known as Buckingham's theorem states: if an equation is dimensionally homogeneous, it can be reduced to a relationship among a complete set of dimensionless products (Langhaar[4]).

By means of the above theorems we have shown that if there are n variables and the rank of the dimensional matrix is r, there will be p dimensionless products of powers of the variables or π-terms where p is given by

$$p = n - r \qquad (7.3.3)$$

Also a functional relation will exist among these π-terms that can be represented as

$$\pi_1 = f(\pi_2, \pi_3, \ldots, \pi_p) \qquad (7.3.4)$$

It should be noted that there are an infinite number of complete sets of π-terms because new complete sets can be formed from any given complete set. However, it is only necessary to find one complete set. Sometimes it is advantageous to form several complete sets and to use the one that has the simplest π-terms.

7.4   METHOD OF MAKING A DIMENSIONAL ANALYSIS

The first and most important step in performing a dimensional analysis of a given problem is to determine the pertinent variables that enter into the problem. This step can influence the final result in many ways. If an insufficient number of variables is included, the final result may be in error. If too many variables are included, the final result, although correct, may contain so many π-terms that the functional relation is too difficult to interpret or investigate.

The second step is to list the important variables and their dimensions, to construct the dimensional matrix, and determine the rank of the matrix. From the number of variables n and the rank of the dimensional matrix r, the number of π-terms p in a complete set is determined by Eq. 7.3.3.

The third step is to form a complete set of π-terms. These π-terms can be found by solving the set of Eqs. 7.3.2. However, with a little experience we can write down a complete set of π-terms by inspection much easier than by solving the equations of 7.3.2. All that is necessary is to make one variable appear exclusively in one π-term, another variable appear exclusively in another π-term, etc., until $n - r$ π-terms have been formed and all the variables have been used.

The fourth and final step is to write the functional relation between a complete set of π-terms. If there is only one π-term, this term is set equal to an arbitrary constant. If there are more than one π-terms, the functional relation is indicated by Eq. 7.3.4. Any one of the π-terms can be expressed as a function of the remaining terms in the set.

The following examples may clarify these procedures. Consider the problem of determining the dependence of the propagation velocity v in a given medium upon the elastic constants E and ν and the mass density of the medium ρ. The variables, their dimensions, and the dimensional matrix are the following:

Variable	Dimensions		Dimensional matrix			
			v	E	ν	ρ
v	LT^{-1}	M	0	1	0	1
E	$ML^{-1}T^{-2}$	L	1	-1	0	-3
ν	1	T	-1	-2	0	0
ρ	ML^{-3}					

The rank of the matrix is two, since all the third-order determinants that can be formed from this matrix have the value zero. Since there are four variables and the rank of the dimensional matrix is two, there are, by Eq. 7.3.3, two π-terms in a complete set. One set of π-terms is

$$\pi_1 = \frac{v^2\rho}{E}$$

$$\pi_2 = \nu$$

The functional relation between these π-terms is

$$\pi_1 = f(\pi_2)$$

or

$$\frac{v^2\rho}{E} = f(\nu) \tag{7.4.1}$$

or

$$v = \sqrt{\frac{E}{\rho}f(\nu)}$$

Thus from a dimensional analysis only, we have shown that the propagation velocity in a medium is proportional to the square root of E/ρ multiplied by some function of Poisson's ratio. From the theory of wave propagation in solid materials, it is shown that the longitudinal propagation velocity in an infinite medium is given by

$$v = \sqrt{\frac{E(1-\nu)}{\rho(1+\nu)(1-2\nu)}} \tag{7.4.2}$$

and that the longitudinal propagation velocity in a long thin rod is given by

$$v = \sqrt{E/P} \tag{7.4.3}$$

Thus it is evident that the method of dimensional analysis gives the correct form of the functional relationship that exists among the variables.

As a second example, consider the problem of determining the dependence of the propagation velocity v in a long thin bar upon the length l of the bar and the resonant frequency f of the bar. The dimensional matrix for this problem is

	v	f	l
L	1	0	1
T	-1	-1	0

The rank of the dimensional matrix is two and there are only three variables; therefore, there is only one π-term which can be written as

$$\pi_1 = \frac{v}{lf}$$

The function relation is

$$v/lf = c$$

where c is a constant

or
$$v = clf \tag{7.4.4}$$

From the theory of wave propagation, the longitudinal velocity in a long thin rod is related to the length and resonant frequency by the equation

$$v = 2lf \tag{7.4.5}$$

Additional examples of the use of dimensional analysis to obtain functional relationships among variables can be found in Chapter 13.

7.5   TRANSFORMING UNITS

The changing of the magnitude of a quantity in one system of units to the equivalent magnitude of the same quantity in another system is a simple process of transforming units. The need for transforming units arises in numerical calculations when the system of units used in measuring the various quantities is not one to be used in the final presentation of the data, or when an equation containing a dimensional constant in one system of units is to be used in another system of units.

Most engineering, physics, and mathematical handbooks have tables of conversion factors for transforming the standard physical quantities from one system of units to another. These tables are indispensable and should be a standard and ready reference for all technologists. The large number of different units used for each of the many different physical quantities makes it impractical to remember all of the different conversion factors.

The methods used in transforming units in equations are illustrated by the following examples:

1. Calculate the velocity of propagation v in ft/sec for a longitudinal wave in a bar having a Young's modulus E of 5.0×10^6 lb/in^2 and a specific gravity of 2.5. The equation relating Young's modulus and density to velocity is

$$v = \sqrt{E/\rho} \tag{7.5.1}$$

The quantities E and ρ must be expressed in the same system of units as the quantity v. Velocity is to be in ft/sec and E is given in lb/in.^{2} and therefore E must be changed to lb/ft^2. There are 144 in.^{2}/ft^2. Thus

$$E = 5.0 \times 10^6 \times 144 \,(\text{lb/in.}^2)(\text{in.}^2/\text{ft}^2) = 720 \times 10^6 \,\text{lb/ft}^2$$

Specific gravity is the ratio of the density of a substance to that of water. Also density must be expressed in force units. Therefore, since 1 ft^3 of water weighs 62.4 lb and the acceleration of gravity is 32.2 ft/sec,^{2} the density is given by

$$\rho = \frac{2.5 \times 62.4}{32.2} \frac{\text{lb}}{\text{ft}^3} \cdot \frac{\text{sec}^2}{\text{ft}} = 4.84 \frac{\text{lb-sec}^2}{\text{ft}^4}$$

Thus

$$v = \sqrt{\frac{E}{\rho}} = \sqrt{\frac{720 \times 10^6}{4.84} \cdot \frac{\text{lb}}{\text{ft}^2} \cdot \frac{\text{ft}^4}{\text{lb-sec}^2}} = \sqrt{149 \times 10^6 \frac{\text{ft}^2}{\text{sec}^2}} = 12{,}200 \frac{\text{ft}}{\text{sec}}$$

2. Calculate Young's modulus in lb/in.^{2} for a sample of rock in the shape of a right circular cylinder having a diameter of 2 cm, a length of 25 cm, and a weight of 200 gm if the longitudinal resonant frequency of the sample is 8000 cps. The equation relating longitudinal resonant frequency to longitudinal velocity is

$$v = 2lf \tag{7.5.2}$$

Combining Eqs. 7.5.2 and 7.4.3 gives

$$E = 4l^2 f^2 \rho \tag{7.5.3}$$

Density is mass per unit volume; therefore

$$\rho = \frac{m}{\pi r^2 l} = \frac{200}{(3.14)(1)^2(25)} \frac{\text{gm}}{\text{cm}^3} = 2.55 \frac{\text{gm}}{\text{cm}^3}$$

Evaluation of Eq. 7.5.3 gives

$$E = 4(25)^2(8000)^2 2.55 \frac{\text{cm}^2\text{-gm}}{\text{sec}^2\text{-cm}^3} = 4.08 \times 10^{11} \frac{\text{gm-cm}}{\text{sec}^2\text{-cm}^2}$$

Since 1 gm-cm/sec^2 is by definition 1 dyne, $E = 4.08 \times 10^{11}$ dynes/cm^2 The conversion factor from dynes/cm^2 to lb/in.^{2} is

$$1 \frac{\text{dyne}}{\text{cm}^2} = 1.45 \times 10^{-5} \frac{\text{lb}}{\text{in.}^2}$$

Therefore

$$E = 5.9 \times 10^6 \frac{\text{lb}}{\text{in.}^2}$$

7.6   MODEL SCALING

The general theory of model design can be developed by means of the functional relations derived from a dimensional analysis of the prototype. If the important variables have been defined and a dimensional analysis made for the prototype, there exists an equation of the form

$$\pi_1 = f(\pi_2, \pi_3, \ldots, \pi_i, \ldots, \pi_p) \tag{7.6.1}$$

As this equation is entirely general, it applies to any similar system; thus it also applies to the model. In order to distinguish the equation for the prototype from that for the model, a subscript m is used to refer to the model equation. Thus

$$\pi_{1m} = f_m(\pi_{2m}, \pi_{3m}, \ldots, \pi_{im}, \ldots, \pi_{pm}) \tag{7.6.2}$$

A prediction equation for π_1, based upon test results on a model where π_{1m} was determined experimentally, is obtained by dividing Eq. 7.6.1 by 7.6.2. Thus

$$\frac{\pi_1}{\pi_{1m}} = \frac{f(\pi_2, \pi_3, \ldots, \pi_i, \ldots, \pi_p)}{f(\pi_{2m}, \pi_{3m}, \ldots, \pi_{im}, \ldots, \pi_{pm})} \tag{7.6.3}$$

If the model were designed and tested such that

$$\pi_2 = \pi_{2m}$$
$$\pi_3 = \pi_{3m}$$
$$\cdots \cdots \cdots$$
$$\pi_i = \pi_{im} \tag{7.6.4}$$
$$\cdots \cdots \cdots$$
$$\pi_p = \pi_{pm}$$

it follows that

$$f(\pi_2, \pi_3, \ldots, \pi_i, \ldots, \pi_p) = f(\pi_{2m}, \pi_{3m}, \ldots, \pi_{im}, \ldots, \pi_{pm}) \tag{7.6.5}$$

Therefore from Eqs. 7.6.3 and 7.6.5

$$\pi_1 = \pi_{1m} \tag{7.6.6}$$

Equations 7.6.4 are the design equations. If all of these designs conditions are satisfied, then the model is true in the sense that it will give an accurate prediction concerning the behavior of the prototype, that is, provided that all the pertinent variables have been included in the original dimensional analysis. If all of the design conditions in Eqs. 7.6.4 are not satisfied, the behavior of the model may be different from that of the prototype.

Equation 7.6.6 is the prediction equation which is valid if the design equations have been satisfied. Specific examples of the use of dimensional analysis for obtaining prediction equations from model studies are given in Chapter 13.

REFERENCES

1. Bridgeman, P. W., *Dimensional Analysis*, Yale University Press, New Haven, Conn., 1931.
2. Dickson, Leonard Eugene, *Elementary Theory of Equations*, Chapter XI, p. 148, John Wiley and Sons, New York, 1914.
3. Durelli, A. J., E. A. Phillips, and C. H. Tsao, *Introduction to the Theoretical and Experimental Analysis of Stress and Strain*, Chapter 12, McGraw-Hill Book Co., New York, 1958.
4. Langhaar, Henry L., *Dimensional Analysis and Theory of Models*, Chapter 4, John Wiley and Sons, New York, 1951.
5. Murphy, Glenn, *Similude in Engineering*, The Ronald Press Co., New York, 1950.

CHAPTER 8

STATISTICAL ANALYSIS
AND EXPERIMENTAL DESIGN

8.1 INTRODUCTION

The role of statistics in experimental research has permeated all fields of science and engineering. Its usefulness as a research tool for designing experiments, collecting experimental data, measuring the magnitude of variation in the data, estimating population parameters with known precision, testing hypotheses, and studying relationships among two or more variables has been well established. All research workers in the field of rock mechanics should have a basic knowledge of statistical analysis and experimental design. The purpose of this chapter is to present some of the basic concepts of statistical analysis and experimental design which are needed for experimental work in rock mechanics. More advanced treatments of statistical analysis and experimental design can be found in any of the excellent texts given in the references at the end of the chapter.

Rock, by its very nature, is a nonhomogeneous material in that its properties and characteristics vary from place to place or from sample to sample even though all the samples come from the same geologic formation. In studying any property or characteristic of rock we are interested not only in the average value of this quantity, but also the variation that can be expected to occur. Besides the variation that occurs because of the nature of the rock, there is the variation produced by the experimental error inherent in all measurement techniques. Errors of measurement are usually of two types, systematic errors and random errors. Systematic errors are of the same sign for each observation, whereas random errors are normally distributed about an average value of zero. Systematic errors cannot be reduced by increasing the number of observations, whereas random errors can be reduced without limit by increasing the number of observations. However, the reduction is proportional to the square root

202

of the number of readings so that there is a practical limit to the reduction that can be achieved.

For efficient experimentation, the variation in the data produced by equipment or experimental errors should be about one order of magnitude smaller than the variation in the data resulting from the property under study. In experimental rock mechanics, it is important to determine the type of error introduced by the measurement technique and to determine if the variation in the data is mainly a result of measurement technique or if it is a property of the rock or the phenomenon under study. If statistical methods are used to design experiments, to collect data, and to analyze data, it is then possible to distinguish between measurement error and variation produced by the property under study.

8.2   DEFINITIONS, TERMINOLOGY, AND NOTATION

The science of statistics uses a number of words that have specific definitions or meanings. It is appropriate therefore to define these words before developing the mathematical concepts of statistics.

A *population* is any finite or infinite set of individuals, objects, or processes that have some common observable and measurable characteristic. A *sample* is a subset of a population. In experimental work, we observe and measure a characteristic of a sample and from these data estimate the characteristic of the population. For example, consider a given rock mass from which 100 specimens have been obtained. These specimens could be measured, weighed, and the density of each computed. These data represent the sample, but they can also be used to estimate the population, that is, the average density of the rock mass from which the sample was taken.

A *parameter* is a constant describing a population, whereas a *statistic* is a quantity describing a sample. To distinguish between these two quantities it is customary to use Greek letters for population parameters and Latin letters for sample statistics.

The densities of the individual specimens of rock described above will not all have the same value. A measure of the variation in these data is the *range* which is the difference between the highest and lowest reading. A better measure of the variation in the data can be obtained from a histogram. A *histogram* is a bar graph of frequency of occurrence. It is made by dividing the range into a number of equal intervals and plotting the number of observations falling in each interval as a function of the midpoints of the intervals. Such a plot is illustrated in Fig. 8.2.1. Histograms or frequency distributions give a good picture of the spread in the data and show the tendency for the data to group about some *central value*.

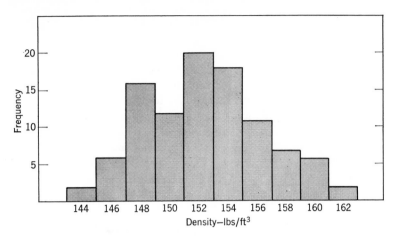

Fig. 8.2.1. Sample histogram representing density of rock.

Many populations are symmetrically distributed about their central value, other populations may be *skewed* or asymmetrically distributed about their central value. A population is said to be normally distributed if its frequency distribution is given by the equation

$$\frac{dN}{N} = (2\pi\sigma^2)^{-\frac{1}{2}}e^{-(\frac{1}{2})[(x-\mu)/\sigma]^2}\,dX \qquad (8.2.1)$$

where dN/N is the fraction of the population in the range X to $X + dX$; μ is the population *mean*, σ is the population *standard deviation*, σ^2 is the *variance*, and e is the base of Napierian logarithms. Mathematical definitions of the quantities μ and σ are given in Sections 8.3 and 8.4, respectively. The quantity $X - \mu$ is the error, or *deviation*. The quantity Z defined as

$$Z = \frac{X - \mu}{\sigma} \qquad (8.2.2)$$

is the deviation measured in units of the standard deviation. Substitution of Eq. 8.2.2 into 8.2.1 gives the *standard normal distribution* as

$$\frac{dN}{N} = (2\pi)^{-\frac{1}{2}}e^{-(\frac{1}{2})Z^2}\,dZ \qquad (8.2.3)$$

The standard normal distribution has a mean of zero and a standard deviation of unity. A graph of Eq. 8.2.3 is shown in Fig. 8.2.2. Further discussion of the standard normal distribution will be given in Section 8.5.

Replication is the repeating of an experiment or measurement under identical conditions. Replication is necessary in experimental design to

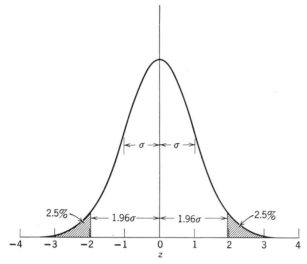

Fig. 8.2.2. Standard normal distribution curve.

provide a reliable measure of experimental error. When performing experiments it is not always possible to control all variables. For example, the order of performing tests is a variable that may affect the final results. Uncontrolled variables can produce bias in experimental data unless their effects are randomized. *Randomization* is the process of arranging experimental conditions so that every possible order has the same likelihood of occurrence. If samples are selected at random from a population, then each sample in the population has an equal chance of being selected.

A statistical hypothesis is a statement about the parameters of a population. A test of a statistical hypothesis is a criterion for determining from the sample results if the hypothesis is to be accepted or rejected. Usually the hypothesis to be tested is the *null hypothesis*, H_0, which is an assumption that there is no difference between two parameters. The alternative hypothesis is that there is a significant difference between the two parameters. Since decisions concerning the acceptance or rejection of null hypotheses are made on the basis of samples, there is always the possibility of making incorrect decisions. The two types of errors we can make when testing hypotheses are illustrated below.

	Accept H_0	Reject H_0
H_0 is true	No error	Type I error
H_0 is not true	Type II error	No error

Thus a type I error is made if a true hypothesis is rejected and a type II error is made if a false hypothesis is accepted. As will be shown later, the probability of making a type I error can be made as small as desired; however, it is usually set at either 1, 5, or 10%.

The capital letter X will be used to designate a measurement of a sample. To distinguish the measurement on the first sample from that on the second and so forth, the subscripts 1, 2, 3, etc., will be used. Thus X_i is the value of quantity X for the ith reading or ith sample. Double or triple subscripts are used to distinguish different groups of samples or different treatments of the samples. Thus X_{ijk} is the value of the quantity X for the kth sample in the jth group for the ith treatment.

The Greek capital letter Σ is used to show summation. Thus

$$\sum_{i=1}^{4} X_i = X_1 + X_2 + X_3 + X_4$$

or

$$\sum_{i=1}^{n} X_i = X_1 + X_2 + \cdots + X_n$$

If each value of X_i is multiplied by some constant a and a summation performed, the result is

$$\sum_{i=1}^{n} a X_i = a X_1 + a X_2 + \cdots + a X_n = a \sum_{i=1}^{n} X_i$$

The summation of any constant n times is simply n times the constant, thus

$$\sum_{i=1}^{n} a = na$$

If a constant is added to each X_i and a summation performed, the result is

$$\sum_{i=1}^{n} (X_i + a) = \sum_{i=1}^{n} X_i + \sum_{i=1}^{n} a = \sum_{i=1}^{n} X_i + na$$

Because all summations will start with the first reading, the use of $i = 1$ below the summation sign is not required and will not be used hereafter.

8.3 MEASURES OF CENTRAL TENDENCY

The most commonly used measure of central tendency is the arithmetic mean or average defined as

$$\bar{X} = \frac{1}{n} \sum_{i}^{n} X_i \qquad (8.3.1)$$

where $\bar{X} = $ arithmetic mean of the samples
$\quad\ X_i = i$th value of the quantity measured
$\quad\ n = $ total number of samples

The quantity $\bar{X}$ is an unbiased estimate of the true population mean μ which is defined by

$$\mu = \frac{1}{N} \sum_{i}^{N} X_i \tag{8.3.2}$$

where N is the total number of items in the population. The sample mean $\bar{X}$ approaches the population mean μ as the number of samples n increases. Note that $\bar{X}$ is derived entirely from the sample and is the average of the sample; however, as the sample is a subset of the population, $\bar{X}$ estimates the population μ.

The simple definition for $\bar{X}$ given above does not describe all the properties that $\bar{X}$ possesses. For example, consider the following problem: Given a set of n numbers X_i ($i = 1, 2, 3, \cdots n$), find a number $\bar{X}$ such that the sum of the squared deviations of X_i from $\bar{X}$ is a minimum. The ith deviation of X_i from $\bar{X}$ is $(X_i - \bar{X})$, the ith squared deviation is $(X_i - \bar{X})^2$ and their sum Q is

$$Q = \sum_{i}^{n} (X_i - \bar{X})^2 \tag{8.3.3}$$

For Q to be a minimum, the rate of change of Q with respect to $\bar{X}$ must be zero and the second derivative of Q with respect to $\bar{X}$ must be positive. Thus

$$\frac{dQ}{d\bar{X}} = \frac{d}{d\bar{X}} \sum_{i}^{n} (X_i - \bar{X})^2 = \sum_{i}^{n} - 2(X_i - \bar{X}) = -2 \sum_{i}^{n} (X_i - \bar{X}) \tag{8.3.4}$$

and

$$\frac{d^2Q}{d\bar{X}^2} = -2 \frac{d}{d\bar{X}} \sum_{i}^{n} (X_i - \bar{X}) = -2 \sum_{i}^{n} - 1 = 2n \tag{8.3.5}$$

Equation 8.3.5 shows that the second derivative is positive; therefore the value of $\bar{X}$ that makes Eq. 8.3.4 identically zero is the required number. Setting Eq. 8.3.4 equal to zero and solving for $\bar{X}$ gives

$$\sum_{i}^{n} (X_i - \bar{X}) = \sum_{i}^{n} X_i - n\bar{X} = 0 \tag{8.3.6}$$

or

$$\bar{X} = \frac{1}{n} \sum_{i}^{n} X_i \tag{8.3.7}$$

From Eqs. 8.3.3, 8.3.6, and 8.3.7, it is evident that the arithmetic mean is that number which makes the sum of the deviations zero and the sum of the squared deviations a minimum.

If the sample size is large and a frequency distribution table or graph has been constructed, the arithmetic mean can be obtained by the following formula

$$\bar{X} = \frac{\sum\limits_{i}^{n} f_i X_i}{\sum\limits_{i}^{n} f_i} \tag{8.3.8}$$

where f_i is the number of readings in the range whose midpoint is X_i and $\sum_i^n f_i$ is the total number of readings.

Assume that n groups of samples have been obtained and that the group means are $\bar{X}_{i.}$. The grand mean of all the readings X_{ij} must be the same as the average of the group means. For the above to be true, it is necessary to weight each mean $\bar{X}_{i.}$ by the number of readings n_i in each group. Thus

$$\bar{\bar{X}} = \frac{\sum\limits_{i}^{n} n_i \bar{X}_{i.}}{\sum\limits_{i}^{n} n_i} \tag{8.3.9}$$

That Eq. 8.3.9 is the same as the grand mean of all the readings can be shown by replacing $\bar{X}_{i.}$ by its equivalent

$$\bar{X}_{i.} = \frac{\sum\limits_{j}^{n_i} X_{ij}}{n_i}$$

Thus

$$\bar{\bar{X}} = \frac{\sum\limits_{i}^{n} \sum\limits_{j}^{n_i} X_{ij}}{\sum\limits_{i}^{n} n_i} \tag{8.3.10}$$

which is the mean of all the readings.

For large sample sizes, involving values of X_i that have several significant figures, it is often convenient to code the data so that they group about a conveniently selected origin. The data may be coded by letting

$$u_i = \frac{X_i - a}{c} \tag{8.3.11}$$

where a and c are constants chosen to give as simple an expression as possible for u_1. The arithmetic mean of the coded data is

$$\bar{u} = \frac{1}{n} \sum\limits_{i}^{n} u_i = \frac{1}{n} \sum\limits_{i=1}^{n} \frac{X_i - a}{c} = \frac{1}{nc} \sum\limits_{i}^{n} X_i - \frac{a}{c} \tag{8.3.12}$$

Equations 8.3.11 and 8.3.12 combine to give

$$\bar{X} = c\bar{u} + a \tag{8.3.13}$$

Thus the true mean can be obtained directly from the mean of the coded data.

8.4 MEASURES OF VARIABILITY

The standard deviation and variance are the most commonly used measures of variability. These quantities are simply related as the standard deviation is the square root of the variance. For a population of mean μ and N entities, the variance is defined as

$$\sigma^2 = \frac{\sum\limits_{i}^{N}(X_i - \mu)^2}{N} \tag{8.4.1}$$

For a sample of n items whose mean is $\bar{X}$, an unbiased estimator of the population variance is the sample variance S^2 given by

$$S^2 = \frac{\sum\limits_{i}^{n}(X_i - \bar{X})^2}{n-1} \tag{8.4.2}$$

The quantity $n-1$, in Eq. 8.4.2, is the number of degrees of freedom that are available for calculating the variance. For n readings there are n degrees of freedom for calculating $\bar{X}$. Having calculated $\bar{X}$ there are only $n-1$ degrees of freedom left for calculating the variance. The correction for degrees of freedom in Eq. 8.4.2 is necessary to assure that S is an unbiased estimator of σ. The quantity $\sum_i^n(X_i - \mu)^2$ is in general greater than the quantity $\sum_i^n(X_i - \bar{X})^2$ as shown below.

$$\sum_i^n(X_i - \mu)^2 = \sum_i^n[(X_i - \bar{X}) + (\bar{X} - \mu)]^2$$

$$= \sum_i^n[(X_i - \bar{X})^2 + 2(X_i - \bar{X})(\bar{X} - \mu) + (\bar{X} - \mu)^2]$$

$$= \sum_i^n(X_i - \bar{X})^2 + 2(\bar{X} - \mu)\sum_i^n(X_i - \bar{X}) + n(\bar{X} - \mu)^2$$

$$= \sum_i^n(X_i - \bar{X})^2 + 2(\bar{X} - \mu)(n\bar{X} - n\bar{X}) + n(\bar{X} - \mu)^2$$

$$\sum_i^n(X_i - \mu)^2 = \sum_i^n(X_i - \bar{X})^2 + n(\bar{X} - \mu)^2 \tag{8.4.3}$$

Thus

$$\sum_i^n(X_i - \mu)^2 > \sum_i^n(X_i - \bar{X})^2$$

Note that the degrees of freedom for $\sum_i^n (X_i - \mu)^2$ in Eqs. 8.4.3 is n, the degrees of freedom for $\sum_i^n (X_i - \bar{X})^2$ is $n - 1$, and the degrees of freedom for $n(\bar{X} - \mu)^2$ is 1. Thus the total number of degrees of freedom on each side of the equation are equal. The quantity $\sum_i^n (X_i - \bar{X})^2$ is referred to as the sum of the squares and its computation is simplified by noting that

$$
\begin{aligned}
\sum_i^n (X_i - \bar{X})^2 &= \sum_i^n (X_i^2 - 2X_i\bar{X} + \bar{X}^2) \\
&= \sum_i^n X_i^2 - 2\bar{X}\sum_i^n X_i + n\bar{X}^2 \\
&= \sum_i^n X_i^2 - \frac{2\sum_i^n X_i}{n} + \frac{\left(\sum_i^n X_i\right)^2}{n}
\end{aligned}
\tag{8.4.4}
$$

$$
\sum_i^n (X_i - \bar{X})^2 = \sum_i^n X_i^2 - \frac{\left(\sum_i^n X_i\right)^2}{n}
$$

Equations 8.4.4 make possible the calculation of the sum of the squares from the quantities $\sum_i^n X_i^2$ and $\sum_i^n X_i$, both of which can be obtained easily with the aid of modern desk calculators.

The sum of the squares for the original data can be obtained easily from coded data u_i where

$$
u_i = \frac{X_i - a}{c}
\tag{8.4.5}
$$

The sum of the squares of the coded data is

$$
\begin{aligned}
\sum_i^n (u_i - \bar{u})^2 &= \frac{n\sum_i^n u_i^2 - \left(\sum_i^n u_i\right)^2}{n} \\
&= \frac{n\sum_i^n \left(\frac{X_i - a}{c}\right)^2 - \left[\sum_i^n \left(\frac{X_i - a}{c}\right)\right]^2}{n} \\
&= \frac{1}{c^2}\left(\sum_i^n X_i^2 - 2a\sum_i^n X_i + na^2\right) \\
&\quad - \frac{1}{c^2 n}\left[\left(\sum_i^n X_i\right)^2 - 2na\sum_i^n X_i + n^2 a^2\right]
\end{aligned}
\tag{8.4.6}
$$

$$
c^2\sum_i^n (u_i - \bar{u})^2 = \frac{n\sum_i^n X_i^2 - \left(\sum_i^n X_i\right)^2}{n}
$$

From Eqs. 8.4.6 it is evident that adding or subtracting a constant from each reading does not change the variance or the sum of the squares.

Assume that n samples have been selected at random from a normal population, that the number of readings for each sample is n_i, that the mean of each sample is $\bar{X}_i$. and that the variance of each sample is S_i^2. The pooled variance S_p^2 is defined as the weighted mean variance thus

$$S_p^2 = \frac{\sum_i^n (n_i - 1)S_i^2}{\sum_i^n (n_i - 1)} \tag{8.4.7}$$

By means of Eq. 8.4.2, Eq. 8.4.7 can be written as

$$S_p^2 = \frac{\sum_i^n \left[\sum_j^{n_i} (X_{ij} - \bar{X}_{i.})^2 \right]}{\sum_i^n n_i - n} \tag{8.4.8}$$

8.5 DISTRIBUTION FUNCTIONS

The normal distribution and the standard normal distribution were defined in Section 8.2. Many populations have a normal or nearly normal distribution. Random experimental errors tend to follow a normal distribution. To study the frequency distribution of any population, it is necessary to have a large number of samples. In general, as the number of samples is increased, the width of the interval in the bar graph or histogram is decreased. Thus as n approaches infinity, the interval width approaches zero and the histogram approaches a frequency distribution. In practice the number of samples or measurements which are obtained are insufficient to establish accurately the true frequency distribution for the population. For these reasons many statistical procedures are based upon the assumption that the population has a normal distribution of mean μ and standard deviation σ.

The equation for the standard normal distribution can be written as

$$Y = (2\pi)^{-1/2} e^{-(1/2)Z^2} \tag{8.5.1}$$

where Z is the deviation of X from μ measured in units of the standard deviation σ. That is,

$$Z = \frac{X - \mu}{\sigma} \tag{8.5.2}$$

The area under the standard normal distribution curve is unity and is shown in Fig. 8.2.2. As the curve is symmetric about $Z = 0$, the area to the left of some value of $-Z$ is equal to the area to the right of the same value of $+Z$. Let the symbol $-Z_{0.5\alpha}$ stand for that value of Z where the area under the curve to left of $-Z_{0.5\alpha}$ is 0.5α and let the symbol $Z_{1-0.5\alpha}$ stand for that value of Z where the area to the left of $Z_{1-0.5\alpha}$ is $1 - 0.5\alpha$. Therefore the area between $-Z_{0.5\alpha}$ and $Z_{1-0.5\alpha}$ is $1 - \alpha$. As the total area is unity, the probability that a sample X will give a value of Z that falls inside

Table 8.5.1. Standard Normal Distribution

Z $\pm \dfrac{X - \mu}{\sigma}$	Area outside $\pm Z$	Y
± 0	1.0000	0.3999
± 0.126	0.9000	0.3958
± 0.253	0.8000	0.3863
± 0.385	0.7000	0.3704
± 0.524	0.6000	0.3477
± 0.674	0.5000	0.3178
± 0.842	0.4000	0.2800
± 1.036	0.3000	0.2331
± 1.282	0.2000	0.1755
± 1.645	0.1000	0.1031
± 1.960	0.0500	0.0584
± 2.576	0.0100	0.0145
± 3.291	0.0010	0.0018
± 3.891	0.0001	0.0002

of the range $-Z_{0.5\alpha}$ to $+Z_{1-0.5\alpha}$ is $1 - \alpha$. This statement can be represented by the following mathematical equation:

$$P(-Z_{0.5\alpha} < Z < Z_{1-0.5\alpha}) = 1 - \alpha \qquad (8.5.3)$$

For Eq. 8.5.3 to be useful, it is necessary to have available the areas under the standard normal distribution curve for various values of the range $-Z_{0.5\alpha}$ to $+Z_{1-0.5\alpha}$. These areas can be calculated from Eq. 8.5.1; tables of these areas are given in most of the references at the end of this chapter. A short summary of the more useful values of Z and areas are given in Table 8.5.1.

Suppose a large number of samples, with each sample containing n readings, are obtained from a population of known mean μ and known standard deviation σ. Let $\bar{X}$ and S represent the mean and standard deviation of each sample. A critical study of the sample means will show that they are normally distributed about μ with a new standard deviation given by

$$\sigma_{\bar{X}} = \frac{\sigma}{\sqrt{n}} \qquad (8.5.4)$$

The quantity $\sigma_{\bar{X}}$ is the population standard deviation for the means determined from n readings. From Eq. 8.5.4, it is evident that the standard deviation of the means can be made as small as desired by increasing the sample size n.

The statistic

$$Z = \frac{\bar{X} - \mu}{\sigma_{\bar{X}}} \qquad (8.5.5)$$

has a standard normal distribution and Table 8.5.1 can be used to compute the probability of finding $(\mu - \bar{X})/\sigma_{\bar{X}}$ in the range $-Z_{0.5\alpha}$ to $+Z_{1-0.5\alpha}$. If $(\mu - \bar{X})/\sigma_{\bar{X}}$ is to lie in the range $-Z_{0.5\alpha}$ to $+Z_{1-0.5\alpha}$, then the following inequality is true

$$-Z_{0.5\alpha} < \frac{\mu - \bar{X}}{\sigma_{\bar{X}}} < Z_{1-0.5\alpha}$$

which can also be written as

$$\bar{X} - \sigma_{\bar{X}} Z_{0.5\alpha} < \mu < \bar{X} + \sigma_{\bar{X}} Z_{1-0.5\alpha}$$

Using Eq. 8.5.4 and writing the inequality as a probability statement gives

$$P\left(\bar{X} - \frac{\sigma}{\sqrt{n}} Z_{0.5\alpha} < \mu < \bar{X} + \frac{\sigma}{\sqrt{n}} Z_{1-0.5\alpha} \right) = 1 - \alpha \qquad (8.5.6)$$

Equation 8.5.6 states that the probability, before the sample is selected, that the random range $\bar{X} - (\sigma/\sqrt{n}) Z_{0.5\alpha}$ to $\bar{X} + (\sigma/\sqrt{n}) Z_{1-0.5\alpha}$ will include the mean of the population is $1 - \alpha$. Thus if α is 0.05 and a random sample of n readings are obtained from a population of mean μ and standard deviation σ, the probability that the random range $\bar{X} - (\sigma/\sqrt{n}) Z_{0.025}$ to $\bar{X} + (\sigma/\sqrt{n}) Z_{0.975}$ will include the population mean is 0.95. That is 0.95 of the time this range will include the population μ. The interval between $\bar{X} - (\sigma/\sqrt{n}) Z_{0.5\alpha}$ and $\bar{X} + (\sigma/\sqrt{n}) Z_{1-0.5\alpha}$ is known as the $(1 - \alpha)100\%$ confidence interval. The value of the end points of this interval are known as the $(1 - \alpha)100\%$ confidence limits.

In the preceding example it was assumed that the population mean and standard deviation were known. In general, neither μ nor σ are known; therefore, we must work only with the quantities $\bar{X}$, S, and n.

The statistic t defined as

$$t = \frac{\bar{X} - \mu}{S/\sqrt{n}} \qquad (8.5.7)$$

has a distribution known as Students' t distribution which is given by

$$Y = C\left(1 + \frac{t^2}{v}\right)^{-(\frac{1}{2})(v+1)} \qquad (8.5.8)$$

where C is a constant dependent upon the number of degrees of freedom v; the value of C is so chosen as to make the total area under the curve equal to unity. The t distribution is symmetric about the origin, but it is non-normal and is strongly dependent upon the number of degrees of freedom used in the calculation of the variance. A table of values of the areas under the t distribution for various degrees of freedom and various α is given in Table 8.5.2. Let $-t_{0.5\alpha(n-1)}$ and $t_{(1-0.5\alpha)(n-1)}$ be those values of t that include $(1 - \alpha)100\%$ of the area under the t distribution curve. If $\bar{X}$ is the mean and S the standard deviation for a random sample of n readings from a normal population of mean μ, then the probability that the statistic $t = [(\mu - \bar{X})/S/\sqrt{n}]$ is in the range $-t_{0.5\alpha(n-1)}$ to $t_{(1-0.5\alpha)(n-1)}$ is $1 - \alpha$. This statement expressed in equation form is

$$P[-t_{0.5\alpha(n-1)} < t < t_{(1-0.5\alpha)(n-1)}] = 1 - \alpha$$

By use of Eq. 8.5.7 and some algebra, this probability statement can be rewritten as

$$P\left[\bar{X} - \frac{S}{\sqrt{n}} t_{0.5\alpha(n-1)} < \mu < \bar{X} + \frac{S}{\sqrt{n}} t_{(1-0.5\alpha)(n-1)}\right] = 1 - \alpha \quad (8.5.9)$$

Equation 8.5.9 states that the probability that the random interval from

$$\bar{X} - \frac{S}{\sqrt{n}} t_{0.5\alpha(n-1)} \qquad \text{to} \qquad \bar{X} + \frac{S}{\sqrt{n}} t_{(1-0.5\alpha)(n-1)}$$

will include the population mean μ is $1 - \alpha$. That is, $(1 - \alpha)100\%$ of the time the random interval

$$\bar{X} - \frac{S}{\sqrt{n}} t_{0.5(n-1)} \qquad \text{to} \qquad \bar{X} + \frac{S}{\sqrt{n}} t_{(1-0.5\alpha)(n-1)}$$

will include the population μ. For example, if $\alpha = 0.5$ and $n - 1 = 9$, the values of $t_{0.5\alpha}$ and $t_{1-0.5\alpha}$ are ± 2.262 (from Table 8.5.2) and the 0.95 confidence limits are $\bar{X} \pm 2.262(S/\sqrt{n})$.

Let $\bar{X}_1$ and $\bar{X}_2$ be the means for two random samples of size n_1 and n_2, and S_1^2 and S_2^2 be the variance for each sample. The pooled variance S_p^2 as given by Eq. 8.4.7 is

$$S_p^2 = \frac{(n_1 - 1)S_1^2 + (n_2 - 1)S_2^2}{n_1 + n_2 - 2} \qquad (8.5.10)$$

If both samples have been selected at random from the same normal population, then the statistic

$$t = \frac{\bar{X}_1 - \bar{X}_2}{S_p\sqrt{1/n_1 + 1/n_2}}$$ (8.5.11)

Table 8.5.2.* Critical Values of the Statistic t

$t_{1-0.5\alpha}$ df	$t_{0.90}$	$t_{0.95}$	$t_{0.975}$	$t_{0.99}$
1	3.078	6.314	12.706	31.821
2	1.886	2.920	4.303	6.965
3	1.638	2.353	3.182	4.541
4	1.533	2.131	2.776	3.747
5	1.476	2.015	2.571	3.365
6	1.440	1.943	2.447	3.143
7	1.415	1.895	2.365	2.998
8	1.397	1.860	2.306	2.896
9	1.383	1.833	2.262	2.821
10	1.372	1.812	2.228	2.764
15	1.341	1.753	2.131	2.602
20	1.325	1.725	2.086	2.528
25	1.316	1.708	2.060	2.485
30	1.310	1.697	2.042	2.457
40	1.303	1.684	2.021	2.423
∞	1.282	1.645	1.960	2.326
$-t_{0.5\alpha}$	$-t_{0.10}$	$-t_{0.05}$	$-t_{0.025}$	$-t_{0.01}$

* Table values are to be prefixed with a negative sign when table is read from bottom. This table of t values was taken from tables of Fisher and Yates: *Statistical Tables for Biological, Agricultural, and Medical Research*, published by Oliver and Boyd Ltd., Edinburgh, and by permission of the authors and publishers.

has a t distribution. Therefore Table 8.5.2 can be used to establish confidence limits where the degrees of freedom to be used are $n_1 + n_2 - 2$. The appropriate probability equation is

$$P[-t_{0.5\alpha(n_1+n_2-2)} < t < t_{(1-0.5\alpha)(n_1+n_2-2)}] = 1 - \alpha$$ (8.5.12)

Assume that two random samples, of n_1 and n_2 readings each, are obtained from the same normally distributed population which has a variance of σ^2 and a mean of μ. For each sample a variance S_1^2 and S_2^2

can be calculated by means of Eq. 8.4.2. Because of the spread in the individual readings, the two variances will differ even though both are an independent estimate of the population variance σ^2. The statistic F defined as (Fisher[6])

$$F = \frac{S_1^{\;2}}{S_2^{\;2}} \tag{8.5.13}$$

has a distribution given by

$$f(F) = CF^{(\frac{1}{2})(v_1-2)}(v_2 + v_1 F)^{-(\frac{1}{2})(v_1+v_2)} \tag{8.5.14}$$

where C is a constant and where $v_1 = n_1 - 1$ and $v_2 = n_2 - 1$ are the respective degrees of freedom that are associated with the variance in the numerator and the denominator. The function $f(F)$ is asymmetric and is always positive for positive values of v_1 and v_2. The value of the constant C is so chosen that the area under the curve represented by Eq. 8.5.14 is always unity. Thus two critical values of F can be calculated from Eq. 8.5.14 for each pair of degrees of freedom such that the probability of the experimentally determined value of F (Eq. 8.5.13) lying between these two critical values of F is $1 - \alpha$. Table 8.5.3 gives some critical values of F for various degrees of freedom in the numerator and denominator and for two values of α. The appropriate probability statement to use with Table 8.5.3 is

$$P[F_{(0.5\alpha)(n_1-1)(n_2-1)} < F < F_{(1-0.5\alpha)(n_1-1)(n_2-1)}] = 1 - \alpha \tag{8.5.15}$$

where

$$F_{0.5\alpha(n_1-1)(n_2-1)} = \frac{1}{F_{(1-0.5\alpha)(n_2-1)(n_1-1)}} \tag{8.5.16}$$

Equation 8.5.16 is necessary as values of $F_{0.5}$ are not given in the standard F tables.

If the larger value of $S_1^{\;2}$ and $S_2^{\;2}$ is always put in the numerator, then the value of F is always greater than unity and the need for the additional calculation by means of Eq. 8.5.16 is eliminated.

In using the statistical tables given in various books, care should be exercised to determine if the same system of notation is used.

8.6 TESTING HYPOTHESES

The null hypothesis technique is the procedure most often used to perform statistical tests on data. This technique consists mainly of assuming that the parameters of two or more populations are equal and then determining if the statistics, calculated from random samples of the populations, have a small probability of having been derived from the same or equal populations. If the probability is smaller than some previously selected value, the hypothesis or assumption is rejected and the alternate hypothesis that the parameters of the populations tested are not equal is accepted. Thus, in general, statistical analysis is used to show that two quantities are not equal or are not from the same population. This type of statistical proof can be obtained with any degree of confidence that is desired. The reverse process of showing that two quantities are equal or are from the same population cannot be achieved with as high a degree of confidence.

The steps in performing a test of a statistical hypothesis are:

1. State the null hypothesis. For example, $\mu_1 = \mu_2$ or $\sigma_1{}^2 = \sigma_2{}^2$.
2. Select the value of α which is the size of the type I error. Typical values of α are 1, 5, or 10%.
3. Select the proper statistic and calculate its value from the experimental data.
4. Determine from tables those values of the statistic which will cause rejection of the null hypothesis, that is, determine the $(1 - \alpha)$ confidence interval from table values.
5. Accept or reject the hypothesis depending on whether the calculated value of the statistic is inside or outside the $(1 - \alpha)$ confidence interval.

The preceding steps in performing a test of a statistical hypothesis will be outlined for several simple problems involving comparison of means and comparison of variances. When stating a hypothesis, we must carefully determine if the hypothesis can be rejected by a variation in one or two directions. If the hypothesis can be rejected by variation in two directions a two tail test is involved and the values of the confidence limits are determined at $\alpha/2$. If the hypothesis can be rejected by a variation in only one direction, a one tail test is involved and the value of the confidence limit is determined at α.

Problem I. Determine if the population mean μ, estimated by n random samples having a mean of $\overline{X}$, is significantly different from some specified number μ_0, assuming that σ^2 is known.

1. $H_0: \mu = \mu_0$, which involves a two tail test.
2. Select α.
3. Calculate from the experimental data the statistic $Z = (\overline{X} - \mu_0)/(\sigma/\sqrt{n})$.
4. Use Table 8.5.1 to find the confidence limits $Z_{0.5\alpha}$ and $Z_{(1-0.5\alpha)}$.
5. Reject hypothesis if Z falls outside, or accept hypothesis if Z falls inside the confidence interval determined in step 4.

Problem II. Determine if the population mean μ, estimated by n random samples having a mean $\bar{X}$ and a variance S^2 is significantly different from some specified number μ_0 when σ^2 is not known.

1. $H_0: \mu = \mu_0$ which involves a two tail test.
2. Select α.
3. Calculate from the experimental data the statistic $t = (\bar{X} - \mu_0)/(S/\sqrt{n})$.
4. Use Table 8.5.2 to find the confidence limits $t_{0.5\alpha(n-1)}$ and $t_{(1-0.5\alpha)(n-1)}$.
5. Reject hypothesis if t falls outside, or accept if t falls inside the confidence interval determined in step 4.

Problem III. Determine if the population mean μ, estimated from n random samples having a mean $\bar{X}$ and a variance of S^2, is greater than some specified number μ_0 when σ^2 is not known.

1. $H_0: \mu \le \mu_0$,which is a one tail test.
2. Select α.
3. Calculate from the experimental data the statistic $t = (\bar{X} - \mu_0)/(S/\sqrt{n})$.
4. Use Table 8.5.2 to find the confidence limit $t_{(1-\alpha)(n-1)}$.
5. Reject hypothesis if t is greater than, or accept hypothesis if t is less than, the confidence limit $t_{(1-\alpha)(n-1)}$ determined in step 4.

Problem IV. Determine if two population means μ_1, and μ_2 are significantly different where μ_1 is estimated by n_1 random samples having a mean of $\bar{X}_1$ and a variance of S_1^2 and μ_2 is estimated from n_2 random samples having a mean of $\bar{X}_2$ and a variance of S_2^2.

1. $H_0: \mu_1 = \mu_2$, which is a two tail test.
2. Select α.
3. Calculate the statistic t from

$$t = \frac{\bar{X}_1 - \bar{X}_2}{S_p\sqrt{1/n_1 + 1/n_2}}$$

where

$$S_p^{\,2} = \frac{(n_1 - 1)S_1^{\,2} + (n_2 - 1)S_2^{\,2}}{n_1 + n_2 - 2}$$

4. Use Table 8.5.2 to find the confidence limits $t_{0.5\alpha(n_1+n_2-2)}$ and $t_{(1-0.5\alpha)(n_1+n_2-2)}$.
5. Reject the hypothesis if t is outside, or accept the hypothesis if t is inside the confidence interval determined in step 4.

Problem V.   Determine if the population mean μ_1 is significantly smaller than the population mean μ_2 where μ_1 is estimated from n_1 random samples having a mean of $\bar{X}_1$ and a variance of S_1^2; and μ_2 is estimated from n_2 random samples having a mean of $\bar{X}_2$ and a variance of S_2^2.

1. $H_0 : \mu_1 \geq \mu_2$, which is a one tail test.
2. Select α.
3. Calculate the statistic t from

$$t = \frac{\bar{X}_1 - \bar{X}_2}{S_p\sqrt{1/n_1 + 1/n_2}}$$

where

$$S_p^{\,2} = \frac{(n_1 - 1)S_1^{\,2} + (n_2 - 1)S_2^{\,2}}{n_1 + n_2 - 2}$$

4. Use Table 8.5.2 to find the confidence limit $t_{\alpha(n_1+n_2-2)}$.
5. Reject the hypothesis if t is less than or accept the hypothesis if t is greater than $t_{\alpha(n_1+n_2-2)}$.

Problem VI.   Determine if two population variances σ_1^2 and σ_2^2 are significantly different where σ_1^2 is estimated by n_1 random samples having a variance of S_1^2, and σ_2^2 is estimated by n_2 random samples having a variance of S_2^2.

1. $H_0 : \sigma_1^2 = \sigma_2^2$, which is a two tail test.
2. Select α.
3. Calculate the statistic F from the experimental data by

$$F = \frac{S_1^{\,2}}{S_2^{\,2}}$$

4. From Table 8.5.3, find the values

$$F_{0.5\alpha(n_1-1)(n_2-1)} \quad \text{and} \quad F_{(1-0.5\alpha)(n_1-1)(n_2-1)}$$

which are the $1 - \alpha$ confidence limits.

5. Reject hypothesis if F falls outside or accept hypothesis if F falls inside the $1 - \alpha$ confidence interval established in step 4.

Problem VII.   Determine if the population variance σ_1^2 is significantly greater than the population variance σ_2^2 where σ_1^2 is estimated by n_1 random samples having a variance of S_1^2 and σ_2^2 is estimated by n_2 random samples having a variance of S_2^2.

1. $H_0 : \sigma_1^2 \leq \sigma_2^2$, which is a one tail test.
2. Select α.

Table 8.5.3. Critical Values of the Statistic F
$$\alpha = 0.05$$

Degrees of Freedom for Numerator

		2	4	6	8	10	20
Degrees	2	19.0	19.2	19.3	19.4	19.4	19.4
of	4	6.94	6.39	6.16	6.04	5.96	5.80
Freedom	6	5.14	4.53	4.28	4.15	4.06	3.87
for	8	4.46	3.84	3.58	3.44	3.35	3.15
Denominator	10	4.10	3.48	3.22	3.07	2.98	2.77
	12	3.89	3.26	3.00	2.85	2.75	2.54
	14	3.74	3.11	2.85	2.70	2.60	2.39
	16	3.63	3.01	2.74	2.59	2.49	2.28
	18	3.55	2.93	2.66	2.51	2.41	2.19
	20	3.49	2.87	2.60	2.45	2.35	2.12

$$\alpha = 0.01$$
Degrees of Freedom for Numerator

		2	4	6	8	10	20
Degrees	2	99.0	99.2	99.3	99.4	99.4	99.4
of	4	18.0	16.0	15.2	14.8	14.5	14.0
Freedom	6	10.9	9.15	8.47	8.10	7.87	7.40
for	8	8.65	7.01	6.37	6.03	5.81	5.36
Denominator	10	7.56	5.99	5.39	5.06	4.85	4.41
	12	6.93	5.41	4.82	4.50	4.30	3.86
	14	6.51	5.04	4.46	4.14	3.94	3.51
	16	6.23	4.77	4.20	3.89	3.69	3.26
	18	6.01	4.58	4.01	3.71	3.51	3.08
	20	5.85	4.43	3.87	3.56	3.37	2.94

* The data in these tables were taken from "Tables of Percentage Points of the Inverted Beta (F) Distribution," M. Merrington, and C. M. Thompson, *Biometrika*, **33**, 1943–1946.

3. Calculate the statistic F from the experimental data by

$$F = \frac{S_1^{\,2}}{S_2^{\,2}}$$

4. From Table 8.5.3 find the value of $F_{(1-\alpha)(n_1-1)(n_2-1)}$.

5. Reject hypothesis if F is greater than the table value found in step 4, otherwise accept the hypothesis.

8.7 ANALYSIS OF VARIANCE

The procedures outlined in Section 8.6 work very nicely if there are only two means and two variances involved. When several means and several variances are involved, the number of t and F tests required to check all means or all variances increases rapidly as the number of means or variances increases. In addition, the reliability of either the t or F test decreases as it is used over and over again on the same sets of data. A more elegant procedure for handling the problem of several means or several variances is by the method of analysis of variance. This method is capable of testing for significant differences when using various experimental designs such as (1) one-way classification with replication, (2) two-way classification without replication, (3) two-way classification with replication, (4) three-way classification without replication, (5) Latin square design, and (6) specialized factorial designs. It is not possible to discuss the use of analysis of variance techniques in each of the above categories in the space of a single chapter. Therefore, the use of analysis of variance will be described for only the one-way classification, the two-way classification without replication, and the simple Latin square design.

One-Way Classification with Replication

The one-way classification is described more easily by reference to Table 8.7.1. There is one variable of classification tested at c levels. The

Table 8.7.1. One-Way Classification with Replication

		Classification			
1	2	i	c		
X_{11}	X_{21}	X_{i1}	X_{c1}		
X_{12}	X_{22}	X_{i2}	X_{c2}		
.					
.					
.					
.					
.					
X_{1j}	X_{2j}	X_{ij}	X_{cj}		
X_{1n_1}	X_{2n_2}	X_{in_i}	X_{cn_c}		
Total	$T_{1.}$	$T_{2.}$	$T_{i.}$	$T_{c.}$	$T_{.c.}$	Grand total
Mean	$\bar{X}_{1.}$	$\bar{X}_{2.}$	$\bar{X}_{i.}$	$\bar{X}_{c.}$	$\bar{X}_{..}$	Grand mean

readings are designated as X_{ij} where i stands for the level and j stands for each test at a given level. The number of tests per level is n_i and may vary for the different levels.

The total sum of the squares which would be used to compute the total variance S^2 for all the readings is given by

$$\sum_i^c (n_i - 1)S_T^{\,2} = \sum_i^c \sum_j^{n_i} (X_{ij} - \overline{\overline{X}}_{..})^2 \tag{8.7.1}$$

The total sum of the squares can be divided into two parts, the sum of the squares of the means about the grand mean and the sum of the squares of each reading in a given column about the column mean. To show that this is true, add and subtract $\overline{X}_{i.}$ to the quantity in the parenthesis on the right-hand side of Eq. 8.7.1 and expand as shown next.

$$\sum_i^c \sum_j^{n_i} (X_{ij} - \overline{\overline{X}}_{..})^2 = \sum_i^c \sum_j^{n_i} [(X_{ij} - \overline{X}_{i.}) + (\overline{X}_{i.} - \overline{\overline{X}}_{..})]^2$$

$$= \sum_i^c \sum_j^{n_i} [(X_{ij} - \overline{X}_{i.})^2 + 2(X_{ij} - \overline{X}_{i.})$$

$$\times (\overline{X}_{i.} - \overline{\overline{X}}_{..}) + (\overline{X}_{i.} - \overline{\overline{X}}_{..})^2]$$

$$= \sum_i^c \sum_j^{n_i} (X_{ij} - \overline{X}_{i.})^2 + \sum_i^c n_i(\overline{X}_i - \overline{\overline{X}}_{..})^2 \tag{8.7.2}$$

The computing forms for each of the sum of the squares are

$$\sum_i^c (n_i - 1)S_T^{\,2} = \sum_i^c \sum_j^{n_i} (X_{ij} - \overline{\overline{X}}_{..})^2 = \sum_i^c \sum_j^{n_i} X_{ij}^{\,2} - \frac{\left(\sum_i^c \sum_j^{n_i} X_{ij}\right)^2}{\sum_i^c n_i} \tag{8.7.3}$$

$$\sum_i^c (n_i - 1)S_p^{\,2} = \sum_i^c \sum_j^{n_i} (X_{ij} - \overline{X}_{i.})^2 = \sum_i^c \sum_j^{n_i} X_{ij}^{\,2} - \sum_i^c \left[\frac{\left(\sum_j^{n_i} X_{ij}\right)^2}{n_i}\right] \tag{8.7.4}$$

$$(c - 1)S_m^{\,2} = \sum_i^c n_i(\overline{X}_{i.} - \overline{\overline{X}}_{..})^2 = \sum_i^c \left[\frac{\left(\sum_j^{n_i} X_{ij}\right)^2}{n_i}\right] - \frac{\left(\sum_i^c \sum_j^{n_i} X_{ij}\right)^2}{\sum_i^c n_i} \tag{8.7.5}$$

In the preceding equations $S_T^{\,2}$ is the total variance, $S_p^{\,2}$ is the pooled variance or the variance within the groups, and $S_m^{\,2}$ is the variance of the means.

All three of these variances are an estimate of a population variance σ^2 provided that all the entries in Table 8.7.1 are random samples of the same population. If the different columns are from different populations having significantly different means, then the variance of the means will be significantly larger than S_p^2. Thus the two quantities S_p^2 and S_m^2 can be tested for significant differences. The appropriate statistic to use is

$$F = \frac{S_m^2}{S_p^2} \tag{8.7.6}$$

with $(c - 1)$ degrees of freedom in the numerator and $\sum_i^c (n_i - 1)$ degrees of freedom in the denominator.

Table 8.7.2. Analysis of Variance for One Variable

Source	Sum of Squares	df	Variance	Estimate of
Means	$\sum \dfrac{T_{i.}^2}{n_i} - \dfrac{T_{..}^2}{N}$	$n - 1$	S_m^2	$\sigma^2 + n\sigma_m^2$
Within	$\sum \sum X_{ij}^2 - \sum \dfrac{T_{i.}^2}{n_i}$	$N - n$	S_p^2	σ^2
Total	$\sum \sum X_{ij}^2 - \dfrac{T_{..}^2}{N}$	$N - 1$		

Note that the above is a one tail test as the null hypothesis is that $\sigma_m^2 \leq \sigma_p^2$. Thus we are interested only in proving that S_m^2 is significantly greater than S_p^2. If S_m^2 is less than S_p^2, then it is most likely that all entries are from the same population.

Using the notation that

$$\sum_j^{n_i} X_{ij} = T_{i.}, \qquad \sum_i^c \sum_j^{n_i} X_{ij} = T_{..}$$

and $N = \sum_i^c n_i$, the analysis of variance calculations are usually summarized as illustrated in Table 8.7.2.

Two-Way Classification without Replication

The two-way classification without replication is a standard factorial design which involves two variables each tested at several levels. Table 8.7.3 illustrates the method of assembling the data. The number of columns is c and the number of rows is r. The i subscript stands for the different levels in the first variable and varies from 1 to c. The j subscript stands for the different levels in the second variable and varies from 1 to r. Thus the total number of readings is rc. The column totals are represented by $T_{i.}$, the row totals by $T_{.j}$, and the grand total by $T_{..}$.

Table 8.7.3. Two-Way Classification

First Variable

Levels	1	2	$\cdots$	i	$\cdots$	c	Total	Mean
1	X_{11}	X_{21}		X_{i1}		X_{c1}	$T_{.1}$	$\bar{X}_{.1}$
2	X_{12}	X_{22}		X_{i2}		X_{c2}	$T_{.2}$	$\bar{X}_{.2}$
.								
.								
.								
j	X_{1j}	X_{2j}		X_{ij}		X_{cj}	$T_{.j}$	$\bar{X}_{.j}$
.								
.								
.								
r	X_{1r}	X_{2r}		X_{ir}		X_{cr}	$T_{.r}$	$\bar{X}_{.r}$
Total	$T_{1.}$	$T_{2.}$		$T_{i.}$		$T_{c.}$	$T_{..}$	
Mean	$\bar{X}_{1.}$	$\bar{X}_{2.}$		$\bar{X}_{i.}$		$\bar{X}_{c.}$		$\bar{X}_{..}$

(Second Variable — left axis label)

Assuming that all the data came from the same population, the total sum of the squares used to compute the total variance $S_T^{\,2}$ is given by

$$(cr - 1)S_T^{\,2} = \sum_i^c \sum_j^r (X_{ij} - \bar{\bar{X}}_{..})^2 \tag{8.7.8}$$

The total sum of the squares can be shown to be composed of the sum of the squares for the rows plus the sum of the squares for the columns plus a residual sum of the squares. The sum of the squares in Eq. 8.7.8 can be written as

$$\sum_i^c \sum_j^r (X_{ij} - \bar{\bar{X}}_{..}) = \sum_i^c \sum_j^r [(\bar{X}_{.j} - \bar{\bar{X}}_{..}) + (\bar{X}_{i.} - \bar{\bar{X}}_{..})$$
$$+ (X_{ij} - \bar{X}_{.j} - \bar{X}_{i.} + \bar{\bar{X}}_{..})]^2$$
$$= \sum_i^c \sum_j^r [(\bar{X}_{.j} - \bar{\bar{X}}_{..})^2 + (\bar{X}_{i.} - \bar{\bar{X}}_{..})^2$$
$$+ (X_{ij} - \bar{X}_{.j} - \bar{X}_{i.} + \bar{\bar{X}}_{..})^2] \tag{8.7.9}$$

The computing forms for each of the sum of squares are the following.
for row sum of squares

$$(r - 1)S_r^{\,2} = \sum_i^c \sum_j^r (\bar{X}_{.j} - \bar{\bar{X}}_{..})^2 = \sum_j^r \frac{T_{.j}^{\,2}}{c} - \frac{T_{..}^{\,2}}{cr} \tag{8.7.10}$$

for column sum of squares

$$(c - 1)S_c^{\,2} = \sum_i^c \sum_j^r (\overline{X}_{i.} - \overline{\overline{X}}_{..})^2 = \sum_i^c \frac{T_{i.}^{\,2}}{r} - \frac{T_{..}^{\,2}}{cr} \qquad (8.7.11)$$

for residual sum of squares

$$(c - 1)(r - 1)S_R^{\,2} = \sum_i^c \sum_j^r [(X_{ij} - \overline{X}_{.j}) - (\overline{X}_{i.} - \overline{\overline{X}}_{..})]^2$$

$$= \sum_i^c \sum_j^r X_{ij}^{\,2} - \sum_i^c \frac{T_{i.}^{\,2}}{r} - \sum_j^r \frac{T_{.j}^{\,2}}{c} + \frac{T_{..}^{\,2}}{cr} \qquad (8.7.12)$$

and for total sum of squares

$$(rc - 1)S_p^{\,2} = \sum_i^c \sum_j^r (X_{ij} - \overline{\overline{X}}_{..})^2 = \sum_i^c \sum_j^r X_{ij}^{\,2} - \frac{T_{..}^{\,2}}{cr} \qquad (8.7.13)$$

The analysis of variance computations are made as shown in Table 8.7.4.

Table 8.7.4 Analysis of Variance for Two Variables with No Replication

Source	Sum of Squares	df	Variance	Estimate of
Column means	$\sum \frac{T_{i.}^{\,2}}{r} - \frac{T_{..}^{\,2}}{cr}$	$c - 1$	$S_c^{\,2}$	$\sigma^2 + r\sigma_c^{\,2}$
Row means	$\sum \frac{T_{.j}^{\,2}}{c} - \frac{T_{..}^{\,2}}{cr}$	$r - 1$	$S_r^{\,2}$	$\sigma^2 + c\sigma_r^{\,2}$
Residuals	$\sum\sum X_{ij}^{\,2} - \sum \frac{T_{i.}^{\,2}}{r}$ $- \sum \frac{T_{.j}^{\,2}}{c} + \frac{T_{..}^{\,2}}{cr}$	$(c-1)(r-1)$	$S_R^{\,2}$	
Total	$\sum\sum X_{ij}^{\,2} - \frac{T_{..}^{\,2}}{cr}$	$rc - 1$		

The residual sum of the squares has $(c - 1)(r - 1)$ degrees of freedom and is used to compute the residual variance $S_R^{\,2}$, which is a good estimate of the experimental error if there is no interaction between the row and column effects. In the two-way classification it is assumed that if there is a row or column effect, then these effects are additive. Thus the entry in each cell is given as

$$\mu_{ij} = \mu + c_i + r_j \qquad (8.7.14)$$

where μ is the population mean, c_i is the column effect for the ith column, and r_j is the row effect for the jth row.

Two hypotheses can be tested when using the two-way classification without replication: (1) the column effects are zero and (2) the row effects are zero. The statistic to use for (1) is

$$F = \frac{S_c^2}{S_R^2} \qquad (8.7.15)$$

where the degrees of freedom are $(c - 1)$ for the numerator and $(c - 1)(r - 1)$ for the denominator. The statistic to use for (2) is

$$F = \frac{S_r^2}{S_R^2} \qquad (8.7.16)$$

where the degrees of freedom are $(r - 1)$ for the numerator and $(c - 1)(r - 1)$ for the denominator.

Latin Square

The Latin Square is a special experimental design involving three variables. It provides a relatively simple method for determining which of three variables produce a significant variation in a phenomenon under study when the levels of the three variables are changed. Its design will not measure interaction and should only be used if it can be assumed that interaction does not exist.

Each variable in the Latin Square is tested at the same number of levels. The number of cells is determined by the number of levels used. For n levels of the three variables there are n^2 cells. If the levels of the three variables for a 3 × 3 Latin Square are indicated by I, II, III for the first variable, by 1, 2, 3 for the second variable, and A, B, C for the third variable, the resulting experimental design is as illustrated in Table 8.7.5a. Let the subscripts i, j, k of X_{ijk} represent the levels for the first, second, and third variables, respectively. The entries in the nine cells for a 3 × 3 Latin Square are as shown in Table 8.7.5b. Since there are three variables, there are four sets of totals and means—the columns, rows, diagonals, and grand totals and means. The totals, means, and sum of squares are

Table 8.7.5. Latin Square for Three Levels

	a Variables			*b* Measurements		
	I	II	III			
1	A	B	C	X_{111}	X_{212}	X_{313}
2	C	A	B	X_{123}	X_{221}	X_{322}
3	B	C	A	X_{132}	X_{233}	X_{331}

computed as follows:

Column Totals

$X_{111} + X_{123} + X_{132} = T_{1..}$

$X_{212} + X_{221} + X_{233} = T_{2..}$

$X_{313} + X_{322} + X_{331} = T_{3..}$

Row Totals

$X_{111} + X_{212} + X_{313} = T_{.1.}$

$X_{123} + X_{221} + X_{322} = T_{.2.}$

$X_{132} + X_{233} + X_{331} = T_{.3.}$

Diagonals Totals

$X_{111} + X_{221} + X_{331} = T_{..1}$

$X_{212} + X_{322} + X_{132} = T_{..2}$

$X_{313} + X_{123} + X_{233} = T_{..3}$

Column Means

$\overline{X}_{1..} = \dfrac{T_{1..}}{3}$

$\overline{X}_{2..} = \dfrac{T_{2..}}{3}$

$\overline{X}_{3..} = \dfrac{T_{3..}}{3}$

Row Means

$\overline{X}_{.1.} = \dfrac{T_{.1.}}{3}$

$\overline{X}_{.2.} = \dfrac{T_{.2.}}{3}$

$\overline{X}_{.3.} = \dfrac{T_{.3.}}{3}$

Diagonal Means

$\overline{X}_{..1} = \dfrac{T_{..1}}{3}$

$\overline{X}_{..2} = \dfrac{T_{..2}}{3}$

$\overline{X}_{..3} = \dfrac{T_{..3}}{3}$

Grand total

$$= T_{1..} + T_{2..} + T_{3..} = T_{.1.} + T_{.2.} + T_{.3.} = T_{..1} + T_{..2} + T_{..3} = T$$

Grand mean $= \dfrac{T}{9}$

Column sum of squares (CSS)

$$= \frac{T_{1..}^2 + T_{2..}^2 + T_{3..}^2}{3} - \frac{T^2}{9}$$

Row sum of squares (RSS)

$$= \frac{T_{.1.}^2 + T_{.2.}^2 + T_{.3.}^2}{3} - \frac{T^2}{9}$$

Diagonal sum of squares (DSS)

$$= \frac{T_{..1}^2 + T_{..2}^2 + T_{..3}^2}{3} - \frac{T^2}{9}$$

Total sum of squares (TSS)

$$= X_{111}^2 + X_{212}^2 + X_{313}^2 + X_{123}^2 + X_{221}^2$$

$$+ X_{312}^2 + X_{132}^2 + X_{233}^2 + X_{331}^2 - \frac{T}{9}$$

The analysis of variance calculations are made as indicated in Table 8.7.6.

Table 8.7.6. Analysis of Variance for 3×3 Latin Square

Source	Sum of Squares	df	Variance	F Test
Columns	CSS	2	$\dfrac{CSS}{2}$	$\dfrac{CSS/2}{ESS/2}$
Rows	RSS	2	$\dfrac{RSS}{2}$	$\dfrac{RSS/2}{ESS/2}$
Diagonals	DSS	2	$\dfrac{DSS}{2}$	$\dfrac{DSS/2}{ESS/2}$
Residuals	ESS	2	$\dfrac{ESS}{2}$	
Total	TSS	8		

Since the total sum of squares or total degrees of freedom must be equal to the sum of the column, row, diagonal, and residual, the residual sum of squares or degrees of freedom can be found by subtraction. The residual mean square or variance is an estimate of the experimental error or population variance. This quantity is used to perform F tests with the column, row, and diagonal variances. The column, row, and diagonal variances must be large compared to the residual variance if the variables represented by the columns, rows, and diagonals have a significant effect upon the phenomenon under study.

8.8 STATISTICS OF A STRAIGHT LINE

The problem of curve fitting or finding a functional relationship between two variables is fundamental to many research problems. The simplest type of curve is that of a straight line represented by

$$Y = \alpha + \beta X \tag{8.8.1}$$

where $Y =$ the dependent variable
$X =$ the independent variable
$\beta =$ the slope
$\alpha =$ the intercept or the value of Y at $X = 0$

Examples of approximate linear relationships often found in the study of rock behavior are (1) stress-strain relations, (2) compressive strength as a function of confining pressure, and (3) arrival time as a function of travel distance for a seismic wave. Many functional relationships between two variables are not linear, but by making suitable transformation of variables a linear relationship is obtained. For example, consider a power law relation between the two variables ϵ and d given as

$$\epsilon = kd^n$$

Taking the logarithm of both sides of this equation gives

$$\log \epsilon = \log k + n \log d$$

Therefore the transformation $Y = \log \epsilon$ and $X = \log d$ will result in a linear equation similar to Eq. 8.8.1 where the new constants are $\alpha = \log k$ and $\beta = n$. Many other types of functional relationships can be converted to linear relationships by suitable transformations.

The method for finding the best straight-line fit to a set of experimental data, which shows a linear grouping, a homogeneous variance, and a normal distribution about some central value given by Eq. 8.8.1, is discussed in this section. This method can usually be applied successfully to transformed data provided these data have a homogeneous variance and tend to be normally distributed about some central value given by Eq. 8.8.1.

Let Eq. 8.8.1 represent the postulated relationship between the two variables X and Y. Let

$$\hat{Y}_i = a + bX_i \qquad (8.8.2)$$

represent the estimate of the postulated relationship obtained from a sample of n data points (X_i, Y_i), where X_i is selected and Y_i is observed and $(i = 1, 2, \ldots : , n)$. See Fig. 8.8.1. Unbiased estimates of α and β are a and b, where a and b are determined from the sample data by the method of least squares. The deviation of the observed Y_i from the estimated $\hat{Y}_i$ is $(Y_i - \hat{Y}_i)$ and the sum of the squares of the deviation Q is given by

$$Q = \sum_i^n (Y_i - \hat{Y}_i)^2 = \sum_i^n (Y_i - a - bX_i)^2 \qquad (8.8.3)$$

The quantity Q must be minimized with respect to both a and b. Thus

$$\frac{\partial Q}{\partial a} = \sum_i^n - 2(Y_i - a - bX_i) = -2 \sum_i^n Y_i + 2na + 2b \sum_i^n X_i = 0 \quad (8.8.4)$$

$$\frac{\partial Q}{\partial b} = \sum_i^n - 2X(Y_i - a - bX_i) = -2 \sum X_i Y_i + 2a \sum_i^n X_i + 2b \sum_i^n X_i^2 = 0$$

$$(8.8.5)$$

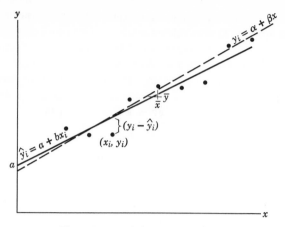

Fig. 8.8.1. Statistics of straight line.

Solving Eqs. 8.8.4 and 8.8.5 for a and b gives

$$a = \frac{\sum_i^n Y_i \sum_i^n X_i^2 - \sum_i^n X_i \sum_i^n X_i Y_i}{n \sum_i^n X_i^2 - \left(\sum_i^n X_i\right)^2} \qquad (8.8.6)$$

$$b = \frac{n \sum_i^n X_i Y_i - \sum_i^n X_i \sum_i^n Y_i}{n \sum_i^n X_i^2 - \left(\sum_i^n X_i\right)^2} \qquad (8.8.7)$$

Solving Eq. 8.8.4 for a gives

$$a = \bar{Y} - b\bar{X} \qquad (8.8.8)$$

where $\bar{X}$ and $\bar{Y}$ are the mean value points for the data. Combining Eq. 8.8.2 with Eq. 8.8.4 gives

$$\sum(Y_i - \hat{Y}_i) = 0 \qquad (8.8.9)$$

From Eqs. 8.8.3, 8.8.8, and 8.8.9, it is evident that the value of a and b determined by the method of least squares (1) determines a straight line that passes through the mean point of the data, (2) makes the sum of the deviations zero, and (3) makes the sum of the squared deviations a minimum.

The variance of the data about the estimate is defined as

$$S_E^2 = \frac{\sum_i^n (Y_i - \hat{Y}_i)^2}{n - 2} = \frac{\sum_i^n (Y_i - a - bX_i)^2}{n - 2} \qquad (8.8.10)$$

The quantity $n - 2$ is the degrees of freedom which is 2 less than n because two quantities a and b have been calculated from the data. The quantity S_E is known as the standard error of the estimate and is an unbiased estimate of the population standard deviation σ_E. Expansion of the summation in Eq. 8.8.10 gives the following forms for calculating $S_E{}^2$:

$$(n - 2)S_E{}^2 = \sum_i^n Y_i^2 - a \sum_i^n Y_i - b \sum_i^n X_i Y_i \qquad (8.8.11)$$

or

$$(n - 2)S_E{}^2 = \sum_i^n (Y_i - \bar{Y})^2 - \frac{\left[\sum_i^n (X_i - \bar{X})(Y_i - \bar{Y})\right]^2}{\sum (X_i - \bar{X})^2} \qquad (8.8.12)$$

The quantity $\sum_i^n (X_i - \bar{X})^2$ is the sum of the squares of the X_i deviations and is related to the sample variance of X_i by

$$(n - 1)S_x{}^2 = \sum_i^n (X_i - \bar{X})^2 = \sum_i^n X_i^2 - \frac{(\sum X_i)^2}{n} \qquad (8.8.13)$$

The quantity $\sum (Y_i - \bar{Y})^2$ is the sum of the squares of the Y_i deviations and is related to the sample variance of Y_i by

$$(n - 1)S_y{}^2 = \sum_i^n (Y_i - \bar{Y})^2 = \sum_i^n Y_i^2 - \frac{\left(\sum_i^n Y_i\right)^2}{n} \qquad (8.8.14)$$

The quantity $\sum_i^n (X_i - \bar{X})(Y_i - \hat{Y})$ is the sum of the products and is related to the sample covariance by

$$(n - 1)S_{xy}{}^2 = \sum_i^n (X_i - \bar{X})(Y_i - \bar{Y}) = \sum_i^n X_i Y_i - \frac{\left(\sum_i^n X_i\right)\left(\sum_i^n Y_i\right)}{n}$$

$$(8.8.15)$$

By means of Eqs. 8.8.14 and 8.8.15, Eq. 8.8.7 can be written as

$$b = \frac{\sum_i^n (X_i - \bar{X})(Y_i - \bar{Y})}{\sum_i^n (X_i - \bar{X})^2} \qquad (8.8.16)$$

By means of Eqs. 8.8.12, 8.8.16, and 8.8.10, we can show that the total sum of squares for Y is composed of two parts, the sum of squares about the estimate line and the sum of squares due to the variation of Y with X, thus

$$\sum_i^n (Y_i - \bar{Y})^2 = \sum_i^n (Y_i - \hat{Y}_i)^2 + b \sum (X_i - \bar{X})(Y_i - \bar{Y}) \qquad (8.8.17)$$

Equation 8.8.17 provides a means for determining if a significant correlation exists between X and Y; that is, if b is significantly different from zero, then the mean sum of squares due to a variation of Y with X must be large compared to the mean sum of squares about the estimate line. Thus the appropriate F statistic is

$$F = \frac{b\left[\sum_{i}^{n}(X_i - \bar{X})(Y_i - \bar{Y})\right]}{\sum_{i}^{n}(Y_i - \hat{Y}_i)^2/(n-2)} \qquad (8.8.18)$$

where F has one degree of freedom in the numerator and $n - 2$ degrees of freedom in the denominator. The appropriate degrees of freedom are found by reference to Eq. 8.8.17. The quantity $\sum(Y_i - \bar{Y})^2$ has $(n - 1)$ degrees of freedom and the quantity $\sum(Y_i - \hat{Y}_i)^2$ has $(n - 2)$ degrees of freedom. As the total degrees of freedom on each side of the equation must be the same, the quantity $b[\Sigma(X_i - \bar{X})(Y_i - \hat{Y})]$ has only one degree of freedom.

The variance of the sample slope b denoted by S_b^2 is obtained from the following equation;

$$S_b = \frac{\sum_{i}^{n}(Y_i - \hat{Y})^2}{(n-2)\sum(X_i - \bar{X})^2} = \frac{S_E^2}{\sum_{i}^{n}X_i^2 - (\sum X_i)^2/n} \qquad (8.8.19)$$

The $(1 - \alpha)$ confidence limits for the population slope β are given as

$$\begin{aligned} L_1 &= b - t_{0.5\alpha(n-2)}S_b \\ L_2 &= b + t_{(1-0.5\alpha)(n-2)}S_b \end{aligned} \qquad (8.8.20)$$

Thus another method for testing that b is significantly different from zero or from any given value β_0 is by the statistic

$$t = \frac{b - \beta_0}{S_b} \qquad (8.8.21)$$

The variance of the sample intercept a denoted by S_a^2 is obtained from the equation

$$S_a^2 = S_E^2\left[\frac{1}{n} + \frac{\bar{X}^2}{\sum(X_i - \bar{X})^2}\right] = S_b^2 \sum \frac{X_i^2}{n} \qquad (8.8.22)$$

Equation 8.8.22 shows that S_a^2 increases as the center of the data moves farther away from the origin. The $(1 - \alpha)$ confidence limits for the population intercept α are given as

$$\begin{aligned} L_1 &= a - t_{0.5\alpha(n-2)a}S \\ L_2 &= a + t_{(1-0.5\alpha)(n-2)}S_a \end{aligned} \qquad (8.8.23)$$

Therefore to determine if a is significantly different from some given value α_0, the appropriate statistic is

$$t = \frac{a - \alpha_0}{S_a} \qquad (8.8.24)$$

If there are two independent samples from which the two estimate equations

$$\begin{aligned} Y_1 &= a_1 + b_1 X_1 \\ Y_2 &= a_2 + b_2 X_2 \end{aligned} \qquad (8.8.25)$$

have been determined and it is desired to test the null hypothesis $H_0 : \beta_1 = \beta_2$ or $H_0 : \alpha_1 = \alpha_2$, then it is necessary to assume that the two samples came from populations having a common variance. An estimate of the common variance is

$$(S_E{}^2)_p = \frac{\sum\limits_{i}^{n_1}(Y_{1j} - \hat{Y}_{1j})^2 + \sum\limits_{j}^{n_2}(Y_{2j} - \hat{Y}_{2j})^2}{n_1 + n_2 - 4} =$$
$$\frac{(n_1 - 2)S_{E_1}{}^2 + (n_2 - 2)S_{E_2}{}^2}{n_1 + n_2 - 4} \qquad (8.8.26)$$

where $S_{E_1}{}^2$ and $S_{E_2}{}^2$ are computed by Eq. 8.8.11. The variance for the difference in the slope is computed by

$$S_{b_1 - b_2}^2 = (S_E{}^2)_p \left[\frac{1}{\sum\limits_{j}^{n_1}(X_{1j} - \bar{X}_{1.})^2} + \frac{1}{\sum\limits_{j}^{n_2}(X_{2j} - \bar{X}_{2.})^2} \right] \qquad (8.8.27)$$

The variance for the difference in the intercepts is computed by

$$S_{a_1 - a_2}^2 = (S_E{}^2)_p \left[\frac{1}{n_1} + \frac{\bar{X}_{1.}{}^2}{\sum\limits_{j}^{n_1}(X_{1j} - \bar{X}_{1.})^2} + \frac{1}{n_2} + \frac{\bar{X}_{2.}{}^2}{\sum\limits_{j}^{n_2}(X_{2j} - \bar{X}_{2.})^2} \right] \qquad (8.8.28)$$

For the null hypothesis $H_0 : \beta_1 = \beta_2$, the appropriate statistic is

$$t = \frac{b_1 - b_2}{S_{b_1 - b_2}} \qquad (8.8.29)$$

where the number of degrees of freedom are $n_1 + n_2 - 4$. For the null hypothesis $H_0 : \alpha_1 = \alpha_2$, the appropriate statistic is

$$t = \frac{a_1 - a_2}{S_{a_1 - a_2}} \qquad (8.8.30)$$

where the number of degrees of freedom are $(n_1 + n_2 - 4)$.

8.9 STATISTICAL TESTS FOR NON-NORMAL DISTRIBUTIONS

Some experimental data because of the nature of the phenomenon under study have a non-normal or nonsymmetric distribution. The use of statistical procedures based upon normal distributions are not justified for other distributions and other means must be found to make statistical analyses of these data.

Tchebycheff's inequality offers one method for making statistical inferences concerning populations having any distributions. Assume that n readings have been obtained from a population of unknown distribution and that the mean and the mean squared deviations have been calculated by

$$\bar{X} = \frac{1}{n} \sum_i^n X_i \tag{8.9.1}$$

$$S^2 = \frac{\sum (X_i - \bar{X})^2}{(n-1)} \tag{8.9.2}$$

Let p be that fraction of the total readings where the absolute value of $|X_i - \bar{X}|$ is less than kS, where k is some number greater than unity. Then $1 - p$ will be that fraction of the total readings where the absolute value of $|X_i - \bar{X}|$ is greater than kS. Let D_{1i} and D_{2i} be the respective deviations. Then the following equations can be written

$$D_{1i} = |X_i - \bar{X}| < kS$$
$$D_{2i} = |X_i - \bar{X}| > kS$$

Squaring both equations and summing gives

$$\sum_i^{np} D_{1i}^2 = \sum_i^{np} (X_i - \bar{X})^2 < k^2 S^2 np \tag{8.9.3}$$

$$\sum_i^{n(1-p)} D_{2i}^2 = \sum_i^{n(1-p)} (X_i - \bar{X})^2 > k^2 S^2 n(1-p) \tag{8.9.4}$$

Addition of these two equations and the use of Eq. 8.9.2 gives

$$\sum_i^{np} D_{1i}^2 + \sum_i^{n(1-p)} D_{2i}^2 = \sum_i^{np} (X_i - \bar{X})^2 + \sum_i^{n(1-p)} (X_i - \bar{X})^2$$

$$= \sum_i^n (X_i - \bar{X})^2 = (n-1)S^2 \tag{8.9.5}$$

Since $nS^2 > (n-1)S^2$ and from Eqs. 8.9.5 and 8.9.4, we obtain the following inequality:

$$nS^2 > (n-1)S^2 > \sum_i^{n(1-p)} D_{1i}^2 > k^2 S^2 n(1-p) \tag{8.9.6}$$

Simplification of the first and last terms of Eq. 8.9.6 gives

$$1 > k^2(1 - p)$$

$$\frac{1}{k^2} > (1 - p) \qquad (8.9.7)$$

$$p > \left(1 - \frac{1}{k^2}\right)$$

Therefore since $(1 - p)$ is by definition that fraction of the total readings where the absolute magnitude of the deviations is greater than kS, the probability of having a reading outside of the range $(\bar{X} \pm kS)$ is less than $1/k^2$—this is Tchebycheff's inequality. For the 95% confidence limits, $1/k^2 = 0.05$ and $k = 4.46$. This value should be compared to the value of $Z = 1.96$ which gives the 95% confidence limits for a normal distribution. Considerable information is lost by not having a normal distribution; the confidence limits that result by the above method are quite wide. Therefore it is necessary to have large differences in two sets of data before a significant statistical difference can be obtained.

A more efficient method for handling populations with distributions other than normal is to use means of small subgroups of the population. If subgroups of n readings each are drawn from a nonnormal population, the means of these subgroups will be more nearly normal than the individual readings regardless of the skewness of the original population distribution. The larger the value of n, the more nearly the distribution of the means approaches that of a normal distribution. However, the effect is very pronounced even with n as low as 3 or 4. If subgroup means are used as the population, the standard statistical procedures for normal distribution can usually be applied with safety and confidence.

REFERENCES

1. Cramer, H., *The Elements of Probability and Some of Its Applications*, John Wiley and Sons, New York, 1955.
2. Dixon, Wilfrid J. and Frank J. Massey, Jr., *Introduction to Statistical Analysis*, McGraw-Hill Book Co., New York.
3. Fisher, R. A., *Statistical Methods for Research Workers*, Oliver and Boyd, Ltd., Edinburgh and London, 1954.
4. Freund, John F., Paul E. Livermore, and Irwin Miller, *Manual of Experimental Statistics*, Prentice-Hall, Englewood Cliffs, New Jersey, 1960.
5. Goulden, C. H., *Methods of Statistical Analysis*, 2nd Edition, John Wiley and Sons, New York, 1952.
6. Ostle, Bernard, *Statistics in Research*, The Iowa State College Press, Ames, Iowa, 1956.
7. Parrott, Lyman G., *Probability and Experimental Errors in Science*, John Wiley and Sons, New York, 1961.
8. Wilson, E. B., Jr., *An Introduction to Scientific Research*, McGraw-Hill Book Co., New York, 1952.

CHAPTER 9

INSTRUMENTATION

9.1 INTRODUCTION

Because of the relative importance of the experimental process in rock mechanics investigations, it is correspondingly important to measure with reasonable accuracy the properties of rocks and the stress, strain, and deformation in their structures. Although it is almost axiomatic that any kind of measurement should be made with reasonable accuracy, this statement has a particular significance in rock mechanics because, unlike many of the older established branches of engineering in which instruments and measuring procedures have experienced the test of time, most of the instrumentation for rock mechanics is of comparatively recent origin. In fact, relatively few instruments are commercially available for measurements in rock. Instead, investigators have either designed and built new equipment or modified existing equipment to meet the needs of particular problems. Thus rock mechanics instrumentation generally lacks any universal acceptance or standardization. Because of this individuality only the more functional aspects of this class of equipment is described, usually the constructional detail can be found in the appropriate references cited at the end of this chapter. The descriptions are further restricted to those instruments and methods that have demonstrated satisfactory performance in actual use.

Many of the rock mechanics instruments reported in the literature were adapted or built to operate in a specific environment or under a special set of conditions. The engineer is cautioned against using any of these instruments in a different environment or under other operating conditions without first calibrating or otherwise checking it under the intended service conditions. Also, a number of instruments have been described that lack an adequate calibration or indication of accuracy so that, in service, a satisfactory response cannot be insured. As a general rule the conditions under which an instrument has been calibrated should be thoroughly explored and compared with the intended service conditions before it is placed in operation.

9.2 DEFINITIONS AND CONCEPTS

A number of words and concepts are used in the description and specification of instruments and procedures for measuring stress, strain displacement, and deformation that have specific meanings. To avoid ambiguity, these words and concepts are defined or explained in the following paragraphs.

Deformation Sensitivity

The *sensitivity* of deformation measuring instruments is defined as the smallest increment of the measured quantity that can be read on the indicating scale.* For an accurately ruled scale read with the unaided eye, the sensitivity is about 0.01 in. However, many instruments employ some kind of *magnification* to improve the sensitivity, which may be mechanical, optical, electronic amplification, etc. If the smallest increment of length that can be read on the scale of a deformation device is d, and the magnification is m, then the *deformation sensitivity* of the device, S_d is

$$S_d = \frac{d}{m} \qquad (9.2.1)$$

Thus, for a ruler, $d = 0.01$, $m = 1$, and $S_d = 0.01$; for a conventional micrometer the scale length is about 60 in. (length of scale per turn times the number of turns per inch); hence the mechanical magnification of a micrometer is about 60, and the deformation sensitivity is approximately 0.00016 in.

Strain Sensitivity

If a deformation measuring device is used as a strain gage, that is, if the deformation is divided by the *gage length L*, the strain sensitivity of the instrument S_s is defined as

$$S_s = \frac{d}{mL} \qquad (9.2.2)$$

Thus for a 2-in. micrometer ($L = 2$ in.), the strain sensitivity would be 0.00008 in./in. (or simply 0.00008).

* This definition applies only to deformation measuring instruments. For transducers the *transfer function* is the relationship between the magnitude of the input quantity Q_i to the output quantity Q_o, and the differential quotient $dQ_o/dQ_i = S$ is the sensitivity of the device.

Table 9.2.1. Breaking Stress and Strain for Some Common Rock Types[1]

Rock Type[2]	Compressive Strength Range (psi × 10³)		Modulus of Elasticity[3] Range (psi × 10⁶)		Breaking Strain Range[4] (microstrains)[5]	
	Minimum	Maximum	Minimum	Maximum	Minimum	Maximum
Marble (6)	6.7	34.5	7.2	11.9	930	4300
Limestone (46)	0.7	29.8	0.4	12.0	1630	8000
Granite (13)	23.0	42.6	3.1	10.0	4030	8660
Diorite (4)	26.2	39.8	8.0	14.6	2070	3370
Gneiss (12)	22.2	36.4	3.5	9.7	2300	7790
Sandstone (43)	4.8	34.1	0.9	7.3	3150	12,000
Dolomite (7)	11.0	53.0	2.8	11.3	2170	4640
Marlstone (13)	8.1	28.2	1.8	7.0	1800	6610
Dibase (8)	22.6	46.6	8.9	13.9	2540	4010
Basalt (6)	17.3	52.2	8.3	12.4	2540	5980
Greenstone (9)	17.7	45.5	6.9	15.2	1540	3800
Shale (12)	10.9	33.5	1.6	9.9	1720	9250
Siltstone (6)	5.0	45.8	4.5	10.8	890	4910

[1] From *U.S. Bureau of Mines Rep. of Investigations* 4459, Part I (S. L. Windes, 1949); 4727, Part II (S. L. Windes, 1950); 5130, Part III (B. E. Blair, 1955); and 5244, Part IV (B. E. Blair, 1956).
[2] The number following the rock type indicates the number of groups of specimens tested. Each group is a petrographically distinct variation of the designated rock type.
[3] Modulus of elasticity determined either in uniaxial compression or sonically (bar velocity method).
[4] Breaking strain calculated from the group uniaxial compressive strength divided by the modulus of elasticity.
[5] Microstrain ($\mu\epsilon$) = 1 μ in/in.

238

Range

In selecting a strain gage or deformeter for a specific measurement, the *range* of instrument must be considered, which is defined as the maximum quantity the instrument can measure. Dial indicators are available graduated to 0.0001 in. divisions with a 0.5 in. range. If the indicator is mounted on the end of a 100 in. bar the range would still be 0.5 in.

In Table 9.2.1 the maximum and minimum compressive breaking strains for a number of common rock types are given. There is over a ten-fold variation between the maximum and minimum values; therefore, the selection of either deformation or strain gages must be considered in relation to the specific rock type under study to insure the proper sensitivity and range selection.

Reproducibility and Accuracy

Because of friction, lost motion, etc., a sensitive instrument will not give exactly the same value in a succession of readings, that is, the readings are not exactly reproducible. The most probable value of an experimentally determined quantity is the arithmetic mean, which is the sum of the individual readings divided by the number of readings. If the number of readings is large, the average approaches the "true value" of the measurement. However, neither the sensitivity nor the reproducibility is any indication of the accuracy of an instrument. The accuracy of an instrument can be determined by measuring or comparing against a known standard.

Some instruments are calibrated by the manufacturer and are sold with a calibration certificate; others such as a micrometer are supplied with a standard (or secondary standard). Instruments developed in a research laboratory are usually calibrated against a primary or secondary standard.

The accuracy of an instrument is often expressed as a percentage of the maximum measuring range, or some arbitrary value within the measuring range of the instrument. Thus, if a 1 in. micrometer has been calibrated by making a succession of readings on a 1 in. calibration standard and found by the proper statistical methods to be accurate within ± 0.0002 in., then the percentage error would be $\pm 0.02\%$, for a 1 in. measurement. For instruments having a uniform scale division such as a micrometer, the magnitude of the error (not the percentage error) is usually constant. Hence for a measurement of 0.5 in. the accuracy would be $\pm 0.04\%$, and in the measurement of a fine wire 0.002 in. in diameter, the error would be $\pm 10\%$.

Kinds of Error

There are two kinds of instrument error: *random* or *accidental*, and *systemic*. Random errors, with equal probability, may be positive or negative with respect to the true value, and may be caused by factors such as friction or lost motion. Because they are random these errors will tend to average out if a sufficient number of readings are taken and the data are treated statistically. Systemic errors are consistently either positive or negative with respect to the true value, and they can be evaluated only by calibration against a standard. Systemic errors are caused by factors such as physical damage, loss of calibration, temperature, humidity, low batteries, etc. In addition to the errors that are inherent to an instrument other errors may be introduced by the observer. Error in judgement or carelessness in reading a meter may cause a random error, whereas personal bias may cause a systemic error. The instrument and personal errors combine to give the measurement error. Mistakes in recording data or in making calculations usually cannot be detected except by repeating the calculation or measurement.

Absolute and Relative Measurement

In this book, *absolute measurement* is used to mean a measurement made with respect to the zero in the scale of measurement, and the value so obtained is called the *absolute value*, whereas a *relative measurement* is made with respect to some arbitrary and usually unknown initial value. For example, absolute pressure is measured with respect to a vacuum, that is, zero pressure, whereas the air pressure in a pneumatic tire is relative because it is measured with respect to atmospheric pressure.

Strain (or Stress) Relief

The distinction between absolute and relative measurement has a special significance in rock mechanics. Unlike most engineering structures which are made from materials that are initially unstressed, structures in rock are made in a material that is initially under stress. If, say, a strain gage is installed on a mine pillar and the mined area around the pillar increased, thereby increasing the load on the pillar, the gage will indicate this increase. This strain measurement is relative because the initial state of strain in the pillar was unknown. However, if this pillar is cut free from the mine roof, the pillar strain will be reduced to zero, and the gage will indicate the absolute decrease in strain. Furthermore, this measured strain (with the sign reversed) is the absolute strain that existed in the pillar before it was cut free. The process of cutting a part of a strained structure free from its surroundings is referred to as *strain-relief*. If a stress-strain relationship for the pillar rock is known, the absolute stress can be calculated and the process is referred to as *stress-relief*. In fact, the terms strain-relief and stress-relief are used almost synonymously. Most procedures for measuring absolute deformation, stress, or strain in a continuum require a partial or complete strain-relief (Section 14.3).

Drift or Time Stability

If any measurement is made over an extended period, the measuring instrument should not drift during the measuring period. *Drift* refers to the change in the reading of an instrument that is not caused by a change in the effect being measured or by factors normally controlled during the course of the measurement. Drift may be caused by many factors, such as temperature, moisture and humidity, creep in component parts of the instrument (springs, etc.), and deterioration of vacuum tubes or batteries. In measuring the change in a small quantity, such as the change in strain in rock, this specification may be difficult to satisfy, especially if the time between the initial and final measurement is long. The drift in an instrument from all causes should be less than the required minimum increment of measurement.

9.3 ENVIRONMENTAL FACTORS

Most laboratory equipment, such as stress, strain, and deformation measuring instruments, is designed to operate without special precaution in a laboratory environment. If some control is necessary it is usually limited to comparatively simple procedures; for example, resistance-strain gages are usually temperature compensated and protected against atmospheric moisture. However, the equipment used in underground measurements is almost always subject to the most adverse environmental conditions, which may include such factors as a wide temperature range, high humidity and moisture, dust and dirt, gassy and corrosive atmosphere, and excessive vibrations and mechanical shock. Moreover, electric power, when available, is usually subject to large and sometimes sudden variations in voltage and it may contain transients produced by switching and other causes.

Although it is almost axiomatic that instruments for underground use should be designed or adapted to tolerate adverse environmental factors, the range of conditions is usually so large and the number of factors so great that the construction of an instrument that would function properly under all conditions is almost a physical impossibility, at least within practical cost limits. The alternative, and probably the more practical procedure, is to make a preliminary survey of the environmental conditions at the test site and to construct or modify the instrumentation accordingly.

A number of the more frequently encountered unfavorable environmental conditions found in underground workings follow next together with suggested precautions and remedial procedures.

Temperature

The rock and air temperature in underground mines ranges from 0° to 140°F. The rock temperature is usually relatively constant, even over periods of a year or more, but the daily air temperature variation may be as much as 20°F, and seasonal changes of 40 to 50°F are not uncommon. Therefore gages or instruments mounted in or on rock are subject to only small temperature changes, whereas equipment exposed to the air experiences much larger changes. Most electronic equipment, especially transistorized amplifiers or solid state devices, should be temperature stabilized. Some types of electronic and electrical apparatus may not function properly or may be subject to early failure at temperatures over 100°F. Resistance and photoelastic strain gages must be temperature compensated or corrected. Extensometers or other deformation measuring devices should be temperature corrected, or temperature effects may be minimized by making the principal components out of invar steel. Provisions for heat dissipation must be made in many types of electrical equipment when operated at elevated ambient temperatures.

Water and Moisture

With the exception of salt mines and some nonmetallic mineral mines, the atmospheric moisture in most underground workings is usually near the dew point, that is, near 100% humidity. Sometimes there is standing water on the floor or a drip from the roof or side walls and, in some instances, the water is relatively acid. Needless to say water and moisture are detrimental to most equipment, especially to electronic, electrical, and optical apparatus. A part of the precaution to prevent moisture problems can be made before taking equipment underground by providing protective coatings and proper seals. However, this type of treatment usually is not sufficient to protect a complete measuring system.

When equipment is taken from the cool surface to warmer underground sites, special precautions should be observed since moisture will condense under these conditions on the cooler surfaces. Camera lenses should always be checked for condensation before taking underground pictures. To prevent condensation in nonoperating (cool) equipment, heaters, such as electric light bulbs, can be placed in instrument cases and left turned on. Instrument cables should be water-proofed with multiple layers of rubber tape or other sealing materials like liquid neoprene rubber. When possible, O-rings should be used to seal encased mechanical parts as, for example, in the borehole deformation gage described in Section 9.4.

Dust and Dirt

Dust and dirt affect practically all classes of scientific equipment, therefore instruments should be transported in dust and dirt-proof cartons, or in containers such as polyethene bags (special care should be exercised during underground transport and handling). Amplifiers, strain-gage bridges, and other electronic and electrical equipment may be kept in boxes with dust seals. However, if the equipment generates much heat, ventilation blowers with dust filters are necessary to keep the equipment cool. Filters are especially important if the dust is corrosive, like that in a salt mine. For more permanent installations, a small sheet metal room should be constructed with a filtered air supply or possibly an air conditioning system. Wood construction should be avoided underground because of the fire hazard.

Some underground atmospheres contain methane or other hydrocarbon gases in sufficient concentration to be explosive. In this instance "permissible" equipment conforming to the U.S. Bureau of Mines regulations should be employed. Certificates of permissibility can be obtained from the United States Bureau of Mines, Pittsburgh, Pennsylvania.

Some mine atmospheres may also contain small concentrations of hydrogen sulfide or nitrous oxide (as a biproduct of explosives). These gases may be absorbed in moisture condensed on equipment to form a highly corrosive product. Usually if instruments are kept dry, damage from this cause can be minimized.

Electric Power Supplies

In most mines and other underground installations, alternating-current power is available, although in some of the older mines, especially coal mines, only direct current is used. Before taking equipment underground the voltage, frequency, and capacity of the electrical power should be determined. However, it is usually difficult to ascertain beforehand any information about the stability of the power or if the alternating-current wave form is relatively good. Voltage fluctuations may affect unregulated equipment, and the presence of switching transients or poor wave form will affect most electronic amplifiers, voltmeters, oscilloscopes, etc. Sometimes poor power supplies can be corrected by employing regulators and filters. If the power demands are small it may be more expedient to use battery-operated equipment.

Shock and Vibration

Because the transportation and handling of equipment in mines usually subjects equipment to strong shock and vibration, instruments, meters, etc., should be built to withstand relatively high shock and vibration levels. Sensitive optical or mechanical systems or components should be shock mounted, and extra precaution should be exercised in packing equipment for transportation to underground sites. After equipment is installed, it may be subject to strong air-borne vibrations from blasting. Even weak ground and air-borne vibration from distant blasting or the operation of mining machinery may be picked up by geophones, or may affect amplifiers if they are microphonic.

9.4   DEFORMATION MEASURING INSTRUMENTS

Instruments for measuring deformation, that is, comparatively small changes in length, include axial and diametral borehole gages, deformeters (or extensometers), fluid levels, and surveying instruments. The change in length measurements may range from less than 100 microinches in a borehole measurement to a number of feet in the case of surface subsidence. If deformation measurements are made in or on the surface of a solid body and the change in length is divided by the distance between the measuring points, the result is the average strain between these points. Thus it is evident that all strain gages are inherently deformation measuring devices. However, custom has established them as a separate class of instrument and they are treated accordingly in this book.

Surveying Instruments

Surveying instruments such as levels, theodolites, tacheometers, and transits have been used to measure lateral and vertical movement, both on the surface over mined areas and in underground workings. In some instances this movement is substantial, as, for example, in the subsidence areas over block caving operations where large voids are created on the surface, or over deposits mined by longwall methods, following which the surface caves or flexes into the stoped area. For these larger deformations an accuracy of 0.05 ft is usually sufficient, and this accuracy can be achieved with most conventional surveying equipment without difficulty. Surveying instruments have also been used for measuring the comparatively small lateral and vertical displacements over mined areas preliminary to failure. These displacements may be of the order of a fraction of an inch, and accurate leveling, chaining, and triangulation measurements are required. Precision levels and theodolites are available with a sensitivity of about 2 sec of arc, which would correspond to a displacement of approximately 0.001 ft at a distance of 100 ft. However, accuracies of this order usually cannot be achieved, partly because of the fact that these displacements occur over periods ranging from days to years, and the measurement must necessarily include the error caused by repeated setups of the instrument at the triangulation stations. This source of error can be reduced by erecting concrete monuments at the triangulation station on which a mounting plate is permanently attached for accurately positioning the surveying instrument. Another source of error results from a seasonal movement of the sighting pins or targets caused by freezing and thawing or variable moisture conditions in the soil. This error may be reduced by anchoring the sighting pins at a sufficient depth below the surface.

Fluid Levels

In some underground areas, such as abandoned stopes, it is not safe or otherwise feasible to make vertical displacement measurements with an instrument requiring an observer in the area. However, fluid levels obviate this requirement and at the same time provide an accurate and comparatively inexpensive means of measuring either large or small vertical displacements. This type of measurement is particularly useful in distinguishing between roof and floor movement. A basic version of this type of level

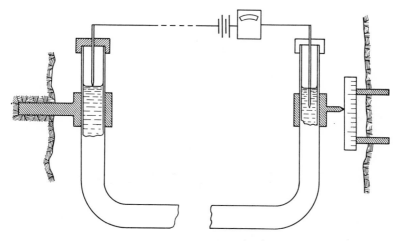

Fig. 9.4.1. Manometer level.

consists of a fluid-filled tube (water is satisfactory) with vertical end-section (Fig. 9.4.1), thus simulating a manometer. One end-section which contains an electrical contact with the fluid is secured to the point of measurement. The other end section, whose elevation is adjustable, is placed in the reference area. The tube is initially filled with fluid until electrical contact is made at the measurement end and the position of the reference end is recorded. If the point of measurement is displaced, the reference end is adjusted until the electrical contact is again established, and the displacement is read on the reference scale. Fluid levels have been constructed with vernier scales or micrometer adjustments at the reference end, and sensitivities of the order of 0.004 in. over a reference-to-measuring point distance of 150 ft have been claimed. A possible source of error may result from temperature variations between the two ends of the manometer, which can cause corresponding variations in the length of the fluid column. A brief summary of the various fluid levels developed during the preceding 30 years is given in a report by Meisser.[1]

Deformeters

Excluding rulers and tapes, the mechanically simplest deformation measuring devices are deformeters, also called extensometers, of which convergence gages are a special type. This class of instruments usually consists of a length-sensing device, such as a vernier scale, micrometer head, or a dial indicator attached to the end of a rod of fixed or sometimes incrementally adjustable length. The sensitivity of the instrument is generally chosen to meet the measurement requirements. Thus as the roof-to-floor convergence in coal of other relatively weak flat bedded rock may

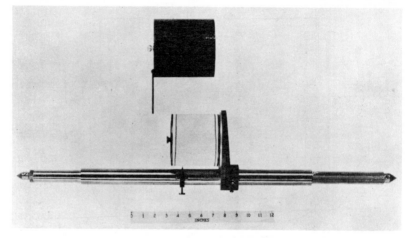

Fig. 9.4.2. Recording convergence gage. (After Greenwald et al.)

amount to several inches, a convergence gage may consist of no more than a steel scale read to an index on the end of a rod. The sensitivity of such an instrument would be about 0.01 in., the magnification unity, and the range, several inches. Greenwald,[2] et al. described a recording type convergence gage (Fig. 9.4.2) consisting of two spring-actuated telescoping tubes, to one of which was attached a recording stylus, and to the other a clock-motor driven recording drum. In a hard rock mine, the roof-to-floor convergence may amount to only a fraction of an inch at failure, and a convergence gage for this type of measurement would require a sensing element such as a micrometer head or dial indicator to provide a satisfactory sensitivity.

If the ends of a deformeter can be permanently anchored, then an accuracy comparable with the sensitivity of the gage can be achieved. Moreover, this type of installation will have an excellent time stability, especially if temperature corrections are made for the expansion or contraction of the bar. However, in many instances permanent installation is impractical because of possible damage to the instruments from mining operations. Merrill[3] described an installation in which deformation measurements were made between studs grouted in the roof and floor. The

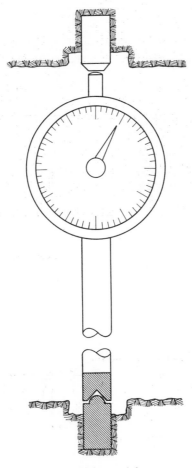

Fig. 9.4.3. Dial gage deformeter.

floor stud had a hemispherical end that mated with a conical end on the extensometer (Fig. 9.4.3), and the roof stud had a small flat area against which the movable end of a dial indicator contacted. This instrument was reported to reproduce to ±0.001 in. when the instrument was repeatedly removed and reinserted. The same type of instrument has been used for measuring the axial deformation of mine pillars. Merrill and Morgan[4] described an extensometer for use in an area that was inaccessible during

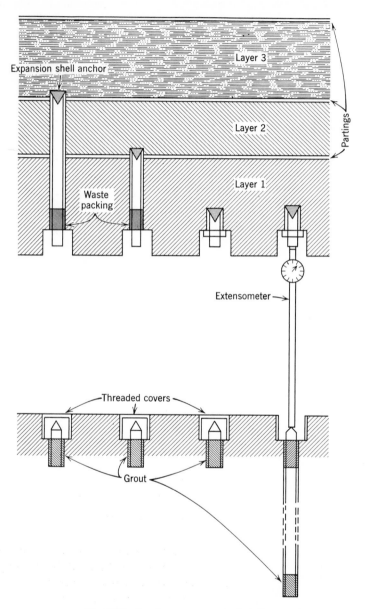

Fig. 9.4.4. Roof sag measurement station.

part of a test. The sensing element on this instrument was a linear differential transformer (Section 9.4 and Table 9.5.1), the output from which could be read on a meter outside of the test area. The base length of the instrument was 45 in., and the deformation sensitivity was reported to be 0.001 in.

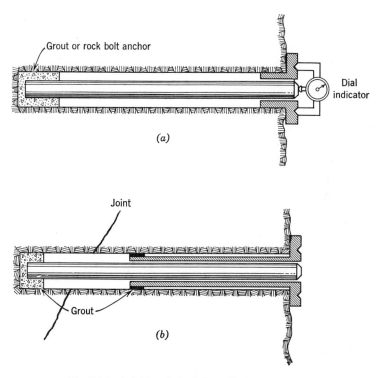

(a)

(b)

Joint

Grout

Grout or rock bolt anchor

Dial indicator

Fig. 9.4.5. Axial borehole deformation gages.

With some modification, extensometers can be anchored in boreholes so as to permit the measurement of deformation within a body of rock. Merrill[5] measured both the roof-to-floor convergence and the differential sag (Section 14.10) within roof strata by anchoring rods at various depths in the roof and measuring the change in the distance to deeply anchored rods in the floor (Fig. 9.4.4). Conventional rock-bolt anchors are usually satisfactory for securing the reference pins in the roof and floor. The lateral deformation in pillars and the side walls of stopes has also been measured by anchoring a rod in a borehole at a prescribed depth and measuring the deformation to a short tube grouted in the collar of the hole (Fig. 9.4.5a) or by anchoring a rod at one point in a hole and a tube at a second point (Fig. 9.4.5b). Several deformation indicators have been developed for measuring the displacement between the end of the rod and the tube, for example see Potts.[6] Usually these displacements are only fractions of an inch so that dial indicators or a micrometer head should be used as the length-measuring device. If this device can be permanently attached to the tube, deformations can be measured with an accuracy of about ±0.001 in. In some installations it is not expedient to leave the indicating device on the tube because of possible damage. In these instances it is necessary to develop a means of accurately reinstalling this device on the tube, a procedure that usually lowers the accuracy of measurement by severalfold.

Borehole Deformation Gages

In an elastic rock, a change in the secondary principal stresses (Section 1.7) in the plane normal to the axis of a borehole will produce a diametral deformation in the hole. Measurement of this deformation can be used to determine the change in these secondary principal stresses. Moreover, if the change in the secondary principal stresses is produced by a complete stress relief, then the measured diametral deformation can be used to determine the absolute magnitude and direction of the secondary principal stresses (Section 14.3). Thus, as a means for determining in situ rock stress, the measurement of the deformation in a borehole is of special significance to structural rock mechanics.

The magnitude of the diametral deformation in rock is small. For example, consider a 1 in. hole in a rock having a compressive strength of 5000 psi and a modulus of elasticity of 3×10^6 psi. If the hole is subjected to a uniaxial compression the maximum diametral deformation would be 0.005 in. (from Eq. 14.3.5). The maximum diametral deformation in a stronger rock would be comparable because the modulus of elasticity of rock generally increases with its strength. Thus a diametral deformation gage should have a range greater than 0.005 in. and a sensitivity equal to approximately 1% of this value, or 0.00005. If a borehole gage is to measure the change in the diametral deformation occurring over an extended period, say, the time required to produce a significant increase in a mined area, the instrument should have a good time stability, that is, it should not drift (Section 9.2). The drift rate of a diametral deformation gage should be determined at the maximum value of the deformation to be measured. Usually the drift rate of an instrument increases as the measured deformation increases. A satisfactory test for determining the drift rate is as follows (see Obert, Merrill, and Morgan[7]):

1. Place the instrument in a proving ring, load to a specified deformation, and record the initial load and deformation value.
2. Increase and maintain the load on the proving ring at a value such

that the difference in the deformation is comparable with that expected in field measurement. Record the deformation during this loaded interval. Also, during this interval subject the instrument to the range of environmental conditions expected in the field.

3. After a specified time (as nearly equal to the expected time of field measurement as possible) decrease the load on the proving ring to its initial value and record the deformation immediately and for a short period thereafter.

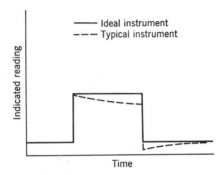

Fig. 9.4.6. Calibration to determine drift.

4. A typical and an ideal deformation versus time chart is shown in Fig. 9.4.6 from which the total drift and the drift rate for any specified time can be determined.

Jacobi and Brändle[8] described a borehole deformation gage manufactured by H. Maihak, in which the diametral sensing contacts act through a system of levers to produce a change in the tension, and hence the frequency of a vibrating wire. This change in frequency is compared with the frequency of a calibrated vibrating wire in an indicating unit. The sensitivity of this gage and indicating unit is excellent and it is claimed to have a good time stability and to be unaffected by environmental factors. However, the gage must be preloaded (to place the sensing contacts against the borehole wall) after it is inserted in the borehole, a procedure that requires a special placement tool to actuate the preloading mechanism within the gage; hence, the mechanical system in the gage is rather complicated. The range of the gage is given as 0.004 in., but it may be reset during the course of a measurement. The vibrating wire transducer in this gage has a low compliance (that is, it is stiff) so that a considerable contact force is required to actuate the sensing mechanism, a factor that in softer rocks may cause some creep at the contact points.

Hast[9] developed a diametral deformation gage for use in a borehole approximately 1 in. in diameter. This gage utilizes a magnetostrictive transducer as a length-sensing device. The gage is so constructed that a change in the diameter of a borehole produces a proportional change in the strain in a nickel core around which a coil is wound. Because of the magnetostrictive effect, the permeability of the core, and hence the impedance of the coil, is strain dependent. The gage is calibrated by measuring the impedance versus deformation relationship in boreholes in

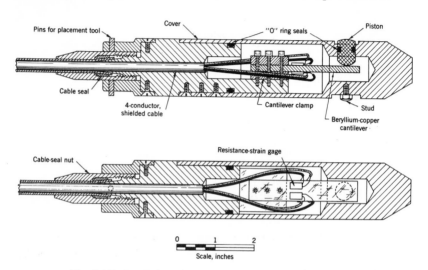

Fig. 9.4.7. Borehole deformation gage. (After Obert et. al.[7])

metal blocks having known elastic moduli. The indicated sensitivity of this gage is good. Like the Maihak gage, the Hast gage must be preloaded after insertion in the borehole. Also, the deformation measuring range is small and hence the gage must sometimes be reset during the course of a measurement. The compliance of the transducer in the Hast gage is low and the contact pressure high, a condition that makes it necessary to calibrate the gage in a block of the rock under investigation. Hast pointed out that a high contact pressure may be a desirable feature because soft or otherwise inelastic areas may be detected at the time the gage is pre-loaded.

Obert, Merrill, and Morgan[7] described a diametral deformation gage for use in a 1½ in. diameter borehole. The transducing element in this instrument is a beryllium-copper cantilever on which four strain gages are mounted and connected as a four-arm bridge (Fig. 9.4.7). This cantilever is actuated by a piston that contacts the walls of the hole. The maximum piston contact force is 30 lb; thus the gage has a relatively high compliance and no correction is necessary for contact force. The sensitivity of the gage when used in conjunction with a conventional strain indicator is approximately 19 microinches, which in a rock with a modulus of elasticity of 3 × 10⁶ psi corresponds to a change in stress of 13 psi. This instrument is stable over long periods and is comparatively insensitive to temperature or other environmental factors. The deformation measuring range of this gage is 0.044 in., which is large enough so that it does not have to be reset during the course of a measurement; in fact, the range is so large that normal variations in the initial borehole diameter can usually be accommodated as well. However, if the variation in the initial hole diameter is excessive, a stud with a proper head thickness can be selected at the time the gage is being placed so that the limits of the measuring range are not exceeded. Because of this large measuring range the gage placement problem is simplified.

Linear Variable Differential Transformers

The linear variable differential transformer is a transducer for converting small displacements into changes in electrical current. These devices are available in a variety of sizes ranging from less than $\frac{1}{2}$ in. to several inches in length, and with displacement measuring ranges from 0.005 to several inches. The transformer consists of a primary coil P and two secondary coils S_1 and S_2 wound in opposite directions, within which there is a movable iron core C, Fig. 9.4.8a. If a low frequency a-c current (60 to 20,000 cps) is supplied to the primary, and if the iron core is centered, the induced voltages in the secondary coils E_1 and E_2 are equal and opposite; hence the output voltage E will be zero. If the core is displaced in either direction from the center position, the unbalanced voltage induced in the secondaries is proportional to the core displacement (Fig. 9.4.8b). As such, the differential voltage in the secondaries is independent of the frequencies (within the specified operating frequency range) but dependent on the voltage applied to the primary. Linear variable differential transformers possess many desirable qualities; they are mechanically simple, relatively unaffected by ambient temperature and humidity changes, and their measurement range is large. When used with a sensitive vacuum-tube voltmeter, sensitivities of the order of 10^{-5} in. can be achieved. On the other hand, the output from the transformer is proportional to the input voltage and hence an accurately regulated voltage supply must be used. Because these devices will permit remote operation, they are particularly suitable for use as a length-sensing element in deformeters designed for installation in areas that become inaccessible or unsafe during the course of measurement.

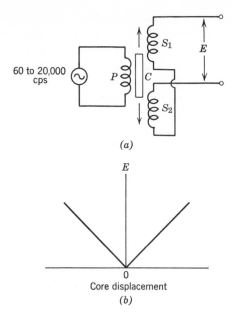

60 to 20,000
cps

S_1

E

P C

S_2

(a)

E

0
Core displacement
(b)

Fig. 9.4.8. Circuit (a) and response (b) for a linear differential transformer.

9.5   STRAIN-MEASURING INSTRUMENTS

In general, strain gages are extremely sensitive devices, for example, the compressive breaking strain of some rocks is as small as 0.001 in./in. (Table 9.2.1), and tensile breaking strains are even smaller. For a strain gage to have a sensitivity of 1% of this compressive breaking strain, the smallest increment of measurement should be about 10 microstrains (microstrain = strain $\times$ 10^{-6}, denoted by $\mu\epsilon$). Unlike extensometers and convergence gages that are usually improvised to meet the needs of a particular experiment, strain gages require a constructional precision outside the capabilities of the average investigator. Fortunately, a wide variety of strain gages are manufactured commercially and at a nominal price in relation to the sensitivity of the device. Because of the extreme sensitivity of most strain gages, special techniques are required in their use. This art is extensive and outside the scope of this book. However, most strain gage manufacturers issue very complete instructions on the use of their product, and because this information is so generally available, only a brief description of the basic strain-gage types that have been used in rock mechanics investigation is presented, together with some special precautions and instructions in relation to the use of these gages on rock. In Table 9.5.1 a number of the more frequently used strain-measuring

Table 9.5.1. Characteristics of a Number of Commercially Available Strain Gages

Type of Gage	Strain Sensitivity Microstrains	Magnification	Gage Length Range, inches	Measuring Range, inches
Bonded resistance wire*	2–4	2500–5000	$\frac{1}{8}$–6	0.0012–0.12
Unbonded resistance wire*	5	2000	1	0.003
Tuckerman (optical)	4	2500	$\frac{1}{4}$–4	0.0045 (1 in. gage)
Differential transformer	10	—	$\frac{1}{2}$–6	0.050–3.0
Vibrating wire (Maihak)	1	10,000	4	0.004
Photoelastic (circular)	40	—	$1\frac{1}{4}$ D	—
Photoelastic (linear)	70	—	1–2	—
Dial indicator (Starrett)	50	750	—	0.25

* With conventional strain indicator.

devices are listed, together with their strain sensitivity, magnification, range of available gage lengths, and measuring range.

Bonded and Unbonded Resistance-Strain Gages

Resistance-strain gages together with their companion indicators provide an accurate means of measuring strain. The experimental procedure is comparatively simple and the necessary equipment is relatively inexpensive. Although there are a number of books explaining the fundamentals and the art of using resistance-strain gages, special techniques for using these on rock surfaces not considered in these references are required. To provide a basis for considering these problems, the fundamental concepts involved in this process are reviewed briefly.

Resistance-strain gages can be divided into two general classes, bonded and unbonded. Bonded gages are made either by cementing a grid of fine wires to a paper or bakelite backing, or by depositing a metallic grid on a plastic (epoxy) film. This gage element is cemented to the surface on which the strain measurement is to be made, hence the designation, bonded gage. If the cementing operation is properly executed, any strain produced on the surface will be transmitted to the wire or foil grid.* Unbonded wire gages are made by stretching a wire or grid of wires between two contacts attached to the test area. The wires must be pretensioned if the gage is to respond to both tensile and compressive strains. Unbonded wire gages are commercially available as encased units in which the strain is transmitted to the wire through knife edges in contact with the surface.

The principle of the resistance-strain gage is quite simple: if a wire is lengthened, its cross-sectional area is decreased and its electrical resistance

* Bonded gages are also available that employ piezo-resistive crystals (single crystals of silicon).

increased proportionately. The change in resistance per unit resistance $\Delta R/R$ is proportional to the change in length per unit length $\Delta L/L$, that is,

$$\frac{\Delta R}{R} = G \frac{\Delta L}{L} \qquad (9.5.1)$$

where G is the *gage factor*. The magnitude of G depends on the type of resistance wire used in the gage. For constantan (advance) wire, which is

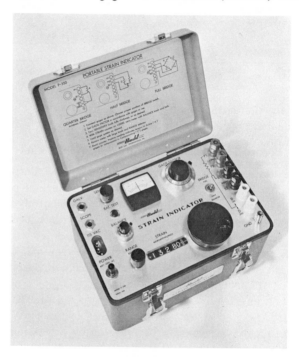

Fig. 9.5.1. Strain gage indicator. (Courtesy Budd Co.)

commonly used because of its low temperature-resistivity coefficient, the value of G is approximately 2.0. Thus the resistance-strain gage is a transducer for converting change of strain to a change in resistance. The indicating device used in conjunction with a strain-gage element is a resistance bridge calibrated in strain units (Fig. 9.5.1). This bridge must be relatively sensitive, for example, to have a strain sensitivity of 5 micro-strains (one part in 200,000); a strain indicator when used with a conventional 120 ohm gage element (gage factor = 2.0) must have a resistance sensitivity of 0.0012 ohm. The measurement of changes in resistance of this magnitude creates a number of problems; the most frequently encountered is that due to moisture. For example, if a leakage resistance due to moisture of 1.2 megohms develops across the terminals of a gage element or the leads to the indicator, an error of approximately 50 micro-strains would result. Note that this error results from a *change* in the leakage resistance during the measurement period—a fixed resistance of 1.2 megohms across a 120 ohm gage element would only produce an error of about 0.005%. Thus moisture becomes a serious problem in effecting measurements over extended periods with resistance-strain gages.

The resistance of metals (including strain-gage wires) is temperature dependent, and any change in resistance caused by temperature cannot be distinguished from a change caused by strain. However, this effect can be eliminated by employing temperature compensated gages or a temperature compensating gage in an adjoining arm of the indicating bridge. This gage, referred to as a dummy, should be cemented to an unstrained piece of the same rock on which the active gage is mounted (to avoid differences in thermal expansion) and placed in the proximity of the active gage.

The problem of moisture-proofing, resistance-strain gages bonded to metals or other nonporous materials is relatively simple. Usually if the attached gage is coated with one of the waxes recommended by the gage manufacturer, a satisfactory protection will be achieved. However, most rocks are porous to the degree that moisture will migrate to the gage-rock interface in an amount sufficient to cause a shunting resistance across the gage and in some instances to affect the bonding properties of the cement. If the rock is being tested in the laboratory, this difficulty can be eliminated by carefully drying the rock at slightly above room temperature and then bonding the gage to the rock with an epoxy cement hardened by heating with an infrared lamp. Finally the gage should be coated with a protective wax. The use of resistance-strain gages on in situ rock is much more difficult. If the above procedure is followed, a satisfactory bond may be achieved for a short period, that is from 24 hr to several days. However, the use of resistance-strain gages in direct contact with rock is not recommended for longtime underground measurement.

Another factor seriously affecting the long-time stability of bonded gages, is that most cements used for attaching gages are plastics that creep when strained. Over long periods this creep causes a reduction in the strain transferred to the wires in the gage element, and hence an error in the measured strain. The effect can be minimized by using epoxy-backed, foil-type gages cemented to the test surface with an epoxy cement cured by several cycles of heating and cooling. However, this procedure is not suitable for underground application. Also epoxy or other cements may permeate the rock below the gage to the extent that the elastic properties of the rock are altered.

Vibrating Wire Strain Gage

The vibrating wire or acoustic-strain gage utilizes the principle previously described in Section 9.4, namely, that the vibration frequency of a wire depends on the tension in the wire. Jacobi and Brändle[8] described a strain gage, the transducing element in which is virtually identical to that in the vibrating wire, borehole-deformation gage. The change in length between two knife edges in contact with the specimen acts through a lever to cause a change in length and therefore tension and frequency of a vibrating wire. The indicator used with this gage is the same as that used with the vibration wire, borehole-deformation gage. This type of gage is extremely sensitive, stable, and not subject to moisture or electrical problems. However, the gage has a very small measuring range—about 0.004 in.

Tuckerman Optical Strain Gage

The Tuckerman strain-measuring system consists of two principal components, an autocollimating telescope and an optical strain gage (Fig. 9.5.2). A simplified functional explanation of the system is as follows: light from the source L emerges from the autocollimator as parallel light

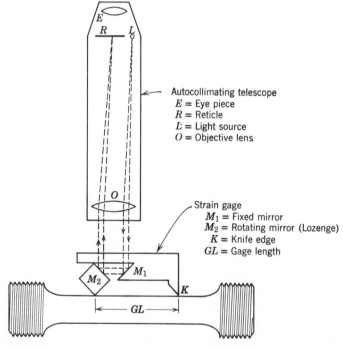

Fig. 9.5.2. Tuckerman optical strain gage (not to scale).

and is directed to the fixed mirror M_1, reflected to the rotating mirror M_2, then reflected to the autocollimator, and finally focused on the reticle R, where the image is observed through the eyepiece E. If the specimen is strained (lengthened or shortened), the mirror M_2 is caused to rotate, thereby changing the angle between the incident and reflected light from the strain gage, and correspondingly causing a proportional displacement of the image on the reticle. Since the displacement of the image on the reticle depends only on the angle between the incident and reflected light from the strain gage, the autocollimator may be placed (or hand-held at any convenient distance or position with respect to the strain gage without introducing any significant error.

The Tuckerman gage has several advantages over cemented types of gages for measurement of strain on rock surfaces. First, because the gage is mechanically fastened to the surface, a procedure taking only a few seconds, it can be used immediately after installation; second, the gage can be used on comparatively rough surfaces; third, there is no tendency for the gage to creep on the surface during the measurement period, and fourth, the gage can be used on wet surfaces or on surfaces at elevated temperature (up to 500°F). The sensitivity of a 1 in. (gage length) gage with a 0.2 in. rotating mirror is 4 microstrains and the deformation range is 0.02 in.

Photoelastic Strain Gages

In Chapter 12 procedures are described for determining the magnitude and direction of stresses at points in a photoelastic model. The stress-dependent pattern produced in the model results from the transmission of polarized light through the birefringent model material. As such the method cannot be used for determining the strains in an opaque body. Zandman and Wood[10], developed a technique for cementing a birefringent material with a reflective backing to the surface of an opaque body. Polarized light incident on this patch will produce patterns in the photoelastic material dependent on the strain transmitted to the patch from the body. From an analysis of these patterns the direction and magnitude of the strain can be determined; hence the patch is in effect a strain gage. In a later modification the reflective backing was eliminated by cementing the patch to the surface with a reflective cement (an epoxy resin loaded with aluminum powder).

Photoelastic strain gages utilizing the back-reflecting principle are commercially available. One version (Fig. 9.5.3a) consists of a thin plate of birefringent material, coated on the under side with a metallic reflecting film (Fig. 9.5.3b). The plate is prestressed during manufacture "freezing in" a series of fringes. These fringes, which shift when the plate undergoes longitudinal strain, are made visible by a polarizing film on the outer face. When cemented to a surface, these gages indicate strain in the direction of

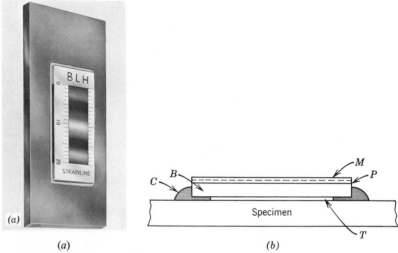

(a)   (b)

Fig. 9.5.3. (*a*) Linear photoelastic gage. (Courtesy Baldwin Lima Hamilton Corp.)
(*b*) *B*—Reflective backed birefringent plate. *C*—Cement, *P*—Polarizer, *M*—Mask,
T—Rubber tape.

the gage orientation. They may be used in either ordinary daylight or
artificial light without any accessory equipment. The sensitivity of this
gage is less than 70 microstrains. Figure 9.5.4 is a circular photoelastic
strain gage consisting of a birefringent layer containing a frozen circular

Fig. 9.5.4. Circular photoelastic gage. (Courtesy Budd Co.)

fringe, a reflective backing, and a polarized layer on the front surface. When the gage is cemented to a surface and illuminated with either daylight or artificial light, it will permit the determination of the direction and magnitude of the principal strains. The direction of the principal strains can be obtained by inspection of the fringe pattern, but evaluation of the magnitudes requires a more detailed analysis. Magnitudes can be measured within ± 40 microstrains, directions within $\pm 5°$.

Bonded photoelastic-strain gages are subject to some of the errors inherent to bonded resistance-strain gages, namely errors due to creep or ageing of cement, or to the cement permeating the rock.

9.6 RIGID INCLUSION STRESS GAGES

A number of procedures have been developed for determining the stress in an elastic body but, in general, strains or deformations are measured and stresses are calculated by using relationships involving the elastic constants of the body material. Thus the analysis of stress is in most instances an analysis of strain. However, the rigid inclusion stress gage, when used within specified operational limits, is an exception to this generalization in that this instrument provides a direct means of measuring the stress in a body. Moreover, the measurement does not require any specific information about the elastic properties of the body material. This latter point can be used to distinguish between stress and strain, or deformation gages; that is, a stress gage should measure stress in a body, and a strain gage should measure strain in a body without requiring any knowledge of the elastic properties of the body material. Some so-called stress gages require a calibration in *each* rock type in which measurement is made. Although it may not be immediately evident, this calibration procedure usually involves the elastic properties of the rock (otherwise any material could be used in the calibration), and instruments of this type are not in the true sense stress gages.

The theoretical basis for the rigid inclusion stress gage is a consequence of Sezawa and Nishamura's[11] treatment of the stress distribution in a circular inclusion in a plate subjected to biaxial stress. In this treatment the plate and circular inclusion are assumed to be isotropic, homogeneous, linear-elastic, and of different elastic moduli. Also, the bond across the inclusion-plate boundary must be mechanically perfect. On the basis of their results, Coutinho[12] derived relationships between the biaxial principal stresses applied to the plate $P_r Q_r$ and the stresses $P_b Q_b$ developed in a rigid inclusion in the plate:

$$P_r = KP_b + kQ_b$$
$$Q_r = KQ_b + kP_b$$

(9.6.1)

The constants K and k depend on the ratio of the elastic moduli of the plate and inclusion E_r/E_b and on their relevant Poisson's ratios v_r and v_b.* These relationships are shown graphically in Fig. 9.6.1 from which it is evident that the effect of Poisson's ratio is small, and for $E_b > 2E_r$ the stress in the inclusion is virtually proportional to the stress in the plate and independent of the elastic modulus of either. Coutinho suggested

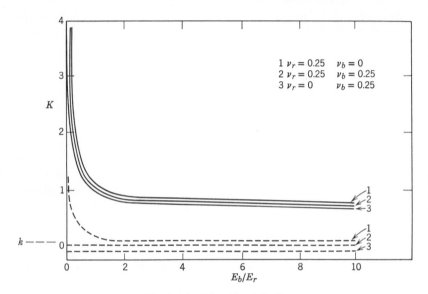

Fig. 9.6.1. (After Coutinho[12])

that on the basis of this relationship a gage for measuring stress could be developed.

Utilizing this principle, Wilson[13] developed a gage for measurement in rocks having a modulus of elasticity of 7×10^6 psi or less. The construction requires a brass rod ($E_b = 13.7 \times 10^6$ psi) to be split lengthwise and two resistance-strain gages (connected in series) cemented with an epoxy resin in a thin cavity in one of the halves (Fig. 9.6.2). The two halves are then cemented together with epoxy resin and the assembly machined with a $1°$ taper. In service the gage is pressed and epoxy cemented into a

* The relationships for K and k given by Wilson[13] are:

$$K = \frac{(1 + v_b)(3 - 4v_b)}{8(1 - v_r)(1 + v_r)} \frac{E_r}{E_b} + \frac{5 - 4v_r}{8(1 - v_r)}$$

$$k = \frac{(1 + v_b)(1 - 4v_b)}{8(1 - v_r)(1 + v_r)} \frac{E_r}{E_b} + \frac{4v_r - 1}{8(1 - v_r)}$$

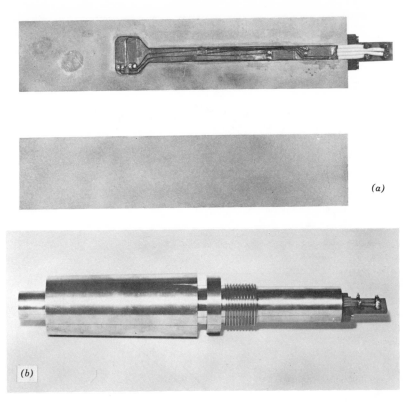

Fig. 9.6.2. Rigid inclusion gage. (After Wilson.[13]) (*a*) Two halves ready for joining. (*b*) Assembled gage.

1° tapered hole in the rock. This procedure insures a satisfactory gage-to-rock bond and at the same time compressively prestresses the gage so that it can respond to tensile stress.

The gage response was checked by cementing it in a biaxially loaded slab of rock. It was found that if the slab was loaded in the direction normal to the plane of the strain gages, the response in the direction of the strain gages was greater than that calculated from the value of Poisson's ratio for brass. A factor was empirically determined to correct this cross-sensitivity.* When so corrected the empirically determined response of the gage was in good agreement† with that calculated from theory (Fig. 9.6.3).

---

* A number of strain-measuring devices require correction for cross-sensitivity, and all instruments of this type should be checked for this effect.
† The procedure for determining the theoretical response of this gage corrected for cross-sensitivity is given in the report by Wilson.[13]

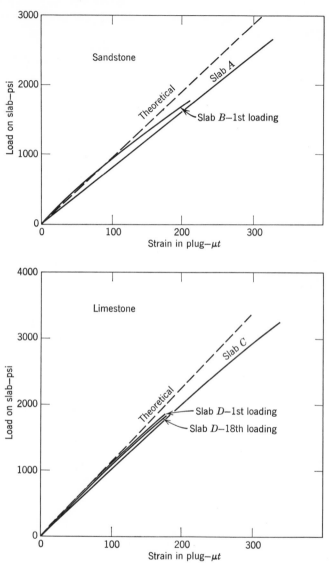

Fig. 9.6.3. Load-strain response for rigid inclusion gage. (After Wilson.[13])

Although the gage response is relatively independent of the elastic moduli of various rocks (provided $E_b > 2E_r$), a better agreement with theory can be obtained if the constants K and k in Eqs. 9.6.1 are calculated using even approximate values for the modulus of elasticity of the rock. Wilson found that the rigid inclusion gage will respond properly to a compressive stress even if it is not prestressed, but the gage will not respond to tensile stresses unless it is prestressed. Other tests showed that relatively accurate results can be obtained in fractured rock (with E_r between 0.5×10^6 and 3×10^6 psi) in which a $1°$ tapered hole cannot be reamed by cementing the gage in an oversize hole with an expanding Portland cement (E-cement $= 1.5 \times 10^6$ psi).

The rigid inclusion-type gage is limited in its application because it is unidirectional, that is, it will respond only to the component of stress in the direction in which it is oriented. Thus if it is desired to measure the maximum compressive stress in rock, it may be assumed that the direction of the maximum stress is vertical and the gage oriented accordingly. In coal measure rocks, the direction of the maximum compressive stress is probably vertical, but in other rock formations this assumption may not be valid. Also, the gage cannot be used to measure absolute stress by the overcoring procedure because the prestress in the gage would normally split the overcore. Because the gage is cemented in the hole it must either be sacrificed or recovered by some overcoring procedure. Despite these disadvantages the rigid inclusion gage is one of a few types of instruments that can be used for determining changes in stress in rocks that tend to creep or that are otherwise inelastic.

Potts and Tomlin[14] described a tapered, rigid inclusion-type gage which fits into a split tapered sleeve (Fig. 9.6.4). This gage and sleeve assembly can be pretensioned in a straight (nontapered) hole. The transducing element in this gage utilized a thin, oil-filled cavity within the body of the gage. When the cavity is deformed, the displaced oil activates a diaphragm

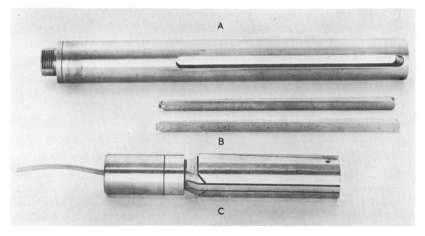

Fig. 9.6.4. Rigid inclusion gage with tapered mounting sleeve. (After Potts and Tomlin.[14])

on which resistance-strain gages are mounted. This strain is measured with a conventional strain indicator. Potts and Tomlin reported that if accurate results are desired, the gage must be calibrated in each rock type in which it is employed.

9.7 HYDRAULIC CELLS—FLATJACKS

Hydraulic cells are thin-walled, fluid-filled metal bladders or containers (Fig. 9.7.1) designed to withstand a hydraulic pressure of several thousand pounds per square inch when properly confined by cementing or grouting

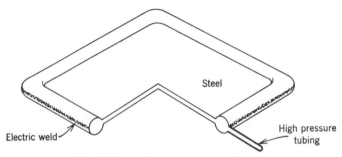

Fig. 9.7.1. Section through flat-jack.

them in a borehole or slot in rock. These cells are used for a variety of purposes in rock mechanics investigations (Chapter 14) including the measurement of force (dynamometers), as jacks for loading mine pillars, for applying a force to in situ rock,* as pressure transmitters, and as pressure sensors†.

These cells are fabricated by forming two identical halves out of malleable sheet metal and brazing or welding the halves together (Fig. 9.7.1). Copper, brass, or mild steel has been used for this purpose. Panek and Stock[15] have constructed cells ranging from 2 × 8 in. for use in boreholes to 48 × 60 in. for slots. The maximum pressure that cells will withstand depends on a number of factors: the type and thickness of the body metal, the brazing or welding technique, and even more important it depends on the quality of the grouting material and the care exercised in preventing the formation of voids in the grout during installation of the

---

* Flat, thin hydraulic cells, referred to as flatjacks, were first developed for prestressing concrete or otherwise loading parts of concrete structures.

† Panek and Stock[15] make a distinction between jacks which apply a pressure to an area but are not capable of deforming if the area is deformed and therefore not capable of maintaining a uniform load over the area, and pressure transmitters which can deform so as to apply a uniform load to a correspondingly deforming area. Pressure sensors also have this same deforming capability.

cells in slots or boreholes. A typical 12 × 16 in. cell made of $\frac{1}{16}$ in. mild steel with an electric welded joint, and grouted in a slot with an expanding-type, Portland cement-sand grout withstood a pressure of 8500 psi before failing, and then the cause of failure was attributed to the rock rupturing.[15] Although the pressure in the cells is temperature dependent, the variation in temperature in rock is usually so small that this source of error can be neglected. Except for the possibility of leaks which are not difficult to prevent, hydraulic cells have good stability (do not drift) over long periods and they are particularly well suited for use in an underground environment.

9.8 DYNAMOMETERS—LOAD CELLS

Dynamometers or load cells are instruments for measuring force or load. In rock mechanics investigations they have been used for measuring the load on props and chocks, the tension developed in rock bolts, and the pressure (load on a given area) developed in backfill or in caved areas (self-fill) in worked-out stopes.

As previously noted (Section 9.7), thin, flat hydraulic cells can be used as dynamometers if they are properly constrained. For example, this type of cell has been used to measure the pressure in hydraulic backfill which is sufficiently fluid to provide the necessary constraint. Because the bladders are stiff and offer some resistance to external pressure, they should be calibrated, preferably in an identical ambient constraint.

Several laboratories have fabricated load cells using as the transducing element a resistance-strain gage mounted on a metal annular ring, (Fig. 9.8.1a). Barry et al.[16] developed a cell for measuring the tension in rock bolts utilizing this type of transducer. Eight resistance-strain gages were bonded to the exterior surface of the annular ring and connected as a four-arm bridge (Fig. 9.8.1b). With this arrangement the gages are temperature compensating. By securing end plates to the annular ring this transducing element is converted into a load cell suitable for use in measuring prop loads or the pressure in backfill. Usually in this type of instrument the gages are mounted on the internal surface of the ring, thereby making them less vulnerable to moisture or mechanical damage. This same principle is used in commercially available dynamometers.

9.9 INSTRUMENTS FOR VISUAL OR PHOTOGRAPHIC OBSERVATIONS IN BOREHOLES

Stratascope

The stratascope is a device for visually or photographically examining the interior of a borehole in rock to locate and observe partings, fractures, and layers of competent and incompetent rock.

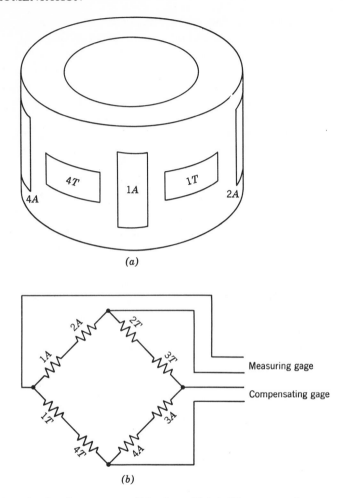

Fig. 9.8.1. (*a*) Annular ring dynamometer. (After Barry.[16]) 1*A*, 1*T* etc., are strain gages. (*b*) Four arm bridge for temperature compensation.

The instrument can be used quantitatively to measure displacements across partings, joints, and fractures, and from repeated measurements to determine the change in these displacements. The instrument consists of a low-power telescope with a calibrated reticle, a light source for illuminating the interior of the hole, and a pair of prisms for directing the light (Fig. 9.9.1). Provision is made for attaching a camera. An extendable stratascope developed by the U.S. Bureau of Mines is made in 5-ft sections to permit observation in holes from 5 to 20 ft long. The diameter of the tubes is 2 in. so that the device can be used in a BX diamond drill hole.

Diamond drill holes are more satisfactory than percussion or auger drill holes because the interior surfaces are relatively smooth, free from scoring, and rifling. Also diamond drilling does not produce as much chipping at the edges of partings, fractures, and joints—a factor that tends to obscure

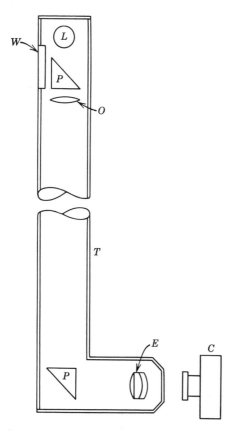

Fig. 9.9.1. Stratoscope. *L*—Light. *W*—Window and reticle. *P*—Prism. *O*—Objective lens. *T* = 5 ft–20 ft Extendable tube. *E*—Eyepiece. *C*—Camera.

these mechanical defects. In the photograph of a fracture, Fig. 9.9.2, the reticle divisions are 0.05 in. apart, and as an estimate can be made to about $\frac{1}{5}$ division, the sensitivity is about 0.01 in. The Bureau of Mines has also developed a smaller diameter model designed to fit in an AX diamond drill hole. This instrument is 1 in. in diameter and it has a maximum length of 10 ft. It utilizes a telescopic gun sight having a variable magnification from 2× to 4×.

Fig. 9.9.2. Stratascope photograph. Reticle divisions = 0.05 in.

Borehole Cameras

The borehole camera is an important instrument for exploring and evaluating the geology of a proposed mine or site for an underground installation, especially if the rock contains weathered zones or other defects of geologic origin which may not be discernible from an examination of exploratory cores, or because the core recovery was poor. Also, the dip and strike of subsurface geological formations can be established which is usually not possible from an examination of cores.

Burwell and Nesbitt[17] described an optical borehole camera that can be used to photograph the interior of a 3-in. diameter (NX) diamond drill hole. In this camera a light source illuminates the wall of the borehole through a cylindrical window in the wall of the instrument. The reflected light from the wall of the hole again passes through the cylindrical window, and is reflected from the conical mirror into the camera. The instrument is synchronized so that as it is lowered or raised in the hole it will take a picture for every inch of travel; thus the entire length of the hole can be photographed. The instrument contains a compass which is also photographed so that the azimuthal orientation of the hole can be established. In this instrument the light intensity is adequate for both black and white and color photography.

The wall of the hole as viewed from the conical mirror by the camera appears as an annular ring. To improve the visualization of the hole, a projector was developed that projects the image on the film off of an identical conical mirror onto the interior of a cylindrical ground glass screen.

Viewed from the outside of the cylindrical screen, the wall of the hole appears undistorted.

Borehole Television Cameras

Borehole television cameras have been developed by the International Eastman Company, Hannover, West Germany, and by the Lawrence Radiation Laboratory, Livermore, California. Both cameras are $2\frac{1}{2}$ in. in

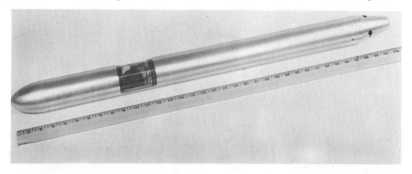

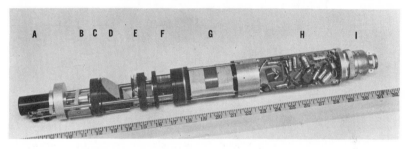

Fig. 9.9.3. Borehole television camera. (After Short.[18])

diameter and will operate in holes 3 in. in diameter or larger. The Eastman camera includes both an inclinometer for measuring deviation from the vertical and a compass. Thus the deviation of a hole from the vertical can be plotted and the strike and dip of geological formation determined. The Eastman camera is side viewing with a viewing angle of 52°. Two Lawrence Radiation Laboratory cameras have been described by Short.[18] One camera is side viewing with a viewing angle of 120°, and the other camera is end viewing (Fig. 9.9.3).

The resolution obtained with television cameras appears to be as good as that obtained by photographic methods. Television borehole cameras have the advantage over photographic cameras in that the image can be viewed directly. However, television borehole cameras will only reproduce in black and white.

9.10   SEISMIC EQUIPMENT

In Chapter 14 seismic methods are described for detecting incipient rock failure, delineating subsurface subsidence, determining the in situ elastic properties of rock, and for determining the stress in rock. These methods depend on the detection of low intensity seismic pulses created either by fracture or movement in rock, or by the detonation of small

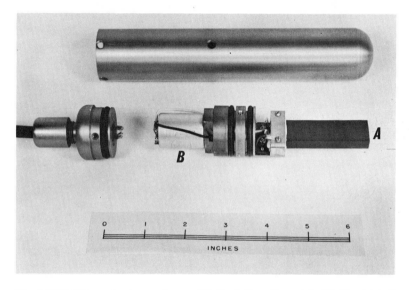

Fig. 9.10.1. Micro-seismic geophone. (A) Rochelle salt crystal. (B) Crystal-to-line transformer. (After Obert and Duvall.[19])

explosive charges. Conventional seismic equipment such as that used in geophysical exploration is not sufficiently sensitive for this purpose. However, the response of this type of equipment can be improved by increasing the electrical output of the seismic detectors (geophones) and/or by increasing the amplification of the geophone output.

Obert and Duvall[19] developed a high-sensitivity geophone and amplifier, collectively referred to as micro-seismic equipment, for the detection of these low intensity pulses. The mechanical-electrical transducing element in the geophone is a Rochelle salt piezo-electric crystal (bimorph bender) mounted as a cantilever (Fig. 9.10.1). The natural frequency of this undamped transducer is approximately 1000 cps, which is in the part of the spectrum of these seismic pulses containing relatively strong component frequencies. Consequently, this geophone responds predominantly to a frequency of, or close to, 1000 cps. This response does not affect the utility of the instrument because in its intended service the shape of the seismic pulse is of no concern. A cantilever is a mass-spring type of mechanical system, hence the output from the transducer is proportional to the interial force, and, for frequencies below its natural frequency this geophone is essentially an accelerometer. At its natural frequency the output from the transducing element is >30 volts/g (where g is the acceleration of gravity) which is about one-hundred times the output from accelerometers used in geophysical work.

The threshold of detection of any transducer is determined by the combined electrical "noise" inherent to the transducer and amplifying system, and the spurious electrical noise picked up from external sources by the geophone, geophone cable, and the amplifying system. With proper electrical shielding and amplifier design, the combined electrical noise (equivalent electrical noise at input to amplifier) for this type of equipment in service can be kept below 10 microvolts. Thus the threshhold of detection for an equipment with this specification would be about 10 microvolts, which for a geophone with an output of 30 volts/g would correspond to an acceleration threshhold of approximately 1/3 micro-g. The specification for a low-noise amplifier for use with this geophone is not within the scope of this book. However, in comparison with the amplifiers used with conventional geophysical-type accelerometers, the voltage gain should be about one hundred times greater.

REFERENCES

1. Meisser, O., "Seismiche Methode in der Gebirgsmechanik," *Ländertreffen des Internationalen Büros für Gebirgsmechanik*, No. 1 (1961).
2. Greenwald, H. P., E. R. Maize, I. Hartmann, and G. S. Rice, "Studies of Roof Movement in Coal Mines," *U.S. BurMines Rept. Invest.*, **3355** (1937).
3. Merrill, R. H., "Design of Underground Mine Openings, Oil Shale Mine, Rifle, Colo.," *U.S. BurMines Rept. Invest.*, **5089** (1954).
4. Merrill, R. H., and T. A. Morgan, "Method of Determining the Strength of a Mine Roof," *U.S. BurMines Rept. Invest.*, **5406** (1958).
5. Merrill, R. H., "Roof-Span Studies in Limestones," *U.S. BurMines Rept. Invest.*, **5348** (1957).
6. Potts, E. L. J., "Underground Instrumentation," *Sec. An. Symp. on Rock Mechanics, Colo. Sch. of Mines Quart.*, **52**, No. 3 (1957).
7. Obert, L., R. H. Merrill, and T. A. Morgan, "Borehole Deformation Gage for Determining the Stress in Mine Rock," *U.S. BurMines Rept. Invest.*, **5978** (1962).
8. Jacobi, O. and E. Brändle, *Electric Remote Measuring Instruments*, Glückauf, **92**, No. 1314 (1956).
9. Hast, N., "The Measurement of Rock Pressures in Mines," Sveriges Geol. Undersokn., Årsbok **52**, No. 3 (1958).
10. Zandman, F. and M. R. Wood, "Photostress," *Prod. Eng.* 1956 (Sept).

11. Sezawa, K. and G. Nishimura, "Stresses Under Tension in a Plate with a Heterogeneous Insertion," *Rep. Aero. Res. Inst., Tokyo,* **6** (1931).

12. Coutinho, A., "Theory of an Experimental Method for Determining Stresses not Requiring an Accurate Knowledge of the Modulus of Elasticity," *Internat'l Assn. of Bridge and Structural Eng. Cong.,* **9** (1949).

13. Wilson, A. H., "A Laboratory Investigation of a High Modulus Borehole Plug Gage for the Measurement of Rock Stress," *Fourth Symp. on Rock Mech. Penn. State Univ.,* 1961.

14. Potts, E. L. J. and N. Tomlin, "Investigation into the Measurement of Rock Pressures in the Mines and in the Laboratory," *Internat'l Conf. on Strata Control, Rept. D-4,* 1960.

15. Panek, L. A. and J. A. Stock, "Development of a Rock Stress Monitoring Station Based on the Flat Slot Method of Measuring Existing Rock Stress," *U.S. BurMines Rept. of Invest.,* **6537** (1964).

16. Barry, A. J., L. A. Panek, and J. S. McCormick, "Use of a Torque Wrench to Determine the Load in Roof Bolts, Part III, Expansion Type ¾-in. Bolts," *U.S. BurMines Rept. of Invest.,* **5080** (1954).

17. Burwell, E. B. and R. H. Nesbitt, "NX Borehole Camera," *Mining Eng.,* **6** (1954).

18. Short, N. M., "Borehole TV Camera Gives Geologists Inside Story," *UCRL Rpt.,* **7133** (1962).

19. Obert, L. and W. I. Duvall, "Micro-Seismic Method of Determining the Stability of Underground Openings," *BurMines Bull.,* **573** (1957).

CHAPTER 10

MECHANICAL PROPERTIES AND BEHAVIOR OF ROCK

10.1 INTRODUCTION

One factor necessary for designing or evaluating the stability of any kind of structure is a knowledge of the mechanical properties of its material, that is, a knowledge of how the material deforms and/or fails under the action of applied forces. The composition of most structural materials such as metals and concrete is uniform and reproducible to the degree that their mechanical properties in service are virtually the same as those measured in the laboratory; hence, a rational structural design can be made on the basis of published mechanical property values, or data obtained from the producers of structural materials.

However, the same equivalence does not exist for rock. The composition of even the more common rock types is highly variable: for example, sandstones may be bonded with silica or calcite, or otherwise cemented; or the quartz content in granite may vary by a factor of three or more. In addition, in situ rock is affected by geological actions such as faulting or jointing, that (*usually*) are followed by chemical processes that produce alteration and decomposition. Unless mechanical property tests are conducted at a scale such that the test specimen includes these defects in normal proportion, the results will not be representative of the in situ rock. The specimen size that satisfies this requirement is generally too large to be tested in the laboratory because of the physical limitations of test equipment. The alternative is to test in situ. This procedure is limited by difficulties encountered in preparing an area (specimen) of a shape such that the test results will be interpretable and also in applying a force of sufficient magnitude to an area of this size.

As a consequence of these limitations, virtually no laboratory tests have been made on specimens large enough to contain defects of geological

275

origin, and only a few in situ tests have been made under conditions that permit a satisfactory interpretation. Thus, except for the instances in which laboratory tests have been made on specimens taken from underground operations in uniform geologically undisturbed rock, the data relevant to the mechanical properties and behavior of in situ rock are meager. On the other hand, a large number of tests have been made on small, comparatively uniform, rock specimens to determine their mechanical properties and behavior. These tests can be divided into two general classes: static or time-independent tests, and time-dependent or creep tests. Static mechanical property tests are designed to measure the deformation and fracture of rock specimens subjected to a variable uniaxial, biaxial, or triaxial load. They furnish numerical strength and elastic property values and thus provide the quantitative aspects to rock testing. Static tests are considered to be time-independent because they are effected in a short period and at comparatively low rates of loading which can be applied in a testing machine. Creep tests usually are made by applying a constant uniaxial or triaxial load to a rock specimen and measuring the corresponding deformation as a function of time. Generally, these tests are more qualitative than quantitative since they describe behavioral characteristics rather than provide numerical values.

In this chapter the mechanical properties and behavior of both ideal and real materials are discussed with emphasis on brittle materials, especially rock. The deformational behavior of rock is dependent among other things on the state of stress in the rock. Although many tests have been made at high stress levels simulating conditions deep within the earth's crust, this book will be generally restricted to examining the mechanical properties and behavioral results obtained within the stress range normally experienced in engineering practice.

10.2 DEFINITIONS AND CONCEPTS

The terms and concepts used in describing the mechanical properties and behavior of materials can be defined or explained by considering, first, the stress-strain diagram for an idealized ductile metal (Figure 10.2.1). For an applied stress less than σ_a the strain is proportional to the stress and if the stress is removed, the strain will vanish. In this stress range, the material is *linear-elastic* (sometimes referred to as *perfectly elastic* or *Hookean*). The theory of elasticity (see Chapter 3) assumes linear-elasticity and is valid only for materials that are linear-elastic. The point *A* is called the *proportional limit*. For an applied stress greater than σ_a, the strain is no longer proportional to the stress, and for this range of stress the material behaves *inelasticically* where *inelasticity* refers to behavior beyond the

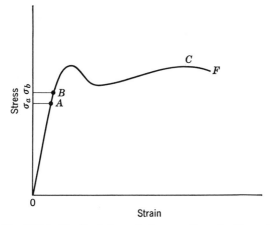

Fig. 10.2.1. Idealized stress-strain curve for a ductile material.

linear-elastic range. If the applied stress is increased above σ_a, a value σ_b will be reached above which the strain will no longer vanish when the stress is removed; that is, there will be a *permanent deformation*. Point B is referred to as the *yield stress*. The region on the stress-strain curve from O to B is called the *elastic region*. For most substances the proportional limit and the yield stress virtually coincide; hence, for brevity, only the term yield stress will be used. Also it will be assumed that for any stress greater than the yield stress the substance will not only behave inelastically, but that a permanent deformation will result.

Deformation in the inelastic region is termed *plastic* or *viscous deformation*. A material that possesses a yield stress is *perfectly plastic* if it will

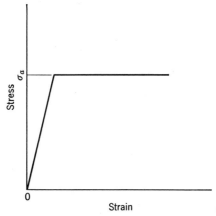

Fig. 10.2.2. Stress-strain curve for a perfectly elastoplastic material.

not sustain a stress greater than the yield stress, but will deform without limit under this stress unless constrained. Moreover, this deformation is irrecoverable. Figure 10.2.2 illustrates the behavior of an elastoplastic material. If the stress-strain curve rises in the region above the yield stress as indicated in the region from A to F (Fig. 10.2.3), the material has been *strain hardened*. If the stress σ_c is first applied and then removed, the strain will decrease to ϵ_d, where ϵ_d is the *permanent set* or *permanent deformation*. If the stress σ_c is reapplied, the stress-strain curve will return to F, and the

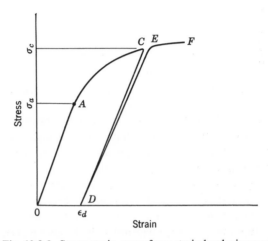

Fig. 10.2.3. Stress-strain curve for a strain-hardening material.

slope of CD will be about equal to the slope DE. Thus, in effect, strain hardening has increased the yield stress from σ_a to σ_c, but has not affected the elastic constant of the material. If the stress is increased further, the material will continue to deform until failure (by necking) occurs at F (Fig. 10.2.1). The stress-strain curves in Figs. 10.2.1, 10.2.2, and 10.2.3 are characteristic of *ductile* materials. Materials, such as cast iron and many rocks, usually terminate by fracture at or only slightly beyond the yield stress. These materials are referred to as *brittle*.

10.3 STATIC MECHANICAL PROPERTIES OF ROCK

Static mechanical property tests provide the strength and elastic property data necessary for the evaluation of rock structure problems. The mechanical properties determined by static testing include the uniaxial (unconfined) compressive, tensile, shear, and flexural strengths; the triaxial, compressive, and shear strength, and the elastic constants (modulus of elasticity and Poisson's ratio) obtained from uniaxial, biaxial, and triaxial stress-strain diagrams*. The U. S. Bureau of Mines[1] and the Department of Mines and Surveys, Canada,[2] have proposed standardized tests for determining some of the properties of mine rock and the American Society for Testing and Materials[3] has recommended procedures for testing building stone. However, there are no generally accepted industry standards for determining the static mechanical properties of rock. As previously mentioned, static tests are considered to be time-independent because they are usually made in a conventional testing machine at rates of loading of the order of $10^1 - 10^2$ psi/sec. However, a finite time, usually of the order of several minutes, is required to execute these tests, hence the results are time-independent only because time is not taken into account in the test procedure. From a structural standpoint deformational behavior measured at higher rates of loading, such as impact or explosive loading ($10^4 - 10^{10}$ psi/sec), is of minor importance and is not considered in this book; although tests made over long periods at slowly increasing rates ($10^{-1} - 10^{-4}$ psi/sec) or at constant loads are relatively important. Time-dependent behavior is discussed in Section 10.4.

Table 10.3.1 gives the static mechanical properties for many of the more common rock types. The equipment and procedures used in these tests are described in Chapter 11. For each rock type the sampling procedure was as follows[1]. Diamond drill cores or pieces of quarry or mine rock were selected, usually from a number of geographical areas; this stock was then divided into groups on the basis of geologic or petrographic examination. From each group a number of specimens were prepared, tested, and the results averaged. The maximum and minimum average values of these groups were given, and if the average group values clustered to the degree that 50% or more lay within a limited range, this range was also given.

The mechanical property data in Table 10.3.1 show the extreme variability of most rock types: for example, the crushing strength of granite ranges from 23,000 to 42,600 psi; for limestone from 5300 to 37,600 psi; and for quartzite from 21,100 to 91,200 psi. Considered in another way, rock with an average crushing strength of 25,000 psi could be any one of fourteen of the fifteen rock types listed in Table 10.3.1. Thus the geological designation does not provide a satisfactory basis for classifying rock according to its mechanical strength. For example, a sandstone could be a weak rock with a crushing strength less than 5000 psi, or a comparatively strong rock with a crushing strength greater than 30,000 psi. A more detailed petrographic identification might limit the range of mechanical property values for a specified rock, but at the same time it would also

---

* Physical property tests for determining the hardness of rock have been proposed but the results are of no concern in the treatment of structural problems. Also, the apparent density and porosity are usually measured by established procedures; for example, see references 1 and 2.

Table 10.3.1. Static Mechanical Properties of Rock[1]

Rock Type and Number of Test Groups[2]	Compressive Strength of Test Group × 10³ psi			Tensile Strength of Test group × 10³ psi		Modulus of Rupture of Test Group × 10³ psi			Static Modulus of Elasticity of Test Group × 10⁶ psi[3]		Dynamic Modulus of Elasticity of Test Group × 10⁶ psi[4]		
	Max.	Min.	>50% of Data within	Max.	Min.	Max.	Min.	>50% of Data within	Max.	Min.	Max.	Min.	>50% of Data within
Amphibolite (13)	74.9	30.4	—	—	—	7.4	4.0	—	—	—	15.1	6.7	—
Basalt (9)	52.0	11.8	—	—	—	6.6	2.1	—	—	—	12.4	5.9	—
Dibase (10)	51.8	23.2	40–50	—	—	8.0	4.5	5.0–5.5	—	—	13.9	10.2	—
Diorite (11)	48.3	22.5	25–35	—	—	7.3	2.0	—	—	—	6.12	3.6	4.0–5.0
Dolomite (10)	52.0	9.0	—	—	—	3.8	2.5	—	—	—	12.3	3.2	—
Gneiss (15)	36.4	22.2	25–35	—	—	3.1	1.2	2.0–3.0	—	—	15.1	3.5	7.0–10
Granite (17)	42.6	23.0	25–35	—	—	3.9	1.2	2.0–3.0	—	—	11.9	1.5	—
Greenstone (11)	45.5	16.6	—	8.1	4.1	6.7	1.7	—	11.0	2.5	15.2	3.4	10–13
Limestone[5] (46)	37.6	5.3	20–30	—	—	5.2	0.4	—	9.0	6.9	14.1	1.2	—
Marble (8)	34.5	6.7	30–35	—	—	3.3	1.7	—	12.0	4.2	—	—	—
Marlstone (15)	28.2	10.4	10–20	—	—	4.8	0.4	—	4.8	0.6	7.0	1.5	—
Quartzite (11)	91.2	21.1	—	—	—	6.4	1.2	—	—	—	—	—	—
Sandstone (48)	34.1	4.8	10–20	2.8	1.0	3.6	0.6	0.6–1.0	7.3	1.4	8.0	0.8	—
Shale (18)	33.5	10.9	—	—	—	4.2	0.3	—	7.6	1.7	9.9	1.5	1.5–2.5
Siltstone (8)	45.8	5.0	—	—	—	5.0	1.1	—	—	—	9.3	1.0	—

[1] From *U.S. Bureau of Mines Reports of Investigations* 3891, 4459, 4727, 5130, and 5244.
[2] The number following rock type indicates the number of petrographically distinct groups tested in uniaxial compression. A smaller number of groups were tested for the other listed mechanical properties.
[3] Tangent modulus of elasticity at the midstrength point.
[4] Determined by the resonant bar velocity method.
[5] Excluding chalk and coral rock.

decrease the probability of a given rock type agreeing with one of the identified types (presuming that only a part of all combinations of petrographic variables are tested). Thus, as opposed to testing materials for the purpose of obtaining or duplicating an equivalent material at a future date, testing rock without some specific requirement for the test results has very little purpose.

Bodies of rock as they occur in nature are divided by partings, bedding planes, joints, fractures, faults, and other planes of weakness. Hence, a body of rock can be considered to be made up of closely packed constituent units or building blocks such as the parallelepipeds formed by a three-dimensional system of joints. The mechanical properties of a specimen cut from one of these parallelepipeds generally would be representative not only of this element but of adjoining elements as well. If a mechanical process involves pieces of rock whose dimensions are less than those of the rock unit, as in the crushing or grinding tests, the mechanical properties of the pieces from the rock should relate to the process under examination. On the other hand, if the investigation is concerned with phenomena occuring at a scale greater than the rock unit size, as in blasting studies or in the evaluation of rock structures, the mechanical properties of specimens cut from a unit may poorly approximate the properties of the megascopic rock. Hence, the investigator should fully consider his objectives before he expends any great effort in measuring rock properties.

Uniaxial Compressive Strength

The unconfined compressive strength test is effected by uniaxially loading a right cylinder or prism of rock to failure. The specimen will fail either in shear as indicated by the characteristic cones formed on the shear planes or by splitting axially.* The compressive strength so obtained depends to some degree on the specimen shape, size, moisture content, rate of loading, smoothness of bearing plates, and other factors. However a reliable result can be obtained if the factors suggested in Section 11.7 are observed. The uniaxial compressive strength test prescribed in Section 11.7 is similar to that employed by most rock mechanics laboratories; hence, compressive strength data from these laboratories are usually comparable.

Tensile Strength

The direct tensile strength is obtained by axially loading a cylinder of rock to failure in tension (Section 11.6). If care is exercised in centering

---

* The factor that determines whether a specimen will fail in shear or by splitting is the degree of constraint offered by the bearing platens. The greater the friction between the platen and specimen end, the more likely the specimen will fail in shear.

the tensile load on the axis of the specimen and in preventing the grips from applying a radial stress to the rock, the specimen should fail in plane (tensile) stress. The specimen-to-specimen deviation in tensile strength is larger than that for compressive strength, a result that is due in part to an inability to exactly center the axial load, and in part to the fact that failure occurs on the weakest cross-sectional plane (bedding plane, parting, recemented joint), that is, on the weakest link. This effect is particularly noticeable in testing bedded rock. If the number (including zero) of planes of weakness vary from specimen to specimen, the deviation in tensile strength will vary accordingly; hence, the deviation is dependent on both the sampling procedure and on the specimen size. Of course the lowest value may be a better measure of the tensile strength of the sampled rock than the average value.

The tensile strength of rock also has been determined by applying a compressive load to a cylindrical specimen lying on its side (sometimes referred to as the indirect or Brazilian test, Section 11.6). Theoretically under these conditions a uniform tensile stress should develop across the diametral plane connecting the points of loading; tests have shown that the specimen will fail on this plane. Although this test is easier to effect than the uniaxial tension test, the state of stress in the specimen is highly complex. Compressive stresses exist in both directions parallel to the diametral plane. Also, crushing (prior to failure) occurs along the contact loading lines, which further complicates the stress distribution in the contact loading zone. How these factors affect the tensile failure strength has not been fully evaluated, especially for heterogeneous and anisotropic materials.

From a structural standpoint the tensile strength test is not of too great an importance. Except in bedded roof there are few instances in which tensile stresses of significant magnitude can exist underground; in situ measurements have indicated that, in general, underground stresses are compressive, even in places where elastic theory predicts otherwise.

Flexural Strength

The flexural strength, or modulus of rupture test, is effected by supporting a beam at its ends and loading the midpoint of the span to failure (Section 11.8). The outer fiber tensile strength of the material is calculated by means of Eq. 11.8.1 or 11.8.2.* As a maximum tensile stress occurs on the under surface at the midpoint, only a small part of the specimen is under test. Thus as opposed to the uniaxial tensile test in which the length of the specimen is subjected to a uniform tensile stress, the

---

* In deriving Eqs. 11.8.1 and 11.8.2 a linear stress distribution across the vertical plane was assumed (Section 5.6).

flexural test usually gives higher values because of the weakest link condition, that is, because the probability of a plane of weakness occurring at the midpoint is small. Despite this limitation, the flexural strength of a pneumatic loaded mine roof (Section 14.10) was found to be in reasonable agreement with the laboratory determined flexural strength of rock specimens cored from the roof. The agreement probably was good because the flexural test is an approximate model of a single layer mine roof.

Unconfined Shear Strength

A number of methods have been proposed for measuring the unconfined shear strength of rock (Section 11.9). With the exception of the torsional shear tests, these methods involve shearing a rock specimen between knife edges. However, at the point of contact the knife edges cause a stress concentration in the specimen and, according to Everling,[4] failure is due to a crack that initiates at this point which propagates in a direction that does not correspond to the direction of maximum shear. In some of these tests stresses are induced in the specimen by the clamps that hold the specimen. Also, stresses due to bending may be experienced. The torsional test is subject to both the clamping and bending difficulties. As a result of these factors, the shear stress values obtained by different procedures are dependent on the specimen size and other test conditions to the degree that no unconfined shear test is considered satisfactory.

Wuerker[5] obtained the shear strength of rock from consideration of Mohr's representation of the uniaxial compressive and tensile strength results (Fig. 10.3.1). However, since the results of uniaxial tensile strength

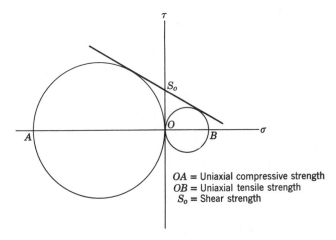

OA = Uniaxial compressive strength
OB = Uniaxial tensile strength
S_o = Shear strength

Fig. 10.3.1. Mohr's representation of tensile, compressive, and shear strengths.

tests are not too reliable and as the tangent curve may not be a straight line, the accuracy of this procedure is questionable.

Triaxial Compressive and Shear Strength

The triaxial compressive and shear strength test consists of applying a hydraulic pressure to the exterior surface of a cylinder specimen and at the same time axially loading the specimen in compression in increasing

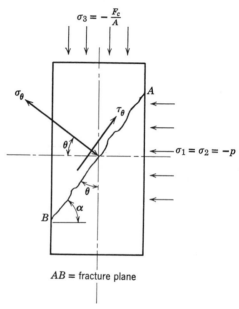

$$AB = \text{fracture plane}$$

Fig. 10.3.2. Principal stresses applied to a triaxial compressive test specimen.

increments to failure (Section 11.10). Although there is no standardized or generally accepted procedure, most laboratories perform this test in a sufficiently similar manner so that the test results are comparable. The greatest difference in procedure is in the size of test specimen, although a length-to-diameter ratio of two is generally used. The hydraulic pressure p, the axial load F_c, the end area of the specimen A, and θ, the angle of the failure plane with respect to the axial direction, or its compliment α, the angle of fracture, are recorded (Fig. 10.3.2). The principal stresses in the specimen are

$$\sigma_3 = -\frac{F_c}{A}, \quad \text{and} \quad \sigma_1 = \sigma_2 = -p.$$

From these data a Mohr's stress circle (Section 1.4) can be constructed in the $\sigma\tau$ plane as shown in Fig. 10.3.3 (recalling the convention that compressive stresses are negative). If the validity of the Coulomb-Navier theory of failure is assumed (Section 10.5), and if a radius is drawn at the angle 2θ on the circle to the point P, the projection of this point on the σ and τ axes will give the magnitude of the normal and shear stress acting on the fracture plane. The *angle of internal friction* ϕ is formed by the tangent to the circle at P and the direction of the σ axis, and the coefficient of internal friction μ can be calculated from $\tan 2\theta = 1/\mu$. Also, $\phi = 90° - 2\theta$ and $\tan 2\theta = \cot \phi$. Thus, from a triaxial test on a single

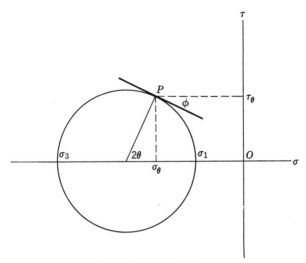

Fig. 10.3.3. Mohr's circle for a triaxial compressive test.

specimen, it is possible to obtain a value for the normal and shear stress acting on the plane of failure, and the angle and coefficient of the internal friction for the material.

Usually in determining the properties of a specific rock type a group of specimens is tested. Unlike the uniaxial compressive strength test in which specimens fracture at approximately the same axial load, in the triaxial test the axial load at failure depends on the radial load. Figure 10.3.4 shows the relationship between the axial and radial loads for a group of specimens. This effect is characteristic of brittle materials and was first reported by von Karman in 1911. For many rock types the $\sigma_1\sigma_3$ relationship is approximately linear, and if so it can be expressed by

$$\sigma_3 = C_0 + \sigma_1 \tan \beta \qquad (10.3.1)$$

where C_0 is the uniaxial compressive strength for a specimen having a length-to-diameter ratio of 2, and the angle β is a constant. If the $\sigma_1\sigma_3$ relationship is not linear, other expressions may be found by curve fitting.

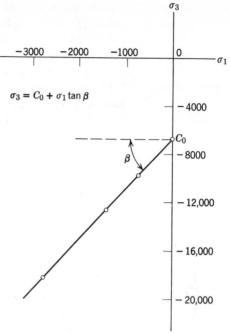

Fig. 10.3.4. Relationship between the axial and radial stress at failure in a group of triaxial specimens.

The $\sigma_1\sigma_3$ data for a group of specimens can also be plotted on the $\tau\sigma$ plane as a family of Mohr's circles (Fig. 10.3.5). The common tangents to these circles (line AB and CD) are referred to as *Mohr's envelope*. The significance of this envelope is considered in Section 10.5. The enveloping curves are approximately linear for many rock types and in this case the envelope can be expressed by

$$\tau_\theta = \pm(S_0 - \sigma_\theta \tan \phi) \qquad (10.3.2a)$$

or, as $\tan \phi = \cot 2\theta$,

$$\tau_\theta = \pm(S_0 - \sigma_\theta \cot 2\theta) \qquad (10.3.2b)$$

where σ_θ and τ_θ are the normal and shear stresses acting on the failure plane, S_0 is the shear strength of the rock, and $\tan \phi$ is the slope of the envelope curves.

For the more general case when the envelope is curved, as in Fig. 10.3.5, Balmer[6] has shown that the envelope can be expressed parametrically in σ_1 and σ_3. From triangle AHD (Fig. 10.3.6)

$$\left(\frac{\sigma_1 + \sigma_3}{2} - \sigma_\theta\right)^2 + \tau_\theta{}^2 = \left(\frac{\sigma_1 - \sigma_3}{2}\right)^2 \qquad (10.3.3)$$

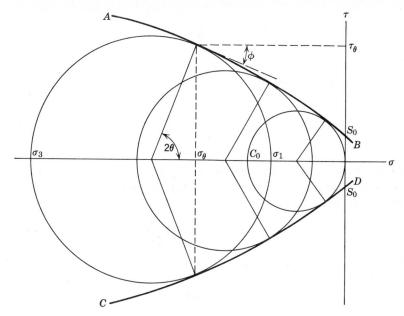

Fig. 10.3.5. Mohr's envelope determined from the failure stresses in a group of triaxial specimens.

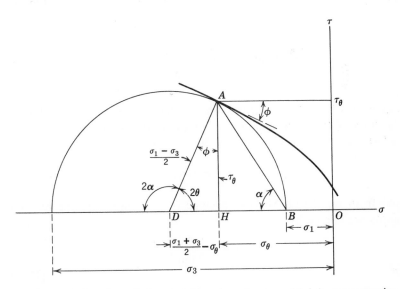

Fig. 10.3.6. Designation of the stresses and angles in a Mohr's representation.

By taking the partial derivative of σ_3 with respect to σ_1 and solving for σ_θ

$$\sigma_\theta = \sigma_1 + \frac{\sigma_3 - \sigma_1}{\dfrac{\partial \sigma_3}{\partial \sigma_1} + 1} \tag{10.3.4}$$

Substituting Eq. 10.3.4 in Eq. 10.3.3 and solving for τ_θ

$$\tau_\theta = \frac{\sigma_3 - \sigma_1}{\dfrac{\partial \sigma_3}{\partial \sigma_1} + 1}\left(\frac{\partial \sigma_3}{\partial \sigma_1}\right)^{\frac{1}{2}} \tag{10.3.5}$$

Equations 10.3.4 and 10.3.5 parametrically represent Mohr's envelope in terms of σ_1 and σ_3. From these two equations and triangle AHB in Fig. 10.3.6, the following relation is obtained

$$\tan \alpha = \frac{\tau_\theta}{\sigma_\theta - \sigma_1} = \left(\frac{\partial \sigma_3}{\partial \sigma_1}\right)^{\frac{1}{2}} \tag{10.3.6}$$

The slope of Mohr's envelope is

$$\frac{\partial \tau_\theta}{\partial \sigma_\theta} = \tan \phi \tag{10.3.7}$$

As $2\alpha = \phi + 90$, $\tan \phi = -1/\tan 2\alpha$, and by means of the identity $\tan 2\alpha = 2 \tan \alpha/(1 - \tan^2 \alpha)$, Eqs. 10.3.7 and 10.3.6 can be combined to give

$$\frac{\partial \tau_\theta}{\partial \sigma_\theta} = \tan \phi = \frac{\tan^2 \alpha - 1}{2 \tan \alpha} = \frac{\dfrac{\partial \sigma_3}{\partial \sigma_1} - 1}{2\left(\dfrac{\partial \sigma_3}{\partial \sigma_1}\right)^{\frac{1}{2}}} \tag{10.3.8}$$

Equation 10.3.8 shows that the slope of the Mohr's envelope $\partial \tau_\theta/\partial \sigma_\theta$ and the slope of the axial-radial stress curves $\partial \sigma_3/\partial \sigma_1$ are related. If the axial-radial curve is linear (Eq. 10.3.1), then

$$\frac{\partial \sigma_3}{\partial \sigma_1} = \tan \beta \tag{10.3.9}$$

and from Eq. 10.3.8 Mohr's envelope is also linear and the respective slopes are related by

$$\tan \phi = \frac{\tan \beta - 1}{2 \tan^{\frac{1}{2}} \beta}$$

or

$$\tan \beta = \frac{\sin \phi + 1}{\sin \phi - 1} \tag{10.3.10}$$

For a linear Mohr's envelope the relationship between C_0 and S_0 is obtained by letting $\sigma_1 = 0$ in Eqs. 10.3.1, 10.3.2a, 1.3.11, and 1.3.12. Thus

$$\sigma_3 = C_0$$

$$\tau_\theta = -S_0 + \sigma_\theta \tan \phi$$

$$\sigma_\theta = \frac{\sigma_3}{2}(1 - \cos 2\theta) \qquad (10.3.11)$$

$$\tau_\theta = \frac{\sigma_3}{2} \sin 2\theta$$

Table 10.3.2. Physical Properties of Some Typical Foundation Rocks*

	C_0	$\tan \phi$	S_0	$\tan \beta$
Andesite	18,900	5.7	4060	1.0
Basalt	24,600	7.4	4500	1.2
Diorite	12,200	9.2	2000	1.4
Gneiss	16,100	10.0	2500	1.4
Granite	22,300	11.8	3250	1.6
Graywacke	8,700	6.5	1700	1.1
Limestone	15,300	12.6	2150	1.6
Sandstone	13,000	7.0	2450	1.1
Schist	10,000	15.7	1260	1.9
Shale	10,300	19.7	1160	2.1
Siltstone	4,000	7.7	720	1.2
Tuff	430	4.9	100	0.9

* *Bureau of Reclamation Report* SP-39, August 1953.

Combining these equations and using the relation $2\theta + \phi = 90°$ gives

$$C_0 = -\frac{2S_0 \cos \phi}{1 - \sin \phi} \qquad (10.3.12)$$

Thus the angles and constants in Eqs. 10.3.1 and 10.3.2 are related.

In Table 10.3.2 values of C_0, S_0, $\tan \beta$, $\tan \phi$ are given for a number of common rock types for which the $\sigma_1 \sigma_3$, or $\tau_\theta \sigma_\theta$ relationship are linear (Eqs. 10.3.1 and 10.3.2). Note that for this case there are only two independent constants, either C_0 or S_0, and one of the angles θ, ϕ, or β.

In view of the fact that it is difficult to perform an in situ uniaxial compressive test, the problem of performing a triaxial test in situ is even more difficult. As a result very limited data are available on in situ triaxial strengths of rock. However, some information can be inferred from the behavior of the supporting pillars in German potash mines. For example, in one mine the depth is 3300 ft and the extraction 50%. Thus the average pillar load would be approximately 6600 psi (Section 17.3). However, the measured uniaxial compressive strength of this potash ore is only 4000–5000 psi. Simply on the basis of the uniaxial strength, we would expect instability. However, the pillar height-to-width ratio in these mines is approximately $\frac{1}{4}$. Owing to the lateral constraint on both the top and bottom of these pillars, a radial constraint is produced which places the core of the pillar in a state of triaxial stress. Stability is achieved because this lateral constraint places the core of the pillar in triaxial compression and hence the average pillar strength is increased to a value greater than 6600 psi.

Elastic Constants

Within the stress range experienced in most mines, a large part of rocks tested in uniaxial compression is relatively elastic; that is, in a stress-strain cycle, as the stress is removed, the strain is recovered. However, only a part of these rocks exhibit anything approaching linear elasticity. Also, the elastic behavior of a given rock type from different geographical areas varies considerably; for example, the stress-strain diagrams shown in Figs. 10.3.7a and b are for two granites tested in uniaxial compression to failure. The Georgia granite was virtually linear-elastic; the Colorado granite departed from linear elasticity with the modulus of elasticity increasing at low stress and decreasing at high

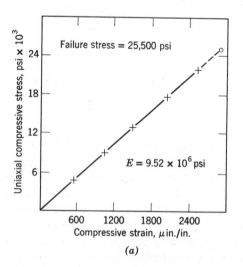

(a)

Fig. 10.3.7a. Stress-strain curve for Georgia granite.

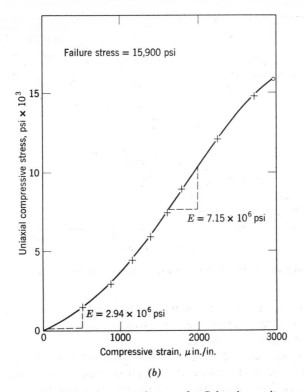

Failure stress = 15,900 psi

$E = 7.15 \times 10^6$ psi

$E = 2.94 \times 10^6$ psi

Uniaxial compressive stress, psi $\times 10^3$

Compressive strain, μ in./in.

(b)

Fig. 10.3.7b. Stress-strain curve for Colorado granite.

stress. An increase in the modulus of elasticity with stress is characteristic of many sedimentary rocks, especially those found adjoining coal measures.

Figure 10.3.8 is a family of stress-strain curves for a granite tested in triaxial compression.[7] In this test the radial pressure p_0 is held at some fixed value and the axial stress σ_z varied (Section 11.11). For higher values of confining pressure the stress-strain curves become more nearly linear and the area within the stress-strain loop (hysteresis loop) decreases. The elastic behavior exemplified by the Colorado granites, Figs. 10.3.7b and 10.3.8, has been attributed by Nishihara[8] to the presence of microscopic cracks in the rock that reversibly open at low stress and close at high stress. Cracks observable under low magnification were detected in a limestone for which the stress-strain curve was strongly concave upward as shown in Fig. 10.3.9, curve *B*.[9] Stress-strain measurements made on areas of the same specimen which did not contain micro-cracks were virtually linear-elastic as shown in Fig. 10.3.9, curve *A*. Because nonlinear-elastic effects tend to decrease with increasing confinement, they are

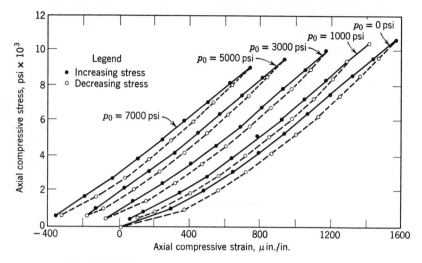

Fig. 10.3.8. Stress-strain curves for a triaxially loaded granite.

probably not as pronounced in situ; in fact, it has been suggested that a part of these micro-cracks may be produced in coring or in otherwise relieving the stress in in situ rock.[9]

The static modulus of elasticity (Young's modulus) of a number of rock types is given in Table 10.3.1. These values were obtained for specimens tested in compression, and the tangent value for the midrange

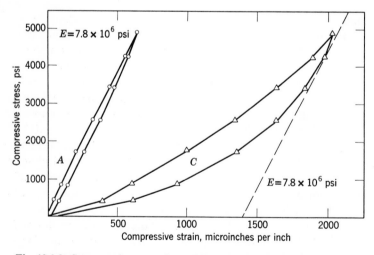

Fig. 10.3.9. Stress-strain curves from different areas on a limestone specimen.

stress is quoted.* For comparison, elastic constant data from dynamic tests (pulse velocity and resonant bar velocity methods) are given. Static elastic constant data from in situ (underground) tests are meager. Merrill and Morgan[10] pneumatically loaded a layer of roof rock overlying a large opening (50 ft × 100 ft). From the measured tensile strain at the midpoint of the span and the applied stress calculated from the pneumatic pressure (considering the roof layer as a beam), it was determined that the initial (zero stress) elasticity of the layer was $E = 11.0 \times 10^6$ psi as compared with a laboratory determined static value (in tension) of $E = 10.0 \times 10^6$ psi. However, as the in situ stress was increased the modulus of elasticity decreased progressively to $E = 2.6 \times 10^6$ as the failure stress was approached. A decrease in the modulus of elasticity with increasing tensile stress is characteristic of most rocks.

10.4 TIME-DEPENDENT PROPERTIES OF ROCK

In Section 6.2 the characteristics of a number of viscoelastic models made up of various combinations of springs and dashpots were described. Either the deformational response of a model under constant stress or the variation in the stress under constant strain was considered as a function of time. Equations relating stress and strain to time were given, in which one constant appeared for each elastic or viscous element. The utility of these phenomenological models depends on how closely a model behavior can be made to correspond to the behavior of a real material. When the correspondence is good, the elastic and viscous constants can be evaluated and as such they specify the behavior of the material. In this section we present experimental results illustrating the time-dependent behavior of rock. In a few instances, the elastic and viscous constants have been determined, but generally this evaluation is difficult; hence, the results are more qualitative than quantitative. A wide variety of tests have been performed on a large number of rock types, but no attempt has been made to standardize any time-dependent tests.

Figure 10.4.1 is an idealized creep curve for rock that shows in addition to the transient phase (from $t = 0$ to $t = t_1$) and the steady-state phase (from $t = t_1$ to $t = t_2$), a tertiary phase (from $t = t_2$ to $t = t_3$) which is terminated at $t = t_3$ by failure. Figures 10.4.2, 10.4.3, and 10.4.4 are creep curves for salt, marble, and granodiorite. Although the test conditions are dissimilar, they indicate that the transient creep period is

---

* The tangent modulus of elasticity is the slope of the stress-strain curve at some specified stress, that is, $(\Delta\sigma/\Delta\epsilon)_{\sigma=1000 \text{ psi}}$.
  The secant modulus of elasticity is the stress divided by the strain at some specific stress, that is, $(\sigma/\epsilon)_{\sigma=1000 \text{ psi}}$ (Fig. 11.11.1).

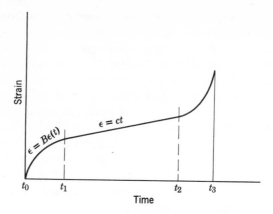

Fig. 10.4.1. Idealized creep curve.

short, usually less than one day; the duration of the steady-state phase is comparatively long but dependent on the magnitude of the applied load; and the tertiary phase is very brief, in fact for most rock types and especially the more brittle rocks, the duration is too short to permit observation. In tests in which the tertiary phase is absent, rock may rupture at a time when the strain rate is either constant or decreasing.

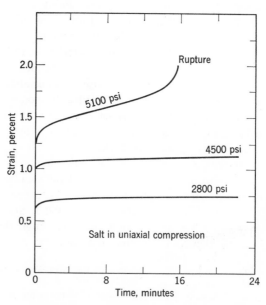

Fig. 10.4.2. Creep curves for salt; from Höfer.[16]

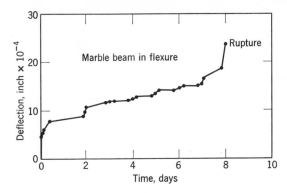

Fig. 10.4.3. Creep curve for marble. (After Phillips.[11])

Hence, Phillips[11] concluded that a decreasing rate of strain does not necessarily insure stability.

It has been found that the creep curve for a number of materials can be expressed by a relationship of the form

$$\epsilon = A + Bt + C\epsilon(t) \tag{10.4.1}$$

where A is the elastic strain at $t = 0$; Bt is the steady-state creep; and $C\epsilon(t)$ is the transient or primary creep. A term for tertiary creep is not included because laboratory studies indicate that its duration is usually so short that once it is initiated failure cannot be arrested. Griggs[13] reported that the transient creep in a Solenhofen limestone specimen loaded for 550 days in uniaxial compression at approximately 50% of its crushing strength was logarithmic (Fig. 10.4.5) and could be expressed by

$$\epsilon = (6.1 + 5.2 \log t)10^{-5}$$

A term for steady-state creep was either absent or negligible. However, under confined pressure Griggs found that the creep rate in this limestone

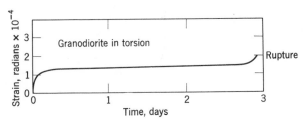

Fig. 10.4.4. Creep curve for granadorite. (After Lomitz.[12])

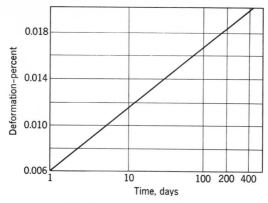

Fig. 10.4.5. Creep of Solenhofen limestone under load of 20,000 psi. (After Griggs.[13])

included a steady-state term, namely,

$$c = (1.52 + 0.90 \log t + 0.50t)10^{-2}$$

and the departure from strictly logarithmic creep was observable within 10 min after start of the test.

Evans[14] presented experimental creep curves for several rocks (Fig. 10.4.6). These curves are similar to the creep curves for the exponentially decreasing creep term in the equation for the Burger's model, that is, $\epsilon = C[1 - e^{-kt}]$. Iida and Kumazawa[15] also recorded creep curves

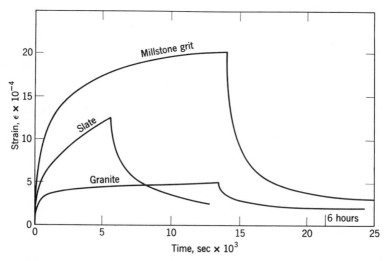

Fig. 10.4.6. Creep under constant stress and recovery curves. (After Evans.[14])

similar to those for the Burger's model. However, for limestone, sandstone, and granite a fifth (elastic) element had to be added to the model to obtain agreement between the model and specimen behavior.

In relation to the design of rock structures the duration of the transient creep period is too short to consider. However, immediately following an underground blast there is evidence of the effects of transient creep as manifest by small rock falls, the development of fractures, and both audible and micro-seismic noises (Section 14.12). These effects usually decrease rapidly with time so that 15 to 60 min after blasting the disturbed area

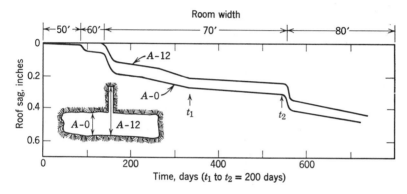

Fig. 10.4.7. Roof sag in experimental room in oil shale. (After Merrill.[16])

returns to normal. This afterworking immediately following blasting presumably relieves high stress concentration on the newly created rock surfaces. Steady-state creep has been measured over long periods in the roof and pillars of underground mines.

Data presented in a report by Merrill[16] show that the roof sag rate in an experimental room in oil shale (kerogenaceous marlstone) was constant over a 200-day period (see Fig. 10.4.7 from t_1 to t_2; also Section 14.11). Höfer[17] measured the lateral deformation in pillars in a number of German potash (halite plus sylvite) mines (Section 14.11). In general the deformation rate was constant for measurement periods ranging from 63 days to almost 10 years, and for strain rates up to 4% per year (Fig 10.4.8).

Contrary to the conclusion reached on the basis of laboratory tests that the duration of the tertiary phase is short, micro-seismic investigations indicate that there is a period of accelerated movement in in situ rock preceding failure (Section 14.12). This period may range from a few hours for a fall of rock of several tons to several weeks for a large body of rock, such as a 250-ton roof fall or the failure of a mine pillar. Thus it

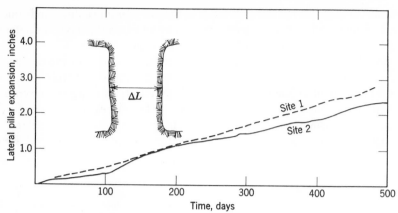

Fig. 10.4.8. Creep in potash ore pillars. (After Höfer.[17])

can be inferred that in situ rock exhibits in varying degree transient, steady-state, and tertiary creep.

Creep rate is dependent on the magnitude of the applied stress, as illustrated in Fig. 10.4.9. It has been found experimentally that at given time t, usually in the transient phase, the creep rate may be expressed by

$$\dot{\epsilon} = D\sigma^n$$

or

$$\epsilon = Dt\sigma^n$$
(10.4.2)

The values of n for a number of rock types tested uniaxially at room temperature is given in Table 10.4.1. However, the dependence of the creep

Table 10.4.1. Values of the Exponent n in the Equation $\dot{\epsilon} = D\sigma^n$ for Creep in Various Rocks at Room Temperature*

Rock Type	Maximum Strain $\times\ 10^{-3}$	Maximum Stress psi $\times\ 10^3$	Exponent n
Slate	1	51	1.8
Granite	1	50	3.3
Alabaster (in water)	5	2.8	2.0
Limestone	7	20	1.7
Halite	20	4.2	1.9
Gabbro	0.01	1.4	1.0
Granodiorite	0.2	1.4	1.0
Granite	3	14	3.0
Shale	3	1.4	2.7

* From a compilation by E. C. Robertson[18]

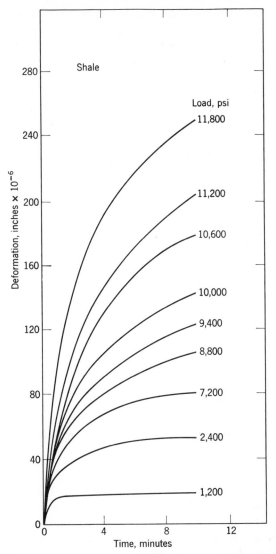

Fig. 10.4.9. Time versus deformation curves for different applied loads. (After Phillips.[11])

rate on stress is complex and Eqs. 10.4.2 may not be valid for all ranges of strain or for creep occurring in the steady-state phase. As previously noted, Höfer[17] measured lateral pillar deformation in a number of potash mines over a period of several years. Figure 10.4.10 shows the relationship between the average pillar stress, calculated from the depth of the workings and extraction, and the strain rate. This result shows that in potash ore there is a sharp transition in the strain rate occurring at an average pillar stress of 6000 psi, below which the strain rate is almost zero and above which the strain rate becomes proportional to the stress.

The creep rate also depends on the rock temperature. From an engineering standpoint, for most rock types this effect is usually negligible within the range of temperatures encountered in mining. However, creep in salt and possibly other evaporite minerals is temperature dependent

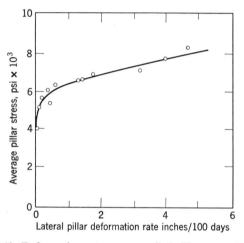

Fig. 10.4.10. Deformation rate versus applied pillar stress. (After Höfer.[17])

to the degree that in some cases this effect might be a factor in design considerations. The constants A, B, and n in the relationship for time-dependent creep

$$\epsilon = A + Bt^n \qquad (10.4.3)$$

were experimentally evaluated by LeComte[19] for different confining pressures and temperatures. Although most of the measurements were made at a temperature and confining pressure higher than that encountered in mining, the results indicate that creep is temperature-dependent to the extent that within the range of temperatures encountered in mining (approximately 50 to 110°F) significant differences might be expected. LeComte also found that the confining pressure reduces the creep rate.

As pointed out in Section 10.2, if a ductile metal is cyclicly strained at a stress value above the yield point, it will strain-harden, that is, it will become harder; the yield point will be raised, but the modulus of elasticity will remain unchanged. Robertson[20] found that cyclic triaxial-loading affects Solenhofen limestone similarly in that if the confining pressure is raised after each loading cycle, the yield stress can be increased without affecting the elastic properties (Fig. 10.4.11). However, as opposed to metal, the rock did not harden, hence Robertson preferred to call the process strain-strengthening. Moreover, it was found that if the strengthened specimen was allowed to stand unloaded for 24 hrs, the yield point decreased to very nearly its original value. Strain-strengthening is evident in varying degree in limestone and other sedimentary and metamorphic

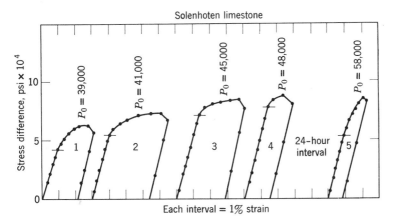

Fig. 10.4.11. Shift in yield point of cyclicly loaded Solenhofen limestone specimen. (After Robertson.[19]) Cycles 1 through 4 consecutive with no time interval between.

rocks, but is absent or negligible in most igneous or more siliceous rocks. Many rock types do not possess a sufficiently defined yield point to evaluate this phenomenon.

10.5 THEORIES OF FAILURE

Experimental results presented in Section 11.10 show that an unconfined brittle rock will fracture when subjected to a certain stress, but under confinement the same rock will withstand a much greater stress before it fractures or deforms excessively. Thus failure, which may occur either by fracture or excessive deformation, depends upon the mechanical condition in the rock, a fact that has been shown experimentally to be true for most brittle materials. In designing or evaluating rock structures or for that matter, in solving most mechanical problems, it is desirable to have a theory of failure based on theoretical or experimental results specifying how the strengths of materials are affected by the state of stress in the specimen, time, temperature, and other factors. Although reasonably valid theories of fracture exist for ductile materials such as metals and plastics, the subject is not as advanced for complex polycrystaline substances such as rock. Of the various theories of failure that have been proposed, only those that have been found on the basis of experiment to be reasonably valid for rock are considered.

Coulomb-Navier Theory of Failure

The maximum shear stress theory proposed by Coulomb postulates that failure will occur in a material when the maximum shear stress at a point in the material reaches a specific value S_0, which is referred to as the shear strength. If σ_1, σ_2, and σ_3 are the principal stresses in a material*, the maximum shear stress is given by

$$\tau_{max} = \tfrac{1}{2}(\sigma_1 - \sigma_3)$$

Thus Coulomb's theory predicts that failure will occur when the shear stress has the magnitude

$$\tfrac{1}{2}(\sigma_1 - \sigma_3) = S_0 \qquad (10.5.1)$$

and that the failure plane will bisect the angle between the maximum and minimum principal stresses. For example, in a triaxial test (Fig. 10.3.2) the plane of failure should be at 45° to the axial load—a conclusion that is not borne out by experiment. Rather, this direction varies from one rock type to another and if $\sigma_3 < \sigma_1$, the plane of failure is usually less than 45° from the direction of the larger compressive stress, and in uniaxial tension the plane of failure is at about 90° from the direction of the applied stress. Moreover, this theory implies that the shear strength in tension and compression are equal, a condition that is not even roughly approximated for brittle materials.

Navier modified this theory by assuming that the normal stress acting across the plane of failure increases the shear resistance of the material by an amount proportional to the magnitude of the normal stress. Considering the two-dimensional case, if σ_θ and τ_θ are the normal and shear stresses acting on the failure plane, then this modified theory stipulates that failure will occur when the magnitude of the shear stress acting on the failure plane reached a value

$$|\tau_\theta| = S_0 - \mu\sigma_\theta \qquad (10.5.2)$$

As $\mu\sigma_\theta$ is analogous to the frictional force on an inclined plane due to a normal reaction, μ is referred to as the *coefficient of internal friction*.

---

* Throughout this discussion the convention that tensile strength is positive is followed, and that $\sigma_1 > \sigma_2 > \sigma_3$.

The normal and shear stress are (Eqs. 1.3.11 and 1.3.12)

$$\sigma_\theta = \frac{\sigma_1 + \sigma_3}{2} + \frac{\sigma_1 - \sigma_3}{2} \cos 2\theta$$

and

$$\tau_\theta = -\frac{\sigma_1 - \sigma_3}{2} \sin 2\theta$$

Thus Eq. 10.5.2 can be written as

$$S_0 = \mu\sigma_\theta + |\tau_\theta| = \mu\frac{\sigma_1 + \sigma_3}{2} + \frac{\sigma_1 - \sigma_3}{2}(\sin 2\theta + \mu \cos 2\theta) \quad (10.5.3)$$

Equation 10.5.3 has a maximum value when $dS_0/d\theta = 0$, that is when

$$\tan 2\theta = \frac{1}{\mu} \quad (10.5.4)$$

and the magnitude of S_0 at this angle is, from Eqs. 10.5.3 and 10.5.4,

$$S_0 = \frac{\sigma_1}{2}[\mu + (\mu^2 + 1)^{1/2}] + \frac{\sigma_3}{2}[\mu - (\mu^2 + 1)^{1/2}] \quad (10.5.5)$$

For $\mu = 0$ Eq. 10.5.3 gives $\theta = 45°$, which is the condition for maximum shear stress failure (Eq. 10.5.1). For $\mu > 0$, $\theta < 45°$, and for $\mu = \infty$, $\theta = 0°$. Hence, as μ increases, the plane of failure moves toward the direction of the algebraically least stress (compression being negative).

In this treatment only the magnitude of τ_θ is considered; hence changing the sign of θ does not change τ_θ and S_0 is symmetric about $\theta = 0°$. The plane of failure passes through the axis of the intermediate principal stress and makes an angle of $\pm\tan^{-1}(1/\mu)$ with the least principal stress.

A criterion for failure can be expressed in terms of T_0 and C_0, the tensile and compressive strength of the material, by first making $\sigma_1 = T_0$ and $\sigma_3 = 0$ in Eq. 10.5.5 so that

$$\frac{T_0}{2}[\mu + (\mu^2 + 1)^{1/2}] = S_0 \quad (10.5.6)$$

which corresponds to failure in tension, and then by making $\sigma_3 = -C_0$ and $\sigma_1 = 0$ so that

$$\frac{C_0}{2}[(\mu^2 + 1)^{1/2} - \mu] = S_0$$

which corresponds to failure in compression. Thus

$$\frac{C_0}{T_0} = \frac{(\mu^2 + 1)^{1/2} + \mu}{(\mu^2 + 1)^{1/2} - \mu} \quad (10.5.7)$$

and from Eq. 10.5.6

$$\frac{\sigma_1}{T_0} - \frac{\sigma_3}{C_0} = 1 \qquad (10.5.8)$$

In Fig. 10.5.1 Eq. 10.5.8 is represented by the line CD. As $\sigma_1 > \sigma_3$ the only values of σ_1 and σ_3 for which the material will not fail lie within the angle BCD.

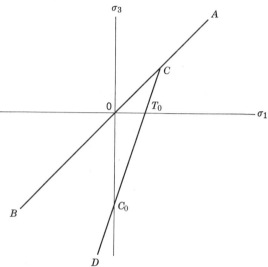

Fig. 10.5.1. Limiting states of stress for brittle failure according to Coulomb-Navier theory.

This theory predicts that the compressive strength is greater than the tensile strength, but that the ratio of the strengths is not as large as indicated in Table 10.3.1. For example, from Eq. 10.5.7, with $\mu = 1$, that is, for $\theta = 22.5°$, $C_0/T_0 = 5.8$, whereas for measured strengths this ratio usually ranges from 10 to 50. Since μ is constant, the angle of failure given by $\frac{1}{2} \tan^{-1}(1/\mu)$ is the same for tension and compression failure. In compression the angle of failure is relatively constant for most rock types (Section 11.10), but in tension the failure surface is usually normal to the direction of the tensile stress. Jaeger[21] points out that "the theory assumes shear fracture, so that T_0 should not be the actual (brittle) tensile strength but the value at which shear failure in tension would take place if in fact brittle failure did not occur in practice before this value is reached." The difference in the appearance of the fracture surfaces created in tension and shear also indicate that the mechanism of failure is not the same in the two cases.*

* The characteristics of fracture surfaces are discussed in Nadai, *Theory of Flow and Fracture of Solids*, Vol. 1, McGraw-Hill, 1950.

Mohr's Theory of Failure

Mohr's theory of failure postulates that a material will fracture or begin to deform permanently when the shear stress τ_θ on the fracture or slip plane has increased to a value which depends on the normal stress σ_θ acting on the same plane, or when the largest tensile principal stress has reached a limiting value T_0. Thus at failure either

$$\tau_\theta = f(\sigma_\theta)$$

or $$\sigma_1 = T_0$$ (10.5.9)

The functional relationship $\tau_\theta = f(\sigma_\theta)$ must be determined experimentally and is in fact represented by the curves AB and CD in Fig. 10.3.5. Since these curves are the envelope to the Mohr's circles for the values of $\sigma_3\sigma_1$ at failure, their physical significance is as follows: for any state of stress represented by a Mohr's circle lying completely within the envelope the material will not fail, whereas if any part of the circle lies outside of the envelope the critical stresses will be exceeded. For the circle tangent to the envelope the material will fail on the plane making the angle θ with respect to the least principal (largest compressional) stress as in Fig. 10.3.2. Mohr's theory further implies that the intermediate principal stress σ_2 has no influence on failure. This is evident from a consideration of Mohr's representation of the three-dimensional state of stress (Fig. 10.5.2) in which the normal and shear stresses on any plane

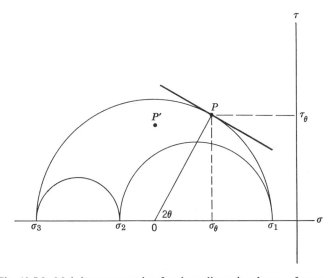

Fig. 10.5.2. Mohr's representation for three-dimensional state of stress.

can be represented by a point P' lying within the area bounded by the $\sigma_2\sigma_1$, $\sigma_3\sigma_2$, and $\sigma_3\sigma_1$ circles. For a given normal stress, it is evident that the maximum normal shear stress will be represented by P on the $\sigma_3\sigma_1$ circle, and that the value of the intermediate stress cannot affect τ_θ.

Thus this theory of failure not only specifies the state of stress at failure, but predicts the direction of the failure plane. The slip planes

Fig. 10.5.3. Trona specimen subjected to triaxial load showing slip lines.

at failure in a trona specimen are shown in Fig. 10.5.3, and Fig. 10.5.4 is Mohr's representation for a group of trona specimens. The measured angle of the failure plane is $\theta = 28 \pm 3°$. For specimens tested in triaxial compression, τ_θ increases monotonically with σ_θ. Hence Mohr's theory implies that a material will not fail in hydrostatic compression, a consequence that is consistent with experimental fact. The envelope curves if projected into the $+\sigma$ quadrants do not predict the correct angle of failure in tension. In these quadrants the maximum stress theory is assumed; that is, failure will occur if σ_1, σ_2, or σ_3 reaches a critical value T_0, and the failure plane will be normal to the direction of failure stress.

For the special case when the envelope curves are straight lines

$$\tau_\theta = \pm(S_0 - \sigma_\theta \tan \phi) \tag{10.5.10a}$$

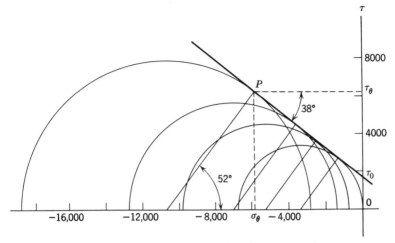

Fig. 10.5.4. Mohr's circles and envelope for trona specimens.

or from Eq. 10.5.4 and recalling that $\tan \phi = \cot 2\theta = \mu$

$$\tau_\theta = \pm(S_0 - \mu\sigma_\theta) \qquad (10.5.10b)$$

Hence, for this case the Coulomb-Navier and Mohr's theories are identical.

Griffith's Theory of Brittle Fracture

The Coulomb-Navier and Mohr's theories of failure are concerned with the mechanisms of fracture and yield occurring on a macroscopic scale and, except for the fact that the mechanisms of shear and tensile fracture are identified with the appearance of shear and tensile fracture surfaces, they make no attempt to account for the cause of failure on an internal or microscopic basis. The tensile strength of single crystals has been computed on the basis of interatomic forces with the result that the theoretical values are from 100 to 1000 times larger than observed values. However, it is known from observation that crystalline materials contain microfractures, and Griffith[22] hypothesized that stress concentrations develop at the end of these cracks causing the crack to propagate and ultimately contribute to macroscopic failure. For a thin elastic strip of unit thickness containing an elliptical hole oriented with its long axis perpendicular to an applied tensile stress (Fig. 10.5.5), the maximum stress σ_{max} at the apex of the ellipse depends on ρ, the radius of curvature at the apex, and $2c$, the length of the crack. Inglis[23] determined this stress to be

$$\sigma_{max} = 2\sigma_0 \sqrt{\frac{c}{\rho}} \qquad (10.5.11)$$

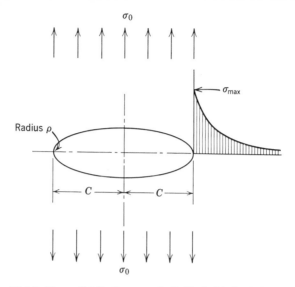

Fig. 10.5.5. Stress distribution at end of elliptical hole. (After Inglis.[23])

where σ_0 is the applied mean stress. The radius of curvature at the end of the ellipse may be of the order of intermolecular spacing.

Griffith computed the difference of energy in the strip with and without the hole to be

$$W_e = \frac{\pi c^2 \sigma_0^{\,2}}{E} \qquad (10.5.12)$$

The surface energy resulting from the formation of the crack is

$$W_s = 4cT \qquad (10.5.13)$$

where T is the surface tension. Hence the elliptical hole has decreased the total energy by

$$W = W_e - W_s = \frac{\pi c^2 \sigma_0^{\,2}}{E} - 4cT \qquad (10.5.14)$$

Instability will result and the crack will propagate if $\partial W/\partial c = 0$, that is, if the total energy becomes the maximum. Hence, from Eq. 10.5.14

$$\sigma_0 = \sqrt{\frac{2ET}{\pi c}} = T_0 \qquad (10.5.15)$$

where T_0 is the tensile strength of the material. This relationship was verified by Griffith from tests on internally pressurized glass tubes and spherical shells containing cracks made with a diamond. If in the two-dimensional case randomly distributed and oriented cracks occur throughout the body, the criterion for fracture is as follows:

If $\sigma_1 > \sigma_3$ and $3\sigma_1 + \sigma_3 < 0$ fracture will occur when

$$(\sigma_1 - \sigma_3)^2 = -8T_0(\sigma_1 + \sigma_3) \qquad (10.5.16a)$$

and at an angle given by

$$\cos 2\theta = -\frac{1}{2}\left(\frac{\sigma_1 - \sigma_3}{\sigma_1 + \sigma_3}\right) \qquad (10.5.16b)$$

If $\sigma_1 > \sigma_3$ and $3\sigma_1 + \sigma_3 > 0$, fracture will occur when

$$\sigma_1 = T_0 \qquad (10.5.17a)$$

and at an angle

$$\theta = 0 \qquad (10.5.17b)$$

Murrell[24] showed that the fracture criterion expressed by Eqs. 10.5.16a and 10.5.16b correspond to a Mohr's envelope at failure given by

$$\tau_\theta{}^2 + 4T_0\sigma_\theta = 4T_0{}^2 \qquad (10.5.18)$$

Thus Griffith's comparatively simple model of a brittle material containing micro-cracks of a specified length and a surface energy proportional to the surface tension, leads to failure criterion represented by a parabolic Mohr's envelope. Although some of the sedimentary rocks (limestone, sandstone and coal measure rocks) have a curved Mohr's envelope, it is common for the more brittle rocks such as granite and quartzite to have a straight Mohr's envelope in compression. Moreover, as the stress concentration around a Griffith crack is calculated on a basis of elastic theory this mechanism of failure should be time-independent and hence would not account for variation in strength with stress or strain rate. McClintock and Walsh[25] modified Griffith's theory by assuming that in compression Griffith's cracks close and a frictional force develops across the crack surface. Failure occurs when

$$\mu(\sigma_3 + \sigma_1 - 2\sigma_c) + (\sigma_1 - \sigma_3)(1 + \mu^2)^{1/2} = 4T_0\left(\frac{1 - \sigma_c}{T_0}\right)^{1/2} \quad (10.5.19)$$

where μ is the coefficient of friction for the crack surface and σ_c is the stress normal to the crack required to close it. Brace[26] pointed out that σ_c is small and can be neglected. Hence Eq. 10.5.19 becomes

$$\mu(\sigma_3 + \sigma_1) + (\sigma_1 - \sigma_3)(1 + \mu^2)^{1/2} = 4T_0 \qquad (10.5.20)$$

If $C_0 = -\sigma_3$ and $\sigma_1 = 0$ are the conditions for simple compression, the ratio of the compression to tensile strength is

$$\frac{C_0}{T_0} = \frac{4}{(1 + \mu^2)^{1/2} - \mu} \qquad (10.5.21)$$

Substituting Eq. 10.5.21 in 10.5.20

$$\frac{\sigma_1(1 + \mu^2)^{\frac{1}{2}} + \mu}{C_0(1 + \mu^2)^{\frac{1}{2}} - \mu} - \frac{\sigma_3}{C_0} = 1 \tag{10.5.22}$$

Thus for compression the McClintock and Walsh modification of Griffith's theory predicts a linear relation between σ_1 and σ_3, as does the Coulomb-Navier theory.

10.6 EFFECT OF PLANES OF WEAKNESS

On a megascopic scale most rock contains planes of weakness that affect its structural properties. These planes may be the result of joints, fractures, faults, bedding, or partings. Adler[27] described graphical methods for treating materials for which the failure envelope of the weakness plane(s) is known. First consider a material containing a plane of weakness, the fracture envelope for which has been determined by direct shear or triaxial tests, line AB in Fig. 10.6.1a. Let this material be subject to principal compressive stresses, σ_1 and σ_3, with the plane of weakness oriented at an angle γ with respect to the σ_3-axis, Fig. 10.6.1b. If the angle 2γ in Fig. 10.6.1a is such that the radius OC touches or intersects the failure envelope (between D and E) failure will result; for all other values of 2γ there will be no failure. Or, if the orientation and the failure envelope of the weakness plane is known, the values of σ_1 and σ_3 for which there will be no failure, can be determined from the construction in Fig. 10.6.2. Let AB be the failure envelope and γ' be the angle of the plane of weakness. Any Mohr's circle drawn with OC as a radius will give the limiting values of σ_1 and σ_3.

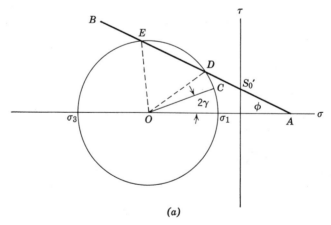

Fig. 10.6.1a. Representation of a plane of weakness AB.

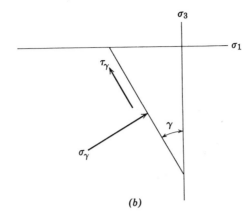

Fig. 10.6.1b. Normal and shear stress on plane of weakness.

If the material contains a number of planes of weakness with identical failure envelopes and if the stress circle intersects the failure envelope, the range of γ for which failure will not occur is restricted, as illustrated in Fig. 10.6.3 for three equally spaced fracture planes, OA, OB, OC. Thus it is evident that as the number of planes of weakness increases, the radius of the stress circles must be correspondingly decreased if they are to represent conditions that will not produce failure. For a highly fractured rock the

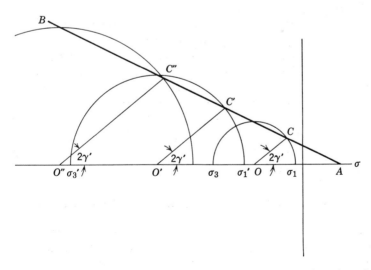

Fig. 10.6.2. Determination of permissible principal stresses when orientation of the plane of weakness and its failure envelope are known.

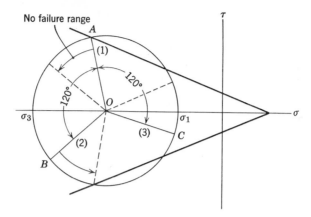

Fig. 10.6.3. Mohr's construction for a rock containing planes of weakness OA, OB, and OC 120° apart.

permissible stress circles, that is, circles for which there will be no failure, must lie virtually within the envelope.

The use of this graphical procedure presumes a knowledge of the fracture criterion for the planes of weakness, that is, in Fig. 10.6.1a

$$|\tau_y| = S_0' - \mu\sigma_y \qquad (10.6.1)$$

where S_0' is the shear strength of the plane of weakness, $\mu = \tan\phi$ is the coefficient of internal friction of the plane, and τ_y and σ_y are the shear and normal stresses acting on the plane. To evaluate this relationship a number of specimens of a given rock type with variously oriented but otherwise identical natural weakness planes (joints, partings, etc.) must be tested triaxially. To prepare a group of specimens to this specification for laboratory tests would be a difficult task. However, Jaeger[28] prepared triaxial specimens with simulated joint planes cut in prescribed directions. The types of joint planes were (1) plaster filled, (2) bare surfaces ground flat, and (3) natural surfaces across which shear failure had taken place. Measured coefficients of friction were in the range 0.5 to 0.8 corresponding to $\phi = 27°$ to 39° (Fig. 10.6.1a).

10.7 EFFECT OF MOISTURE AND PORE PRESSURE

Besides depending on the state of stress, strain, and stress or strain rate, the mechanical properties of rock also are affected by moisture and pore pressure. In an underground operation the moisture and pore pressure range is usually limited and difficult to vary. As a consequence the effects of these factors have received little study. In the laboratory the

effect of moisture and pore pressure has been more thoroughly investigated, but the problem of relating the results of small-scale tests to
full-scale structures has received only limited study.

Effect of Moisture

Most underground rock contains moisture ranging from less than 1 %
for some of the evaporite minerals to over 35% for porous sandstones,
and in the majority of mines and underground workings the rock is
nearly saturated. In some instances water will drip or flow from joints,
fractures, and faults; in some rocks, especially limestone, large quantities
of water will enter the workings through water courses or channels.
The presence of moisture or water (at atmospheric pressure) can have
several effects. If the water is migrating, there can be a mechanical action
in which softer materials such as clay or other decomposition products
are washed from joints or faults. Chemical action may dissolve some of the
more soluble mineral components resulting in a change in the mechanical
properties of the rock. For example, Griggs[12] found that the constant-
load creep rate in wet gypsum (alabaster) was more than twenty-five
times that for dry gypsum.

Chemical hydration or dehydration (gain or loss of water by hydration)
in certain rocks may cause a volumetric change accompanied by a change
in mechanical properties. Alternate wetting and drying will cause some
rocks to expand and contract and thus will affect their properties.
Hartmann[29] found that draw slate expanded 0.42% in four days when
placed in an atmosphere having a 95% humidity, and contracted 0.30%
in the same period when the humidity was then lowered to 30% (Fig.
10.7.1). This expansion or contraction was substantially reduced if the

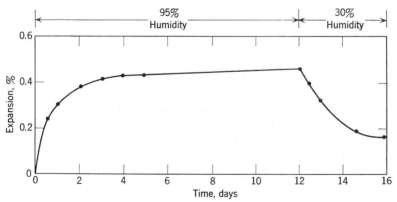

Fig. 10.7.1. Effect of moisture on the expansion of shale. (After Hartmann and
Greenwald.[29])

specimen was painted with a moisture-resistant coating. Hartmann also showed that if a draw slate specimen was uniaxially constrained ($\epsilon_{axial} = 0$) and then placed in water, an axial stress of 300 psi developed in 4 hr. Also, exposure to a varying humidity or alternate wetting and drying will cause some rocks to separate on bedding planes or, in more extreme cases, to completely disintegrate. Hence, in an underground operation where the humidity is seasonably or otherwise variable, deterioration of surface rock may be expected, especially in sedimentary rocks containing clay or other decomposition products. Some estimate of these effects can be

Table 10.7.1. Effect of Moisture Content on Compressive Strength

Moisture Condition	Ratio of Oven-dried and Saturated Compressive Strength to Air-dried Compressive Strength						
	Marble	Lime-stone	Granite	Sand-stone 1	Sand-stone 2	Slate	Average
Oven-dried	1.01	1.03	1.07	1.01	1.18	1.06	1.06
Air-dried	1.00	1.00	1.00	1.00	1.00	1.00	1.00
Saturated	0.96	0.85	0.92	0.90	0.80	0.85	0.88

obtained by subjecting cores from exploratory drilling to an atmosphere in which the humidity is varied.

Moisture also affects the compressive strength and the elastic properties of rock, although the reason for this effect is not understood. For example, the modulus of elasticity in granite and marble (determined by the bar velocity method, Section 11.12) increased by 38% and 40% respectively as the moisture content in the rock was increased from dry to saturated.[1] The effect of moisture on the compressive strength is less pronounced, as indicated in Table 10.7.1. In general the strength decreased with an increase in moisture, but since this effect is small, for most engineering purposes it can be disregarded, especially if the compressive strength of the rock is measured at a moisture content corresponding to that occurring in situ. A part of this decrease in strength with increase in moisture content may be due to an inability of the pore water to migrate freely (within the time limits of the test), and hence a pore pressure may develop as the load on the specimen is increased.

Effect of Pore Pressure

The effect of pore pressure on the strength of rock has been considered theoretically by Hubbert and Rubey[30] using as a model for rock a material made up of an aggregate of spherical grains. This material, in a triaxial test is subjected to a pore pressure, p_0, and principal stresses σ_3 and $\sigma_2 = \sigma_1$. The pore pressure reduces the principal stresses, so that the effective principal stresses $\sigma_1{}'$ and $\sigma_3{}'$ are

$$\sigma_3{}' = \sigma_3 - p_0$$

and   (10.7.1)

$$\sigma_1{}' = \sigma_1 - p_0$$

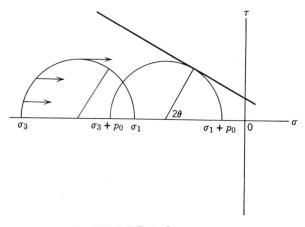

Fig. 10.7.2. Effect of pore pressure.

Thus the effect of the pore pressure is to displace the Mohr's circle in Fig. 10.7.2 by an amount p_0, but the stress difference $\sigma_3 - \sigma_1$ and the slope of the envelope remain unchanged. The Coulomb-Navier condition for failure becomes

$$\tau_\theta = S_0 - \mu(\sigma_\theta - p_0)$$
$$\sigma_1 + p_0 = T_0$$
(10.7.2)

from which it is evident that the effect of the pore pressure reduces the normal stress on the failure plane by an amount p_0.

Murrell[24] considered the effect of uniform internal pressure p_0 in a Griffith crack, and on this basis derived a modified failure criterion, which for the two-dimensional case is for $\sigma_1 > \sigma_3$ and $3\sigma_1 + \sigma_3 + 4p_0 \leq 0$

$$(\sigma_1 - \sigma_3)^2 + 8T_0(\sigma_1 + \sigma_3 + 2p_0) = 0 \qquad (10.7.3)$$

and the fracture angle θ is

$$\cos 2\theta = -\frac{1}{2}\frac{\sigma_1 - \sigma_3}{\sigma_1 + \sigma_3 + 2p_0} \qquad (10.7.4)$$

If $3\sigma_1 + \sigma_3 + 4p_0 \geq 0$, then

$$\sigma_1 + p_0 = T_0$$

and   (10.7.5)

$$\theta = 0$$

In drilling stress-relief holes (Section 14.3), holes for instruments, or in core drilling for rock samples, significant water pressures have not been reported within the limits of the stress concentration zone surrounding underground openings. In blocked-off holes that normally drain water, the static pressure was less than 100 psi, even in mines operating at depths of 2000 to 3000 ft. Presumably the drainage through joints, fractures, faults, partings, and water courses is such that the water pressure gradient can only increase slowly with the distance from the surface of the opening, although the gradient may depend on the areal extent of the openings. For example, a tunnel, shaft, or single opening may have a larger gradient than an extensively mined area. Thus the most likely condition in the rock surrounding an underground opening is a moist or nearly saturated rock with a comparatively small pressure gradient developing behind the rock surface.

REFERENCES

1. Obert, L., S. L. Windes, and Wilbur I. Duvall, "Standardized tests for Determining the Physical Properties of Mine Rock," *U.S. BurMines Rept. of Invest.*, **3891** (1946).
2. Hardy, H. R. Jr., "Standardized Procedures for the Determinations of the Physical Properties of Mine Rock under Short-Period Uniaxial Compression," *Tech. Bull.* **8,** *Dept. of Min. and Tech. Sur.*, Ottawa (Dec. 1959).
3. American Society for Testing Materials, Toughness of Rock, ASTM Designation D3-18, A.S.T.M. Standards, 1942, Part II.
4. Everling, G., "Zur Definition der Schubfestigkeit," *Gluckauf Bergmannische Zeitschrift*, **98,** No. 18 (1962).
5. Wuerker, R. G., "The Shear Strength of Rocks," *Mining Eng.*, **11,** No. 10 (1959).
6. Balmer, G., "A General Analytic Solution for Mohr's Envelope," *A.S.T.M. Proc.*, **52** (June 1952).
7. Obert, L., "Triaxial Method for Determining the Elastic Constants of Stress Relief Cores," *U.S. BurMines Rept. of Invest.*, **6490** (1964).
8. Nishihara, "Stress-Strain-Time Relation of Rocks," II, *Doshisha Kogaku Kaishi*, **VIII,** No. 3 (Nov. 1957).
9. Obert, L., "Effects of Stress Relief and Other Changes in Stress on the Physical Properties of Rock," *U.S. BurMines Rept. of Invest.*, **6053** (1962).
10. Merrill, R. H. and T. A. Morgan, "Method of Determining the Strength of a Mine Roof," *U.S. BurMines Rept. of Invest.*, **5406** (1958).
11. Phillips, D. W., "Tectonics of Mining," Part II, *Colliery Eng.* (August 1948).
12. Lomnitz, C., "Creep Measurement in Igneous Rocks," *J. Geol.*, **64,** No. 5 (1956).
13. Griggs, David, "Creep of Rocks," *J. Geol.*, **XLVII,** No. 3 (April-May 1939).
14. Evans, R. H., "The Elasticity and Plasticity of Rocks and Artificial Stone," *Leeds Philosophical and Literary Soc. Proceedings*, **3,** part 3 (1936).

15. Iida, K. and M. Kumazawa, "Viscoelastic Properties of Rocks," *J. Earth Sci, Nagoya Univ.*, **5**, No. 1 (1957).

16. Merrill, R. H., "Design of Underground Mine Openings, Oil Shale Mines, Rifle, Colo.," *U.S. BurMines Rept. Invest.*, **5089** (1954).

17. Höfer, K., "Beitrag zur Frage der Standfestigkeit von Bergfesten im Kalibergbau," *Freiberger Forschungsh.*, **A100**, 1958.

18. Robertson, E. C., "Viscoelasticity of Rocks, International Conference on the State of Stress in the Earth's Crust," *Rand Corporation Memo RM* 3853 (May 1963).

19. LeComte, P., "Creep and Internal Friction in Rock Salt," *Harvard Univ. Doctoral Dissertation*, 1960.

20. Robertson, E. C., "Experimental Study of the Strength of Rocks," *Bull. Geol. Soc. Am.*, **66** (October 1955).

21. Jaeger, J. C., *Elasticity, Fracture and Flow*, Methuen and Co., London (1962), op cit p. 78.

22. Griffith, A. A., "The Phenomena of Rupture and Flow in Solids," *Phil. Trans.*, **221** (1921).

23. Inglis, C. E., *Trans. Inst. Naval Architects, London*, **60** (1913).

24. Murrell, S., "A Criterion for Brittle Fracture of Rocks and Concrete Under Triaxial Stress and the Effect of Pore Pressure on the Criterion," *Proc. Fifth Rock Mechanics Symposium*, Pergamon Press (1963).

25. McClintock, F. A. and J. Walsh, "Friction on Griffith's Cracks in Rock Under Pressure," *Int. Cong. Appl. Mech.*, Berkeley, Cal. (1962).

26. Brace, W., "Brittle Fracture of Rocks," *Int. Conf. on State of Stress in the Earth's Crust, Reprint of Papers, Rand Corp. Memo* 3583 (May 1963).

27. Adler, L., "Failure in Geological Materials Containing Planes of Weakness," *Trans. AIME, SME*, **226** (March 1963).

28. Jaeger, J. C., "The Frictional Properties of Joints in Rock," *Geofis. Pura Appl.*, **43** (1959).

29. Hartmann, I. and H. Greenwald, "Effect of Changes in Moisture and Temperature on Mine Roof," *U.S. BurMines Rept. Invest.*, **3588** (Oct. 1941).

30. Hubbert, M. K. and W. W. Rubey, "Role of Fluid Pressure in Mechanics of Overthrust Faulting," *Bull. Geol. Soc. Amer.*, **70** (Feb. 1959); **71** (May 1960).

CHAPTER 11

MECHANICAL PROPERTY TESTS

11.1 INTRODUCTION

In this chapter equipment and test procedures are described for measuring the mechanical properties of rock. These tests provide the mechanical property data required for the evaluation of rock structure and other rock mechanics problems. Probably because of its relative importance this subject has received more study than any other phase of rock mechanics. A variety of apparatus and procedures has been developed, many rock types have been tested, and a large volume of mechanical property data has been published. These efforts show that the test results depend not only on the properties of the rock but also on the manner and environment in which the rock is tested. Factors that affect these measurements include specimen size, specimen shape, flatness and smoothness of specimen surfaces, friction between loading platens and the specimen, rate of loading, the state of stress in the specimen, and the temperature and moisture content in the rock.

It is common practice in materials testing to select a set of conditions so that the combined effects of these extraneous factors are relatively small and hence the measured results are, to as large a degree as possible, indicative of the material properties. Although these effects can be minimized, they cannot be eliminated—a fact that suggests the need for standardizing the procedures used in measuring the mechanical properties of rock so that the results from different laboratories can be appraised or compared. Some attempts have been made in this direction; the American Society of Testing and Materials[1] has prescribed tests for measuring the compressive and flexural strength of building stone; the U.S. Bureau of Mines[2] proposed a number of standardized procedures for measuring the physical properties of mine rock; and the Department of Mines and

318

Technical Surveys, Canada[3] and the South African Institute of Mining and Metallurgy[4] have adopted similar procedures. The procedures prescribed by these laboratories are for static tests in which the specimens are subject to uniaxial loading.* No standardized tests for measuring the triaxial compressive or shear strength have been proposed although the essential aspects of the procedures presently in use are similar. Also, a variety of methods have been employed for measuring the time-dependent properties of rock, but no generally accepted testing techniques have evolved. As a consequence there is only a limited basis for equating time-dependent test results.

11.2   SAMPLING PROCEDURE

The purpose in specifying a sampling procedure is to insure the selection of an adequate group of test specimens from each mechanically different rock type that forms an essential part of a structure, model, or is otherwise related to the project under investigation. The sampling procedure should be guided by the project objectives; for example, in selecting a horizon or area for an underground chamber, long lengths of exploratory drill core may be sampled so that variations in rock properties can be evaluated and a site with optimum properties delineated. On the other hand, there is usually no choice in selecting a mining horizon in a mineral deposit, and the sampling procedure is generally directed at obtaining the properties of the rock that eventually will comprise the roof, floor, side walls, pillars, or other specific parts of the structure.

Assuming that in relation to sampling the broad objectives of a project have been considered, the next step is to identify macroscopically the rock by geological type (sandstone, limestone, granite, shale) and within each type, to subdivide the rock according to grain or crystal size; macroscopically discernible differences in mineral composition; attitude of bedding planes, partings, joints, or other planes of weakness; degree of alteration; porosity; and possibly color. The extent of the sampling, that is, the number of specimens taken from each of these classifications, will depend on the requirement for each type of mechanical property test, the desired accuracy, the stock of rock available for test (often the amount of exploratory core available for mechanical property testing is limited because of drilling costs or because the first priority on cores from mineral deposits is usually for chemical analysis), and the cost and

---

* The U.S. Bureau of Mines includes in their standardized tests a dynamic (sonic) method for determining the elastic constants of rock. Dynamic tests are usually associated with the propagation of, or the effects produced by, stress waves, and are not considered time dependent in the sense that the tests described in Section 11.12 are.

time required for testing. A preliminary analysis of the test results may indicate that the results from one or another of the subdivisions are not significantly different and the groups may be combined. On the other hand, the analysis may disclose significant differences within a given group of specimens and a further subdivision may be required. Most rock is anisotropic and, if the core stock and sampling procedure permits, a group of specimens should be obtained from the three mutually perpendicular directions. Usually these directions are oriented with respect to some petrographic property of the rocks such as bedding, schistosity, cleavage, or fabric. In bedded rock the greatest difference in properties occurs in specimen taken perpendicular and parallel to the bedding, and generally this type of rock is sampled only in these two directions.

A petrographic examination may disclose mineral components that are soluble or that expand or soften in water, as for example, bentonites or other clays, and thin-section or other mineralogic studies are usually recommended, especially for sedimentary rocks. However, microscopic examinations in themselves are not satisfactory bases for classifying rock mechanically, for some rocks that are microscopically similar have significantly different mechanical properties.

11.3   PREPARATION OF SPECIMENS

Most mechanical property test specimens are cut from diamond drill cores from exploration drilling, underground diamond drilling performed specifically to obtain core for test purposes, or from cores drilled in the laboratory from blocks of rock selected from underground areas. Cores from exploration or underground drilling are preferred to those from blocks of rock that may have been subjected to high intensity stress waves from blasting. Cores or blocks of rock should be transported to the laboratory in moisture-proof containers, as loss of moisture affects the properties of some rock.

A core preparation laboratory should contain the following principal pieces of equipment:* (1) a coring drill (Fig. 11.3.1a); (2) a diamond cut-off saw (Fig. 11.3.2); (3) a grinding lap; and (4) a surface grinder (Fig. 11.3.3). A conventional drill press powered with a one horse power motor can be adapted to cut cores up to $2\frac{1}{8}$ in. diameter (NX) provided that thin-wall bits are employed. There are two types of thin-wall bits: detachable, and integrally attached to the core barrel (Fig. 11.3.1b). Also, cores with a diameter larger than $2\frac{1}{8}$ in. can be drilled with thin-wall bits, but for most

* A part of this equipment is described in more detail in U.S. Bureau of Mines Report of Investigations RI 3891[2]. These principal pieces of equipment are also required in preparing rock models.

(a)

(a)

Fig. 11.3.1a. Laboratory coring drill.

rock types a more powerful drill is required. Water or air admitted through a swivel in the core barrel should be used to clear drill cuttings and cool the bit. Cores from most nonsiliceous rocks can be sawed with an abrasive blade, using air as a coolant. Diamond blades are necessary for cutting siliceous rocks and water is normally used as a coolant. A grinding lap is a flat brass rotating disk on which a fine abrasive is distributed. For core diameters $2\frac{1}{8}$ in. D or less, a lap can be used for grinding flat end surfaces on specimens, although producing a sufficiently flat surface by this method

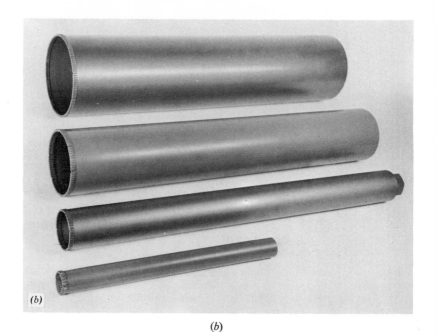

(b)

Fig. 11.3.1b. Thin-wall diamond bits and barrels.

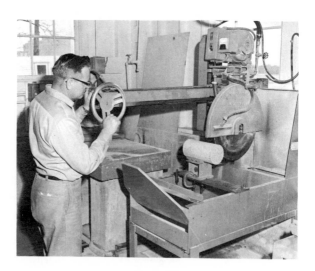

Fig. 11.3.2. Diamond cut-off saw.

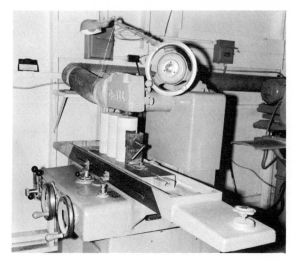

Fig. 11.3.3. Surface grinder.

is an art. Conventional surface grinders provide about the only practical means of preparing flat surfaces larger than $2\frac{1}{8}$ in. *D*. Rock types at least as hard as granite can be surfaced with these machines.

The degree of flatness required in preparing the bearing surface of specimens to be loaded in compression is not generally appreciated. For example, because the breaking strain of some rock is as low as 0.001 (Table 9.2.1), the deviation in the bearing surfaces of a compressive strength specimen 1 in. long must be small compared with 0.001 in. Although this is the extreme example, a deviation of 0.001 in. in the flatness of the bearing surfaces of most compressive tests specimens would cause a significant error.

Within the range of specimen size used in the standardized tests proposed by the Bureau of Mines, the test results were relatively independent of the specimen size. However, these tests were performed on comparatively fine-grained rocks. Many rock types such as salt, potash ores, pegmatites and some granites have crystal sizes ranging from $\frac{1}{4}$ in. to more than 1 in. For these rock types the ratio of the crystal to specimen size should be investigated to establish a ratio below which the test results are relatively unaffected by specimen size.

The moisture in test specimens is another factor that should be controlled. Although the compressive strength is not too sensitive to the moisture in a specimen, the modulus of elasticity of marble (determined by the bar velocity method) increased by more than 30% as the moisture in the specimen was increased from dry to saturated[2]. It has been determined that small changes in the moisture content of nearly dry specimens do not strongly affect rock properties. Hence, specimens are tested by most laboratories after they have been air-dried at room temperature for two weeks. Additional moisture can be removed from air-dried specimens by placing them in a vacuum or in a desiccator. Specimens should not be dried in an oven (especially not above 212°F) as heating may cause an irreversible change in rock properties.

Although air-dry testing may be convenient and satisfactory for measuring the properties of rock used in laboratory models, it may not be a realistic specification for determining the properties of rock in underground workings. With the exception of the evaporite minerals, most underground rocks are comparatively wet. However, the exact moisture content for in situ rock is difficult to ascertain because of evaporation from, or condensation on underground rock surfaces, changes in the circulation of underground water caused by mining, or because water must be used in coring or cutting rock specimens.

Some shales and rocks containing clay will disintegrate if allowed to dry. Usually the disintegration of diamond drill cores can be prevented by wrapping the cores as they are drilled in a moisture-proof material such as aluminum foil or chlorinated rubber, or sealing them in moisture-proof containers. Mud shales and rock containing bentonites (for example tuff) may soften if the moisture content is too high. Most of the softer rocks can be cored or cut using compressed air to clear cuttings and to cool the bit or saw.

11.4 STATISTICAL TREATMENT OF DATA*

Most physical tests involve tabulation of a series of readings, with computation of an average said to be representative of the whole. The question arises as to how representative this average is as the measure of the characteristic under test. Three important factors introduce uncertainties in the result:

1. Instrument and procedural errors.
2. Variations in the sample being tested.
3. Variations between the sample and the other samples that might have been drawn from the same source.

If a number of identical specimens were available for test, or if the test were nondestructive and could be repeated a number of times on the same

* Excerpted from "Standardized Tests for Determining the Physical Properties of Mine Rock," by Leonard Obert, S. L. Windes, and Wilbur Duvall, *U.S. Bureau of Mines Report of Investigations* 3891 (1946).

specimen, determination of the procedural and instrument errors would be comparatively simple, because in such a test the sample variation would be zero. Periodic tests on this specimen or group of specimens could be used to check the performance of the test procedure and equipment. However, as most of the tests used in determining the mechanical properties of rock are destructive, the instrument and sample variations cannot be separated. In this case the best procedure is to prepare a number of groups of as nearly identical specimens as possible, and use the variation of one of these groups as an indication of the performance of the test.

Table 11.4.1. Compressive Strength of Quarried Rocks

Rock Type	Number of Specimens	Compressive Strength, Psi	Standard Deviation Percent
Marble	12	30,000	9.8
Limestone	12	10,900	5.8
Granite	12	33,200	3.5
Sandstone	12	10,400	5.0
Greenstone	8	39,000	5.8

The most commonly used measure of the central tendency of a sample of n specimens is the arithmetic mean or average, $\bar{X}$, defined as (Eq. 8.3.1).

$$\bar{X} = \frac{1}{n} \sum_{i}^{n} X_i \qquad (11.4.1)$$

where X_i is the measured value of the ith specimen. The standard deviation of the sample S is a measure of the sample variability (Section 8.4.2), where

$$S^2 = \frac{1}{n-1} \sum_{i}^{n} (X_i - \bar{X})^2 \qquad (11.4.2)$$

Table 11.4.1 gives the average compressive strength and standard deviation of a group of specimens carefully prepared from uniform quarried rocks. For compression test groups of this size (12 specimens), these standard deviations are about the minimum that can be achieved.

When a series of readings is made on each of two groups of specimens and their average values are different, the question arises if this difference is significant. If the test groups are small, as is generally the case in testing rock, and if the distribution of the readings about their average is unknown or skewed, Tchebycheff's theorem (Section 8.9) can be used to delimit the range in which a predetermined minimum percentage of the data will fall.

From Eq. 8.9.7 at least $100(1 - 1/k^2)\%$ of the observations will fall in the range $\bar{X} \pm kS$. Thus, for $k = 3$, 89% of the data will fall in the range $\bar{X} \pm 3S$. If there are two groups of data with average values $\bar{X}_1$ and $\bar{X}_2$ with $\bar{X}_1 < \bar{X}_2$, and if $\bar{X}_1 + kS_1$ does not overlap $\bar{X}_2 - kS_2$ where $k > 3$, the difference in the values is significant. That is, for $k = 3$, $\bar{X}_2 - \bar{X}_1 \geq 3(S_1 + S_2)$ (Fig. 11.4.1).

If the number of tests exceeds ten the individual values should be studied to determine if a normal distribution of the data exists. Divide the range of the data (highest value minus the lowest value) in about five equal

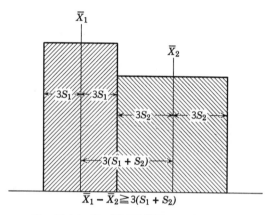

Fig. 11.4.1. Significant difference in means.

intervals and plot the number of readings against the midpoint of the interval. If there is a strong tendency for the data to group about the central interval, the data most likely have a normal distribution. If such is the case, then the procedures outlined in Section 8.6 can be used to test for significant differences between averages.

11.5 APPARENT SPECIFIC GRAVITY—APPARENT POROSITY

Apparent specific gravity ASG is defined as the ratio of the weight of a specimen of given *exterior* volume to the weight of an equal volume of water. Thus apparent specific gravity does not take into account voids within the specimen. For specimens of regular shape (prisms or cylinders), the exterior volume V_c can be calculated from linear measurements; hence

$$\text{ASG} = \frac{W_r}{V_c \gamma_w} \tag{11.5.1}$$

where W_r is the dry weight of the specimen and γ_w is the unit weight of water (0.0361 lb/in.³). For specimens of irregular shape the exterior volume can be determined from the saturated weight of the specimen W_s, and the weight of the saturated specimen when it is immersed in water W_v, by

$$V_c = \frac{W_s - W_v}{\gamma_w} \qquad (11.5.2)$$

Hence from Eq. 11.5.1,

$$ASG = \frac{W_r}{W_s - W_r} \qquad (11.5.3)$$

Table 11.5.1. Specific Gravity and Apparent Porosity of Typical Rocks

Rock Type	Average Specific Gravity	Apparent Porosity (AP) Percent
Marble	2.87	0.6
Limestone	2.37	11.0
Granite	2.66	0.9
Sandstone	2.06	16.0
Slate	2.74	1.0
Greenstone	3.02	0.7

The apparent porosity AP is defined as the ratio of the open pore space to the exterior volume of the specimen. This quantity can be calculated from the dry and saturated specimen weights by

$$AP = \frac{W_s - W_r}{V_c \gamma_w} \qquad (11.5.4)$$

Representative ASG and AP data for several rock types are given in Table 11.5.1. Because some of the voids in rock may not be interconnected, the true porosity generally will be greater than the apparent porosity. It is extremely difficult to saturate rock completely. Immersion in water even for long periods is usually not sufficient to achieve saturation. A better method is to evacuate a vessel containing the rock, and then to flood the vessel.

11.6   TENSILE STRENGTH

The "direct" tensile strength test is effected by loading a cylindrical or prismatic specimen in tension to failure (Fig. 11.6.1a and b). The tensile strength is given by $T_0 = F_t/A$, where F_t is the tensile load at failure and A is the cross-sectional area of the specimen. At least in concept this is an ideal mechanical property test because if the normal precautions

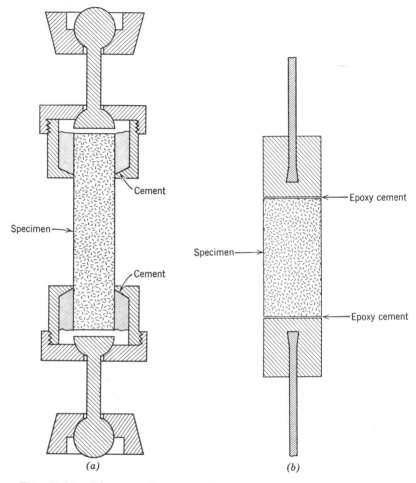

Specimen

Cement

Cement

Epoxy cement

Specimen

Epoxy cement

(a)

(b)

Fig. 11.6.1a. Direct tensile strength tests.

Fig. 11.6.1b. Tensile strength tests. (After Fairhurst.[5])

are observed, the specimen should fail in plane tensile stress and, for a homogeneous isotropic material, the plane of failure should be normal to the axis of the specimen. The precautions are twofold: first, the applied tensile load must be uniformly distributed over the end of the specimen and parallel to its axis; second, the grips, or method of holding the specimen, must not produce significant lateral stresses in the specimen. To minimize the effects of grips, the central section of metal tensile strength specimens are usually machined to a smaller diameter. With rock this is a difficult operation and other methods of holding the specimen have been developed.

Obert[2] et al. cemented the ends of the specimen in cups (Fig. 11.6.1a) and applied the tensile load through ball and socket joints. Fairhurst[5] employed an even simpler method, in which the ends of a cylindrical specimen were cemented with an epoxy resin to flat caps and the tensile load applied through flexible cables (Fig. 11.6.1b).

It is known that even small scratches on the surface of metal specimens will reduce the tensile strength appreciably, and some investigators fine-grind or polish the cylindrical surface of rock specimens to minimize this effect. However, most rock contains planes of weakness, incipient cracks, or other mechanical defects, and failure usually occurs at the point of these defects, a factor that causes a comparative large deviation in the tensile strength measurements from a group of specimens. Also, as the length of tensile strength specimens is increased, the probability of including a weaker defect is increased, and hence the average tensile strength decreases with the size of test specimens.

No specification for a direct tensile strength determination has been proposed. However, a satisfactory determination can be made by testing a group of 10 or more air-dried specimens, using the methods illustrated in Fig. 11.6.1a or b. Core diameter from $1\frac{1}{8}$ in. to $2\frac{1}{8}$ in. and a specimen length of 4 in. between grips should give a result reasonably independent of specimen size. A rate of loading of 500 psi/min would be consistent with general practice.

In the indirect tensile strength test a cylindrical test specimen is placed horizontally between the bearing plates of a testing machine and loaded to failure in compression (Fig. 11.6.2). If the diameter of the specimen is D and the length L, and a line load F is applied along the length of the specimen at AB and CD, a uniform tensile stress

$$\sigma = \frac{2F_c}{\pi DL}$$

should develop across the plane $ABCD$ (Fig. 11.6.2). If F_c is the load at failure, the tensile strength T_0 is

$$T_0 = \frac{2F_c}{\pi DL} \qquad (11.6.1)$$

This test was proposed by Carneiro[6] and independently by Akazawa[7] and is referred to as the Brazilian test. However, in addition to the tensile strength that develops across the diametral plane, a vertical compressive stress occurs in the plane, which varies from $-\dfrac{6F_c}{\pi DL}$ at the center to $-\infty$ at the line of contact with the platens (Section 4.6). This compressive strength causes high shear stresses and local crushing along the loading line. To prevent this type of failure Hondros[8] distributed the line load over an area by applying the load to a strip of width $D/12$ (Fig. 11.6.3).

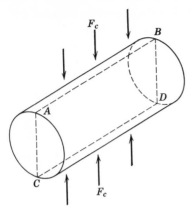

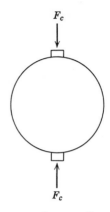

Fig. 11.6.2. Indirect tensile strength test.

Fig. 11.6.3. Indirect tensile strength test with distributed load.

With this procedure the uniform tension across the diametral plane is unaffected except in the proximity of the loaded area where a horizontal compressive stress develops. Thus failure should initiate in the interior of the specimen, and should not be affected by surface scratches or irregularities. Mitchell[9] found that a layer of cardboard inserted between the specimen and the platen will serve the same purpose. For a uniform concrete Akazawa[7] found the ratio of the indirect tensile strength to flexural strength was constant, although the flexural strength (outer fiber tensile strength) was approximately 2.4 times the indirect tensile strength. The indirect tensile strength should also be greater than the direct tensile strength because in the latter test the specimen has a better opportunity to select the point of minimum strength.

Although several investigators have proposed standardized procedures for measuring the indirect tensile strength of concrete, no similar procedure has been proposed for testing rock.

11.7 UNIAXIAL COMPRESSIVE STRENGTH

The uniaxial compressive strength of rock or other brittle materials is determined by loading a cylindrical or prismatic specimen to failure in a compression machine (Fig. 11.7.1). The compressive strength C_0 is given by

$$C_0 = \frac{F_c}{A} \tag{11.7.1}$$

where F_c is the applied compressive load at failure and the A is the cross-sectional area of the specimen.

The compressive strength is the most commonly determined property of rock and the procedure used by most laboratories in making this measurement is virtually equivalent. In concept the compressive strength test is deceptively simple, but in reality there are a number of factors that can affect the test results significantly, such as the flatness of the bearing surfaces, the specimen size and shape, the moisture content in the specimen, the effect of friction between the bearing platens and the specimen, the alignment of the swivel head, and the rate of loading.

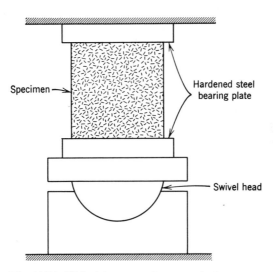

Fig. 11.7.1. Uniaxial compressive strength test.

As noted in Section 11.3, the flatness of the specimen ends (bearing surfaces) is critical. In general the deviation from flatness should be small compared to the deformation produced by the axial load at failure. This degree of flatness is difficult to achieve by lapping even if jigs are used to hold the specimens, but uniformly flat surfaces can be obtained by surface grinding. Parallelness between end surfaces should be within $\pm 1°$, a specification that is not difficult to realize.

The alignment of the spherical bearing head used to accommodate for the lack of parallelism between the ends of the specimen is one of the more critical factors in compression testing. If the diameter of a spherical head is too large in relation to the specimen diameter or if the spherical surfaces are not properly ground or lubricated, or if the specimen is not accurately centered, the measured compressive strength values may be substantially less (as much as 50%) than those obtained using a proper technique. For testing drill core specimens with diameters between $1\frac{1}{8}$ in. and $2\frac{1}{8}$ in. a

3 in. diameter spherical head lubricated with oil and graphite and used on the lower bed of the testing machine* is relatively easy to align.

When a specimen is loaded in compression, the specimen and bearing plates expand laterally. If the specimen is rock and the bearing plates steel, this difference in expansion creates lateral stresses in the ends of the specimen. The magnitude of these stresses depends on the coefficient of friction between rock and steel and on their elastic constants.† To minimize any change in the smoothness of the bearing plates (that would affect the coefficient of friction between the rock and bearing plate) hardened steel-bearing plates with ground surfaces are generally employed. Capping material such as lead foil, leadite, blotting paper, and Portland and gypsum cement mortars have been used to correct for lack of flatness in the specimen ends. In most instances these capping materials have a lower modulus of elasticity than the rock and, in addition, tend to flow plastically. In either case this softer material in contact with the rock produces tensile stresses in the ends of the specimen that lower its compressive strength or cause it to split in an axial plane at a relatively low applied load. Consequently, capping materials are generally avoided in testing rock.

The compressive strength of rock increases with the rate of loading. For high rates of loading as, for example, that produced by stress waves from blasting, this increase in strength may be appreciable. Also, for very long loading periods a substantially reduced strength may be obtained. However, for rates of loading of 100, 200, and 400 psi/sec the difference in the measured compressive strength of rock was found to be negligible.[2]

Obert[2] et al. reported that for cylindrical specimens the effect of specimen size is less than the normal intragroup variation, provided that the ratio of the specimen length-to-diameter ratio is kept constant. However, the length-to-diameter ratio L/D has a significant affect on the crushing strength which can be corrected by the following relation:‡

$$C_0 = \frac{C_p}{0.778 + 0.222 \dfrac{D}{L}} \qquad (11.7.2)$$

where C_0 is the compressive strength of a specimen of the same material

---

* Use of the spherical head on the lower platen of the compression machine makes it easier to retain the lubricant on the spherical surfaces.

† If the modulus of elasticity of the bearing plate is greater than that of the specimen, the lateral stress in the ends of the specimen is compressive (assuming that both materials have approximately the same Poisson's ratio.)

‡ Equation 11.7.2 is identical to the L/D relationship for prismatic specimens given in the ASTM standards.[1]

having a $1:1$ length-to-diameter ratio, and C_p is the compressive strength of a specimen for which $2 > (L/D) > \frac{1}{3}$.

The standardized procedure proposed by the U.S. Bureau of Mines[2] specifies testing a group of at least ten air-dried specimens. The diameters of the specimens can range from $\frac{7}{8}$ to $2\frac{1}{8}$ in.; their length-to-diameter ratio should be either unity, or the measured compressive strength should be corrected by Eq. 11.7.2, and the rate of loading should be 100 psi.

11.8   FLEXURAL STRENGTH TEST (MODULUS OF RUPTURE)

The flexural strength or modulus of rupture is a measure of the outer fiber tensile strength of a material. This property is determined by loading either cylindrical or prismatic specimens in a three-point loading device

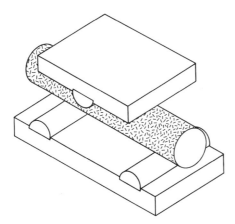

Fig. 11.8.1. Flexure strength test.

(Fig. 11.8.1) to failure. The flexural strength R_0 for a cylindrical specimen of diameter D is

$$R_0 = \frac{8F_c L}{\pi D^3} \tag{11.8.1}$$

where F_c is the applied compressive load at failure and L is the length between the bearing (knife) edges of the lower plate (Fig. 11.8.1). For a prismatic specimen of thickness a and width b,

$$R_0 = \frac{3F_c L}{2ba^2} \tag{11.8.2}$$

In the loading device used for testing prismatic specimens two of the bearing edges must be self-aligning (pivoted) to permit uniform loading if the specimen surfaces are not exactly parallel. Self-aligning bearing edges are not required for testing cylindrical specimens. A satisfactory specification for testing specimens obtained from diamond drill core is as follows. The distance between bearing edges L should be 5 in. A test group of ten or more specimens at least 6 in. long and from $\frac{7}{8}$ to $2\frac{1}{8}$ in. in diameter, air dried at room temperature for two weeks should be tested. The ends of the specimens do not need to be surfaced. The specimen should be loaded to failure at a uniform rate of 500 psi/min.

Although the flexural strength is a measure of the *outer-fiber* tensile strength of a material, the value obtained by this procedure is higher than determined in direct tension. This higher value is presumed to result from the fact that only a small area (or point) on the opposite side of the specimen directly under the point of loading is subject to the maximum tensile strain. Hence, the probability of a defect (plane of weakness or incipient fracture) occurring at or near this point is less than that for an equivalent defect occurring in the length of a tensile specimen.

11.9   UNCONFINED SHEAR STRENGTH

A number of procedures have been devised for measuring the unconfined shear strength of rock. Four of these methods, illustrated in Fig. 11.9.1, are referred to as the single shear test, double shear test, punch shear test, and torsional shear test. For the single shear test the shear strength S_0 is given by

$$S_0 = \frac{F_c}{A} \tag{11.9.1}$$

and for the double shear test by

$$S_0 = \frac{F_c}{2A} \tag{11.9.2}$$

where A is the cross-sectional area of the specimen and F_c is the force in the direction of the plane A necessary to cause failure.

For the punch test

$$S_0 = \frac{F_c}{2\pi r a} \tag{11.9.3}$$

where a is the thickness of the specimen and r the radius of the punch, and for the torsional test

$$S_0 = \frac{16M_c}{\pi D^3} \tag{11.9.4}$$

where M_c is the applied torque at failure and D is the diameter of the cylinder.

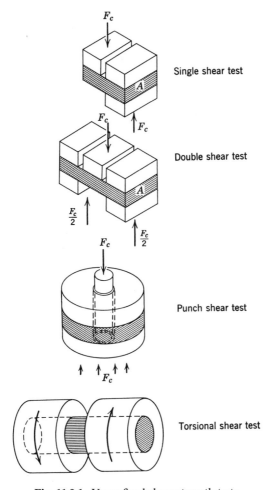

Fig. 11.9.1. Unconfined shear strength tests.

Unconfined shear tests are subject to several inherent faults. In the single, double, and punch shear tests, large stress concentrations develop along the shear, or punch and die edges. A fracture (or crack) presumably initiates along this line and propagates across the cross section of the specimen in a direction that does not necessarily correspond to the direction of the maximum shear stress.* Also, the specimen may bend with the applied force causing tensile or compressive stresses to develop on opposite surfaces of the specimen. Finally, stresses are induced by the clamping

* See Everling[10] for a discussion of unconfined shear failure.

action of the grips. In the torsional shear test both the clamping of the grips and possible bending affect test results. Collectively these factors affect the measured shear strength to the degree that erratic and inconsistent results are obtained by these methods. Usually the measured shear strength is not inversely proportional to the cross-sectional area as implied by Eqs. 11.9.1, 11.9.2, and 11.9.3. Although a number of investigators have described unconfined shear tests in detail, there is virtually no information available to show how test results depend on rate of loading, specimen size, shape, and the other factors that should be controlled in a standardized procedure. Also, only limited test results have been published for unconfined shear strengths determined by these methods.

11.10   TRIAXIAL COMPRESSIVE AND SHEAR STRENGTH

The triaxial compressive and shear strength of rock is of special importance in calculating the bearing capacity of foundation rock for surface structures and, in a less definite manner, in determining the strength of mine pillars and other parts of underground structures. The triaxial test is effected by applying a constant hydraulic pressure $-p$ to the curved surface of a cylindrical specimen and at the same time applying to the ends of the specimen a compressive axial load which is incrementally or continuously increased until the specimen fails. The specimen is usually jacketed with a rubber or metal-foil sleeve to prevent penetration of the hydraulic fluid into the pore space in the specimen. Otherwise the pore pressure must be taken into account as indicated in Section 10.7. If F_c is the axial load at failure, the principal stresses in the specimen at failure are (Fig. 11.10.1)

$$\sigma_3 = -\frac{F_c}{A} \quad \text{and} \quad \sigma_1 = \sigma_2 = -p$$

where A is the end area of the specimen. The angle of the fracture plane with respect to the axis of the specimen θ, or its complement α, the angle of fracture, is also determined.

Thus the triaxial compressive strength of a specimen is not a single value but a pair of values, σ_1 and σ_3. To determine the triaxial compressive strength of a given material it is customary to test a group of about six specimens, each at a different value of the radial pressure. The resulting $\sigma_1\sigma_3$ values are used to establish a functional relationship, $\sigma_3 = f(\sigma_1)$, and it is this relationship that specifies the triaxial compressive strength of the material (Section 10.3). The data from this group of specimens can also be used to determine the triaxial shear strength of the material. Construct a Mohr's circle for each pair of $\sigma_1\sigma_3$ values and draw an envelope curve $\tau_\theta = f(\sigma_\theta)$ tangent to these circles. The intercept of the envelope with the τ-axis is the triaxial shear strength of the material S_0 (Fig. 10.3.5).

The data from a group of triaxial specimens cannot be averaged in the same statistical manner used in treating the results from other static tests because the triaxial result is a single functional relationship. However, a better functional relationship between σ_3 and σ_1, or τ_θ and σ_θ, can be obtained by fitting the curve by the method of least squares. Methods for fitting linear curves are given in Section 8.8, and a method for fitting quadratic curves is described by Balmer.[11]

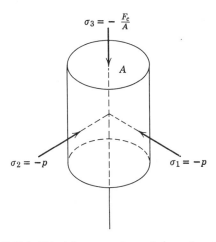

Fig. 11.10.1. Triaxial compressive and shear strength test.

Various triaxial equipments have been developed, most of which were designed for testing specimens at a much higher stress level than that encountered in underground mining. A simple and relatively inexpensive equipment designed specifically for testing diamond drill core specimens in the stress range equivalent to that experienced in mining at depths to 10,000 ft was developed by Obert.[12] The pressure jacket and bearing plate detail is given in Fig. 11.10.2, and the complete assembly in Fig. 11.10.3.

The specimen preparation is virtually the same as that for compression specimens (in fact, if $p = 0$, σ_3 is the uniaxial compressive strength of a specimen with a L/D ratio of two). Although the effects of specimen size, shape, moisture content, and rate of loading have not been studied in the same detail as for the uniaxial compressive strength test, the similarity in the uniaxial and triaxial procedures is such that the effects of these factors should be about the same. An L/D ratio of two and a rate of loading between 20 and 100 psi/sec conform to the practice of most laboratories.

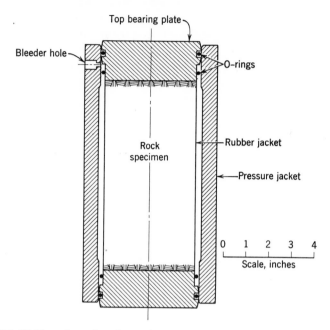

Fig. 11.10.2. Half section of rock specimen, pressure jacket, and bearing plate assembly.

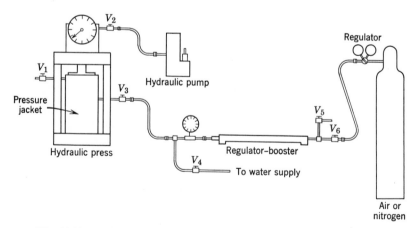

Fig. 11.10.3. Triaxial equipment—schematic diagram of hydraulic system.

11.11 STATIC ELASTIC CONSTANTS

Of the five elastic constants defined in Section 3.2; namely, the modulus of elasticity E, Poisson's ratio ν, the bulk modulus K, the modulus of rigidity G, and Lame's constant λ, only two are independent for an isotropic material. Hence, if any two elastic constants of a material are measured, the other three can be evaluated using the relationships given in Table 3.2.1. The two elastic constants most commonly determined by static tests are

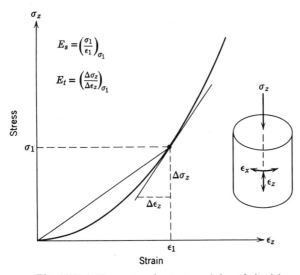

Fig. 11.11.1. Tangent and secant modulus of elasticity.

the modulus of elasticity and Poisson's ratio. The procedure generally employed for evaluating these constants is to subject a prismatic or cylindrical specimen to an increasing and/or decreasing axial compressive or tensile stress σ_z, and to measure the corresponding change in strain parallel and normal to the direction of the applied stress, ϵ_z and $\epsilon_x = \epsilon_y$ (Fig. 11.11.1). The tangent and secant modulus of elasticity E_t and E_s are defined by

$$E_t = \frac{\Delta \sigma_z}{\Delta \epsilon_z}, \qquad E_s = \frac{\sigma_z}{\epsilon_z} \qquad (11.11.1)$$

Except for linear-elastic, or quasi linear-elastic materials the stress at which these fractions are evaluated should be specified as indicated in Fig. 11.11.1. Poisson's ratio is given by

$$\nu = -\frac{\epsilon_x}{\epsilon_z} = -\frac{\epsilon_y}{\epsilon_z} \qquad (11.11.2)$$

If this ratio varies with the stress, the stress at which the determination is made also should be specified.

The static modulus of elasticity also has been determined from the bending in beams and cantilever.[5] However, because of the nonuniform stress distribution in a beam or cantilever, and because only a small part of the specimen is under either a maximum tensile or compressive stress this method is not considered as reliable as the uniaxial procedure. The axial strain on the surface of specimens subjected to either tensile or

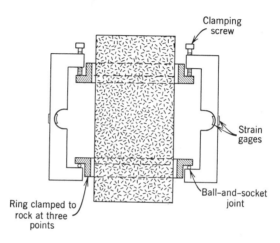

Fig. 11.11.2. Compressometer for measuring axial strain. (After Fairhurst.[5])

compressive loads has been measured with: (1) both optical and resistance-strain gages mounted on the specimen surface at the midplane normal to the specimen axis. Two gages placed at diametrically opposite points must be used to compensate for possible asymmetrical loading; (2) with a compressometer, developed by Leeman and Grobbelaar[13] (Fig. 11.11.2). This device is attached to the specimen at approximately the quarter-planes so that it measures the strain over the central half of the specimen; or (3) with dial indicators mounted to measure between the bearing platens. The last procedure has been found to be the least accurate because of two suspected causes—the lack of complete flatness of the end surfaces of the specimen may cause an excessive deformation at low stress; distortion of the platens (convex bending) as the load is applied. In this respect it should be noted that for a rock with a modulus of elasticity of 10×10^6 psi, a 10-in. long specimen loaded to 10,000 psi deforms only 0.010 in. Hence a measurement error of 0.001 in. would cause a 10% error in the measurement of the axial deformation.

The measurement of the strain on the surface of cylindrical or prismatic specimens is also subject to some error. The stress distribution in a specimen loaded in uniaxial tension or compression depends on the specimen shape and constraint offered by the bearing platens. If the constraint is such that the radial displacement at the perimeter of the end cross-section is zero, Filon[14] showed theoretically that for a specimen with a length-to-diameter ratio of $\pi:3$ the stress in the axial direction at the midplane normal to the specimen axis is approximately 10% higher on the axis, and 5% lower on the exterior surface than the applied axial stress. Thus the strain measured on the surface at the midplane should be about 5% low, and the calculated modulus of elasticity about 5% high. However, if hardened and ground steel platens are used, the slippage at the ends reduces the end constraint substantially. For an aluminum test specimen the radial deformation at the end was found to be only 20% less than that of the midplane.* Also if the specimen length-to-height ratio is two or more this effect is further reduced. Hence it it presumed that the error due to this cause is substantially less than 5% and can be disregarded for practical engineering purposes.

For many rock types the slope of the initial or virgin increasing and/or decreasing stress-strain curve is less than the slope of the curve for subsequent cycles. The cause of this difference has not been definitely determined, although several investigators have attributed it to the opening of micro-cracks at the time the specimen is separated from its surroundings and its confining stresses are relieved. Since the difference between the slope of the stress-strain curve for the initial and subsequent cycles may be large, and because the difference between successive cycles usually decreases, it has become the common practice to subject the specimen to a number of loading cycles before recording the data cycle. Also, if this practice is observed, the data cycle should be specified, for example, sixth cycle, decreasing load.

The elastic constants also can be determined under triaxial loading by making a comparatively simple modification of one of the bearing plates used in the triaxial equipment shown in Figs. 11.10.2 and 11.10.3. This modification permits the use of resistance-strain gages on the exterior surface of the specimen. The modified bearing plates and a test specimen are shown in Fig. 11.11.3a and b. The relationship between the applied radial and axial pressure p_0 and p_x, (from Eq. 4.8.5, with $p_i = 0$ and $a = 0$) is

$$\epsilon_z = \frac{2\nu p_0}{E} - \frac{p_z}{E} \qquad (11.11.3)$$

* Unpublished measurement made by authors.

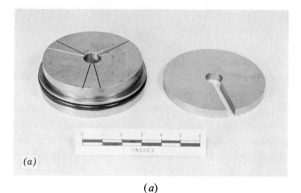

(a)

(a)

Fig. 11.11.3a. Bearing plate, with provision for bringing out strain gage leads.

where v is Poisson's ratio. With p_0 constant, the incremental change in the stress and strain is

$$E = \frac{\Delta p_z}{\Delta \epsilon_z} \tag{11.11.4}$$

A modified triaxial apparatus was devised by Obert[15] for the measurement of the triaxial elastic constants of stress-relief cores obtained by the

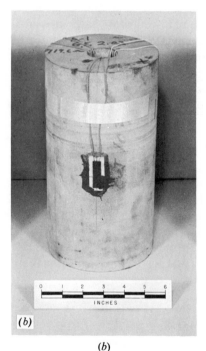

(b)

(b)

Fig. 11.11.3b. Triaxial test specimen with strain gages attached.

overcoring method (Section 14.3). These cores contain a central hole, and hence the test specimen is a thick-wall cylinder (Fig. 11.11.4). If, for a given value of the radial pressure $-p_0$, the applied axial stress $-p_z$ is varied and the corresponding axial strain ϵ_z measured, the modulus of elasticity E can be determined from Eq. 4.8.10, namely

$$\epsilon_z = \frac{2\nu b^2 p_0}{E(b^2 - a^2)} - \frac{p_z}{E} \tag{11.11.5}$$

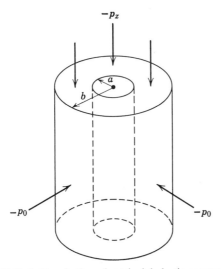

Fig. 11.11.4. Terminology for triaxial elastic constants test.

where a is the radius of the hole, b is the outside radius of the specimen, and ν is Poisson's ratio. With $p_0 = 0$ the uniaxial modulus of elasticity is

$$E = \frac{p_z}{\epsilon_z} \tag{11.11.6}$$

and for $p_0 = $ constant, the triaxial modulus of elasticity is

$$E = \frac{\Delta p_z}{\Delta \epsilon_z} \tag{11.11.7}$$

In addition, if the diametral deformation of the hole U is measured with a borehole deformation gage, then from Eq. 4.8.11

$$U = -\frac{4(1 - \nu^2)ab^2 p_0}{E(b^2 - a^2)} - 2a\nu\epsilon_z \tag{11.11.8}$$

or

$$U = -\frac{4p_0 b^2 a}{E(b^2 - a^2)} + \frac{2\nu a p_z}{E}$$

For $p_0 = 0$

$$U = -2\nu a \epsilon_z \quad \text{or} \quad \nu = -\frac{U}{2a\epsilon_z}$$

and for $p_0 = \text{constant}$

$$\nu = -\frac{1}{2a}\left(\frac{\Delta U}{\Delta \epsilon_z}\right) \tag{11.11.10}$$

Eliminating ϵ_z from Eqs. 11.11.5 and 11.11.8

$$p_0 = -\frac{E(b^2 - a^2)U}{4ab^2} + \frac{\nu(b^2 - a^2)p_z}{2b^2} \tag{11.11.11}$$

For $p_z = 0^*$

$$E = -\frac{4ab^2 p_0}{U(b^2 - a^2)} \tag{11.11.12}$$

and for $p_z = \text{constant}$

$$E = -\frac{4ab^2}{(b^2 - a^2)}\frac{\Delta p_0}{\Delta U} \tag{11.11.13}$$

Thus Eqs. 11.11.7 and 11.11.13 provide independent ways of determining the modulus of elasticity. Also, Poisson's ratio can be calculated from Eq. 11.11.10. Figure 10.3.8 presents the axial stress-strain results from a granite specimen, that characteristically show that rock becomes more linear-elastic and the area under the hysteresis loop decreases as the degree of confinement is increased. Because most in situ rock is under triaxial stress (except on surfaces), this procedure makes it possible to determine the elastic constants under a state of stress simulating that experienced in situ.

11.12 DYNAMIC ELASTIC CONSTANTS

The dynamic elastic constants of a solid material can be determined indirectly by measuring the propagation velocities in the material. For an isotropic solid there are two types of body or free-medium waves: a longitudinal or compressional wave which travels with a velocity V_p; and a shear or transverse wave which travels with a velocity V_s. These velocities are related to the elastic constants by

$$V_p = \left[\frac{Eg(1 - \nu)}{\gamma(1 + \nu)(1 - 2\nu)}\right]^{\frac{1}{2}} \tag{11.12.1}$$

$$V_s = \left[\frac{Gg}{\gamma}\right]^{\frac{1}{2}} \tag{11.12.2}$$

* This is the biaxial case, that is, $\sigma_z = 0$, $\sigma_x = \sigma_y = -p_0$. Fitzpatrick[16] developed a biaxial equipment for measuring the modulus of elasticity of stress-relief cores, but because the biaxial load produces a uniform tensile strain in the axial direction, and because the tensile strain strength of rock is low, elastic constant determinations can be made by this method only at comparatively low stress levels.

where E is the modulus of elasticity, G is the modulus or rigidity, ν is Poisson's ratio, g is the acceleration of gravity, and γ is the unit weight of the material. Note that the elastic constant G can be determined from V_s without a knowledge of ν but that the value of ν is needed to determine the elastic constant E. The relationship

$$G = \frac{E}{2(1 + \nu)} \tag{11.12.3}$$

provides the necessary equation for calculating both ν and E. Thus

$$E = \frac{V_s^2 \gamma}{g}\left[\frac{3(V_p/V_s)^2 - 4}{(V_p/V_s)^2 - 1}\right] \tag{11.12.4}$$

$$\nu = \frac{1}{2}\left[\frac{(V_p/V_s)^2 - 2}{(V_p/V_s)^2 - 1}\right] \tag{11.12.5}$$

Also from the relationship

$$K = \frac{E}{3(1 - 2\nu)}$$

the bulk modulus K can be determined from

$$K = \frac{\gamma}{g} V_s^2 \left[\left(\frac{V_p}{V_s}\right)^2 - \frac{4}{3}\right] \tag{11.12.6}$$

These equations are valid only if the material is isotropic, homogeneous, and linear-elastic. Some indication of the degree of anisotropy can be determined by measuring the propagation velocities in mutually perpendicular directions.

Both the longitudinal and shear velocities in a prismatic or cylindrical rock specimen can be determined by placing a piezo-elastic crystal (usually barium titanate) in mechanical contact with one end of the specimen (Fig. 11.12.1). A high voltage pulse of short duration (say, 800 volts for 1 to 10 μsec) is applied to the crystal. If the crystal is x-cut it will produce a compressional pulse in the specimen; if it is y-cut it will produce a shear pulse

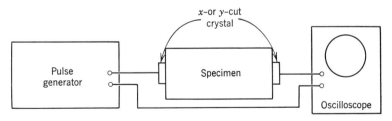

Fig. 11.12.1. Seismic pulse apparatus for determining elastic constants.

in the specimen. A companion x- or y-cut crystal placed in mechanical contact with the other end of the specimen will act as a mechanical-electrical transducer (pickup) and, on arrival of the mechanical pulse, it will generate an equivalent electrical pulse. By measuring the time required for the compressive or shear pulse to travel the length of the specimen, the

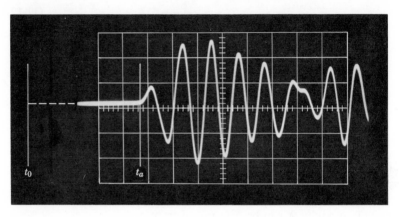

Fig. 11.12.2. Oscilloscope recording of seismic pulse. Traveltime $= t_a - t_0 = \Delta t_p =$ 48.44 μsec.

corresponding longitudinal or shear propagation velocity can be determined, that is,

$$V = \frac{L}{\Delta t_p} \quad \text{and} \quad V_s = \frac{L}{\Delta t_s} \qquad (11.12.7)$$

where L is the length of the specimen, Δt_p is the traveltime for the compressional pulse, and Δt_s is the traveltime for the shear pulse. The traveltime can be measured with an oscilloscope having a calibrated sweep rate, by initiating the sweep with the driving pulse and recording the arrival of the pulse from the pickup (Fig. 11.12.2).

Several variations in the pulse technique have been developed. For example, Fairhurst[5] fabricated metal crystal holders which also serve as bearing plates (Fig. 11.12.3) that permit the specimen to be axially loaded and pulsed at the same time. This procedure makes it possible to measure the modulus of elasticity at any specified uniaxial load. Also, by various methods, such as employing xy-cut crystals, the compressional and torsional pulses can be produced simultaneously. The oscilloscope display of a longitudinal and shear wave arrival is shown in Fig. 11.12.4.

The elastic constants also can be determined by measuring the resonant frequency of a vibrating bar or rod. This procedure is sometimes referred

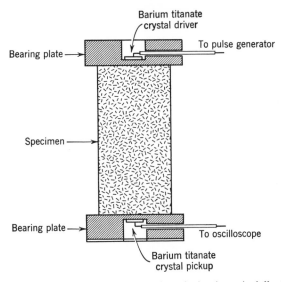

Fig. 11.12.3. Apparatus for measuring seismic velocity in uniaxially loaded rock specimens. (After Fairhurst.[5])

to as the bar velocity or resonant frequency method. Bars or rods have three possible modes of vibration, namely, longitudinal, transverse (bending), and torsional. For the longitudinal mode, the longitudinal bar velocity V_b is related to the fundamental longitudinal frequency f_b by

$$V_b = 2f_b L \qquad (11.12.8)$$

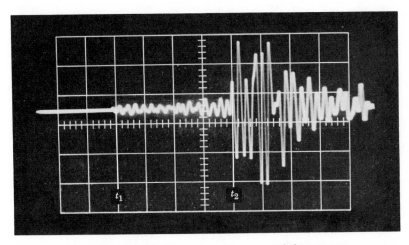

Fig. 11.12.4. Arrival of longitudinal wave t_1 and shear wave t_2.

where L is the length of the specimen. The modulus of elasticity is given by

$$E = V_b^2 \frac{\gamma}{g} = 4f_b^2 L^2 \frac{\gamma}{g} \tag{11.12.9}$$

Comparison of Eq. 11.12.9 with Eq. 11.12.1 shows that the longitudinal bar velocity and longitudinal, free-medium velocity are related by

$$V_p = V_b \left[\frac{(1 - \nu)}{(1 + \nu)(1 - 2\nu)} \right]^{\frac{1}{2}} \tag{11.12.10}$$

The validity of Eqs. 11.12.8 and 11.12.9 requires that the specimen is homogeneous, isotropic, linear-elastic, and the diameter or cross-sectional dimensions are small in comparison with the specimen length, that is, for a circular bar of radius r, $r/L < 1/10$.

If $r/L > 1/10$, a correction can be made as follows:

$$V_b = V_{b'} \left(1 + \frac{\pi^2 r^2 \nu^2}{4L^2} \right) \tag{11.12.11}$$

and

$$V_{b'} = 2f_{b'}L$$

where r is the radius of the rod, and $f_{b'}$ and $V_{b'}$ are the measured bar frequency and bar velocity.

The shear velocity in a bar V_s is given by

$$V_s = 2f_s L \tag{11.12.12}$$

where f_s is the fundamental torsional frequency of the bar. Also,

$$G = V_s^2 \frac{\gamma}{g} = 4f_s^2 L^2 \frac{\gamma}{g} \tag{11.12.13}$$

hence, the shear velocity in a bar and in a free medium are identical (Eq. 11.12.5).

Poisson's ratio can be calculated from

$$\nu = \frac{E}{2G} - 1 = \frac{V_b^2}{2V_s^2} - 1 \tag{11.12.14}$$

An apparatus and procedure was developed by Obert et al.[2] for measuring both the longitudinal and torsional frequencies of rock specimens cut from diamond drill cores (Fig. 11.12.5). Vibration in the specimens is produced by cementing to one end of the core a soft iron pole-piece that interacted with an electromagnetic driver unit (Fig. 11.12.6). An identical pole-piece cemented to the other end of the core interacted with an equivalent electromagnetic pickup to sense the vibration amplitude. To determine the resonant frequency the oscillator is tuned to give the maximum reading on the voltmeter.

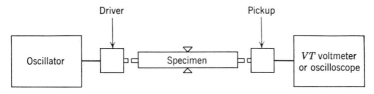

Fig. 11.12.5. Resonant frequency apparatus for measuring longitudinal and torsional vibration in rock cores.

A small correction must be made for the added weight of the pole-pieces, given by

$$f_b = f_b'\left(1 + \frac{W'}{W}\right)$$

$$f_s = f_s'\left(1 + \frac{k^2 W'}{W}\right)$$

where f_b = longitudinal frequency without pole-pieces
$\quad\ \ f_b'$ = longitudinal frequency with pole-pieces
$\quad\ \ f_s$ = torsional frequency without pole-pieces
$\quad\ \ f_s'$ = torsional frequency with pole-pieces
$\quad\ W'$ = weight of both pole-pieces
$\quad\ W$ = weight of specimen
$\quad\ \ k$ = ratio of radius of gyration of pole-pieces to that of specimen

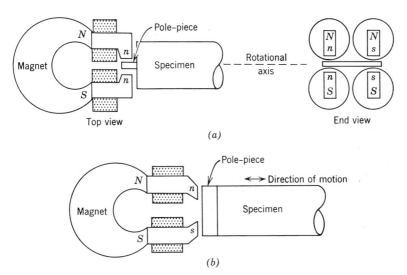

Fig. 11.12.6. Driver and pick-up heads. (*a*) Torsional. (*b*) Longitudinal.

For materials that are homogeneous and isotropic such as glass and metals the resonant frequency method of determining the elastic constants is accurate and comparatively easy to effect. For less homogeneous materials such as rock the method is not as satisfactory because in a bar vibrating in either its fundamental longitudinal or torsional mode the strain reaches a maximum at the midpoint and is always zero at the ends of the bar. Hence a mechanical defect (fracture, joint, etc.) will have a greater effect if it is at or near the center than if it is at or near either end. Also the effects of anisotropy become evident if Poisson's ratio is calculated from the longitudinal and torsional bar velocity. In fact, if Poisson's ratio is calculated from Eq. 11.12.14, negative values or values greater than 0.5 are sometimes obtained, which on the basis of elastic theory is impossible. This anomaly results from the fact that in the longitudinal vibration the strain is in the axial direction, whereas in torsional vibration the shear strain is in any plane parallel to the specimen axis. Despite these possible sources of error, about one-third of rocks tested by this method give reasonable values for Poisson's ratio.

For best results both the resonant frequency and pulse method should be used to determine, on the same sample of rock, the propagation velocities V_b, V_p, and V_s. If both methods are used, two independent measurements of V_s are obtained, one from the resonant torsional frequency and one from the shear pulse traveltime. These two independent shear velocity determinations should agree within the experimental error of the measurements if the rock sample does not have any major defects near its center section. The average value of $\bar{V}_s$ is used to calculate G by means of Eq. 11.12.13. The longitudinal bar velocity V_b is used to calculate E by means of Eq. 11.12.9. The longitudinal pulse velocity V_p and the longitudinal bar velocity V_b are used to calculate Poisson's ratio by means of the equation

$$ \nu = \frac{1}{4}\left\{ \left[\left(9 - \frac{V_b^2}{V_p^2} \right) \left(1 - \frac{V_b^2}{V_p^2} \right) \right]^{1/2} - \left(1 - \frac{V_b^2}{V_p^2} \right) \right\} \quad (11.12.15) $$

Equation 11.12.15 is obtained by solving Eq. 11.12.10 for ν. If both V_b and V_p are obtained by producing an axial strain along the length of the same cylindrical bar, both the velocities are related to the same elastic constant E in the axial direction. The difference between the longitudinal pulse velocity and the bar velocity is a result of the average Poisson's ratio in the plane normal to the axis of the bar. Thus even if the rock is anisotropic, Eq. 11.12.15 will give a good average value for ν.

11.13 CREEP TESTS

Apart from the slow change that occurs during periods of excavation, the stress in the structural parts of underground openings is otherwise virtually constant. Hence, the inelastic rock property of concern in the evaluation of underground structures is the creep rate under constant load. At great depth, or with rock that is plastic, the effects of creep are more readily evident, and in some instances they may become the limiting factor in mining or maintaining an underground opening. Despite the importance of this subject only a comparatively few studies have been made of time-dependent deformational behavior of mine rock subjected to the range of load conditions experienced underground, and the procedures employed in these tests differ markedly. For example, Iida and Wada[17] and Lomnitz[18] measured the creep in rock cylinders subjected to a constant torsional shear stress; Griggs[19] measured the creep in specimens subjected to a constant compressive stress; and Phillips[20] studied the creep in loaded beams and cantilevers. Despite the differences in these procedures two conditions must be satisfied in creep testing; first, the apparatus must be able to provide a constant load over long periods, which may be from hours to years; and second, the deformation or strain-measuring instruments must not drift (Section 9.2) during the measuring period.

The first requirement has been satisfied by developing the load by means of dead weights, as illustrated in Fig. 11.13.1a, b, and c. For bending or torsional tests the weights can be applied directly, but for compression testing a system of levers is usually used to develop the required stress. Also, relatively constant loads can be maintained in spring loaded systems provided the stiffness of the spring loading system is low compared to that of the specimen. Since the advent of 0-ring seals, constant loads can also be developed by hydraulic systems. A hydraulic jack or ram utilizing this type of seal will maintain a constant load over long periods provided the system is maintained at a constant temperature and that the deformation rate in the specimen is small. A simple means of developing regulated pressure can be achieved with a regulator-booster (Fig. 11.13.2). This device multiplies the pneumatic pressure (which can be obtained from a tank of compressed nitrogen or air) by the ratio of the pneumatic-to-hydraulic piston areas. Because of the large volume in the pneumatic cylinder a small displacement of the piston will not change the pneumatic pressure significantly. Hence, the hydraulic pressure will be maintained approximately constant even though the jack or ram may move to compensate for small displacements in the specimen. Because the pneumatic chamber in this device acts as a gas thermometer it must be kept at a constant temperature.

If the specimen size is large and the applied load of such a magnitude that the ultimate deformation is of the order of 0.1 in. or greater, dial indicators, or mechanical strain gages, with a sensitivity of 0.0001 in. will be sufficiently sensitive, and this class of instrument is relatively free from drift. Also optical-mechanical strain gages such as the Tuckerman gage

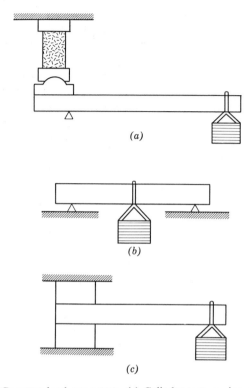

Fig. 11.13.1. Constant load creep tests. (*a*) Cylinder compression. (*b*) Beam tension and compression. (*c*) Cantilever tension and compression.

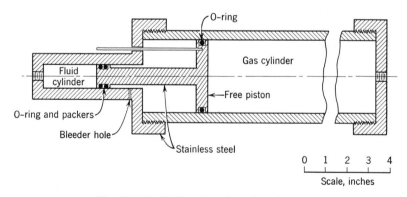

Fig. 11.13.2. Half section of regulator-booster.

(Section 9.5) are relatively free from drift. Electromechanical transducers usually are less stable over long periods, especially devices that must be cemented to the deforming body such as bonded resistance-strain gages. It is difficult to prevent the bonding cement from creeping at high stress, and this creep cannot be distinguished from creep in the specimen. Finally, if a load is applied to a specimen through a leverage system and the indicated deformation magnified by the same lever (Fig. 11.13.1a), the loading system must be checked for creep. This may be done by substituting a hardened steel specimen for the rock specimen provided, of course, that the applied stress does not approach the yield stress for the steel.

REFERENCES

1. *ASTM Standards*, Part II, 1942, Sections:

    D3-18. "Toughness of Rock," p. 427.
    C20-40. "Apparent porosity, Water absorption, Apparent Specific Gravity, and Bulk Density of burned refractory products," p. 278.
    C97-36. "Absorption and Apparent Specific Gravity of Natural Building Stone," p. 177.
    C140-41T. "Compressive Strength of Natural Building Stone," p. 1102.
    C99-36. "Flexural Testing of Natural Building Stone," p. 179.
2. Obert, L., S. L. Windes, and W. I. Duvall, "Standardized Tests for Determining the Physical Properties of Mine Rock," *U.S. BurMines Rept. Invest.*, **3891** (1946).
3. Hardy, H. R., "Standardized Procedures for the Determination of the Physical Properties of Mine Rock Under Short-Period Uniaxial Compression, *Dept. of Mines and Surveys Tech. Bull.*, No. 8, Canada.
4. Denkhaus, H. G., "Uber Festigkeitsuntersuchungen in der Gebirbsmechanik," *Landertreffen des Internatl. Buros fur Gebirgsmechanik*, Leipsig (1960).
5. Fairhurst, C., "Laboratory Measurement of Some Physical Properties of Rock," *Fourth Symp. on Rock Mech. Penn. State Univ.*, 1961.
6. Carneiro, F., "Une Novelle Methode d'essai pour Deternirer la Resistance a la Traction du Beton," *Reunion des Laboratoires d'Essai de Materiaux*, Paris (June 1947).
7. Akazawa, T., "Tension Test Method for Concrete," *Union of Testing and Research for Materials and Structures*, No. 16 (1953).
8. Hondros, G., "The Evaluation of Poisson's Ratio and the Modulus of Materials of Low Tensile Resistance by the Brazilian (Indirect Tensile) Test with Particular Reference to Concrete," *Australian J. of Appl. Sci.*, **10**, No. 3 (1959).
9. Mitchell, N. B., "The Indirect Tension Test for Concrete," *Mater. Res. and Std.*, **1**, No. 10 (1961).
10. Everling, G., "Zur Definition der Schubfestigkeit," *Gluckauf*, **98**, No. 18 (1962).
11. Balmer, G., "A General Analytic Solution for Mohr's Envelope," *ASTM Proc.*, **52** (1952).
12. Obert, L., "An Inexpensive Triaxial Apparatus for Testing Mine Rock," *U.S. BurMines Rept. Invest.*, **6332** (1963).
13. Leeman, E. R. and C. Grobblaar, "A Compressometer for Obtaining Stress-Strain Curves of Rock Specimens up to Fracture," *J. Sci. Inst.*, **34** (1957).

14. Filon, L. N. G., "On the Elastic Equilibrium of Circular Cylinders Under Certain Practical Systems of Load," *Phil. Trans., Ser. A*, **198** (1902).
15. Obert, L., "Triaxial Method for Determining the Elastic Constants of Stress Relief Cores," *U.S. BurMines Rept. Invest.*, **6490** (1964).
16. Fitzpatrick, J., "Biaxial Device for Determining the Modulus of Elasticity of Stress-Relief Cores," *U.S. BurMines Rept. Invest.*, **6128** (1962).
17. Iida, K. and T. Wanda, "Measurement of Creep in Igneous Rocks," *J. of Earth Sci.*, **8**, No. 1, Nagoya Univ., Japan (1960).
18. Lomnitz, C., "Creep Measurements in Igneous Rocks," *J. Geol.*, **64**, No. 5.
19. Griggs, D. T., "Creep in Rocks," *J. Geol.*, **47**, No. 2 (1939).
20. Phillips, D. W., *Tectonics of Mining, Part II, Colliery Eng.* (August 1948).

CHAPTER 12

PHOTOELASTICITY AND PHOTOELASTIC MODEL STUDIES

12.1 INTRODUCTION

Most optical isotropic solids when stressed become optically anisotropic, a phenomenon that was first noted by David Brewster in 1816. While discovering that glass, when stressed or strained, becomes double refracting, he noticed also that the degree of this optical anisotropy was proportional to the stress or strain. It has since been found that this property is present to some degree in most transparent solids.

The engineering applications of this optical phenomenon were not realized until almost a century later. Shortly after the turn of this century interest in the stress-optical properties of transparent solids grew rapidly and methods were developed for making analytical studies of the stresses in models of engineering structures. The methods proved useful, both as a design technique and as a means of corroborating the results from the mathematical theory of elasticity. Because both optical and elastic theory are necessary for an understanding of this method of making stress determinations, the term photoelasticity was coined to refer to the science and technology of the stress-optical properties of transparent solids.

Today photoelasticity is one of the more useful methods for making quantitative stress analyses in both two and three dimensions. Even though the method is experimental, the assumptions and principles of the theory of elasticity are required to translate the measured optical effects into terms of stress. Thus the interpretation of the experimental results is in terms of ideal elastic materials.

The number of problems that can be solved by the mathematical methods of the theory of elasticity are limited to those involving relatively simple boundary conditions. Such limitations do not exist for the photoelastic method of stress analysis. Consequently the photoelastic method is

355

used in all fields of engineering design. Since the early thirties it has been used to study the stress distributions in models of complicated underground mine openings. More recently it has been extended to the determination of stresses on the surface of in situ rock structures, to the study of primitive stresses in small samples of rock, and to the study of petrofabrics.

12.2 PHYSICAL OPTICS

Because the photoelastic method of stress analysis depends upon optical methods and theory, a brief discussion is given of those phases of physical optics that are fundamental to this method.

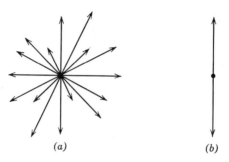

(a)   (b)

Fig. 12.2.1. Vector representation of a plane wave of light. (a) Ordinary light. (b) Plane polarized light.

Light is assumed to be a transverse electromagnetic wave. Such a wave is characterized by its maximum amplitude A_0, frequency f, period T, velocity V, and state of polarization. The equation that relates the variation of the amplitude a with time t and distance z for a simple harmonic plane wave traveling in the positive z direction can be written in the form

$$a = A_0 \sin \left[2\pi f \left(t - \frac{z}{v} \right) \right]$$ (12.2.1)

The amplitude A_0 may be identified with either the electric or magnetic field or the dielectric displacement. Usually the dielectric displacement is shown as the light vector. In free space and in isotropic media the dielectric displacement is proportional to and in the same direction as the electric field.

A plane wave of ordinary light may be resolved into a superposition of numerous elementary waves having the same form, in which A_0 assumes all possible directions normal to the direction of propagation (Fig. 12.2.1a). If, however, the vector A_0 remains constantly in one plane, the light is said to be plane polarized (Fig. 12.2.1b). If the amplitude vector representing ordinary light is resolved into two normal components, these components are also equal. However, if plane polarized light is resolved into normal components, these components will be proportional to the sine and cosine of the angle that the plane of polarization makes with the new direction (Fig. 12.2.1b).

The quantity f in Eq. 12.2.1 is the frequency or the number of vibrations per second. Its reciprocal is the period T or the time for one complete vibration. Each color in the visible spectrum is represented by a different frequency. White light is composed of all the frequencies in the spectrum, whereas monochromatic light is composed of only one frequency.

The quantity v is the velocity of light, which in a vacuum is taken as 186,000 mi/sec or 2.997×10^{10} cm/sec. The relation between the wavelength λ, frequency f, period T, and velocity v is

$$v = \lambda f = \frac{\lambda}{T} \tag{12.2.2}$$

It can be shown experimentally that when monochromatic light goes from one medium to another that both the velocity and the wavelength change but the frequency remains constant. Thus color of light is associated with frequency rather than wavelength even though most textbooks on optics describe color in terms of the wavelength of light in air. The wavelength of the visible spectrum extends from about 3800 to 7600 A, where $1A = 10^{-8}$ cm. The quantity $2\pi f(t - z/v)$ in Eq. 12.2.1 is called the phase of the wave.

If c represents the velocity of light in a vacuum and v the velocity in another medium, then the absolute index of refraction n is defined as

$$n = \frac{c}{v} \tag{12.2.3}$$

Since the velocity of light is a maximum in a vacuum, the absolute index of refraction is greater than unity. The velocity of light diminishes in direct proportion to the increase in density of the medium, and for air and gases the absolute index of refraction is very nearly unity. If n_1 and n_2 are the absolute indexes of refraction for media 1 and 2, then

$$n_1 = \frac{c}{v_1}$$

$$n_2 = \frac{c}{v_2}$$

and

$$\frac{n_2}{n_1} = \frac{v_1}{v_2} = n_{12} \tag{12.2.4}$$

where n_{12} is the relative index of refraction for light traveling from medium 1 to medium 2. Hence for a light ray going from a rarer to a denser medium, the relative index of refraction is greater than unity and for a ray going from a denser to a rarer medium it is less than unity.

If i is the angle of incidence, r is the angle of reflection and R the angle of refraction; the laws of reflection and refraction for ordinary light are

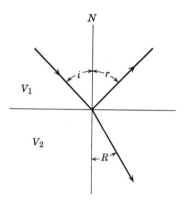

Fig. 12.2.2. Reflection and refraction of light at a boundary between two media.

summarized in the following two equations respectively:

$$i = r \qquad (12.2.5)$$

and

$$\frac{\sin i}{\sin R} = \frac{v_1}{v_2} = n_{12} \qquad (12.2.6)$$

The relations expressed by Eqs. 12.2.5 and 12.2.6 are shown graphically in Fig. 12.2.2. The law expressed by Eq. 12.2.6 is known as Snell's law of refraction.

When v is independent of the orientation of the amplitude of the light vector A_0, the medium is optically isotropic, when v depends on the orientation of the amplitude of the light vector A_0, the medium is optically anisotropic. Crystals, with the exception of those belonging to the cubic system are anisotropic. Although most stress-free transparent solids are isotropic, they become anisotropic when stressed.

Crystals that have two values of v are said to be double refracting because they split an ordinary incident ray of light into two rays—the ordinary ray and the extraordinary ray. Both rays become polarized at right angles to one another on passing through the crystal. For some crystals the absorption of one ray is much greater than for the other. Such crystals are said to be dicroic. Anisotropic crystals thus provide one means of obtaining plane polarized light Other means are by the use of polarizing filters or by reflecting light off of glass plates at their polarizing angle, about 57°. The most efficient method for obtaining plane polarized light is by means of the Nicol prism. Nicol prisms and polarizing filters are always made so that only one of the polarized rays is transmitted. The other ray is either refracted to the side or absorbed.

The properties of anisotropic materials are usually studied by means of interference effects observed in polarized light. Interference effects result from the superposition of two waves, each traveling independent of the other. The resultant wave is the sum of the two. Using Eq. 12.2.1, the sum of two waves of the same frequency, polarized in the same plane and traveling in the same direction but with different velocities, is given by

$$a = a_1 + a_2 = A_1 \sin\left[2\pi f\left(t - \frac{z_1}{v_1}\right)\right] + A_2 \sin\left[2\pi f\left(t - \frac{z_2}{v_2}\right)\right] \quad (12.2.7)$$

Expanding this expression and collecting terms gives

$$a = A \sin(2\pi f t - \delta) \quad (12.2.8)$$

where δ is a new phase constant and the new amplitude A is given by

$$A^2 = A_1^2 + A_2^2 + 2A_1 A_2 \cos\Delta \quad (12.2.9)$$

The phase difference Δ at any given time between the two waves is given by

$$\Delta = 2\pi f\left(\frac{z_1}{v_1} - \frac{z_2}{v_2}\right) \quad (12.2.10)$$

Note that the phase difference is a result of a difference in travel time which occurs from either a difference in path length or a difference in propagation velocity or both.

By use of Eqs. 12.2.2 and 12.2.3, Eq. 12.2.10 can be written as

$$\Delta = \frac{2\pi}{\lambda}(n_1 z_1 - n_2 z_2) \quad (12.2.11)$$

For a constant path length $d = z_1 = z_2$, Eq. 12.2.11 becomes

$$\Delta = \frac{2\pi d}{\lambda}(n_1 - n_2) \quad (12.2.12)$$

The intensity of a light beam is proportional to the square of its amplitude. Therefore if the quantity $d(n_1 - n_2)$ in Eq. 12.2.12 is equal to an integral number of λ, the resultant intensity of the two light beams is given by

$$A^2 = A_1^2 + 2A_1 A_2 + A_2^2 \quad (12.2.13)$$

Thus the resultant amplitude is just $A_1 + A_2$. However, if the quantity $d(n_1 - n_2)$ is equal to $[(2m + 1)/2]\lambda$, where m is an interger, the resultant intensity is

$$A^2 = A_1{}^2 - 2A_1A_2 + A_2{}^2 \qquad (12.2.14)$$

and the resultant amplitude is $A_1 - A_2$. Thus there will be regions of excess energy and others of a deficiency of energy. However, the average energy is still just the sum of the two intensities.

For circular polarized light the amplitude vector is constant but rotates about the light ray. For elliptic polarized light the amplitude vector varies in size as it rotates about the light ray. Elliptical or circular polarized light can be produced from plane polarized light by a device known as a quarter-wave plate. A quarter-wave plate is made from double refracting crystals of such a thickness that the change in phase between the ordinary and extraordinary rays, after passing through the plate, is just a quarter of a wavelength of the light or a phase difference of $\pi/2$. If the incident plane polarized ray of monochromatic light is represented by

$$a = A_0 \sin 2\pi f\left(t - \frac{z}{v}\right)$$

the two waves in the quarter-wave plate can be represented by

$$x = A_1 \sin 2\pi f\left(t - \frac{z}{v_1}\right)$$

$$y = A_2 \sin 2\pi f\left(t - \frac{z}{v_2}\right)$$

where $A_1 = A_0 \cos \theta$ and $A_2 = A_0 \sin \theta$ and θ is the angle that the plane of polarization of the incident wave makes with the plane of polarization of the ordinary wave in the quarter-wave plate (Fig. 12.2.3). After the two

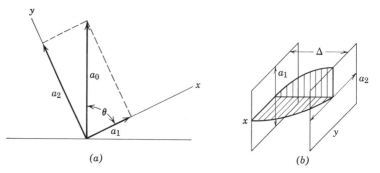

(a)   (b)

Fig. 12.2.3. Resolution of a plane polarized light ray into two plane polarized rays at right angles to each other by means of double refracting material.

waves have passed through the quarter-wave plate their phase difference will be given by Eq. 12.2.12. Thus at any given point z there will be two simple harmonic motions given by

$$x = A_1 \sin \omega t$$
$$y = A_2 \sin (\omega t - \Delta)$$

where $\omega = 2\pi f$.

Eliminating t between these two equations gives

$$\frac{x^2}{A_1{}^2} - \frac{2xy}{A_1 A_2} \cos \Delta + \frac{y^2}{A_2{}^2} = \sin^2 \Delta \qquad (12.2.15)$$

If $d(n_1 - n_2) = \lambda/4$ in Eq. 12.2.12, then $\Delta = \pi/2$, and Eq. 12.2.15 becomes

$$\frac{x^2}{A_1{}^2} + \frac{y^2}{A_2{}^2} = 1 \qquad (12.2.16)$$

which is simply the equation of an ellipse. If $\theta = 45°$, then $A_1 = A_2$, and Eq. 12.2.16 becomes the equation for a circle. Thus circular polarized light can be obtained by passing plane polarized light through a quarter-wave plate oriented so that the plane of polarization of the incident light makes an angle of 45° with the plane of polarization of the ordinary ray in the quarter-wave plate.

If circular polarized light is incident on another polarizer (usually called an analyzer), the emergent light is again plane polarized and the amplitude of the emergent light vector is independent of the orientation of the analyzer with respect to the quarter-wave plate.

12.3 STRESS OPTICAL LAW

Maxwell[1] showed that, for a glass bar under simple tension, the principal axes of stress coincide with the principal axes of optical symmetry and that the changes in the principal indices of refraction are linear functions of the principal stresses. That is, the relation between stress and double refraction is similar to that between stress and strain except that the change in the indices of refraction are used rather than their absolute values. By analogy with Eq. 3.2.1, the relationships between indices of refraction and principal stresses can be expressed as follows:

$$n_a - n_o = C_1 \sigma_1 + C_2(\sigma_2 + \sigma_3)$$
$$n_b - n_o = C_1 \sigma_2 + C_2(\sigma_3 + \sigma_1) \qquad (12.3.1)$$
$$n_c - n_o = C_1 \sigma_3 + C_2(\sigma_1 + \sigma_2)$$

where n_a, n_b, and n_c are the principal indices of refraction, n_o is the original index of refraction of the unstressed material, and C_1 and C_2 are the stress optical coefficients. Equations 12.3.1 are the fundamental relations connecting stress and optical effects. In theory, then, if the principal indices of refraction and the directions of the principal axes of optical symmetry are determined for a stressed transparent body, we can then calculate the state of stress at a point. The experimental measurement of these six quantities presents great difficulty in the general case. Therefore the photoelastic method was at first limited to two-dimensional problems where three of the six quantities (magnitude and direction of the principal stresses) are known a priori.

For plane-stress problems there are two principal stresses P and Q in the plane of the plate. The third principal stress is zero and normal to the plane of the plate. The two principal indices of refraction for plane stress are n_1 and n_2. Thus for plane stress conditions, Eqs. 12.3.1 reduce to

$$n_1 - n_o = C_1 P + C_2 Q$$
$$n_2 - n_o = C_1 Q + C_2 P \tag{12.3.2}$$

Equations 12.3.2 are difficult to use because they require the measurement of absolute retardation or phase change as given in Eq. 12.2.11. However, if relative retardation Δ is measured, then the use of Eq. 12.2.12 and Eq. 12.3.2 gives

$$\Delta = \frac{2\pi \, dC}{\lambda}(P - Q) \tag{12.3.3}$$

where $C = C_1 - C_2$ is the relative stress-optical coefficient and d is the path length or the plate thickness. Equation 12.3.3 shows that if the relative retardation is measured, we can calculate the principal stress difference or maximum shear stress (see Eq. 1.3.10). Finally, if the directions of the principal axes of optical symmetry are determined, these will correspond to the direction of the principal stresses. It should be noted that at a free boundary in the plane of the plate the direction of the principal stresses are normal and tangent to the boundary, and the normal stress is zero. Thus the determination of the relative retardation at a free boundary gives sufficient information to calculate the state of stress at the boundary. Because of this fact the determination of the stress concentration around holes in plates is easily made by the photoelastic method.

12.4   PHOTOELASTIC POLARISCOPE

The basic components of a photoelastic polariscope are shown in Fig. 12.4.1 and the description follows.

a. **Light Source.** Both white light and monochromatic light are required. The mercury vapor lamp with a 5461A filter is usually used as

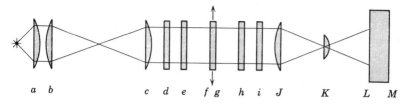

Fig. 12.4.1. Schematic diagram of photoelastic polariscope: (*a*) Light source. (*b*) Condensing lens. (*c*) Field lens. (*d*) Polarizer. (*e*) Quarter-wave plate. (*f*) Straining frame. (*g*) Model. (*h*) Quarter-wave plate. (*i*) Analyzer. (*j*) Field lens. (*k*) Focusing lens. (*l*) Viewing screen. (*m*) Camera.

the monochromatic light source. An incandescent filament lamp can be used as the white light source.

b. **Condensing Lens.** The condensing lens is used to concentrate the light from the source. This lens is of short focal length and is placed near the light source. Its function is to collect the light from the source and bring it to a focus at the focal point of lens (*c*).

c. **Field Lens.** This lens is used to produce a field of parallel light rays. Lenses (*b*) and (*c*) must be so adjusted that the focus of lens (*b*) is at the focal point of lens (*c*) and that the field of light just fills lens (*c*).

d. **Polarizer.** The purpose of the polarizer is to produce plane polarized light. Large field polarizers are made of Polaroid and are placed after the field lens. Small field polarizers are usually Nicol prisms and are placed at the focus of lens (*b*). With the polarizer in the latter position the field lens must be free of any internal stress; otherwise distortions in the fringe pattern of the model will result. It is necessary to have the polarizer in a frame that permits 360° angular adjustment relative to the axis of the loading frame.

e. **Quarter-Wave Plates.** Mica is used as the double refracting material as it is available in large sheets. Also plastic films are used that exhibit the same effects. The purpose of the quarter-wave plate is to produce circular polarized light from plane polarized light.

f. **Straining Frame.** The purpose of the straining frame is to apply a load to the model. A wide variety of designs and sizes is available. Loading in tension or compression should be possible. The loading frame or the model in the loading frame should be adjustable about a vertical axis so that the plane of the model can be set normal to the beam of light. Also the loading frame should be adjustable both in the vertical and horizontal directions so as to position it in the polariscope field.

g. **Model.** The model is made from one of the many photoelastic materials such as Bakelite BT61-893, celluloid, glass, Columbia Resin CR-39, or epoxy resins. For two-dimensional studies the model is a plate about $\frac{1}{8}$ to $\frac{1}{4}$ in. thick. Extreme care should be used in preparing the model so that edge and internal stresses are not developed.

h. **Quarter-Wave Plate.** Same as the quarter-wave plate in (*e*). Both quarter-wave plates should be housed so that they can be set parallel or crossed to one another. It should be possible to remove the quarter-wave plates from the polariscope when plane polarized light is desired.

i. **Analyzer.** The analyzer is identical to the polarizer. It is in this element of the polariscope that the interference effects occur that are finally observed as isoclinic or isochromatic fringes in the image of the model on the viewing screen.

j. **Field Lens.** This lens converges the light beam to the focal point of lens (*k*).

k. **Focusing Lens.** This lens is used to focus the image of the model on the viewing screen or on a photographic film of a camera.

l. **Viewing Screen.** A ground glass or a projecting screen can be used to view the image of the model and the photoelastic fringe patterns.

m. **Camera.** For precise photoelastic studies a camera should be used interchangeably with the viewing screen.

The polariscope can be used with white or monochromatic light and with various combinations of polarizer, quarter-wave plates, and analyzer. Examples of several of these combinations are illustrated in Fig. 12.4.2. To understand how the polariscope works it is necessary to write down the equations representing the vibrations of the light vector at each point in the polariscope.

As an illustration consider the use of monochromatic light with analyzer and polarizer crossed, that is, with their planes of polarization at right angles (Fig. 12.4.3). The amplitude of the emergent light from the polarizer can be represented by

$$a_1 = A_0 \sin \omega t \qquad (12.4.1)$$

Let θ be the angle that the plane of polarization in the polarizer makes with the principal plane of stress at a point in the plane of the model (Fig. 12.4.3). The incident plane polarized light ray splits into two rays as it enters the model. The amplitude of the two light rays in the model at the point under consideration will be the sine and cosine projections of a_1. Also as these rays pass through the model they will develop a phase

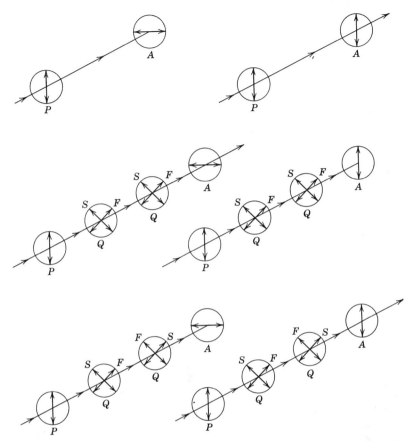

Fig. 12.4.2. Various arrangements for polariscopes (arrow after analyzer means light transmitted). *P*—Polarizer. *Q*—Quarter-wave plate. *A*—Analyzer. *F*—Fast ray. *S*—Slow ray.

difference Δ given by Eq. 12.3.3. Thus the two light rays emerging from the model will be given by

$$a_2 = A_0 \cos \theta \sin (\omega t + \Delta)$$
$$a_3 = A_0 \sin \theta \sin \omega t$$

(12.4.2)

When these two rays enter the analyzer only the sine and cosine components of these rays in the plane of polarization of the analyzer will be transmitted (Fig. 12.4.3). Thus the amplitude of the light transmitted by the analyzer will be given as

$$a_4 = A_0 \sin \theta \cos \theta \sin (\omega t + \Delta) - A_0 \cos \theta \sin \theta \sin \omega t$$

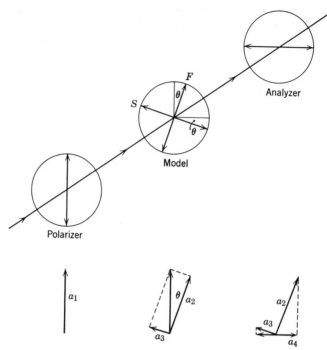

Fig. 12.4.3. Analysis of light vector for polariscope having crossed polarizer and analyzer.

which can be written as

$$a_4 = A_0 \sin 2\theta \sin \frac{\Delta}{2} \cos \left(\omega t + \frac{\Delta}{2} \right) \tag{12.4.3}$$

The intensity I of the light is proportional to the square of the amplitude, thus

$$I = A_0{}^2 \sin^2 2\theta \sin^2 \frac{\Delta}{2} \tag{12.4.4}$$

Equation 12.4.4 shows that the intensity is zero when $\theta = 0$, $\pi/2$, $3\pi/2$, etc., regardless of the value of $\Delta/2$. Also for $\Delta/2 = 0$, π, 2π etc., there will be zero intensity. To remove the dark lines resulting from the critical values of $\Delta/2$, it is necessary only to replace the monochromatic light with white light. As $\Delta/2$ in Eq. 12.3.3 depends on wavelength, the dark lines are replaced by colored patterns when white light is used. The dark lines resulting from the critical values of θ remain, as θ does not depend on the wavelength of the light. Therefore if white light is used, the directions of the principal stresses in the model can be located by observing the position of the dark lines as the polarizer and the analyzer are rotated while they remain crossed. The dark lines or lines of zero intensity observed with given settings of the crossed polarizer and analyzer are called isoclinics. Isoclinics are loci of points where the direction of the principal stresses are constant.

If the analyzer and polarizer are set at 45° to the direction of the principal stresses at a point in the model, then the intensity of the monochromatic light coming from this point and passing through the analyzer is by Eq. 12.4.4

$$I = A_0^2 \sin^2 \frac{\Delta}{2} \qquad (12.4.5)$$

From Eq. 12.3.3 Δ is proportional to $P - Q$. Thus the intensity of the light will vary cyclically through maxima and minima as the load in the model is increased. Knowing the number of cycles or fringe order, the thickness of the model, wavelength of light and the stress optical coefficient of the model material, the quantity $P - Q$ may be computed. The above method for obtaining $P - Q$ is not too satisfactory as it requires a point- by-point survey of the model. A more direct method for obtaining $P - Q$ is described next.

As a second illustration consider the use of monochromatic light, a crossed polarizer, analyzer, and two quarter-wave plates with one on each side of the model oriented so that their planes of polarization are parallel and at 45° to the plane of polarization of the analyzer and polarizer (Fig. 12.4.4).

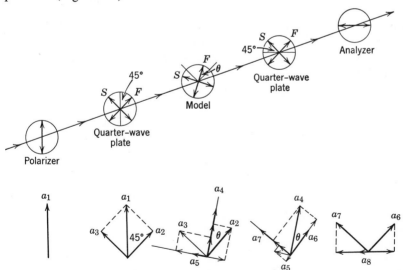

Fig. 12.4.4. Analysis of light vector for crossed polarizer and analyzer and parallel quarter-wave plates.

The amplitude of the emergent light vector from the polarizer is

$$a_1 = A_0 \sin \omega t \tag{12.4.6}$$

The direction of vibration for this light ray is at 45° to the planes of polarization in the quarter-wave plate. As this light ray enters the quarter-wave plate, two rays of equal amplitude will develop, each polarized in a plane at 45° to the plane of polarization of the incident ray. After these rays pass through the quarter-wave plate, a phase difference of $\pi/2$ will have developed. Thus the two emergent rays from the quarter-wave plate will be given by

$$a_2 = 0.707 A_0 \sin \left(\omega t + \frac{\pi}{2} \right)$$
$$a_3 = 0.707 A_0 \sin \omega t \tag{12.4.7}$$

The above two equations represent circular polarized light. As this light enters and passes through the model it will be divided into two rays, the direction of vibrations determined by the angle that the principal stress direction makes with the incident rays. Also a phase difference of Δ will develop which is given by Eq. 12.3.3. Thus the emergent light from the model is represented by

$$a_4 = 0.707 A_0 \cos \theta \sin \left(\omega t + \frac{\pi}{2} + \Delta \right) + 0.707 A_0 \sin \theta \sin (\omega t + \Delta)$$
$$a_5 = -0.707 A_0 \sin \theta \sin \left(\omega t + \frac{\pi}{2} \right) + 0.707 A_0 \cos \theta \sin \omega t \tag{12.4.8}$$

These two rays next pass through the second quarter-wave plate which has the same orientation angle θ with respect to these rays as the first quarter-wave plate. Thus the two emergent rays from the second quarter-wave plate can be represented by

$$
\begin{aligned}
a_6 = {} & 0.707 A_0 \cos^2 \theta \sin (\omega t + \Delta + \pi) \\
& + 0.707 A_0 \sin \theta \cos \theta \sin \left(\omega t + \Delta + \frac{\pi}{2} \right) \\
& + 0.707 A_0 \sin^2 \theta \sin (\omega t + \pi) \\
& - 0.707 A_0 \sin \theta \cos \theta \sin \left(\omega t + \frac{\pi}{2} \right) \\
a_7 = {} & 0.707 A_0 \sin \theta \cos \theta \sin \left(\omega t + \Delta + \frac{\pi}{2} \right) \\
& + 0.707 A_0 \sin^2 \theta \sin (\omega t + \Delta) \\
& - 0.707 A_0 \sin \theta \cos \theta \sin \left(\omega t + \frac{\pi}{2} \right) \\
& + 0.707 A_0 \cos^2 \theta \sin \omega t
\end{aligned}
\tag{12.4.9}
$$

These two rays of light then pass through the analyzer whose plane of polarization is at 45° to the quarter-wave plate principal planes. Thus the light emergent from the analyzer is given by

$$a_8 = (a_6 - a_7) \cos 45°$$

$$a_8 = -0.5A_0 \sin (\omega t + \Delta) - 0.5A_0 \sin \omega t \qquad (12.4.10)$$

$$= -A_0 \cos \frac{\Delta}{2} \sin \left(\omega t + \frac{\Delta}{2}\right)$$

Therefore the intensity of the emergent light is given by

$$I = A_0{}^2 \cos^2 \frac{\Delta}{2} \qquad (12.4.11)$$

Equation 12.4.11 shows that the intensity is independent of the orientation of the principal stress direction at each point in the model; therefore the quarter-wave plates have removed the variation in intensity resulting from the isoclinics. The only variation in intensity is a result of $\Delta/2$, where Δ is given by Eq. 12.3.3. Thus maximum light occurs when $\Delta/2$ is an integral number N of π or

$$\frac{\Delta}{2} = N\pi \qquad (12.4.12)$$

Combining Eqs. 12.4.12 and 12.3.3 gives

$$P - Q = \frac{N\lambda}{dC} \qquad (12.4.13)$$

The number N is the fringe order and is a measure of $P - Q$. The lines of maximum light intensity as given by Eq. 12.4.13 are the isochromatics. Isochromatics are loci of points where the differences between the principal stresses are a constant. The dark lines between the lines of maximum light intensity are the fractional fringe orders $\frac{1}{2}$, $1\frac{1}{2}$, $2\frac{1}{2}$, etc.

The preceding example represents a light field polariscope, for total transmission of light occurs when $\Delta/2$ is zero, that is, when the model is not in the circular polarized light. A dark field polariscope is obtained if the polarizer and analyzer are parallel and the quarter-wave plates are parallel and at 45° to the plane of polarization (Fig. 12.4.2). For such an arrangement the light emergent from the second quarter-wave plate is given by Eq. 12.4.9, and the light emergent from the analyzer is given by

$$a_9 = (a_6 + a_7) \cos 45° = A_0 \sin \frac{\Delta}{2} \cos \left(\omega t + \frac{\Delta}{2} + 2\theta\right) \qquad (12.4.14)$$

Thus the intensity of the emergent light is given by

$$I = A_0^2 \sin^2 \frac{\Delta}{2} \tag{12.4.15}$$

and the field is dark when $\Delta/2$ is zero or when the model is not in the circular polarized light. Also extinction of light occurs when $\Delta/2$ is an integral number of π or

$$\frac{\Delta}{2} = N\pi \tag{12.4.16}$$

Equation 12.4.16 is identical to Eq. 12.4.12. However, for the dark field polariscope, the fringe order N represents lines of zero light intensity, whereas in the light field polariscope the fringe order N represents lines of maximum light intensity.

It is interesting to note that from Eqs. 1.3.5, 1.3.6, and 12.4.13 that

$$(P - Q)^2 = (\sigma_x - \sigma_y)^2 + 4\tau_{xy}^2 = \left(\frac{N\lambda}{dC}\right)^2 \tag{12.4.17}$$

When $P - Q = 0$, the curves representing $\sigma_x - \sigma_y = 0$ and $\tau_{xy} = 0$ intersect at a point which is of zero fringe order, $N = 0$. Such points are called isotropic points. Note also that if the conditions for an isotropic point are substituted into Eq. 1.3.4 (which gives the directions of the principal stresses), the resulting equation is indeterminate. Thus isoclinics of all orientations pass through an isotropic point.

The arrangements of the polariscope to obtain isoclinics and isochromatics have been briefly described. These two sets of curves determine the direction of the principal stresses and the difference in their magnitude. Additional information is required to determine the values of the principal stresses P and Q except at free boundaries where the magnitude of one of the principal stresses must be zero. More than a dozen methods have been devised to determine the magnitude of P and Q in connection with plane-stress analysis by the photoelastic method. Excellent description of these methods are given by Mindlin[2] and Frocht.[3]

Methods have also been devised to measure the fringe order at any point. A dark field circular polariscope is used to determine the loci for the integral fringe orders 1, 2, 3, etc., and a light field circular polariscope is used to determine the loci for the fringe orders $\frac{1}{2}$, $1\frac{1}{2}$, $2\frac{1}{2}$, etc. At other points in the model it is possible to add or subtract known retardation to make the total retardation correspond to an integral fringe order. This procedure is known as compensation. Some of the more common methods of compensation are the tension or compression strip, Babinet compensator, Friedel's method, and Tardy's method. Some of these methods of compensation also enable determining which principal stress is associated with which principal direction. Excellent description of these methods is given by Durelli.[4]

12.5 PHOTOELASTIC METHOD

The basic steps in the photoelastic method of stress analysis are as follows:

1. Selection of photoelastic material.
2. Construction of model.
3. Calibration of photoelastic material.
4. Loading the model.
5. Determination of isoclinics.
6. Determination of isochromatic fringe patterns.
7. Determination of stress trajectories throughout the model, if desired.
8. Determination of maximum shear stress distribution throughout the model.
9. Determination of the principal stresses throughout the model, if desired.
10. Presentation of data in graphical or tabular form.

The ideal photoelastic material has not been discovered. However, there are several materials that are suitable for photoelastic stress analysis, for example, Bakelite BT 61-893, celluloid, CR-39, or epoxy resin. The desirable properties of a photoelastic material are the following:

1. Optical transparency.
2. Optical sensitivity to stress—high stress optical coefficient.
3. Linear stress strain relationship with high proportional limit.
4. Optical and elastic isotropy and homogeneity.
5. Freedom from creep.
6. Absence of permanent set.
7. Freedom from initial internal stresses.
8. Freedom from aging and edge effects.
9. Machinability.
10. Reasonable cost.

For plane-stress problems a two-dimensional model is made from a plate of photoelastic material. The thickness of the plate should not exceed $\frac{1}{4}$ in. and should be considerably less if small openings or small radii of curvature exist in the model. For large field polariscopes (8 to 10 in. diameter), the model can be of about the same width or larger. Extreme care should be used in preparing the model. A coolant should be used for all machine milling, drilling, and grinding. The plate of photo-elastic material should be checked for internal fringe patterns and edge effects both before and after the model is made. The model should be tested as soon after preparation as possible so that aging effects will not develop. Models that show initial internal fringes or edge fringes should not be used for test purposes until these fringes are removed by an annealing process. A calibration strip should be cut from the same piece of material as the model. Pure bending, simple tension, or a circular disc loaded diametrically can be used to determine the stress optical coefficient for the photoelastic model material. At least three independent determinations should be made.

The calibration equation for a simple tension strip is

$$C = \frac{N\lambda A}{dL} = \frac{N\lambda}{Pd} \qquad (12.5.1)$$

where L is the total load and A is the cross-sectional area of the tension strip at its center and P is the uniaxial stress. Calibration tests should be made for different fringe orders and a plot made of retardation or fringe order versus tensile stress. The resulting curve should be a straight line; its slope F is given by

$$P = FN$$

The slope F is a constant for any particular model material and light source. This constant F is defined as the model fringe value and is given by

$$F = \frac{\lambda}{dC} \qquad (12.5.2)$$

Thus when N is observed in the model, the value of $P - Q$ will be given by

$$P - Q = FN \qquad (12.5.3)$$

Usually photoelastic models are loaded in tension rather than compression because alignment problems are simplified. While the model is loaded the polariscope should be set for visual observation of the isochromatics so that the zero fringe order can be identified and isotropic points noted. A photograph of the fringe pattern is made at a given load value. The quarter-wave plates are removed and the isoclinics for each 5, 10, or 15° of rotation are either sketched or photographed.

From the preceding data the stress trajectories can be sketched and the maximum shear stresses computed. At free boundaries the maximum principal stresses can be determined.

To determine the principal stresses at points other than on free boundaries requires additional tests. Probably the most direct method is to determine $P + Q$ by the lateral extensometer method. Equation 3.4.1 in Chapter 3 shows that the lateral strain for a plate in plane stress is proportional to the sum of the normal stresses. As the sum of the normal stresses is a constant at any point Eq. 3.4.1 can be written as

$$\epsilon_z = - \frac{\nu}{E}(P + Q) \tag{12.5.4}$$

Therefore a determination of ϵ_z at any point in the model will determine $P + Q$ if both ν and E of the photoelastic material are known. To determine ϵ_z requires that the thickness of the model plate be measured at each point both before and after the load is applied.

Various types of lateral extensometers have been devised to make these measurements. Extensometers with large optical or mechanical magnifications are required since strains of the order of 10 μ in./in. should be measurable.

12.6   THREE-DIMENSIONAL PHOTOELASTICITY

Three-dimensional photoelasticity is based upon the fact that it is possible to produce frozen stress patterns in photoelastic materials. A frozen stress pattern is one that remains in the material after the load is removed. At room temperature the optical effects produced by the application of stress to photoelastic materials disappear upon removal of the stresses. The same is true at high temperatures. However, if the load is applied at high temperature and maintained constant while the temperature of the photoelastic material is cooled slowly to room temperature, the photoelastic fringe patterns become frozen in the material. Removal of the load at room temperature after this annealing process does not remove the fringe patterns. Once the fringe patterns are frozen in a model they are unaffected by changing the shape of the model by cutting, milling, or filing, provided these operations do not raise the temperature of the material to the critical temperature. Because of this fact it is possible to stress three-dimensional models at elevated temperatures, cool them slowly under constant load, remove the load at room temperature, and saw the model into thin plates or small cubes. These plates or cubes are then observed in a standard polariscope for their frozen fringe patterns.

From both theory and experiment it can be shown that stresses in the direction of the propagation of light produce no photoelastic effects upon the light. Thus if the direction of the light is in the z direction, the components of stress having a z subscript have no photoelastic effect, that is, σ_z, τ_{xz}, and τ_{yz} have no stress optical effect. The only components that have a stress optical effect are σ_x, σ_y, and τ_{xy}. These components of stress give rise to the secondary principal stresses for the z direction discussed in Section 1.7. Thus from Eq. 1.7.1 the difference between the secondary principal stresses P' and Q' is given by

$$(P' - Q')_z = \sqrt{(\sigma_x - \sigma_y)^2 + 4\tau_{xy}^2} \tag{12.6.1}$$

Rewriting Eq. 12.3.1 in terms of secondary principal stresses P', Q', and R' gives

$$n_p - n_0 = C_1 P' + C_2(Q' + R')$$
$$n_q - n_0 = C_1 Q' + C_2(P' + R')$$

Subtraction of these two equations gives

$$n_p - n_q = C'(P' - Q') \tag{12.6.2}$$

Therefore from Eq. 12.2.12 the retardation Δ is given by

$$\Delta = \frac{2\pi \, dC'}{\lambda}(P' - Q') \tag{12.6.3}$$

Equation 12.6.3 is the stress optical law for three-dimensional photoelasticity.

The complete determination of the state of stress at a point in a three-dimensional photoelastic model requires considerable experimental ingenuity and laborious mathematical calculations. Good descriptions of the various techniques that are used are described in the references given at the end of this chapter.[2,3,4]

12.7 TWO-DIMENSIONAL PHOTOELASTIC MODELS

A number of problems related to the design and stability of underground openings can be considered with reasonable accuracy on a two-dimensional basis. In Chapters 4 and 5 several problems of this type were treated theoretically. For example, the theoretical stress distribution around a circular or rectangular hole in an infinite elastic plate was determined. This solution approximates the stress distribution around a drift or shaft of the same cross-sectional shape, provided the opening is in an elastic rock and situated at a depth large in comparison with the dimensions of the opening. However, if the problem is one of determining the stress distribution around multiple openings the theoretical approach becomes more difficult. The two-dimensional stress distribution around two, three, and an infinite number of circular holes lying in a row has been determined theoretically, but for other multiple hole shapes and configurations photoelastic models usually provide a simpler means for investigating stress distributions.

Photoelastic model studies made in relation to structural problems in rock can be divided into two broad classes: first, there is the class in which the model is a scaled representation of a proposed or actual, but usually unique prototype structure. Because studies of these models make it possible to evaluate structural problems that are not amenable to mathematical treatment, this procedure is of great value to the engineer. However, since there is virtually an infinite number of possible shapes and configurations, this class of model studies does not lead to any generalizations or functional relationships between the variables involved. Second, hypothetical structures containing one or more idealized openings are considered. In a group of models one quantity, such as the height-to-width ratio of the opening, the number of openings, or the orientation of the opening(s) with respect to the direction of the applied stress, is varied, and a relationship between the stress concentration and the selected variable determined. Several examples of two-dimensional photoelastic model studies are presented in the following paragraphs to illustrate the general utility of this procedure.

The accuracy with which the tangential boundary stress in a two-dimensional photoelastic model can be determined depends primarily on the accuracy with which the fringe order can be measured. An error is produced because of the inability to estimate correctly fractional fringe orders and because of the presence of some residual fringe introduced at the time the model was machined, especially at points of small radius of curvature. The results for larger openings having larger radii of curvature, therefore, are more accurate than those obtained on small openings.

Some idea of the accuracy that can be obtained with small photoelastic models is illustrated by the following example. The theoretical value of the maximum stress concentration around a circular hole in a uniaxially stressed plate whose width is four times the diameter of the hole is 3.22 as compared with 3.00 for a plate of infinite width (Howland[5]). Duvall[6] photoelastically determined the maximum stress concentration for a $\frac{1}{2}$ in. diameter hole in a 2 in. plate, and for a $\frac{3}{8}$ in. hole in a $1\frac{1}{2}$ in. plate to be 3.05 and 3.17, respectively. These results are within 6% of the theoretical value. The accuracy of photoelastic measurement is also indicated by a comparison made by Duvall of the experimentally and theoretically determined maximum stress concentration f_{max} on the boundary of a uniaxially loaded ellipse, as a function of the height-to-width ratio of the ellipse h/w. The maximum concentration occurs at the ends of the axis perpendicular to the direction of the applied stress S_y, and is given by the theoretical expression (see Eq. 5.2.7)

$$f_{max} = \frac{(\sigma_\theta)_{max}}{S_y} = \left(1 + 2\frac{w}{h}\right) \qquad (12.7.1)$$

where σ_θ is the tangential boundary stress. The experimental results are shown in Fig. 12.7.1 for ratios of h/w from 0.5 to 1.5, together with the theoretical curve for Eq. 12.7.1.

The maximum stress concentration on the boundary of a single elliptical, ovaloidal*, and rectangular (with rounded corners) openings subjected to an applied uniaxial load was determined photoelastically[6] for various height-to-width ratios (Fig. 12.7.2). By superposition the maximum stress concentration for any biaxial applied load can be calculated. Thus, if the

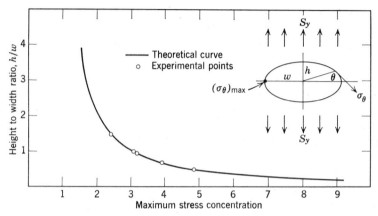

Fig. 12.7.1. Maximum stress concentration on the boundary of ellipses having various height-to-width ratios.

maximum stress concentration is a critical factor in the design of an underground opening that can be approximated by a two-dimensional model, these data are sufficient for selecting a favorable opening shape. This problem is discussed in more detail in Chapters 16 and 17.

The maximum stress concentration on the boundary of most of the single opening shapes considered above could have been derived from the theoretical treatment given in Chapters 4 and 5. However, as previously mentioned, if the opening boundary contains discontinuities, irregularities, or in general, if it is such that it cannot be expressed mathematically, models of one kind or another provide the only means of evaluating the maximum stress concentration and the distribution of stress. For example, Hoek[7] investigated the state of stress around a circular hole containing a crack of variable length by means of photoelastic models.

If the problem involves more than one opening, mathematical treatment usually becomes more difficult, we must resort to models. A photoelastic

* An ovaloid as considered here is a rectangle with semicircular ends; see middle insert, Fig. 12.7.2.

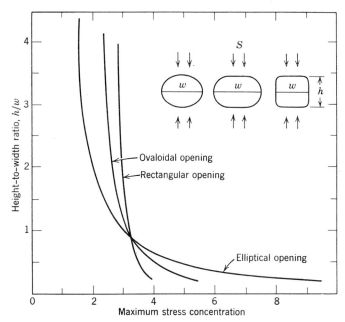

Fig. 12.7.2. Maximum stress concentration on the boundaries of ellipses, ovaloids, and rectangles.

model study of multiple openings also made by Duvall[8] exemplifies the use of this technique to develop generalizations or functional relationships between the variables involved. In this study the maximum stress concentrations in two-dimensional models containing two or more circular or ovaloidal openings lying in a row perpendicular to the direction of the applied stress were determined. A model with five circular openings separated by four pillars is illustrated in Fig. 12.7.3. Figure 12.7.4 shows the isochromatic fringes in this model. The maximum stress concentrations on these openings occur at the points *ABCDE*, and their values for opening-to-pillar width ratios of 1.07 to 4.35 are given in Table 12.1.

Table 12.1 Stress Concentration in a Model Containing Five Circular Openings.

Ratio of Opening-to-Pillar Width	Stress Concentration at Positions				
	A	*B*	*C*	*D*	*E*
1.07	3.29	3.29	3.29	3.29	3.29
2.21	3.63	3.72	3.89	4.03	4.03
2.96	3.53	4.08	4.22	4.39	4.39
4.35	3.96	5.12	5.22	5.28	5.28

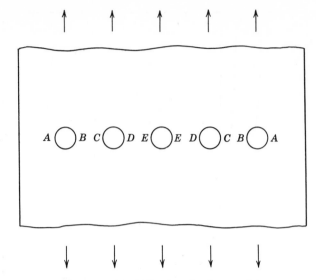

Fig. 12.7.3. Model for five circular openings.

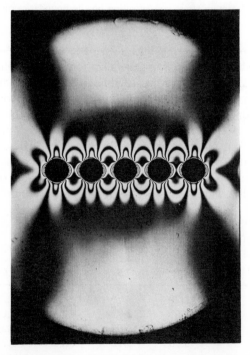

Fig. 12.7.4. Isochromatic fringes around five circular openings in a plate.

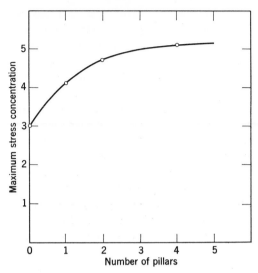

Fig. 12.7.5. Maximum stress concentration on the boundary of circular openings separated by pillars.

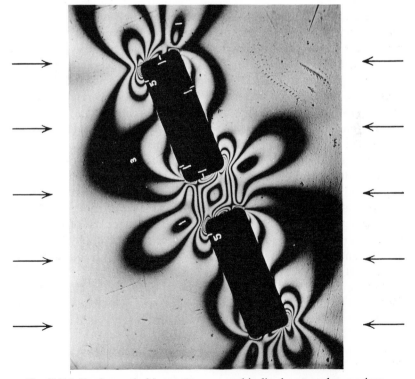

Fig. 12.7.6. Isochromatic fringe pattern around inclined rectangular openings.

From photoelastic data for one, two, three, and five circular opening models (or zero, one, two and four pillar models with an opening width-to-pillar width ratio of 4.0), a relationship was developed between the maximum stress concentration on the boundary of the openings and the number of pillars, which is shown graphically in Fig. 12.7.5. The implication derived from this relationship is that for a system of long parallel openings the maximum stress concentration in the pillars does not increase significantly when the number of pillars is greater than four.

Correspondingly, Panek[9] studied the stress distribution around and in the pillar between two rectangular openings (with round corners). The angle of inclinations of the opening with respect to the direction of the applied stress, the ratio of the opening height-to-width, and the ratio of the pillar height-to-width to the opening height-to-width were varied. The isochromatic fringe pattern of two vertically loaded models (Fig. 12.7.6) shows the shift in the shear stress concentration at the corners of the rooms.

12.8   EPOXY AND GELATIN MODELS

In Section 12.6 a technique was described for freezing stresses in a plastic model so that by cutting the frozen model into plates or rectangular prisms the stress distribution in the model could be analyzed in three dimensions. The material generally used in these models is an epoxy resin or similar photoelastic plastic that can be cast or machined to the desired shape, and one in which stresses can be frozen at a moderate temperature (about 175°C).* These models are loaded externally during the freezing process, and they are generally shaped so that the stress field in the model is uniform except in the proximity of the opening or structural shape to be investigated. Hence the stress field in such a model corresponds to that for an underground opening whose dimensions are small compared to its depth, and it is not representative of an opening at or near the surface, or of a partially isolated body or rock such as a layer of roof rock that has become detached from the overlying rock. The latter class of structure or structural element is loaded either entirely or significantly by the force of gravity, and in the analysis of a corresponding model an equivalent body force must be applied.

Because gelatin has a large stress optical coefficient it can be used to form two- or three-dimensional models that will body load by their own weight to the extent that a satisfactory analysis of both isoclinic and

---

* The properties of a number of plastics suitable for three-dimensional photoelastic analysis are given in a report by Leven.[10]

isochromatic fringe patterns can be made. In addition such a model can be externally loaded so that the stress distribution from a combined body and external loading can be determined. Phillippe and Mellinger[10], for example, reported a study of the stress distribution in a dam foundation

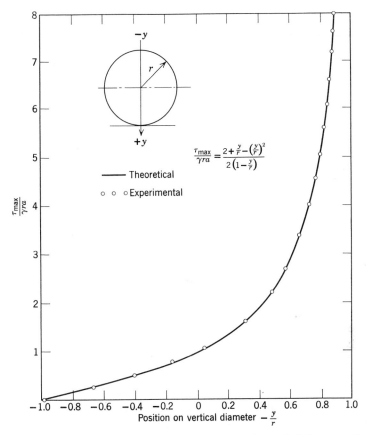

$$\frac{\tau_{max}}{\gamma r a} = \frac{2 + \frac{y}{r} - \left(\frac{y}{r}\right)^2}{2\left(1 - \frac{y}{r}\right)}$$

——— Theoretical

o o o Experimental

Position on vertical diameter $-\frac{y}{r}$

Fig. 12.8.1. Maximum shear stress distribution on the vertical diameter of a body-loaded disk.

using gelatin models. In addition to the body loading due to gravity, a boundary load representing the weight of the overlying embankment was applied to the model by weighting it with lead shot.

The body loading produced in a model by rotating it in a centrifuge (Section 13.6) is proportional to the centrifugal acceleration a given by $a = 4\pi^2 r n^2$, where r is the radius, and n the angular speed of the centrifuge rotor.* Two procedures have been used to observe the fringe patterns in centrifugally loaded models. By using a stroboscopic light source the fringe pattern can be photographed while the model is being rotated. If a model made of epoxy resin (or equivalent) is heated in an oven secured to a centrifuge rotor, and if it is then centrifugally loaded and at the same time allowed to cool, the internal stresses frozen in the model will be those due to the centrifugal body load. The model can then be removed, sectioned, and analyzed. This procedure has been employed by Hoek[7]; to demonstrate the accuracy of the method, the shear stress distribution in a circular disc resting on its edge was measured and compared with the theoretical distribution given by

$$\frac{\tau_{max}}{\gamma r a} = \frac{2 + (y/r) - (y/r)^2}{2[1 - (y/r)]} \qquad (12.8.1)$$

where a is the centrifugal acceleration in multiples of g, γ is the unit weight of the disc, and the dimensions y and r are indicated in the inset in Fig. 12.8.1.

12.9   PHOTOSTRESS TECHNIQUE

In the conventional photoelastic method described in.Section 12.5, plane or circularly polarized light is transmitted through a transparent photoelastic model. Because of the physical limitations imposed by this procedure the method can only be used to study the stress distribution in transparent models. In the photostress method, developed by Mesnager in 1930 and later modified by Zandman and Wood[12], a layer of photoelastic plastic is bonded directly to the surface of the prototype part or structure and plane or circularly polarized light reflected from the back (bonded) surface of the plastic (Fig. 12.9.1). Thus this procedure permits the determination of the stress developed at any point on the surface of an opaque body which, for example, may be a rock model or part of a prototype structure. Apart from the fact that the polarized light passes through the plastic twice, and hence $2d$ insteady of d is used in Eqs. 12.3.3, 12.4.13, and 12.4.14, the photostress method is similar to the conventional photoelastic method.

---

* The stress distribution in a centrifugally loaded model does not exactly correspond to that produced by gravity because a gravity load exists in the direction normal to the direction of the centrifugal acceleration and the finite radius of the centrifugal motor. If $a \gg g$ the first effect may be made negligible; for a centrifuge with a rotor with $r = 48$ in., the maximum difference in the magnitude of the stress at the inner and outer edges of a 4 in. thick model would be approximately 8%. Across the dimensions of the model opening, the difference will be smaller.[9]

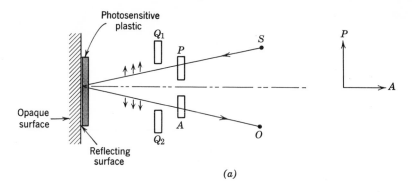

(a)

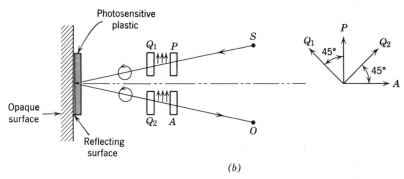

(b)

Fig. 12.9.1. Photo-stress method. (a) Crossed plane polariscope. (b) Circular crossed polariscope.

Photoelastic coatings can be applied in several ways: for example, a transparent fluid (epoxy resin) is available that hardens into a photoelastic plastic. A uniform layer of this fluid can be applied to a reflecting surface, such as a polished metal surface, or to a surface that has been painted with an aluminum or silver paint. Second, this fluid can be cast into a sheet of uniform thickness, and this sheet can be cemented to a polished surface with a transparent cement, or to an unpolished surface with a reflecting cement (usually an acetate cement loaded with aluminum powder). Third, photoelastic plastics are available in thin sheets that can be cut, fit, and attached to reflecting surfaces with a transparent cement, or to a nonreflecting surface with an aluminized cement. The latter materials can be softened by heating so that they will conform to surface

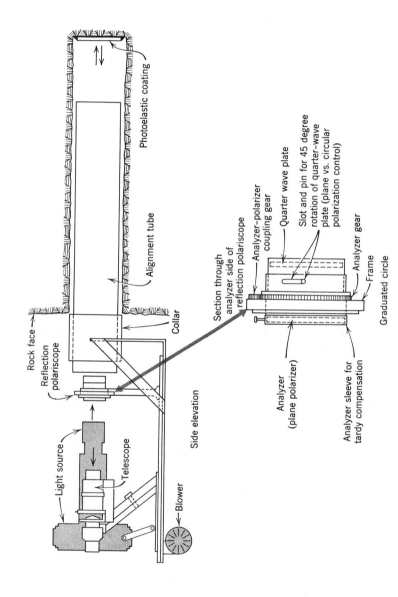

Photoelastic coating

Alignment tube

Rock face

Reflection
polariscope

Collar

Light source

Telescope

Blower

Side elevation

Section through
analyzer side of
reflection polariscope

Analyzer-polarizer
coupling gear

Quarter wave plate

Slot and pin for 45 degree
rotation of quarter-wave
plate (plane vs. circular
polarization control)

Analyzer gear

Frame

Graduated circle

Analyzer
(plane polarizer)

Analyzer sleeve for
tardy compensation

Fig. 12.9.2. Borehole polariscope (private communication from Professor Pincus).

384

contours. The photoelastic strain gages described in Section 9.5 utilize the back-reflecting principle to observe the displacement of a "frozen" fringe. Pincus[12] produced "frozen" rings in photoelastic stock by placing a thin metallic ring on a piece of heated plastic and allowing it to cool. These frozen ring gages serve as a convenient means of ascertaining by visual inspection the direction of the principal stresses; also, if the ring gages are calibrated the magnitude of the principal stresses can be determined from an analysis of the fringe pattern.

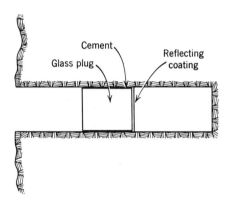

Fig. 12.9.3. Borehole photoelastic gage. (After Hiramatsu, et al.[14])

Back-reflecting photoelastic gages have been used to measure the direction and magnitude of the secondary stresses in underground bodies of rock. The gages were cemented in the ends of boreholes and observed with the borehole polariscope developed by Pincus[13] and shown schematically in Fig. 12.9.2.

Hiramatsu et al.[14] developed a photoelastic borehole gage consisting of a glass plug, the back surface of which was silvered (Fig. 12.9.3). This plug was cemented in a borehole and the change in stress determined from the fringe pattern in the glass, which was observed with a borehole polariscope similar to that in Fig. 12.9.2. Although the stress optical constant of glass is low, the sensitivity of the device was improved by a long optical path twice through the glass. Glass was selected as a photoelastic material because it is stable in a mine atmosphere. Emery[15] utilized the photostress technique by cementing back-reflective plastics directly (in distinction to edge cementing) to the surface of rock specimen to observe the development of fringe patterns around individual grains or crystals as the rock deformed due to time-dependent creep.

REFERENCES

1. Maxwell, J. C., "On the Equilibrium of Elastic Solids," *Trans. Roy. Soc. Edinburgh*, **20**, No. 87 (1950).
2. Mindlin, Raymond D., "A Review of Photoelastic Method of Stress Analysis," I, *J. Appl. Phys.*, **10** (April 1939), and II, **10** (May 1939).
3. Frocht, Max Mark, *Photoelasticity*, Vol 1 and Vol 2, John Wiley and Sons, New York, 1941 and 1948.
4. Durelli, A. J. and W. F. Riley, *Introduction to Photomechanics*, Prentice-Hall, Englewood Cliffs, New Jersey, 1965.
5. Howland, R. O. J., "On the Stresses in the Neighbourhood of a Circular Hole in a Strip Under Tension," *Phil. Trans. Roy. Soc.*, London A **229** (1930).
6. Duvall, W. I., "Stress Analysis Applied to Underground Mining Problems;" Part I, "Stress Analysis Applied to Single Openings," *U.S. BurMines Rept. Invest.*, **4192** (1948).
7. Hoek, E., "Experimental Study of Rock-Stress Problems in Deep-Level Mining," *Experimental Mechanics*, Pergamon Press, New York, 1963.
8. Duvall, W. I., "Stress Analysis Applied to Underground Mining Problems;" Part II, "Stress Analysis Applied to Multiple Openings and Pillars," *U.S. BurMines Rept. Invest.*, **4387** (1948).
9. Panek, L. A., "Stress about Mine Openings," *Dissertation, Columbia University*, New York, 1948.
10. Leven, M. M., "Epoxy Resin for Photoelastic Use," *Symp. on Photoelasticity*, The Macmillan Co., New York.
11. Phillippe, R. R. and F. N. Mellinger, "Theoretical and Experimental Stress Analysis," *Colo. Sch. of Mines Quarterly*, **52**, No. 3 (1957).
12. Zandman, F. and M. R. Wood, "Photo Stress," *Prod. Eng.* (Sept 1956).
13. Pincus, H. U., *An Evaluation of the Capabilities of Photoelastic Coatings for the Study of Strain in Rock*.
14. Hiramatsu, Y., Y. Niwa, and Y. Oka, "Measurement of Stress in Field by Application of Photo-elasticity," *Eng. Res. Inst., Kyoto U.*, **VII**, No. 3 (1957).
15. Emery, C. L., "The Measurement of Strains in Mine Rock," *Mining Research*, George B. Clark, editor, Pergamon Press, New York, 1962.

CHAPTER 13

ROCK MODEL STUDIES

13.1 INTRODUCTION

The use of real materials, such as rock or simulated rock, greatly expands the scope of problems that can be investigated with models. Because this type of model can be loaded to destruction, the strength as well as the stress distribution in the model can be determined, and if the corresponding prototype is of the same rock its strength can also be evaluated. Rock models can be made of a number of components of either the same or different materials, thus simulating more complex prototype structures, such as a laminated mine roof, or a mine pillar in bedded rock. These models have been used to study the properties of different rock types subjected to various conditions of load or states of stress. Rock models, like photoelastic models, have been used to investigate the behavior of unique structures; but probably more important, by varying the properties of the material, dimensions, or applied loads in a group of models, relationships between the relevant variables can be developed.

The testing of rock models presents several problems that should be considered before initiating this type of investigation. For one thing, rock models are comparatively difficult to prepare. Cutting a smooth surface on large blocks of rock requires special equipment (some of which is described in Section 11.3) and cutting anything other than a round hole in the block is even more difficult. If the models are made of more than one component each part usually must be ground (with a surface grinder) to close tolerance, as illustrated in Fig. 13.1.1, which shows a seven-member model of a mine roof in which the laminae are 18 in. long, 3 in. wide, and only $\frac{1}{4}$ in. thick. Because most mine models are tested in compression, and because of the high compressive strength of rock, the capacity of the testing machine must be large. For example, a granite model with a bearing area of 1 sq ft

Fig. 13.1.1. Model of laminated roof C 12 in. span. (After Panek.[7])

if tested to failure would require a testing machine with a capacity of about three million pounds (assuming a compressive strength for granite of 20,000 psi). To lessen this requirement low strength model materials, such as mortars, are sometimes substituted for rock. However, to satisfy similitude requirements, not only the strength but the elastic properties of the synthetic model materials must scale to those of the prototype rock, a condition that is not easy to achieve.

In mathematical or photoelastic models the stress or strain can be determined at any point in the model (which is referred to as a whole field determination), whereas in rock models the strain is usually measured only at discrete points, or more accurately stated, on discrete areas (if for no other reason than the fact that strain gages occupy a finite area).* Hence, the placement of gages should be made judiciously so as not to miss an area in which the strain might be critical. Also, if the object of the model study is to determine the stress distribution, an inordinately large number of gages may be required.

13.2 MODEL-PROTOTYPE SIMILITUDE CONDITIONS

In Chapter 7 it was shown that if a problem can be specified by a number of independent variables a dimensional matrix can be formed, the rank of which subtracted from the number of variables gives the number of dimensionless products, or π-terms in a complete set (Buckingham's π-theorem). Moreover, a functional relationship can be expressed between the π-terms such that

$$\pi_1 = f(\pi_2, \pi_3, \ldots, \pi_q)$$

A function of this type can be written for both the model and prototype (the respective π-terms for which are designated by the subscripts m and

* Brittle or other strain-sensitive coatings provide whole field determinations, but are usually not satisfactory on rock, especially if heat is required to harden the coating.

p), and a relationship between π_{1m} and π_{1p} can be obtained from the ratio

$$\frac{\pi_{1p}}{\pi_{1m}} = \frac{f(\pi_{2p}, \pi_{3p}, \ldots, \pi_{qp})}{f(\pi_{2m}, \pi_{3m}, \ldots, \pi_{qm})}$$

provided that the model is designed and tested such that

$$\pi_{2p} = \pi_{2m}$$

$$\pi_{3p} = \pi_{3m}$$

$$\text{etc.}$$

If these conditions are satisfied, then

$$f(\pi_{2p}, \pi_{3p}, \ldots, \pi_{qp}) = f(\pi_{2m}, \pi_{3m}, \ldots, \pi_{qm})$$

and hence

$$\pi_{1p} = \pi_{1m}$$

To illustrate the application of this procedure, consider a problem in which the stress in the prototype σ_p depends on one of the dimensions of the structure, L_p; the elastic properties of the rock, E_p and ν_p; the external load F_p applied to the structure; and the unit weight of the rock γ_p. The dimensional matrix is*

	σ_p	γ_p	L_p	E_p	F_p	ν_p
M	1	1	0	1	1	0
L	−1	−2	1	−1	1	0
T	−2	−2	0	−2	−2	0

Since all third-order determinants vanish the rank of the matrix is two, hence the number of dimensionless products in a complete set is four, which may be chosen as

$$\frac{\sigma_p}{\gamma_p L_p}, \quad \frac{\gamma_p L_p}{E_p}, \quad \frac{F_p}{E_p L_p^2}, \quad \nu \tag{13.2.1}$$

However, the prototype may be a more complex structure, made of several materials, of a shape involving several dimensions and loaded by more than one force. These additional variables can be expressed as dimensionless products of the original variables; thus, if $\gamma_p{}'$, $\gamma_p{}''$, etc., $E_p{}'$, $E_p{}''$, etc., are the properties of the other materials, then the dimensionless products would be

$$\frac{\gamma_p{}'}{\gamma_p}, \quad \frac{\gamma_p{}''}{\gamma_p}, \quad \text{etc.,} \qquad \frac{E_p{}'}{E_p}, \quad \frac{E_p{}''}{E_p}, \quad \text{etc.}$$

* Since M and T appear in each row at the same power, the number of rows in this matrix could be reduced to two, the dimensions of which would be L and M/T^2. The rank of the matrix would remain unchanged.

Correspondingly, the other forces could be specified by

$$\frac{F_p{'}}{F_p}, \quad \frac{F_p{''}}{F_p}, \quad \text{etc.}$$

and the other dimensions by

$$\frac{L_p{'}}{L_p}, \quad \frac{L_p{''}}{L_p}, \quad \text{etc.}$$

A functional relationship can be written for any one of the π-terms as, for example, for the stress

$$\frac{\sigma_p}{\gamma_p L_p} = f_1\left[\frac{\gamma_p L_p}{E_p}, \frac{F_p}{E_p L_p{}^2}, \frac{L_p{'}}{L_p}, \frac{L_p{''}}{L_p}, \dots, \frac{E_p{'}}{E_p}, \frac{E_p{''}}{E_p}, \dots, \right.$$
$$\left. \nu, \nu', \nu'', \dots, \frac{\gamma_p{'}}{\gamma_p}, \frac{\gamma_p{''}}{\gamma_p}, \dots, \frac{F_p{'}}{F_p}, \frac{F_p{''}}{F_p}, \dots \right] \quad (13.2.2)$$

where f_1 is an undetermined function. Replacing the subscripts p with m gives an equivalent expression for the stress in the model. The condition for model-prototype similitude is that the following design equations are satisfied:

$$\frac{\gamma_m L_m}{E_m} = \frac{\gamma_p L_p}{E_p} \qquad\qquad \frac{F_m}{E_m L_m{}^2} = \frac{F_p}{E_p L_p{}^2}$$

$$\frac{L_m{'}}{L_m} = \frac{L_p{'}}{L_p}, \quad \text{etc.} \qquad \frac{E_m{'}}{E_m} = \frac{E_p{'}}{E_p}, \quad \text{etc.}$$

$$\frac{\nu_m{'}}{\nu_m} = \frac{\nu_p{'}}{\nu_p}, \quad \text{etc.} \qquad \frac{\gamma_m{'}}{\gamma_m} = \frac{\gamma_p{'}}{\gamma_p}, \quad \text{etc.} \qquad (13.2.3)$$

$$\frac{F_m{'}}{F_m} = \frac{F_p{'}}{F_p}, \quad \text{etc.}$$

and for this condition

$$\frac{\sigma_m}{\gamma_m L_m} = \frac{\sigma_p}{\gamma_p L_p} \qquad (13.2.4)$$

From the first of Eqs. 13.2.3

$$\frac{L_p}{L_m} = \frac{\gamma_m E_p}{\gamma_p E_m} \qquad (13.2.5)$$

where L_p/L_m is the *prototype-to-model scale ratio*. From the second of Eqs. 13.2.3

$$\frac{F_p}{F_m} = \frac{E_p L_p{}^2}{E_m L_m{}^2} \qquad (13.2.6)$$

which specifies the requirement for scaling loads. The third of Eqs. 13.2.3 requires a dimensional similitude between the model and prototype; the fourth, fifth, and sixth of Eqs. 13.2.3 require a mechanical property similitude between model and prototype; and the seventh of Eqs. 13.2.3 requires a similar load distribution in model and prototype.

Equation 13.2.4 can be written

$$\frac{\sigma_p}{\sigma_m} = \frac{\gamma_p L_p}{\gamma_m L_m} \tag{13.2.7}$$

and by combining Eqs. 13.2.5 and 13.2.7

$$\frac{\sigma_p}{\sigma_m} = \frac{E_p}{E_m} \tag{13.2.8}$$

By the same reasoning, relations for the strain ϵ and the displacement u can be determined such that

$$\epsilon = f_2(\cdots)$$
$$\frac{u}{L} = f_3(\cdots) \tag{13.2.9}$$

where the terms in f_2 and f_3 include all of the π-terms in Eq. 13.2.2. If Eqs. 13.2.3 are satisfied then

$$\epsilon_p = \epsilon_m$$
$$\frac{u_p}{u_m} = \frac{L_p}{L_m} \tag{13.2.10}$$

Consider a special case, namely, that of a beam of unit width, length L_m, and thickness t_m, made of a material having elastic constants E_m and ν_m, and loaded by its own weight γ_m. If σ_m is the stress at a point in the beam, the dimensional matrix for the beam is

	σ_m	γ_m	L_m	t_m	E_m	ν_m
M	1	1	0	0	1	0
L	-1	-2	1	1	-1	0
T	-2	-2	0	0	-2	0

the rank of which is two, hence there are four dimensionless products, which can be expressed by the functional relationship

$$\frac{\sigma_m}{\gamma_m L_m} = f_1\left(\frac{\gamma_m L_m}{E_m}, \frac{t_m}{L_m}, \nu_m\right) \tag{13.2.11}$$

Also, the strain ϵ and the displacement u can be obtained by the same reasoning, that is,

$$\epsilon_m = f_2(\cdots)$$

$$\frac{u_m}{L_m} = f_3(\cdots) \tag{13.2.12}$$

If this beam is a model of a mine roof, a similar set of equations can be written for the stress, strain, and displacement at the corresponding point in the mine roof. To satisfy similitude requirements, the corresponding arguments in f_1, f_2, and f_3 for the model and prototype must be equal. If this condition is satisfied, the problem remains to determine f_1, f_2, and f_3 by model test.

The problem is simplified if it is known that σ_p, ϵ_p, and u_p are linearly related to the body load γ_p. For this case a material with corresponding characteristics can be chosen for the model, and as σ_m would then be proportional to γ_m and independent of E_m, $\gamma_m(L_m/E_m)$ should not appear in f_1, and only as a factor in f_2 and f_3, that is

$$\frac{\sigma_m}{\gamma_m L_m} = f_1\left[\frac{t_m}{L_m}, \nu_m\right]$$

$$\epsilon_m = \frac{\gamma_m L_m}{E_m} f_2\left[\frac{t_m}{L_m}, \nu_m\right] \tag{13.2.13}$$

$$\frac{u_m}{L_m} = \frac{\gamma_m L_m}{E_m} f_3\left[\frac{t_m}{L_m}, \nu_m\right]$$

and the only requirements for model-prototype similarity are geometrical similitude and equal ν's. Generally the effect of ν is small and can be neglected. Hence, for this case, as the stresses are independent of the deformations, any material that obeys Hooke's law may be used for the model material. Of course, if the model is made from rock from the mine roof the only requirement is geometrical similitude.

13.3   DESTRUCTIVE MODEL TESTING

If the ultimate strength of an underground structure or structural component is calculated using laboratory determined mechanical property data (compressive, shear, and tensile strengths, etc.), it is necessary to assume a criterion of failure because the states of stress in the model (physical property specimen) and prototype (underground structure) are different. This problem is obviated if the model is a scale representation of the prototype, made from the prototype rock, and it is tested to destruction, because in this case the states of stress in both can be made identical. However, because of experimental errors and the variability of rock a number of models must be tested to obtain a statistically reasonable average strength. Unless the shape of the model is simple, model production may become a limiting factor.

Models for destructive testing are made from two kinds of material—either a material the same as in the prototype (which in this case is rock), or a synthetic material, such as mortars, ceramics, or plasters, with properties that scale to those in the prototype (see Section 13.4). Models made from the same material as the prototype satisfy the following conditions:

$$E_m = E_p$$

$$\gamma_m = \gamma_p$$

$$\nu_m = \nu_p$$

If the model-prototype scale factor is L_p/L_m, then

$$\frac{F_p}{F_m} = \frac{L_p^{\,2}}{L_m^{\,2}}$$

and

$$\sigma_p = \sigma_m$$

Moreover, if σ_{Fm} and σ_{Fp} are the stresses at failure in the model and prototype, then

$$\sigma_{Fm} = \sigma_{Fp}$$

To illustrate the utility of destructive testing consider the investigations of Hiramatsu and Oka[1]. Marble and sandstone models containing a circular hole (Fig. 13.3.1a) were subjected to either a uniaxial compressive load F_y, or a biaxial load F_y and $F_x = F_y/4$, which produced the state of stress indicated in Fig. 13.3.1b (assuming the model to be a thin plate). Because the load applied to the model was large compared to the weight of the rock the body load γ was neglected. This is equivalent to assuming that the depth of the prototype opening is large compared to its vertical dimensions. When the uniaxial load was applied tension cracks occurred along or near the lines TT' in the direction of σ_y, and when the biaxial load was applied crushing failure occurred along the line CC'. The stress at points T and C was calculated from elastic theory (Section 4.7) using the second of Eqs. 4.7.7, namely

$$\sigma_\theta = \sigma_y(1 + 2 \cos 2\theta) + \sigma_x(1 - 2 \cos 2\theta)$$

For the uniaxial case $\sigma_x = 0$, and the tangential stress at point C is obtained by letting $\theta = 0°$, and $\sigma_C = 3\sigma_y$; at point T, $\theta = 90°$ and $\sigma_T = -\sigma_y$. For the biaxial case with $\sigma_x = \tfrac{1}{4}\sigma_y$ the tangential stress at point C is $\sigma_C = -2.75\sigma_y$, and at point T, $\sigma_T = 0.25\sigma_y$. A comparison of the calculated failure stress at these points with the laboratory determined uniaxial compressive, tensile, and flexural strengths is given in Table 13.3.1. For

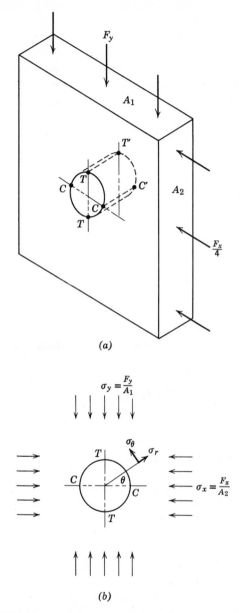

Fig. 13.3.1. Model of a circular hole in plate.

Table 13.3.1*

| Kind of Test | Rock Type | Number of Tests | Failure Begins to Appear | | Theoretical Stress, psi | | Uniaxial Compressive Strength, psi | Uniaxial Tensile Strength, psi | Flexural Strength, psi |
			At	For Average Vertical Stress (σ_v), psi	σ_C	σ_T			
Uniaxial compressive $\sigma_v = \dfrac{F_v}{A}$ $\sigma_h = 0$	Marble	12	Point T	-1680	-5050	1680	$-12{,}800$	750	1820
	Sandstone	16	Point T	-3540	$-10{,}600$	3540	$-22{,}300$	1570	3270
Biaxial compressive $\sigma_v = \dfrac{F_v}{A}$ $\sigma_h = \dfrac{F_h}{4A}$	Marble	15	Point C	-4600	$-13{,}700$	1160	$-12{,}800$	750	1820
	Sandstone	15	Point C	-8600	$-23{,}700$	2170	$-22{,}300$	1570	3270

* From Hiramatsu and Oka[1]

395

the biaxial case the magnitude of the compressive failure stress at C for both marble and sandstone is in agreement with the corresponding uniaxial compressive strength. For the uniaxial case, the magnitude of the tensile failure stress at T is in reasonable agreement with the flexural strength (outer fiber tensile strength), but the uniaxial tensile strength of both rocks was less than half the corresponding tensile failure stress in the model. The better agreement between flexural strength and in situ failure stress

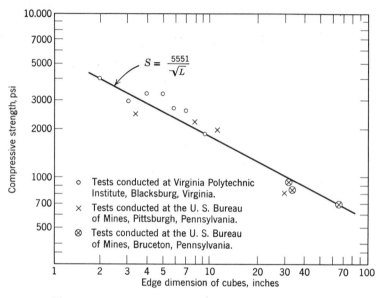

Fig. 13.3.2. Strength of cubical coal specimens. (After Holland.[2])

was also observed in the destructive loading of a limestone and marlstone roof (Section 14.10). Thus for a circular tunnel or drift in an uniaxial or biaxial stress field, these model results indicate that the maximum stress theory is valid, and suggest that any unconstrained or partially constrained body of rock will fail when either its uniaxial compressive strength or flexural strength is exceeded.

Another problem that has been investigated by means of models is the dependence of the strength of mine pillars on size and shape. Generally these studies have been performed to develop a functional relationship between the strength of the model material and model dimensions. For example, in Fig. 13.3.2 the results of several investigations are summarized in which cubical coal specimens of edge dimensions L, ranging from 2 to 65 in, were loaded in compression to failure. For geometrically similar models of a given material the compressive strength C_p would be expected to be independent of the model size, but these results indicate that the pillar strength varies as $L^{-\frac{1}{2}}$. This decrease in strength with size is presumed to result from the fact that the probability of flaws (cracks) in the specimens increases with size. Holland[2] studied the shape factor for model coal pillars of height h and width w, Fig. 13.3.3, (the other lateral dimension

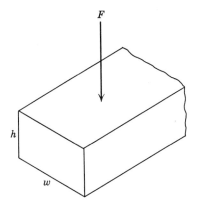

Fig. 13.3.3. Notation for pillar dimensions.

does not affect the pillar strength significantly provided it is greater than w) and found that the compressive strength C_p increases as $w^{\frac{1}{2}}/h$, that is,*

$$C_p = \frac{kw^{\frac{1}{2}}}{h}$$

This increase is presumed to result from the lateral constraint that developed across the end surface of the specimen due to bearing plate-specimen friction. This constraint places the core of short specimens under triaxial compression, thereby increasing the specimen strength.

In the conventional manner of testing model pillars the bearing plate-specimen friction, and hence the lateral end constraint, is indefinite and possibly variable from specimen to specimen. To more nearly simulate the condition under which a mine pillar is loaded, Obert[3] developed the pillar model illustrated in Fig. 13.3.4, in which the ends of the pillar are a continuous part of the roof and floor, and the steel rings produce a uniform and increased lateral constraint (as determined by measuring the strain on the exterior vertical surface of the ring).

---

* For specimens with a height-to-width ratio less than 1, this relationship gives a higher value for the specimen strength than that given by the empirically determined Eq. 11.7.2 for prismatic or cylindrical rock specimens.

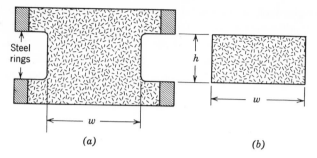

Fig. 13.3.4. (*a*) Model pillar. (*b*) Conventional test specimen, $w/h = 2$.

13.4 SYNTHETIC MODEL MATERIALS

As previously noted, most rock is difficult to cut and shape, especially into multi-opening configurations, and the model size is usually restricted because of the capacity of testing machines. Also, many rocks, such as the metallic mineral-bearing types contain fractures, joints or other inhomogeneities that preclude their use in models. Obviously low strength synthetic materials, such as plasters, mortars, and ceramics, that can be cast into the desired shape would simplify model testing. However, the mechanical properties of synthetic materials must satisfy model-prototype requirements. In a destructive model test in which the effect of body forces can be neglected, if equal model and prototype strain is required, that is, if $\epsilon_m = \epsilon_p$, the pertinent variables are: the compressive, shear, and tensile strengths, C_o, S_o, T_o; the modulus of elasticity E and Poisson's ratio, ν; the applied load F; and the dimensions L, a, b. The following design equations must be satisfied:

$$\frac{F_m}{E_m L_m{}^2} = \frac{F_p}{E_p L_p{}^2} \quad \text{or} \quad \frac{F_p}{F_m} = \frac{E_p}{E_m}\left(\frac{L_p}{L_m}\right)^2$$

$$\frac{C_{o_m}}{E_m} = \frac{C_{o_p}}{E_p} \qquad \frac{S_{o_m}}{E_m} = \frac{S_{o_p}}{E_p}$$

$$\frac{T_{o_m}}{E_m} = \frac{T_{o_p}}{E_p} \qquad \nu_m = \nu_p \qquad (13.4.1)$$

$$\frac{a_m}{L_m} = \frac{a_p}{L_p} \qquad \frac{b_m}{L_m} = \frac{b_p}{L_p}$$

Thus a mortar with a compressive strength one-tenth that of the prototype rock can be used for a model material provided that the shear and tensile strength, and the modulus of elasticity of the mortar are also one-tenth that of the prototype rock, and that their Poisson's ratios are equal (also assuming that both the mortar and prototype rock are linear-elastic). The scale ratio in the first of Eqs. 13.4.1 can then be selected so that the model load is within the capacity of the testing machine. For example, if $E_p/E_m = 10$ and $L_p/L_m = 20$, the ratio of prototype-to-model load would be 4000:1. However, as noted in the preceding section, since the strength of a model may be dependent on its size, it is usually considered good practice to keep the scale ratio as small as possible.

A synthetic model material that in relation to the prototype rock can be made to satisfy all the requirements of Eq. 13.4.1 is probably not attainable; at least no material of this kind has been reported. Usually some compromise is necessary and first consideration should be given to matching the more important properties. Thus, if the shear strength is considered to be the factor that will dominate failure in the prototype, the relationship

$$\frac{S_{0_m}}{E_m} = \frac{S_{0_p}}{E_p}$$

should be satisfied and the other model strengths disregarded. Generally Poisson's ratio will have the least effect on model-prototype similitude.

In Table 13.4.1 plaster of Paris model mixes and their corresponding mechanical properties are specified. Although it is improbable that any specified mix will exactly scale to the corresponding properties of the prototype rock, the range of values given in this table should be large enough to permit a reasonable match.

Denkaus et al.[5] described a ceramic material, whose elastic properties and strength are given in Fig. 13.4.1. This material was developed to scale

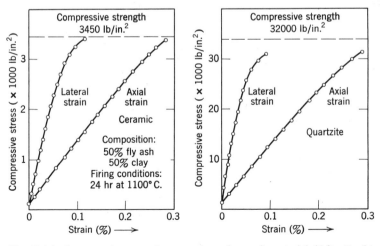

Fig. 13.4.1. Stress-strain curves for quartzite and ceramic material. (After Denkhaus.[5])

Table 13.4.1. Data for Plaster of Paris Mixes[1] (From Barron and Laroque[4])

| Density, lb/ft³ | Proportion by Weight | | | Retarder,[2] gm/liter of Water | Young's Modulus, psi | Compressive Strength, psi | Tensile Strength, psi | Set Time, minutes |
	Water	Plaster	Diatomaceous Earth					
75	60	100	0	0.50	1,000,000	2000	—	—
71	65	100	0	0.25	870,000	1700	274	7
67	70	100	0	0.25	730,000	1500	230	7
62	80	100	0	0.20	600,000	1100	—	11
58	90	100	0	0	500,000	800	145	6.5
53	100	100	0	0	440,000	700	—	10
48	110	100	0	0	368,000	470	103	7
46	120	100	0	0	316,000	400	82	9
43	130	100	0	0	273,000	330	72	9.5
40	140	100	0	0	225,000	280	69	9.5
38	150	100	0	0	190,000	240	—	10
37	160	100	8	0	164,000	200	51	12
36	170	100	8	0	131,000	170	—	12

[1] Poisson's ratio 0.24 and independent of mix.
[2] Sodium citrate.

to the properties of a quartzite. Ceramic material in distinction to plasters and mortars probably will fail more like a brittle material.

In some destructive model tests the shear strength may be more important than either the tensile or compressive strength. Also in nondestructive tests the time-dependent properties of synthetic materials may require evaluation and comparison with prototype rock.

13.5 NONDESTRUCTIVE MODEL TESTING

In a nondestructive test, strains or deformations in the model usually are measured at a succession of loads. If a number of strain gages are employed, the results so obtained generally permit the development of empirical relationships between the pertinent variables. If body loading can be neglected, the variables of concern generally are E, v, F, L, ϵ, and σ, and a complete set of π-terms is

$$\frac{\sigma}{E}, \quad \frac{F}{EL^2}, \quad \epsilon, \quad v \tag{13.5.1}$$

For a geometrically scaled model made from the prototype rock ($E_m = E_p$, $v_m = v_p$), if

$$\frac{L_m^{\,2}}{L_p^{\,2}} = \frac{F_m}{F_p}$$

then

$$\sigma_m = \sigma_p \quad \text{and} \quad \epsilon_m = \epsilon_p \tag{13.5.2}$$

Although these geometrical and force requirements are not difficult to satisfy, as previously noted, it is usually difficult to prepare models from prototype rock, and the required size and capacity of the loading machine may present problems. However, if the prototype rock is assumed to be linear elastic so that the stress, strain, and force are linearly related, the model can be made from any material that satisfies Hooke's law. Or, if the model material is such that σ, ϵ, and u are linearly related to the load, the model results may be applied to any prototype in a correspondingly elastic rock. Thus by a judicious selection of model materials, the geometric and force scale ratios can be kept within reasonable limits. Nondestructive model tests have a further advantage in that the models can be reused, thereby permitting changes in the model dimensions or the loading conditions. This factor generally reduces the time, effort, and cost of preparing and testing models.

Nondestructive model testing is exemplified by a study made by Merrill and Peterson[6] in which the deformation of a borehole in biaxially loaded models made of three relatively elastic materials—marble, limestone, and Hydrostone (a high strength gypsum cement)—was measured.

These measurements permitted a comparison of the deformation of a borehole in a real but quasi-elastic material, with the theoretical deformation of a hole in a thin, linear-elastic, isotropic, homogeneous, infinite plate. Because the model materials were quasi-elastic, the results are indicative of the accuracy with which the stress in an underground body of rock can be determined by the borehole deformation method.

Fig. 13.5.1. Apparatus used for biaxial test. (After Merrill and Peterson.[6])

The models were 5 × 5 × 10 in. prisms with a 1-in. diameter hole at the center, normal to a 5 × 10 in. side (Fig. 13.5.1). This figure also shows the loading presses, and precision air gage used to measure the borehole deformation. Strain gages on the exterior model surfaces were used to check the uniformity of the applied loads. From the measured deformation in the 60° and 45° rosette directions, the magnitude and direction of the applied biaxial stresses were calculated by Eqs. 14.3.1* and compared with the known applied stresses, Table 13.5.1. The difference between the calculated and applied stresses was least for limestone which uniaxial stress-strain tests showed to be virtually linear-elastic, and elastically isotropic. The larger differences for Hydrostone were attributed to anisotropy, although this material was relatively linear-elastic. The differences for marble resulted from both anisotropy and inelasticity.

---

* These equations are for the 60° deformation rosette. Similar equations are given in the Merill and Peterson report for the 45° rosette.

Table 13.5.1. Comparison of the Magnitude and Direction of the Computed and Applied Biaxial Stress

Rock Type	Strain Rosette, degrees	Longitudinal Stress, psi		Lateral Stress, psi		Deviation in Direction of Applied and Calculated Stress, Degrees
		Calculated	Applied	Calculated	Applied	
Hydrostone	45	808	800	177	240	1
Do	60	817	800	203	240	1
Limestone	45	2340	2400	920	1000	2
Do	60	2355	2400	925	1000	2
Marble	45	2200	2400	1180	1000	9
Do	60	2215	2400	1125	1000	14

13.6 CENTRIFUGAL MODEL TESTING

In a body of rock every volume element contributes to the load exerted on the rock thereunder by an amount

$$\gamma = \frac{mg}{V} \qquad (13.6.1)$$

where V is the volume of the element, m is the mass of the rock, and g is the acceleration of gravity. Thus γ is the body force per unit volume (Section 1.1) and the rock is said to be body loaded. In a mine all rock is body loaded, but at sufficient depth the change in load over a distance comparable with the vertical dimensions of most underground openings is small and can be neglected. Hence a surface-loaded model approximates the loading condition for a mine or underground opening at depth. However, the effect of body forces cannot be neglected if the underground structure is at shallow depth. Also, some structural components are loaded only by body forces as, for example, a mine roof that has become detached from the overlying rock. To study the effects of body loading in a model presents a problem. Generally, it is not practical to increase body forces by increasing the unit weight of the model material because the unit weight range is too limited, and because of the difficulties in scaling the other properties of the model rock to those of the prototype. The only other possibility is to increase the acceleration above that of gravity. This can be done by placing the model in a centrifuge in which the acceleration a is given by

$$a = 4\pi^2 r n^2$$

where r is the radius and n the rotational speed (in rps) of the centrifuge rotor. Hence, the ratio of the centrifugal to gravity load is

$$K = \frac{(m/V)a}{(m/V)g} = \frac{4\pi^2 r n^2}{g} \qquad (13.6.2)$$

If γ is the unit weight of the model material, $K\gamma$ can be considered to be the *effective unit weight* of the model material in the centrifuge. In Eq. 13.2.1 if γ_p is the weight of the prototype rock and $K\gamma_m$ the effective weight of the model material, the design equations become

$$\frac{K\gamma_m L_m}{E_m} = \frac{\gamma_p L_p}{E_p}$$

$$\frac{F_m}{E_m L_m{}^2} = \frac{F_p}{E_p L_p{}^2} \tag{13.6.3}$$

$$v_m = v_p$$

and if these relationships are satisfied

$$\frac{\sigma_m}{K\gamma_m L_m} = \frac{\sigma_p}{\gamma_p L_p} \tag{13.6.4}$$

and the conditions for geometric and force scaling become

$$\frac{L_p}{L_m} = K \frac{\gamma_m E_p}{\gamma_p E_m}$$

$$\frac{F_p}{F_m} = \frac{E_p L_p{}^2}{E_m L_m{}^2} \tag{13.6.5}$$

Combining Eqs. 13.6.4 and 13.6.5 gives

$$\frac{\sigma_p}{\sigma_m} = \frac{E_p}{E_m} \tag{13.6.6}$$

Correspondingly, if the model and prototype strains and displacements are ϵ_m and ϵ_p, and u_m and u_p, it follows that

$$\epsilon_m = \epsilon_p$$

$$\frac{u_m}{L_m} = \frac{u_p}{L_p} \tag{13.6.7}$$

If the model is made from the same rock as the prototype, then

$$\gamma_m = \gamma_p$$

$$E_m = E_p$$

and

$$\frac{L_p}{L_m} = K$$

$$\frac{F_p}{F_m} = K^2 \tag{13.6.8}$$

so that

$$\sigma_p = \sigma_m \tag{13.6.9}$$

Moreover, if the model is loaded to failure the model and prototype failure stress will be equal, that is,

$$\sigma_{Fm} = \sigma_{Fp} \qquad (13.6.10)$$

Finally, if the model is made from any linear-elastic material, and if it is assumed that the effects of ν are negligible

$$\frac{\sigma}{K\gamma L} = f_1 \left[\frac{a}{L} \cdots \right]$$

$$\epsilon = \frac{K\gamma L}{E} f_2 \left[\frac{a}{L} \cdots \right] \qquad (13.6.11)$$

that is, only geometrical scaling is required.

Although the design of a model testing centrifuge is principally a problem in mechanical engineering, there are several factors relevant to the size and capacity of the machine for testing rock models that should be considered. First, because the accuracy of centrifugal model testing generally is better if the geometrical scale ratio is not too large, the machine should accommodate a model weighing 50 lb or more. Also, a model of this size will permit the installation of strain gages and other instrumentation without difficulty. On the other hand, for a model of this weight the mechanical system must be rugged; in fact, the rotor should be designed to withstand the unbalanced force if the model should disintegrate. Second, the rotor should be equipped with multiple slip rings to allow the use of strain gages or other instrumentation on the model. If this provision is not made, the centrifuge will be limited mainly to destructive testing. Third, the model acceleration should be at least 1000 g; if not, the limitations on the choice of model materials are too restrictive. Table 13.6.1 gives the general specifications for rock model-testing centrifuges described by Panek[7] and Hoek[8].

Figure 13.6.1 shows the rotor and slip ring assembly in the U.S. Bureau of Mines centrifuge. Panek checked the performance of this centrifuge

Table 13.6.1

	USBM Panek[7]	CSIR Hoek[8]
Radius of rotor	24 in.*	48 in.†
Maximum model weight	50 lb	100 lb
Rotor speed, maximum	2000 rpm	950 rpm est.
Maximum model accel.	2600 g	1200 g
Number of slip rings	12	12

* Rotor operates in a vacuum.
† Rotor operates at atmospheric pressure.

Fig. 13.6.1. Rotor and case-model testing centrifuge.

first by measuring the midspan strain $\epsilon_{L/2}$ in simply supported beams of Indiana limestone subjected to both centrifugal and third-point static loading—a nondestructive test. From these measurements the modulus of elasticity E_c (centrifugal) and E_b (third-point loading) was calculated from the following relationships:

$$E_c = \frac{3F'L^2}{4\epsilon_{L/2}bt^2}$$

and

$$E_b = \frac{FL}{\epsilon_{L/2}bt^3}$$

where

$$F' = K\gamma bt$$

and where b, t, L, and γ are the width, thickness, span, and unit weight of the beam, K is the centrifugal loading factor (defined by Eq. 13.6.2), and $F/2$ is the third-point load. The agreement between these data is shown in

Table 13.6.2. Comparison of Modulii of Elasticity Determined by Centrifugal and Third-Point Loading. (After Panek[7])

			Third-point loading		Centrifugal loading	
Sample	Number of Specimens	Thickness, Inches	$\epsilon_{L/2}$ in./in.	E_s psi $\times$ 10^6	$\epsilon_{L/2}$ in./in.	E_c psi $\times$ 10^6
1	4	$\frac{3}{8}$	143	4.18	200	4.24
2	3	$\frac{5}{8}$	144	3.67	204	3.72
2	3	1	143	4.04	200	4.07
3	3	$\frac{3}{8}$	217	4.16	196	4.34
3	3	$\frac{5}{8}$	185	4.13	184	4.28

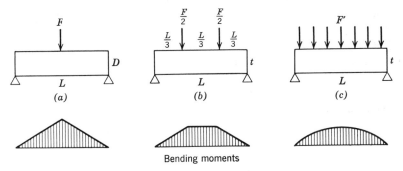

Bending moments

Fig. 13.6.2. Bending moments for (a) center-loaded, (b) third-point-loaded, and (c) body-loaded beams.

Table 13.6.2. Also, in a destructive test the flexural strength of beams of the same material was determined by central, third-point, and centrifugal (body) loading (Fig. 13.6.2a, b, and c). The midspan strain $\epsilon_{L/2}$ for the three cases is given by

$$\epsilon_{L/2} = \frac{8FL}{\pi D^3} \qquad \text{(central loading)}$$

$$\epsilon_{L/2} = \frac{FL}{bt^3} \qquad \text{(third-point loading)}$$

$$\epsilon_{L/2} = \frac{3F'L^2}{4bt^2} \qquad \text{(body loading)}$$

where F and $F/2$ are the static central and third-point loads. The results of this test, given in Table 13.6.3, substantiate a point made in Section 10.3, namely that the strength of a specimen decreases as the length over which the load is distributed increases.

Table 13.6.3. Flexural Strength of Indiana Limestone Determined by Three Methods. (After Panek[7])

	Sample	Number of Specimens	Flexural Strength, psi	Standard Deviation, Percent
Central load	1	7	1340	5.4
	2	9	1590	6.6
Centrifugal load	1	4	1150	8.7
	2	6	1270	4.2
Third-point load	1	—	—	—
	2	6	1100	8.6

The Bureau of Mines centrifuge was built primarily to study rock bolting. In a comprehensive investigation Panek (see references, Chapter 20) performed nondestructive tests on a large number of models of bolted mine roof; from the results, the effect of a number of factors were evaluated including bolt spacing, bolt tension, bolt length, friction between bolted beds, and suspension of roof beds by bolting. These findings are considered in detail in Chapter 20.

REFERENCES

1. Hiramatsu, Y. and Y. Oka, "The Fracture of Rock Around Underground Openings," *Mem. Fac. Eng.*, *Kyoto Univ.*, Japan, **XXI**, Part 2, 1959.
2. Holland, C. T., "Design of Pillars for Overburden Support," Part 1, *Mining Congr. J.* (Mar. 1962) "Part 2," *Mining Congr. J.* (April 1962).
3. Obert, L., "Deformational Behavior of Model Pillars Made From Salt, Trona, and Potash Ore," *Proceedings 6th Symposium on Rock Mechanics*, Mo. School of Mines (October 1964).
4. Barron, K. and G. E. Larocque, "The Development of a Model for a Mine Structure," *Proc. Rock Mech. Symp.*, McGill Univ., Canada (1962).
5. Denkhaus, H. G., F. G. Hill, and A. J. Roux, "A Review of Recent Research into Rockbursts and Strata Movement in Deep-Level Mining in South Africa," *Trans. Inst. Mining Met.*, **68** (April 1959).
6. Merrill, R. M. and J. R. Peterson, "Deformation of a Borehole in Rock," *U.S. BurMines Rept. Invest.*, **RI 5581** (1961).
7. Panek, L. A., "Centrifugal Testing Apparatus for Mine Structure Stress Analysis," *U.S. BurMines Rept. Invest.*, **RI 4883** (1952).
8. Hoek, E., "The Application of Experimental Mechanics to the Study of Rock Stress Problems Encountered in Deep-Level Mining in South Africa," *Int. Cong. Exp. Mech.* (November 1961).

CHAPTER 14

IN SITU MEASUREMENTS

14.1 INTRODUCTION

In Part One relationships between stress, strain, and deformation were derived from a consideration of mathematical models. To obtain these solutions it was necessary to make simplifying assumptions regarding the model material, stress field, and model shape. Consequently the results so obtained are applicable to the solution of real problems only to the degree that the assumed conditions are satisfied in the prototype, and as previously pointed out, in rock mechanics these differences may be great. The study of photoelastic models (Chapter 12) extended the results obtained from theory to bodies of more complicated shape, and the investigation of rock models (Chapter 13) brought us a step closer to reality by employing rock or simulated rock as a model material. Also, some information about structural stability could be obtained from the study of rock models loaded to failure. In this respect it should be noted that the mechanical property tests for measuring the strength of rock can be considered as tests of a simple model.

Despite the advantages afforded by studying rock models it is evident that rock and other materials used in these models do not have the properties of in situ rock. Geological factors such as bedding, fractures, joints, faults, folding, and macroscopic and megascopic inhomogeneities are difficult, if not impossible, to reproduce, especially at reduced scale. Moreover, in model testing the direction and magnitude of the applied stresses are known, whereas in the prototype they usually are not known. Thus neither theory or model studies provide the degree of reality that is required for a good comprehension of the stresses and strains in, and the stability of structures in rock. Our understanding of these problems can be enhanced by making in situ measurements at full scale, or at a scale such that the rock in the test area is representative of the rock in the full scale structure.

The greater part of the investigations described in this chapter deal with direct or indirect methods of measuring stress, strain, or deformation. A point of particular concern is the accuracy with which this type of measurement should be made. If the problem under consideration is one dealing with the stability of a structure in rock, the required accuracy of stress or strain determinations should be consistent with the accuracy with which the strength of the in situ rock is known. Compressive and tensile strength determinations made in the laboratory on samples from uniform, massive rocks, such as thick-bedded sandstones and limestones, marbles, and some granites, show a minimum specimen-to-specimen deviation and it is reasonable to assume that these strength measurements are representative of the in situ rock. Hence, in situ stress or strain measurements in this class of rock should be made with a comparable accuracy. The other extreme would be the highly fractured or jointed rocks that contain, in addition, faults or other mechanical defects of geological origin. For rocks in this category the in situ strength may be so indefinite that an accurate knowledge of absolute stress would be of little value, although the direction of the principal stresses and the ratio of their magnitudes may be useful information (Section 15.4). Some problems do not require a knowledge of the absolute stress but only a determination of the change in stress. For example, it may be important to know at the earliest time whether the stress is increasing or decreasing in some structural component, or whether the mining in one area is affecting stress in an adjoining area. Usually, change in stress determinations should be made with a relatively high accuracy.

Although the literature contains many reports describing underground measurements of one kind or another, there are no generally approved or universally accepted methods. This may be partially due to the extreme variability of the problems that are encountered. In some instances this variability has made it necessary to modify or adapt procedures employed by other branches of engineering, and in others to develop altogether new instruments and methods of measurement. Because of the large number of experimental procedures that have been employed in making in situ measurements, the selection of examples presented in this chapter is limited to those that have introduced or demonstrated new or unique procedures, or produced results of special significance. The details of the equipment used in making these measurements are described in Chapter 9.

14.2 MEASUREMENT OF SURFACE STRAIN

The change in the tangential strain on rock surfaces has been measured with deformeters and strain gages of various types, and the corresponding change in the tangential stress calculated, using a laboratory determined value of the modulus of elasticity. These measured changes in strain have a value in evaluating corresponding changes produced by mining, such as widening a room or reducing the size of a pillar. Because these measurements are usually made over extended periods, mechanical type deformeters and strain gages are usually more satisfactory than bonded resistance or photoelastic strain gages that must be cemented to the rock surface. Also, gages with long base lengths are usually more satisfactory because of the averaging effect. In addition to instrument error this type of measurement is subject to possible errors that result from mounting a gage on a rock surface that may be partially detached from the surrounding rock.

Strain-relief procedures have also been employed in conjunction with surface strain measurements. Lieurance[1] determined the absolute magnitude and direction of surface strains in the walls of a tunnel from measurements on an equiangular strain rosette. The strain rosette was relieved by drilling a ring of overlapping holes around the area. A similar operation is shown in Fig. 14.2.1. The depth of the relief holes should be at least equal to the diameter of the relieved area.

If ϵ_1, ϵ_2, and ϵ_3 are the strains measured in an equiangular strain rosette (Fig. 14.2.2), the magnitude and direction of the principal strains on the surface, ϵ_p and ϵ_q, can be calculated from Eq. 2.5.6, namely

$$\epsilon = \tfrac{1}{3}(\epsilon_1 + \epsilon_2 + \epsilon_3) + \frac{\sqrt{2}}{3}\sqrt{(\epsilon_1 - \epsilon_2)^2 + (\epsilon_2 - \epsilon_3)^2 + (\epsilon_3 - \epsilon_1)^2}$$

$$\epsilon = \tfrac{1}{3}(\epsilon_1 + \epsilon_2 + \epsilon_3) - \frac{\sqrt{2}}{3}\sqrt{(\epsilon_1 - \epsilon_2)^2 + (\epsilon_2 - \epsilon_3)^2 + (\epsilon_3 - \epsilon_1)^2} \quad (14.2.1)$$

$$\theta_{pq} = \frac{1}{2}\tan^{-1}\frac{\sqrt{3}(\epsilon_2 - \epsilon_3)}{2\epsilon_1 - \epsilon_2 - \epsilon_3}$$

where θ_{pq} is the angle from ϵ_1 to ϵ_p or ϵ_q measured in the counter-clockwise direction, and if

$\epsilon_2 > \epsilon_3$	and	$\epsilon_2 + \epsilon_3 < 2\epsilon_1$,	then θ_p is between $0°$ and $45°$
$\epsilon_2 > \epsilon_3$	and	$\epsilon_2 + \epsilon_3 > 2\epsilon_1$,	then θ_p is between $45°$ and $90°$
$\epsilon_3 > \epsilon_2$	and	$\epsilon_2 + \epsilon_3 > 2\epsilon_1$,	then θ_p is between $90°$ and $135°$
$\epsilon_3 > \epsilon_2$	and	$\epsilon_2 + \epsilon_3 < 2\epsilon_1$,	then θ_p is between $135°$ and $180°$

If $\epsilon_1 = \epsilon_2 = \epsilon_3$, it follows that $\epsilon_p = \epsilon_q$, and θ_{pq} is indeterminant. On the other hand, if one of the measured strains is either much larger or smaller than the other two, θ_{pq} can be determined with fair accuracy even though the magnitudes of ϵ_1, ϵ_2, and ϵ_3 may be in considerable error. In some

Fig. 14.2.1. Stress-relieved strain rosette (from *U.S. Bur. Mines Bull. 587*).

instances information about the direction of the principal strains is as important as information about their magnitudes.

Olsen[2] strain-relieved resistance-strain gage rosettes mounted on rock surfaces by overcoring the gages with a large diameter coring bit. Sipperelle and Teichmann[3] employed a similar technique except that a wedge of rock containing the resistance-strain gage rosette was relieved by sawing it from the surface with a diamond saw. As previously pointed out, the resistance-strain gage rosette must be completely water-proofed during the strain-relief operation, a procedure that is usually difficult to effect. Also, the base length of the resistance-strain gage is small and hence the averaging

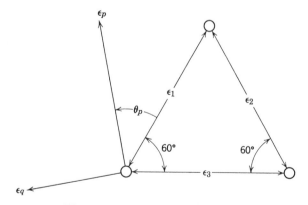

Fig. 14.2.2. Equiangular strain rosette.

effect is usually poor, especially for surface or near-surface rock. In general, attempts to interpret strain rosette measurements made on underground rock surfaces have been inconclusive owing either to lack of sufficient data, or to inconsistencies of an undetermined nature.

14.3 DETERMINATION OF ABSOLUTE STRESS

Six procedures for determining the absolute stress in rock are described, which are referred to as the borehole deformation, flatjack, propagation velocity, hydraulic fracturing, core discing, and static equilibrium methods. The borehole deformation, flatjack, and the propagation methods have been employed by several investigators and some of the more significant results obtained by these procedures are presented in Section 14.5, 14.6, and 14.8. The static equilibrium method proposed by Morgan and Panek[4] is described because the underlying concept is fundamental to an understanding of underground static equilibrium problems.

Borehole Deformation Method

The diametral deformation, that is, the change in diameter of a circular hole in a thin infinite plate, is dependent on the magnitude and direction of the applied stresses in the plane of the plate (Section 4.7). If U is the diametral deformation of a hole of diameter d, and P' and Q' are the secondary principal stresses* normal to the axis of the hole, Eq. 4.7.21 becomes

$$U = \frac{d}{E}[(P' + Q') + 2(P' - Q')\cos 2\theta] \qquad (14.3.1)$$

This equation was derived assuming plane stress, with the stress equal to zero in the direction normal to the plane of the plate. If the surfaces of the plate are constrained so that the strain in the direction normal to the surfaces is zero, the plane-strain equation (Eq. 4.7.27) becomes

$$U = \frac{d(1 - v^2)}{E}[(P' + Q') + 2(P' - Q')\cos 2\theta] \qquad (14.3.2)$$

For a borehole in a three-dimensional, isotropic, linear-elastic medium, the strain in the direction of the borehole generally will not be zero. Assuming plane strain, but with the strain in the axial direction not equal to zero, the contribution to the change in the diameter of the hole will be

---

* Secondary principal stresses are defined in Section 1.7.

$-v\epsilon_z d$, where ϵ_z is the strain in the axial direction of the hole.* By super-position, the total change in the diameter of a borehole will be

$$U = \frac{d(1 - v^2)}{E} \left[(P' + Q') + 2(P' - Q') \cos 2\theta\right] - v\epsilon_z d \quad (14.3.3)$$

Thus if the axial strain and the diametral deformation in three specified directions are measured at a point in a borehole, the direction and magnitude of P', Q', and θ can be determined (assuming that E and v are known). As yet no gage has been developed that will make these four measurements simultaneously. However, in certain instances plane stress can be assumed as, for example, in the center of a small pillar, or in a borehole normal to, and at the surface of any opening; or near the surface of an opening if the rock is fractured so that it cannot sustain an elastic strain normal to the

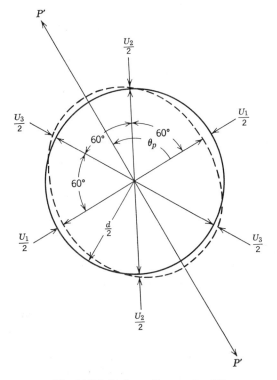

Fig. 14.3.1. Deformation rosette, 60°.

* In a three-dimensional medium the change in the diameter of a borehole due to an axial strain is identical to that in the hole in a thick-wall cylinder, due to a strain in the axial direction (see Section 4.5).

surface. For these cases the axial stress is zero, and P' and Q' can be determined from a set of three diametral measurements U_1, U_2, U_3, and the elastic constants of the material. If the three deformation measurements are made at 60° intervals (Fig. 14.3.1). (the equiangular deformation rosette), the magnitudes and directions of the applied stresses can be determined from*

$$P' = \frac{E}{6d}\left\{(U_1 + U_2 + U_3) + \frac{\sqrt{2}}{2}\right.$$

$$\left. \times \; [(U_1 - U_2)^2 + (U_2 - U_3)^2 + (U_3 - U_1)^2]^{\frac{1}{2}}\right\}$$

$$Q' = \frac{E}{6d}\left\{(U_1 + U_2 + U_3) - \frac{\sqrt{2}}{2}\right.$$

$$\left. \times \; [(U_1 - U_2)^2 + (U_2 - U_3)^2 + (U_3 - U_1)^2]^{\frac{1}{2}}\right\} \quad (14.3.4)$$

$$\theta_p = \frac{1}{2}\tan^{-1}\frac{3(U_2 - U_3)}{2U_1 - U_2 - U_3}$$

where d is the diameter of the borehole, U is positive for an increasing diameter, the angle θ_p is measured from U_1 to P' in the counterclockwise direction, and if

$U_2 > U_3$	and	$U_2 + U_3 < 2U_1$	then θ_p is between 0° and 45°
$U_2 > U_3$	and	$U_2 + U_3 > 2U_1$	then θ_p is between 45° and 90°
$U_2 < U_3$	and	$U_2 + U_3 > 2U_1$	then θ_p is between 90° and 135°
$U_2 < U_3$	and	$U_2 + U_3 < 2U_1$	then θ_p is between 135° and 180°

There are two special cases of interest, namely, for a uniaxial stress field, $Q' = 0$, and $\theta_p = 0°$, Eq. 14.3.4 becomes

$$P' = \frac{EU}{3d} \quad (14.3.5)$$

* To derive Eqs. 14.3.4, in Eq. 14.3.1, let U_1, U_2, and U_3 be the diametral deformations for $\theta = 0°$, $\theta = 60°$, and $\theta = 120°$, and solve the three resulting equations simultaneously. In this solution the direction of P' is given by $\tan 2\theta = -\dfrac{\sqrt{3}(U_2 - U_3)}{2U_1 - U_2 - U_3}$, where in this case θ is the angle measured in the counterclockwise direction from P' to U_1. However, to make U_1 rather than P' the reference for the angular measurement so as to conform to the strain rosette notation (Eq. 14.2.1), and recalling that $\tan \theta = -\tan (180 - \theta)$, the sign on the third of Eqs. 14.3.4 was reversed. Equation 4.7.27 which was derived for plane strain with $\epsilon = 0$, differs from Eq. 4.7.21 only by a factor $1 - v^2$. For $v = 0.25$, $1 - v^2 = 0.94$, hence the difference in the magnitude of P' and Q' calculated on the basis of plane stress or plane strain with $\epsilon = 0$ would be only 6%. The direction of the secondary principal stresses, given by the third of Eq. 14.3.4 is the same for plane stress or plane strain. A set of equations similar to Eqs. 14.3.4 can be derived for a 45° deformation rosette.

where U is the borehole deformation in the direction of P'; for a hydrostatic stress field, that is, $P' = Q' = -p$

$$-p = \frac{U_p E}{2d} \tag{14.3.6}$$

where U_p is the borehole deformation in any direction.

Equations 14.3.4 modified by replacing E with $E/(1 - \nu^2)$ can be used to calculate P' and Q' for plane strain conditions with $\epsilon_z \neq 0$, provided an estimate of the stress σ_z parallel to the axis of the borehole is available. First, the modified Eqs. 14.3.4 are used to calculate estimates of P' and Q'. Second, these values are used as estimates of σ_x and σ_y and the value of ϵ_z is calculated from Hooke's law equation

$$\epsilon_z = \frac{1}{E} [\sigma_z - \nu(\sigma_x + \sigma_y)]$$

Third, the quantity $\nu \epsilon_z d$ is added to the measured values of U_1, U_2, and U_3. Fourth, these new values of the deformations are used in the modified Eqs. 14.3.4 to calculate the more exact values of P' and Q'. A second iteration can be made if desired.

Hast[5] developed a borehole deformation gage (Section 9.4) and a procedure for producing stress-relief at a distance within a body of rock. The essential features of this procedure are as follows (see Fig. 14.3.2):

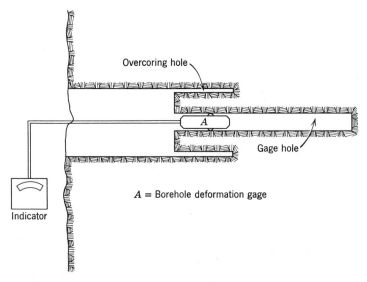

Fig. 14.3.2. Determination of absolute stress by the overcoring procedure.

first, a small diameter hole (approximately 1 in. D) is drilled into an underground rock surface. The borehole deformation gage is inserted in the hole, radially oriented, and an initial reading made. Second, the section of core containing the gage is concentrically overcored with a larger diameter bit, thereby stress relieving the core containing the gage. A second gage reading is taken. The difference between the first and second gage readings is the borehole deformation.* Third, the gage is moved to a point farther in the gage hole, oriented at 60° from the original position and the overcoring process repeated. This procedure is cyclically continued for the 60° rosette positions until the desired range of depths has been investigated. Hast calculated the magnitude and direction of the maximum and minimum stresses in the plane normal to the axis of the gage hole from the borehole deformation data and the modulus of elasticity of the rock, using equations similar to Eqs. 14.3.4. Stress determinations at depths up to 70 ft from the rock face were obtained by this procedure. Details of the drilling and accessory equipment are described in a report by Obert.[6]

Flatjack Method

The flatjack method for measuring the stress in rock was first described by Habib and Marchand.[7] This procedure was effected by first mounting two vibrating-wire strain gages on a rock surface, orienting them to measure in the line of the intended stress determination (Fig. 14.3.3) and then making an initial strain reading. A slot of sufficient area to partially relieve the strain in the proximity of the strain gages is then cut between the strain gages, and a flatjack (Section 9.7) grouted into the slot. After the grout has cured, the flatjack is pressurized to a value such that the strain gages indicate their initial value. For this condition the flatjack pressure is considered equal to the stress normal to the plane of the flatjack that existed in the rock before the slot was cut.

Panek and Stock[8] modified this procedure by replacing the surface-mounted strain gages with copper-foil jacketed resistance-strain gages (Valore type) grouted in slots cut above and below the intended flatjack slot (Fig. 14.3.4). These gages are placed directly over and under the center of the flatjack and oriented so that they will measure the strain in the direction normal to the flatjack. Panek found that this modification improved both the sensitivity and accuracy of the stress determinations. In a later modification the resistance-strain gages were replaced by small

* According to the first two Eqs. 14.3.4 when a concentric core is stress relieved, if the gage hole increases in diameter (U_1, U_2, U_3 positive), P' and Q' are positive, or tensile. However, the relieved stresses are equal but opposite in sign, or compressive. In presenting stress-relief results it is the custom to reverse the sign so that P' becomes the maximum, and Q' the minimum *compressive* stress.

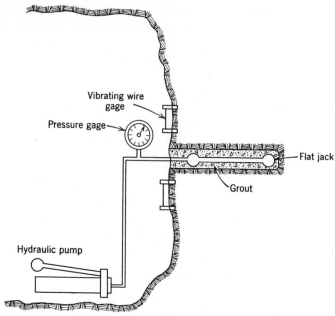

Fig. 14.3.3. Determination of absolute stress by the flat jack procedure.

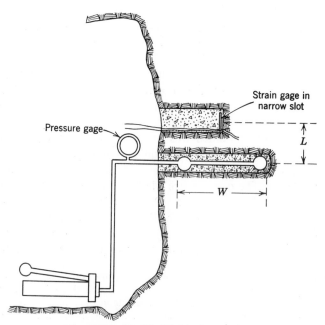

Fig. 14.3.4. Modified flat jack procedure.

hydraulic cells grouted in similarly placed drill holes. It was found that the stability of the hydraulic cells is excellent, that is, their zero drift is virtually nil.

The flatjack method does not require any knowledge of the elastic properties of the rock, and hence it is considered to be a true stress measuring method. However, for the method to give a valid result it is necessary that the gage reading produced by cutting the flatjack slot should be annulled by applying to the surface of the slot (by means of the flatjack) a pressure equal to the preexisting stress. Panek and Stock showed experimentally that this condition is satisfied for a square flatjack of width W if the perpendicular distance from the center of the flatjack to the hydraulic cell L is such that $L/W < \frac{1}{2}$.

Because of the difficulty in cutting deep flatjack slots the method is restricted to near-surface measurements. On the other hand, because of the averaging effect due to the comparatively large area of the flatjack it is less sensitive to local variations in the rock stress. The method is also better adapted to measurement in inelastic rock.

Propagation Velocity Method

The propagation velocity method of determining the in situ stress in rock depends on the passage of stress waves through rock. The longitudinal propagation velocity of a spherical wave V_p in a rock having a unit weight γ, a modulus of elasticity E, and a Poisson's ratio ν, is given by (Section 11.12)

$$V_p = \left[\frac{Eg(1 - \nu)}{\gamma(1 + \nu)(1 - 2\nu)} \right]^{\frac{1}{2}} \tag{14.3.7}$$

where g is the acceleration of gravity. In a linear-elastic material E and ν are independent of the state of stress in the rock, hence V_p is a constant. However, no real material is perfectly linear elastic, and for some materials, including a number of rock types, this departure from linear elasticity is large. For these materials, E and ν, and hence V_p depend on the stress, and this fact is the basis for this seismic method.

To determine in situ stress by this method the propagation velocity is measured between two points in rock. The rock in the measurement area is sampled and a velocity versus stress relationship established by laboratory measurement. From the in situ propagation velocity and the velocity-stress relationship the in situ stress can be determined. It should be noted that this is a nondestructive procedure for determining absolute stress that can be effected without drilling, mining, or otherwise disturbing the rock. Although this procedure is simple in principle, there are a number of difficulties. First, the velocity-stress relationship for some rocks is relatively constant, especially at medium to high stress levels. Hence, for these rocks this method is not applicable. Second, for rocks that do show a significant change of velocity with stress there may be a large specimen-to-specimen deviation in the velocity at a given stress value. If these deviations are comparable with the change of velocity with stress, the accuracy of the method is not satisfactory. Third, in fractured or jointed rock and in some bedded or otherwise heterogeneous rock the reproducibility of the in situ velocity measurements may be poor. As this measurement usually has to be made accurately (because the change in velocity with stress effect may not be too large) this factor may limit the applicability of the method.

Some of the factors that affect the accuracy of absolute stress determinations do not affect in situ change-in-stress determinations to the same degree. For example, even though there may be a considerable specimen-to-specimen variation in the laboratory measured velocity at given stress levels, the *average* change of velocity with stress for a group of specimens may be approximately proportional to the change in velocity with stress in the in situ rock. If this is the case, in situ change of stress can be determined from measured changes in the in situ velocity and the average laboratory determined velocity-stress relationship.

The stress in limestone pillars in a lead mine was determined[9] by measuring the traveltime Δt between the detonation of a blasting cap and the arrival of the detonation generated pulse at a geophone. The velocity was measured parallel to the axis of the pillar. The propagation velocity was calculated from $V_c = L/\Delta t$, where L is the cap-to-geophone distance. The velocity-stress relationship was determined by measuring the fundamental frequency of axially loaded specimens cut from rock samples taken from the pillar and the longitudinal velocity was calculated from the measured frequency.* To check the accuracy of this procedure a mine pillar was cut free from the roof, thereby producing a stress-relief. Thus the zero-stress propagation velocity in the pillar could be compared with the corresponding zero-stress velocity in the laboratory specimens.

Tincelin[10] employed this procedure to study the stress distribution in iron ore pillars. The investigation disclosed that the vertical component of stress near the pillar surface decreases, rather than increases, as predicted by theory. Uhlmann[11] made a similar study in salt and potash pillars. A transition from elastic to plastic propagation was noted as well as indications of incipient failure.

Hydraulic Fracturing

Hydraulic fracturing is a procedure employed by the petroleum industry to enhance well production. A section of a well is sealed off with packers

---

* More refined methods have been developed for measuring the velocity in rock specimens under load as, for example, the pulse technique described in Section 11.12.

(Fig. 14.3.5a) and then the fluid pressure in this section is increased until the surrounding rock is fractured. From a consideration of the stresses produced by the fluid pressure on the walls of the packed-off section, Kehle[12] has shown that inferences can be made regarding the stress field

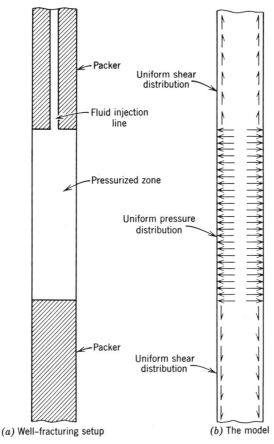

(a) Well-fracturing setup   *(b)* The model

Fig. 14.3.5. Schematic diagram of the pressurized section and the accompanying stresses. (After Kehle.[12])

in the proximity of the well. Fairhurst[13] suggested that this technique could also be used in exploration holes to obtain some knowledge about the state of stress at proposed sites for underground mining, or in the rock surrounding existing openings.

Kehle's treatment of the problem is based on the applied loads indicated in Fig. 14.3.5b. The resulting stresses on the wall of the well are given in

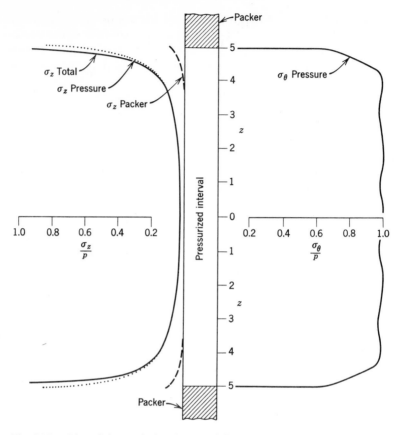

Fig. 14.3.6. Plot of the vertical and tangential stress induced by the well-fracturing operation. (After Kehle.[12])

Fig. 14.3.6, where σ_z and σ_θ are the axial and tangential components of stress. For an applied pressure p, σ_z, and σ_θ are tensile (positive) and the radial stress, $\sigma_r = -p$ is compressive. Because well fracture is initiated in tension, the effects of σ_r can be neglected. Note that in Fig. 14.3.6 at the ends of the packed-off section

$$\sigma_\theta = 0$$
$$\sigma_z = 0.94p$$

(14.3.8)

and along the central part of this section

$$\sigma_\theta = p$$
$$\sigma_z = 0$$

(14.3.9)

The preexisting state of stress on the wall of the well due to the combined gravitational and tectonic stress field can be assumed to be equivalent to that on the boundary of a hole in an infinite plate subjected to biaxial (plane) stress, provided that the principal stresses are parallel and normal to the axis of the well. For a vertical well let R be the vertical, and P and Q the horizontal stresses. From Eq. 4.7

$$\sigma_\theta' = P + Q + 2(P - Q)\cos 2\theta \qquad (14.3.10)$$

where σ_θ' is the tangential stress due to the stress field. The maximum value of the tangential stress $\sigma_{\theta(max)}$ is

$$\sigma'_{\theta(max)} = 3P - Q \qquad (14.3.11)$$

The axial stress due to the stress-field is

$$\sigma_z' = R \qquad (14.3.12)$$

where R, the principal vertical stress, can be approximated from the overburden load, that is, $R = -\gamma h$. By superposition the maximum total stress on the wall of the well is given by combining Eqs. 14.3.8 with Eqs. 14.3.11 and Eq. 14.3.9 with 14.3.12. Thus at the ends of the pressurized section

$$\sigma_{\theta(total)} = 3P - Q$$
$$\sigma_{z(total)} = R + 0.94p \qquad (14.3.13)$$

and along the central part of this section

$$\sigma_{\theta(total)} = 3P - Q + p$$
$$\sigma_{z(total)} = R \qquad (14.3.14)$$

Kehle assumed the validity of Mohr's tensile failure criterion, namely that fracture will occur if the tensile stress is greater than the tensile strength. If the effects of pore pressure are neglected, the well will fracture when p attains a value p_c such that the (tensile) stress given by any of Eqs. 14.3.13 or 14.3.14 is equal to or greater than the tensile strength of the rock, T_0. There are two possibilities for fracture: (1) If $\sigma_{z(total)} = R + 0.94p_c < T_0$, fracture will initiate in the central region when

$$\sigma_{\theta(total)} = 3P - Q + p_c \geq T_0 \qquad (14.3.15)$$

Because the least work is required to propagate a fracture in the direction normal to the maximum (least compressive) stress, the fracture will propagate normal to $\sigma_{\theta(total)}$, that is, in the vertical plane. (2) If $\sigma_{\theta(total)} = 3P - Q + p_c < T_0$, fracture will initiate at an end when

$$\sigma_{z(total)} = R + 0.94p_c \geq T_0 \qquad (14.3.16)$$

and the fracture will propagate in a horizontal plane.

In either case when the fracture has propagated beyond the stress concentration zone around the well, the work required to continue the fracture process is small because of the high stress concentration at the propagating edge of the crack. Hence the pressure $p_s{}^*$ necessary to sustain propagation in the vertical plane is

$$-p_s = P \qquad (14.3.17)$$

and in the horizontal plane is

$$-p_s = R \qquad (14.3.18)$$

The effect of fluid pressure p_0 in the pores of the rock increases the applied principal stresses, and decreases the effective pressure in the packed off region by an amount p_0 (see Section 10.7).† Thus

P should be replaced by $P + p_0$

Q should be replaced by $Q + p_0$

R should be replaced by $R + p_0$

p should be replaced by $p - p_0$

and the conditions for fracture become

(1) if

$$\sigma_{z(\text{total})} = (R + p_0) + 0.94(p_c - p_0) < T_0, \qquad (14.3.19)$$

then fracture will initiate in the central region when

$$\sigma_{\theta(\text{total})} = 3(P + p_0) - (Q + p_0) + (p_c - p_0) \geq T_0 \qquad (14.3.20)$$

and the fracture will propagate in the vertical plane;

(2) if

$$\sigma_{\theta(\text{total})} = 3(P + p_0) - (Q + p_0) + (p_c - p_0) < T_0 \qquad (14.3.21)$$

then fracture will initiate at an end when

$$\sigma_{z(\text{total})} = (R + p_0) + 0.94(p_c - p_0) \geq T_0 \qquad (14.3.22)$$

and the fracture will propagate in a horizontal plane. The pressure necessary to sustain propagation in the vertical plane remains

$$-p_s = P \qquad (14.3.23)$$

and in the horizontal plane

$$-p_s = R \qquad (14.3.24)$$

---

* In well fracturing p_s is referred to as the shut-in pressure.

† Kehle noted that the effective pore pressure depends on the boundary porosity of the rock. This effect may lower p_0 somewhat, but probably not lower than $0.85p_0$. Within the limits of accuracy of this method the effective porosity can be assumed to be 1.0, and the effective pore pressure $1.0p_0$.

Information regarding the principal stresses can be deduced from well fracturing data $(p_c, p_0, p_s, T_0, \gamma, h)$ that satisfy the following conditions:

(1) $0.94(p_c - p_0) < |R + p_0| + T_0$ and $p_s < |R|$

For this condition fracture should initiate in the central region and propagate in a vertical plane. The principal stresses are given by

$$R = -\gamma h$$
$$P = -p_s \quad \text{(from Eq. 14.3.23)}$$
$$Q = 3P + p_0 + p_c - T_0 \quad \text{(from Eq. 14.3.20)}$$

or $$Q = -3p_s + p_0 + p_c - T_0$$

(2) $0.94(p_s - p_0) > |R + p_0| + T_0$ and $p_s > |R|$

The fracture should initiate in the horizontal plane, and the only statement that can be made about the stresses is that

$$3P - Q < T_0 - p_c - p_0$$

(3) $0.94(p_s - p_0) > |R + p_0| + T_0$ and $p_s < R$

The first inequality suggests that a horizontal fracture will initiate, whereas the second inequality implies that the final fracture should be vertical. Assuming this fracture mechanism is correct it follows from the first inequality that

$$\sigma_{z(\text{total})} > \sigma_{\theta(\text{total})}$$

From Eqs. 14.3.19, 14.3.22, and 14.3.23

$$Q > -3p_s - R + 0.94p_0 + 0.06p_c$$

and as $P = -p_s > Q$

$$-p_s > Q > -3p_s - R + 0.94p_0 + 0.06p_c$$

an inequality that bounds Q. If

$$2p_s < -R + 0.94p_0 + 0.06p_c$$

the last inequality cannot hold, as it would imply that $P < Q$. In this case a vertical fracture should initiate and the analysis of the principal stresses should be made according to condition (1).

Thus hydraulic fracturing may provide a means of obtaining information regarding the state of stress in deep exploration holes, and without any knowledge of the elastic constants of the rock. On the other hand, in this analysis it is assumed that the exploration hole is in the direction of one of the principal stresses (R). In an oil field or in other flat-lying sedimentary formations this may be a reasonable assumption but in metamorphic and igneous formations, and especially in the proximity of mineral deposits where the geology is complex, this assumption may be completely invalid.

Core Discing

Sometimes information regarding the subsurface state of stress can be inferred from an examination of exploratory drill cores. Obert and Stephenson[14] reported that core discing, illustrated in Fig. 14.3.7, will occur in a triaxially loaded body of rock that is being cored when

$$S_r > k_1 + k_2 S_z \qquad (14.3.25)$$

Where S_r and S_z are the radial and axial stresses applied to the body of rock, Fig. 14.3.8, k_1 is the value of the radial stress required to produce

Fig. 14.3.7. Discing in drill core.

discing at zero axial stress, and k_2 is approximately 0.7*. Also, it was found that

$$k_1 = -3400 - 2.0S_0 \qquad (14.3.26)$$

where S_0 is the triaxially determined shear strength of the rock. Thus, if discing is observed in a core from a vertical exploratory hole at a depth h, and if it is assumed that the vertical component of stress due to gravity is $S_v = -\gamma h$, then from the geometry in Fig. 14.3.8 the vertical and horizontal stress relate to the applied stresses in Eq. 14.3.25 as follows:

$$S_z = S_v = -\gamma h$$
$$S_r = S_h \qquad (14.3.27)$$

Substituting Eqs. 14.3.27 in Eq. 14.3.25

$$S_h \geq k_1 - k_2 \gamma h \qquad (14.3.28)$$

and from Eq. 14.3.26

$$S_h \geq -3400 - 2.0S_0 - 0.7\gamma h \qquad (14.3.29)$$

---

* The exact values of k_1 and k_2 for a given rock type can be determined from the laboratory test described by Obert and Stephenson.

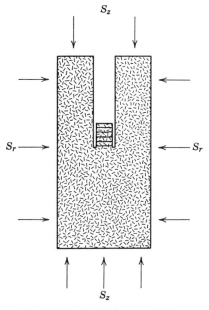

Fig. 14.3.8. Model for core discing.

For example, if a disced core is taken from a depth of 4000 ft in, say, Indiana limestone ($S_0 = 1200$ psi), then

$$S_h \geq -3400 - (2.0 \times 1200) - (0.7 \times 0.09 \times 4000 \times 12)$$

$$S_h = -8000 \text{ psi}$$

Static Equilibrium Method

Morgan and Panek[4] proposed a method for determining the absolute stress in rock, based on a requirement for static equilibrium, namely, that the total load on a sufficiently large underground area must remain unchanged even if an opening or openings are mined in the area. Consider an opening (tunnel or drift) of width w driven through a body of rock at right angles to the xy plane, (Fig. 14.3.9a). Assume that before the opening is driven, the vertical stress σ_y has the preexisting or initial value σ_i. To analyze this two dimensional problem consider a slice parallel to the xy plane, which has the width d and the thickness unity. Before the opening is driven the total force in the y-direction acting on the horizontal plane through 0ψ is equal to

$$\int_0^d \sigma_i \, dx = \sigma_i d \qquad (14.3.30)$$

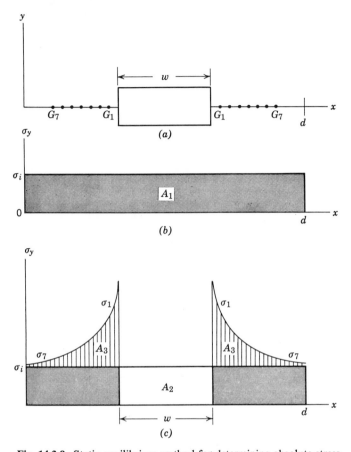

Fig. 14.3.9. Static equilibrium method for determining absolute stress.

which is represented by the area A_1 (Fig. 14.3.9*b*). To satisfy the requirement of static equilibrium for the forces acting in the *y*-direction it is
necessary that the total force remain equal to $\sigma_i d$ after the opening is
created. Driving an opening of width *w* through the rock removes a force
$\sigma_i w$ (area A_2, Fig. 14.3.9*c*). To satisfy the condition of static equilibrium
along 0*d* a force equal to $\sigma_i w$ must now be added to the rock at the sides
of the opening, that is,

$$\sigma_i w = A_2 = 2A_3 \tag{14.3.31}$$

If the rock is reasonably elastic this force will be distributed more or less
as indicated by the curves specified by the ordinates $\sigma_1 \cdots \sigma_7$ (Fig. 14.3.9*c*).

The preexisting stress σ_i can be determined as follows. A number of
stress gages $G_1 \cdots G_7$ are installed at appropriate distances from the

boundary of a proposed opening (Fig. 14.3.9a) and oriented to determine the *change* in the vertical stress that occurs when the opening is made. By plotting these stress measurements as ordinates at the proper spacing, the stress distribution curve at the sides of the opening can be established. By graphical integration the area A_3 under these curves can be determined, and as w is known, σ_i can be calculated from Eq. 14.3.31. Also, the absolute stress distribution at the sides of the opening can be determined from $\sigma_i + \sigma_1$, $\sigma_i + \sigma_2$, . . . , etc.

It should be noted that in this method no assumption is made or required as to the actual stress distribution—it may be like that shown in Fig. 14.3.9 or any of the measured curves shown in Section 14.5. It is not necessary that the rock be elastic, homogeneous, or isotropic; the opening may have any shape—only the width of the opening and the change in stress need to be known. If it is assumed that the stress distribution is symmetrical with respect to the opening, measurements need be made only at one side of the opening.

Because only change of stress measurements are required, rigid or hydraulic type stress gages can be employed. As noted in Sections 9.7 and 14.4, these gages are better adapted for use in inelastic rock than borehole deformation gages.

14.4 DETERMINATION OF THE CHANGE IN STRESS OR STRAIN

Determination of the change in stress or strain is a measurement usually made over an extended period, such as the time required for the superincumbent load to transfer to a pillar as the surrounding area is excavated. This determination can be made either by taking the difference between absolute stress or strain measurements made at the beginning and end of the specified period, or by permanently installing a stress or strain gage, making an initial reading, and then measuring the change that occurs with respect to this reference value. Absolute measurement methods have an advantage over change-in-stress methods in that they are not as subject to time-dependent errors caused by instrument drift and creep in the rock, because the initial and final measurements require only a relatively short time to effect. However, as absolute measurements usually require a stress or strain-relief, the initial and final measurements cannot be made at the same point, a factor that may cause some error.

The use of a permanently installed gage obviates this latter source of error but it is more subject to time-dependent effects. In this respect the choice of gage type is a matter that should be given careful consideration. Temperature corrected or compensated mechanical type deformeters and strain gages generally have the best long-time stability; hydraulic cells, if temperature corrected, are usually satisfactory. Resistance and photo-elastic strain gages are more subject to drift caused by temperature and moisture and the creep in bonding cements used to attach the gage to the rock. Because rigid inclusion and hydraulic-type gages resist the deformation of the borehole in which they are placed, ·they are better suited for long-time measurements in fractured rock or rock that creeps, than high compliance borehole deformation-type gages.

14.5 DETERMINATION OF THE STRESS DISTRIBUTION AROUND UNDERGROUND OPENINGS

Of prime importance is a knowledge of the stress distribution in the walls, pillars, and structural parts of a mine or system of underground openings because the maximum stress on or near the surface of these structural members may be much greater than the magnitude of the stress field. Presumably it is these stress concentrations that are responsible for local, and sometimes general rock failures. In Section 4.7 the stress distribution around a circular opening in a biaxial stress field is derived theoretically. In Chapter 5 the stress distribution around openings of other shapes is presented. In deriving these distributions it was assumed that the opening is in an ideal material, that is, in an isotropic, homogeneous, linear-elastic material. The stress distribution around single and multiple openings has also been determined photoelastically (Chapter 12). In these determinations the photoelastic model material behaves almost as an ideal substance. Also, uniaxially and biaxially loaded models made of relatively uniform elastic rock have been used to study the stress distribution around openings of comparatively simple geometrical shape (Chapter 13).

Stress concentrations occur on or near the surface of all underground openings as a result of the transfer of load from the rock that occupied the opening before mining to the rock surrounding the opening after mining. However, because rock is not an ideal material and boundaries of the openings are not simple geometrical shapes, and because the in situ stress-field is generally not known, the resulting stress distribution can be determined only by measurement, that is, by determining the absolute stress at a number of points in the rock surrounding the opening. Some inferences regarding the stress concentration on the surface of underground openings have been made by taking the ratio of the measured surface or near-surface stress to the calculated gravity stress γh, where γ is the unit weight of the rock, and h is the depth of the opening. This procedure is subject to the errors inherent to surface or near-surface stress measurement

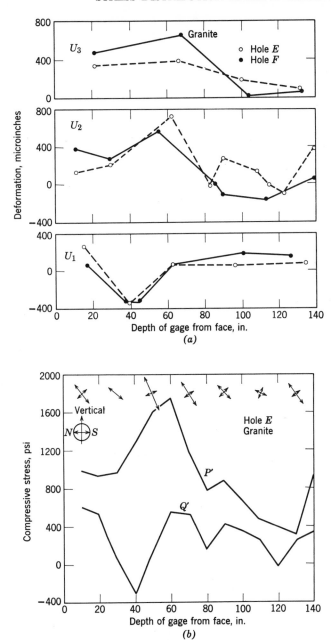

Fig. 14.5.1. (*a*) Borehole deformation measurements. (*b*) Corresponding calculated stress—arrow insets indicate direction and magnitude of P' and Q'.

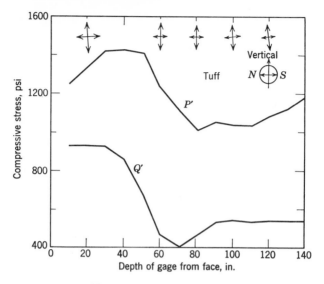

Fig. 14.5.2. Stress distribution in tuff.

and is further complicated if tectonic stresses are present. The borehole deformation-overcoring procedure (Section 14.3) provides a means of determining the stress within a body of rock, and hence for evaluating both the stress distribution near the surface of openings and the stress field. Using this procedure Hast[5] measured the tangential stress distribution in the wall rock of a number of mines in Sweden. His results show that the maximum stress usually does not occur on the surface of the opening as

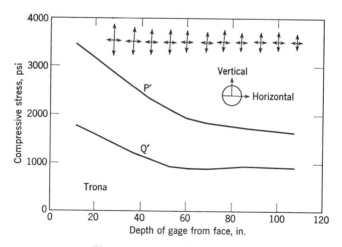

Fig. 14.5.3. Stress distribution in trona.

anticipated by theory, but at a distance from 18 in. to 10 ft or more from the surface, the distance depending on the degree of fracture (either natural or from blasting) present in the rock. Also, most measured stress concentrations were less than two, although concentrations as high as five were recorded.

The borehole deformation-overcoring procedure also was used[15] to measure the stress distribution and stress field at a number of sites and in various rock types. Figure 14.5.1a shows the borehole deformation data (U_1, U_2, U_3 measurement) from pairs of parallel holes 24 in. apart in a competent but blocky and jointed granite. These results show the very erratic but reproducible deformation results for this jointed rock. Figure 14.5.1b shows the corresponding magnitude and direction of the secondary principal stresses in one of these holes. Note that the maximum compressive stress distribution reaches a maximum about 5 ft from the surface but decreases (relaxes) between this point and the surface. Also, note that there is some evidence of tectonic force as indicated by the persistent deviation in the direction of the maximum compressive stress from the vertical. Figure 14.5.2 shows a similar near-surface stress distribution in a comparatively incompent tuff, but in this case the direction of the larger compressive stress was vertical which is indicative of a gravity stressfield. The stress distribution in unfractured relatively elastic trona (an evaporite mineral) is given in Fig. 14.5.3 which, in distinction to granite and tuff, exhibits a distribution more like that of an elastic material. The other extreme occurs in salt (Fig. 14.5.4) which is relatively plastic and in which

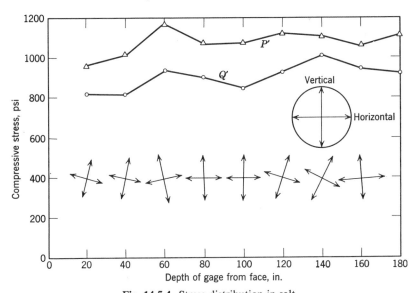

Fig. 14.5.4. Stress distribution in salt.

there is virtually no near-surface stress concentration. Note that in salt the secondary principal stresses are nearly equal, which is characteristic of the more plastic rocks. The variation in the direction of P' and Q' may result from the fact that as P' becomes equal to Q' the direction of these stresses becomes indeterminate.

Stress determinations in trona by Morgan et al.[16] disclosed that this

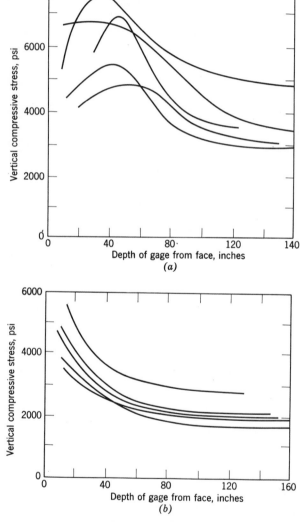

Fig. 14.5.5. Stress distributions in trona.

rock exhibits a transition from an elastic to a plastic behavior at a compressive stress level of about 5000 psi. Figure 14.5.5a presents the vertical tangential stress distributions from five sites. These distributions progressively decrease with the distance from the surface and are similar to the stress distributions for an elastic material. The maximum vertical compressive stress indicated by four of the five curves is less than 5000 psi. In Fig. 14.5.5b stress distribution curves are given for a second group of five sites at four of which the maximum vertical compressive stress exceeded 5000 psi. Note that in the latter group the maximum stress occurs at a point several feet from the surface, and that the stress decreases between that point and the surface. This decrease is presumed to result from the tendency of this mineral to creep at higher stress levels.

14.6 DETERMINATION OF THE STRESS FIELD

The stress field, sometimes referred to as the pre-existing or medium stress, is the state of stress that exists in the rock before mining. The development of procedures for determining this quantity is one of the more important achievements of experimental rock mechanics. In Chapter 4, and in the preceding section it was shown that the stress field in a body is altered by the presence of an opening within the body, and a stress concentration appears in the rock surrounding the opening. The distance from the opening that the stress field is disturbed depends on the shape and size of the opening and on the elastic or inelastic properties of the rock. For example, the stress field is not affected appreciably ($< 10\%$) at three radii from the boundary of a circular tunnel in an elastic rock (Section 4.7) and the corresponding distance from other opening shapes in an elastic rock can be estimated from the distribution curves given in Chapter 5. Usually the influence of the opening extends to a distance proportional to the smaller or smallest cross-sectional dimension of the opening. In an inelastic rock the perturbation in a stress field for the same opening shape may extend several times this distance, owing to the redistribution of stress caused by creep or plastic flow in rock (Section 6.5).

Stress field determinations must be made at a point in the rock outside of the stress concentration zone* by a method for measuring absolute stress, that is, by a procedure involving stress-relief.†

---

* With the exception of stress field determinations made by the static equilibrium method, which are effected by measuring the stress distribution from the opening boundary to the approximate limit of the concentration zone (Section 14.3).

† In theory, the propagation velocity method (Section 14.3) should give an absolute stress determination without a stress-relief. However, the propagation velocity versus stress characteristics of in situ rock cannot be determined with a sufficient accuracy to make this method practical.

Because the borehole deformation-overcoring method provides a means of measuring the stress outside of the stress concentration zone this procedure can be used to determine the stress field, as illustrated in Fig. 14.5.3 in which the secondary stresses P' and Q' tend to level off with the distance from the surface of the opening and approach the stress field values, $P' = 1700$ psi, $Q' = 950$ psi. As P' is virtually vertical its magnitude should agree with the calculated gravity load γh, which in this trona mine is approximately 1650 psi ($h = 1500$ ft; $\gamma = 0.09$ lb/in.^{3}).

In making stress-field determinations it is imperative that the stress measurements be made not only outside of the stress concentration zone surrounding the opening from which the overcoring operation is performed, but outside of the stress concentration zone created by other openings or mined areas that may, by superposition, affect the stress field at the measurement site. The results presented by Morgan, et al.[16] (Fig. 14.5.5) exemplify the superposition effect produced by nearby stoping. The 10 measuring stations from which the data in Fig. 14.5.5a and 14.5.5b were obtained were in the proximity of, but at different distances from extensively mined areas. These stress distribution curves all show the local stress concentrations at the measuring site but, in addition, show the superposition effects caused by nearby stopes, as indicated by range of values at which the stress distribution curves level off (for example, from about 1800 psi for the lowest curve in Fig. 14.5.5a to 4800 psi for the highest curve in Fig. 14.5.5b).

In estimating the extent of the stress concentration zone around mined areas the plastic properties of the rock should be taken into account. As indicated in Section 6.5, the stress concentration zone may extend into a plastic material several times the distance calculated on the basis of elastic theory. The ratio of the secondary principal stress Q'/P' is another factor that reflects the effects of plasticity. For most rock the laboratory (and in situ seismic) determined values of Poisson's ratio range from 0.2 to 0.33, and as the corresponding ratio of the horizontal to vertical components of a gravitational stress field is given (Section 15.4)

$$\frac{S_h}{S_v} = \frac{\nu}{1 - \nu}$$

should lie between $\frac{1}{4}$ and $\frac{1}{2}$. However, measurements made in a number of rock types and at various stress levels indicate that the ratio of S_h to S_v, and more generally of Q' to P', lies between $\frac{1}{2}$ and 1. This higher ratio of S_h/S_v or Q'/P' may result from plastic or viscous flow, which would make the stress field more nearly hydrostatic.

Stress field determinations also make it possible to detect the presence of tectonic stresses or stresses from causes other than the weight of the overlying rock. By comparing the larger measured compressive stress P'

with the calculated gravitational stress $S_v = \gamma h$, or the smaller measured compressional stress Q' with the calculated horizontal stress $S_h = S_v[\nu/(1 - \nu)]$, inferences regarding the nature of the stress field can be made. For example, Hast[5] made stress field measurements in a number of mines in Sweden and found that the horizontal component of stress was persistently larger than the vertical stress, the ratio of the horizontal to vertical stress ranging from 1.5 to 8. On the basis of this condition Hast inferred that the cause was a regional bending in the earth's crust. Obert[15] reported stress field determinations made in mines or underground openings in several rock types in the United States. In most instances the larger compressive stress was vertical or at most not more than 20° from vertical, and the magnitude of the measured vertical stress was within 20% of the calculated gravity value γh. At only one site was the deviation from the gravity condition appreciable. That was in a flat thick-bedded limestone at a depth of 2300 ft, where the horizontal stress in the NS direction was more than twice the vertical stress, and approximately equal to the vertical stress in the EW direction.

14.7 DETERMINATION OF THE STRESS ELLIPSOID

In Section 14.6 inferences and conclusions regarding the stress field were made on the basis of measured secondary principal stresses, because in most instances this is the only information available. However, if the secondary principal stresses in the three mutually perpendicular planes are determined at points sufficiently distant from underground openings to be outside the zone of influence, the ellipsoid for the principal stresses can be determined.

For convenience assume that the stress-relief holes are drilled in the NS, EW, and vertical directions (Fig. 14.7.1) and that these directions are designated as x, y, z, respectively. Also assume that point A on the y axis is outside of the stress concentration zone of any opening, and that the secondary principal stresses measure at A (which lie in the xz-plane) are designated by P_y' and Q_y', and the angle from the z axis to P_y' is γ.*

The normal and shear stresses in the xz-plane are (see Section 1.3)

$$\sigma_z = \frac{P_y' + Q_y'}{2} + \frac{P_y' - Q_y'}{2} \cos 2\gamma$$

$$\sigma_x = \frac{P_y' + Q_y'}{2} + \frac{P_y' - Q_y'}{2} \cos 2\left(\gamma + \frac{\pi}{2}\right)$$

$$\tau_{zx} = -\frac{P_y' - Q_y'}{2} \sin 2\gamma$$

---

* Note that in Eq. 14.3.4, θ_p is from U' to P'. Hence, the angle from the z axis to P_y must be corrected accordingly.

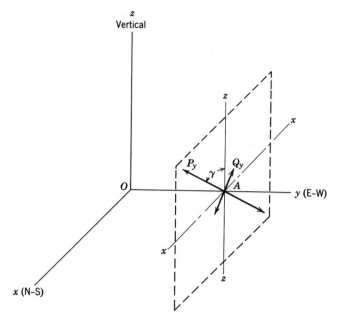

Fig. 14.7.1. Coordinate system for stress ellipsoid.

Correspondingly, the normal and shear stress in the xy and yz-planes are given by

$$\sigma_x = \frac{P_z' + Q_z'}{2} + \frac{P_z' - Q_z'}{2} \cos 2\alpha$$

$$\sigma_y = \frac{P_z' + Q_z'}{2} + \frac{P_z' - Q_z'}{2} \cos 2\left(\alpha + \frac{\pi}{2}\right)$$

$$\tau_{xy} = -\frac{P_z' - Q_z'}{2} \sin 2\alpha$$

$$\sigma_y = \frac{P_x' + Q_x'}{2} + \frac{P_x' - Q_x'}{2} \cos 2\beta$$

$$\sigma_z = \frac{P_x' + Q_x'}{2} + \frac{P_x' - Q_x'}{2} \cos 2\left(\beta + \frac{\pi}{2}\right)$$

$$\tau_{yz} = -\frac{P_x' - Q_x'}{2} \sin 2\beta$$

where P_z' and Q_z' are the secondary principal stresses in the xy-plane, P_x'

and Q_x' are the secondary principal stresses in the yz-plane, and α and β are the angles from the x axis to P_z', and from the y axis to P_x', respectively.

There is a redundancy in the stress measurements in the three orthogonal planes; that is, these data give two values for each of σ_x, σ_y, and σ_z. These pairs of values should be identical if the stress field is uniform in the measurement area and no experimental error exists. Moreover, σ_z, the component in the vertical direction, should be approximately equal to the gravity stress γh.* If these pairs of values are not equal, their difference is an indication of either a variation in the stress field in the measurement area, or the accuracy of the measurement, or both. Furthermore, if the pairs of values do not agree (it is most improbable that they would) some judgment must be made as to how to proceed with the stress ellipsoid determination.

From an engineering standpoint there is one overriding factor that places a limit on the accuracy of stress determinations required for evaluating the stability of underground structures; that is, the corresponding accuracy with which the strengths of the in situ rock is known. Unless large-scale in situ tests can be made, it is probable that the compressive strength cannot be estimated from laboratory results to within $\pm 25\%$. An even larger error would be expected in estimating tensile or shear strengths. Hence, to make it possible to continue with the ellipsoid determination, it will be arbitrarily assumed that if each of these pairs of values σ_x, σ_y, and σ_z agree within 15% from their average value, they can be averaged, and the average value, $\bar{\sigma}_x$, $\bar{\sigma}_y$, and $\bar{\sigma}_z$, used for the ellipsoid determination.†

The next step is to determine the roots σ_1, σ_2, and σ_3, of the cubic equation for the stress ellipsoid (Section 1.5), which are also the axes of the stress ellipsoid

$$\sigma_i^3 - I_1\sigma_i^2 + I_2\sigma_i - I_3 = 0 \qquad (i = 1, 2, 3)$$

where I_1, I_2, I_3 are the invariants of the component normal and shear stresses, namely

$$I_1 = \bar{\sigma}_x + \bar{\sigma}_y + \bar{\sigma}_z$$

$$I_2 = \bar{\sigma}_x\bar{\sigma}_y + \bar{\sigma}_y\bar{\sigma}_z + \bar{\sigma}_z\bar{\sigma}_x - \tau_{xy}^2 - \tau_{yz}^2 - \tau_{zx}^2$$

$$I_3 = \bar{\sigma}_x\bar{\sigma}_y\bar{\sigma}_z - \bar{\sigma}_x\tau_{yz}^2 - \bar{\sigma}_y\tau_{zx}^2 - \bar{\sigma}_z\tau_{xy}^2 + 2\tau_{xy}\tau_{yz}\tau_{zx}$$

* Some local variation in the vertical component of stress may occur, but over any extensive area, the average value should be equal to γh (Section 15.4).
† This method of averaging does not provide for a corresponding adjustment in the shear stresses. More refined methods of averaging might be developed if the quality of the data warrants.

The directions of σ_1, σ_2, and σ_3 are given by

$$\cos(\sigma_i, x) = \frac{A_i}{K_i}$$

$$\cos(\sigma_i, y) = \frac{B_i}{K_i}$$

$$\cos(\sigma_i, z) = \frac{C_i}{K_i}$$

where

$$K_i = [A_i^2 + B_i^2 + C_i^2]^{1/2}$$

$$A_i = (\bar{\sigma}_y - \sigma_i)(\bar{\sigma}_z - \sigma_i) - \tau_{zy}^2$$

$$B_i = \tau_{zy}\tau_{xz} - \tau_{xy}(\bar{\sigma}_z - \sigma_i)$$

$$C_i = \tau_{xy}\tau_{yz} - \tau_{xz}(\bar{\sigma}_y - \sigma_i)$$

In this determination of the stress ellipsoid it was arbitrarily assumed that the xyz-coordinate system corresponds to the NS, EW and vertical direction. If this is not the case the component of stress in, say, the vertical (z') direction can be determined by

$$\sigma_z' = \bar{\sigma}_x \cos^2(z'x) + \bar{\sigma}_y \cos^2(z'y) + \bar{\sigma}_z \cos^2(z'z)$$

$$+ 2\tau_{xy} \cos(z'y) \cos(z'x) + 2\tau_{yz} \cos(z'z) \cos(z'y)$$

$$+ 2\tau_{zx} \cos(z'x) \cos(z'z)$$

where $\cos(z'x)$, $\cos(z'y)$, $\cos(z'z)$ are the angles between the xyz axes and the vertical z' axis. A transformation to any other direction can be made by substituting the proper angles.

14.8   DETERMINATION OF PILLAR STRESS

In a system of underground openings in which no artificial support is used, such as linings, sets, props, etc., the weight of the overlying cover is sustained in part on the side walls of the openings, and in part on areas of unexcavated rock lying within the boundaries of the openings, which are referred to as pillars. These pillars are usually the most important structural element in a system of openings, and a determination of pillar stress is required in the evaluation of their stability.

The preceding section described the use of the borehole-deformation overcoring method for the determination of the stress distribution in the side walls of openings, and the same method was employed[15] to measure the magnitude and distribution of stress in mine pillars. Figure 14.8.1

shows the stress distribution in a vertical limestone pillar. This pillar was 17 ft high, 40 by 60 ft in cross section, and it was situated near the center of an extensively mined area lying at a depth h of 2300 ft from surface. The areal extraction ratio R was 0.6. Note that the direction of the larger compressive stress P' is approximately vertical as would be expected and that the near-surface stress decreased, rather than increased as would be expected on the basis of photoelastic model studies (Section 12.7). As

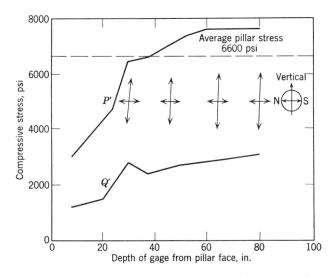

Fig. 14.8.1. Vertical stress distribution in a limestone pillar.

this limestone was relatively elastic it was presumed that this near-surface stress relaxation resulted from the presence of blasting or natural fractures and a loosening of the surface rock. Using $\gamma = 0.092$ lb/in³. for the unit weight of the rock, the average vertical compressive stress in this pillar was calculated to be (see Section 17.2)

$$\bar{\sigma}_v = \gamma h\left(\frac{1}{1-R}\right) = 0.092 \times 12 \times 2300\left(\frac{1}{1-0.6}\right) = 6300 \text{ psi}$$

The average measured vertical stress was determined by assuming that the stress distribution in Fig. 14.8.1 existed over the horizontal midplane of the pillar. By graphical integration over this area a value of $\bar{\sigma}_v = 6600$ psi was obtained. The horizontal stress in the center of a tall pillar (height several times width) should be nearly zero. However, in this short pillar (height-to-width ratio about $\frac{1}{3}$), the relatively large horizontal compressive stress Q' was probably due to two factors: first, the vertical load on a short pillar laterally constrained at the ends should produce some lateral compressive stress in the center of the pillar; second, a compressional component of stress in the floor (which was found to exist) should also contribute to the lateral compressive stress in the pillar.[17]

The borehole-deformation overcoring method was also used to determine the stress distribution in a lead ore (limestone plus galena) pillar that contained widely spaced joints. This pillar was 17 ft high, 24 by 28 ft in

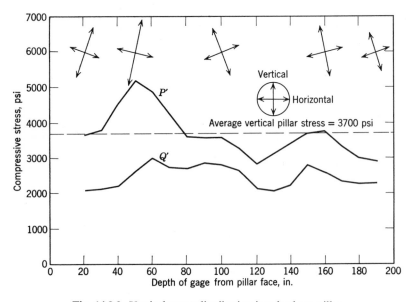

Fig. 14.8.2. Vertical stress distribution in a lead ore pillar.

cross section, and it was in the center of a relatively small mined area lying at a depth of 692 ft. The areal extraction ratio was approximately 0.80. The stress distribution curve (Fig. 14.8.2) shows that the larger compressive stress P' was nearly vertical (within 15°). In the core of the pillar the stress varied in a manner characteristic of a jointed rock. At a point 4 ft from the surface, a stress concentration of about 1.6 occurred, and from this point to surface the stress decreased, probably due to blasting fractures. The average calculated pillar stress was 3700 psi, which compared favorably with the average stress of 3600 psi obtained by integrating the measured stress over the cross-sectional area of the pillar. Thus the inference was made that even over a comparatively small mined area the average pillar stress can be calculated from the deadweight loading, that is, by the use of Eq. 17.2.9 or 17.3.1.

14.9   NATURE OF THE STRESS IN ROCK

In making stress determinations by the stress-relief techniques described in Section 9.3, the deformations experienced by stress-relief cores or bodies or rock may be the result of either or both of two causes. First, and most important, the stress-relieved rock will deform because of the removal of external forces. These external forces include the weight of the overlying cover and tectonic forces of geological origin, such as those associated with folding or faulting. Tectonic forces affect volumes of rock much larger than those involved in stress-relief operations. Second, stress-relieved cores or bodies of rock may deform because internal (body) forces have been disturbed, that is, forces between crystals, crystals and matrix, aggregates of crystals, etc. Since the total deformation in stress-relieved rock is used to calculate external stresses, any contribution to the total deformation due to a readjustment of internal forces would introduce an error. Hence the nature of these internal forces is a matter of concern.

Emery[18] reported that if a photoelastic coating is applied to a specimen cut from a freshly drilled core, fringes will develop after a period of time outlining individual grain or crystal boundaries, indicating that a time-dependent deformation has occurred. In a later report Emery[19] postulated a mechanical model to account for this time-dependent effect, namely, that rock is composed of elastic particles in a viscoelastic matrix. If a piece of rock that has been confined under stress for a long period is stress-relieved, the elastic particles will expand within the matrix causing time-dependent deformations.*

In the process of making stress determinations by the borehole-deformation overcoring method,[15] deformation measurements were continued for periods up to 12 hr after the overcoring operation had been completed. In making these measurements the stress-relieved cores were left intact and the borehole gage was not disturbed. It was found that in most rocks time-dependent deformations occurring after effecting stress-relief were negligible compared to the borehole deformation caused by the stress-relief; hence this effect generally should not cause any significant error.

Also, time-independent deformations may occur in rock as a result of a readjustment of internal stresses. Consider as a model a rock composed of elastic particles under compression in a constraining elastic matrix. A system of this kind might originate in an igneous rock at the time it crystallized if, for example, in a magma the matrix crystallized first and then a second component crystallized and expanded at the same time. If

---

* This elastic-viscoelastic model is similar to a "Newton sponge," which is an elastic sponge saturated with a viscous fluid.

a cut is made through such a material, displacements will occur on the fresh surface provided that a number of the stress zones surrounding these stress particles are intersected. This effect has been observed by altering the shape of a rock specimen on which a photoelastic gage has been attached and noting the instantaneous appearance of a stress pattern. It is conceivable that these internal stresses may have some collective preferred direction, possibly related to the fabric of the rock, and that they affect the manner in which rock fractures or rifts. The deformation produced by the readjustment of time-independent internal stresses may also contribute to the total deformation experienced by a core when external stresses are removed by overcoring. No procedure has been proposed for distinguishing between deformations caused by relief of external stress, and by the readjustment of time-independent internal stress.

14.10   DETERMINATION OF THE IN SITU ELASTIC PROPERTIES AND STRENGTH OF LAMINATED MINE ROOF

Unsupported roof over underground openings can be divided into two classes, arched and laminated. Arched roofs are usually formed in massive rock that is uniform to the degree that the stress or strain in the roof, side walls, and floor of the opening can be approximated from elastic theory (Chapters 16 and 17), or measured in situ using the methods described in Section 14.3. Flat laminated roofs usually are formed in sedimentary rocks that contain partings or other planes of weakness.* Because of discontinuities at these planes of weakness, the roof over these openings cannot be considered as a part of a continuous rock mass. Rather this type of roof generally is composed of one or more lamina of rock that in time becomes detached, thus forming gravity-loaded single or multilamina beams or plates. If the ratio of the areal dimensions of an opening is less than two, the overlying roof lamina should be considered as a plate clamped at four supporting edges; if the ratio is greater than two, it can be treated as a beam clamped at two supporting edges. The shorter of these dimensions is referred to as the *span*.

The in situ modulus of elasticity E of a single-lamina, gravity-loaded roof can be determined by means of the following procedure. Consider a room with a span L less than half the length of the room. From the formula for a beam with built-in ends, the maximum deflection D (roof sag) at the

---

* A distinction is made between laminated and bedded rock. A laminated rock is composed of detached or partially detached (unbonded or partially bonded) rock laminae, which usually form at partings or separations; whereas bedded rock is identified by bands of color and/or mineral difference, but does not necessarily contain partings separations, or other planes or weakness.

center of the room is given by (Eq. 5.6.29)

$$\eta = \frac{\gamma L^4}{32 E t^2} \qquad (14.10.1)$$

where γ is the unit weight of the rock and t the thickness of the roof layer. The deformation of a roof layer from its initial position is difficult to measure because it would be necessary to establish a point of reference in the roof rock before the opening was created. However, if an opening is widened from, say, a width L_1 to a width L_2 the corresponding change in deformation can be measured with relative ease. If this change is $\eta_2 - \eta_1 = \Delta\eta$, then from Eq. 14.10.1

$$E = \frac{\gamma(L_2^4 - L_1^4)}{32 \, \Delta\eta t^2} \qquad (14.10.2)$$

Determination of the In Situ Modulus of Elasticity

Merrill[20] described an investigation in which the in situ modulus of elasticity was determined from roof sag measurements made in an experimental room in a limestone mine. The experimental procedure was as follows: in a test room 10 ft wide and 100 ft-long, reference pins were anchored at various points in both the roof and floor (Fig. 14.10.1). The

Fig. 14.10.1. Experimental room in a limestone mine. (From *Bu Mines Rept. of Inv.* 5348.)

anchorage points in the roof were selected on the basis of stratoscope and core examinations so that a reference pin could be placed in each of the immediate lamina overlying the room. Initial measurements were made between the pairs of roof and floor pins, following which periodic measurements were made as the room was successively widened to 20, 30, 40, and 50 ft. Both the stratoscope observations and differential sag measurements, that is, the difference between the various pairs of roof-to-floor pin measurements indicated that the lowest roof layer had become detached, and hence it deformed as a single member beam. Stratoscope photographs

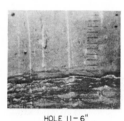 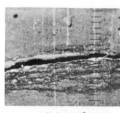 

HOLE 11-6"
Feb. I, 1954 - 10-ft. span

HOLE 11-6"
March 25, 1954 - 20-ft. span

HOLE 11-6"
July 28, 1954 - 30-ft. span

Fig. 14.10.2. Stratoscope photographs of a roof separation. (From *Bu Mines Rept. of Inv.* 5348.)

of this separation are shown in Fig. 14.10.2. From the roof span L, the thickness of the roof layer t, and the unit weight of the rock γ, the modulus of elasticity of the rock was determined by means of Eq. 14.10.2. The value so determined was less than one-half that obtained by testing core specimens in the laboratory. This difference was attributed in part to the fact that the roof deformation could not be measured immediately after the room was widened (for safety reasons); hence some inelastic deformation (creep or plastic flow) was probably included in the deformation measurement when it was made.

Determination of the In Situ Flexural Strength

The maximum tensile stress $(\sigma_x)_{max}$ in a single-lamina, gravity-loaded beam of length L and thickness t occurs at the midpoint of the lower (convex) surface, and is given by Eq. 5.6.29

$$(\sigma_x)_{max} = \frac{\gamma L^2}{2t} \qquad (14.10.3)$$

Equation 14.10.3 can be used for design purposes, that is, to determine a safe span, if in addition to the thickness of the roof lamina, the tensile or flexural strength (outer-fiber tensile strength) of the rock is known. The thickness of the roof lamina can be obtained from cores taken perpendicular to the plane of the roof, and the tensile and flexural strength can be determined in the laboratory from cores taken parallel to the plane of the roof. In determining the thickness of the layer from a core examination it is often difficult to distinguish between breaks in the core caused by mechanical (drilling) action and breaks due to planes of weakness. However, if a number of holes are drilled in the roof, a true plane of weakness will usually be evident from a hole-to-hole persistence in the break.

The in situ flexural strength of the rock in a single-lamina roof can be evaluated by successively widening an experimental room until the roof fails. A test of this type was reported by Merrill[21] in which the flexural strength of a single-layer, oil shale roof 20 in. thick and 80 × 200 ft in extent at failure was calculated from the relationship for a gravity-loaded beam (Eq. 14.10.3). With $\gamma = 0.09$, $L_s = 80$ ft, and $t = 20$ in., the calculated tensile strength is 2050 psi. The laboratory-determined flexural strength of the same rock (measured on cores taken parallel to the bedding) was 3000 psi. A centrifugal model study made by Wright and Bucky[22] using models made from oil shale gave a flexural strength of 1860 psi.

When the experimental room in limestone (Fig. 14.10.1) was widened to 50 ft, it was found from differential sag and stratoscope observations that a 20-in. thick layer had become detached from the overlying roof, the separation amounting to 0.30 in. Rather than continue the widening operation until a failure occurred, a procedure that because of the evident hazard requires extensive safety measures, Merrill and Morgan[23] caused the 20-in. layer to fail by introducing compressed air into the separation. A uniformly distributed load (air pressure) applied to a beam is equivalent to increasing the unit weight γ by an amount p/t, where p is the pneumatic pressure applied to the beam. Thus the maximum tensile stress at the midpoint of the beam becomes, from Eq. 14.10.3,

$$(\sigma_x)_{\max} = \frac{\gamma L^2}{2t} + \frac{pL^2}{2t^2} \qquad (14.10.4)$$

The roof failed at an air pressure of 9.0 psi. From Eq. 14.10.3 the computed tensile strength for $p = 9.0$ psi is 4950 psi, which compared favorably with the laboratory-determined flexural strength of 4650 psi, but unfavorably with the uniaxial tensile strength of 1190 psi. The better agreement between the laboratory-determined flexural strength and the in situ determined tensile strength of the roof layer measured in flexure is probably a consequence of the similitude in the two test procedures.

14.11   MEASUREMENT OF TIME-DEPENDENT DEFORMATION

A number of researchers have measured, over extended periods, the roof-to-floor convergence and the squeeze and closure in tunnels and drifts. However, in most instances props, linings, or other forms of artificial support were employed in the measurement area, and under these conditions the time-dependent behavior of the rock is difficult to evaluate. Only limited data are available pertaining to the time-dependent behavior of completely unsupported mine rock. Höfer[24] measured the lateral deformation in a large number of rib pillars in German potash mines, over periods ranging from 63 days to more than 3 years. The measurements were made with deformeters across the width or half-width of the pillars. Representative results are shown in Fig. 10.4.8. These measurements indicate that, in general, the lateral deformation rate in potash pillars is relatively constant. Strain rates calculated from these data range from less

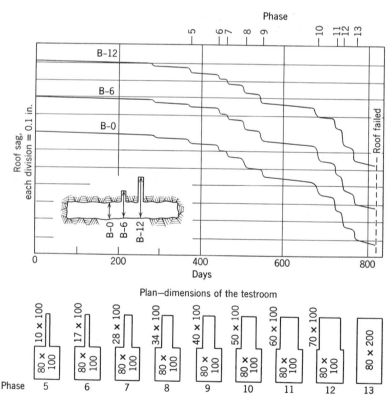

Fig. 14.11.1. Roof-sag results from experimental room in oil shale.

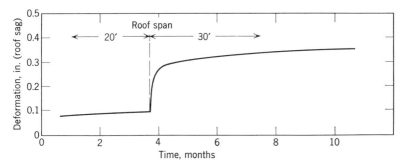

Fig. 14.11.2. Roof-sag in a limestone mine. (From Merrill.[23])

than 1% to more than 15% per year. Höfer showed that the deformation rate increases rapidly when the pillar stress (calculated from the dead weight loading, Eq. 17.2.9) is greater than 5000 psi (Fig. 10.4.10).

In the experimental room tests in oil shale reported by Merrill,[21] (Section 14.10) roof sag measurements were made over a two-year period during which an extension to the experimental room was successively widened, thereby incrementally increasing the stress in the roof layers. Figure 14.11.1 shows the roof sag between pins anchored in the floor and at different elevations in the roof. These curves show the almost instantaneous elastic deformation that occurred each time the room was widened, and the nearly constant sag rate between widenings. Curve B–O is for the deformation between the floor and the exposed surface of roof. The test was concluded when the lowest layer, which was 20 in. thick, failed. It should be noted that in the 50 days preceding failure there was no indication of tertiary creep, that is, a phase in which the strain rate increased.

A curve showing the time-dependent deformation of a limestone roof following the widening of the room is presented on an expanded time scale in Fig. 14.11.2. This curve is similar to the creep curve for the Burger's model (Section 6.2) in that when the roof stress was increased as a result of increasing the room width from 20 to 30 ft an immediate elastic deformation occurred, followed by a transient phase in which the strain rate gradually decreased asymptotically approaching a constant rate.

14.12   MICRO-SEISMIC   METHOD   FOR DETERMINING ROCK STABILITY

Audible rock noises are generally interpreted by men working underground as a warning of imminent danger from rock failure. However, preceding the period during which audible noises are produced there is usually a longer period in which subaudible noises are generated in rock under stress. Subaudible noises are seismic pulses of short duration and extremely small amplitude. They can be detected with a sufficiently sensitive geophone and amplifier (Section 9.10). Obert and Duvall[25] reported that these micro-seismic pulses could be produced in the laboratory by subjecting specimens of rock to a compressive load, and that the production of these pulses, that is, the micro-seismic rate, increases with the load. It is believed that the micro-seismic pulses generated in the laboratory result from the failure of intercrystalline bonds or possibly the failure of individual crystals, but in a mine these pulses may also be caused by movement along fractures or joint interfaces. Whatever the cause, the fact that micro-seismic disturbances are generated in rock under stress, and that the micro-seismic rate increases as failure becomes imminent is the basis for this method of detecting and delineating areas of potential danger.

The micro-seismic pulses are self-generated with random intensities and at random times. Moreover, these pulses are in themselves an indication of rock failure, hence the method does not have to consider the state of stress in the rock, the strength of the rock, or any criterion of failure other than that indicated by the magnitude of the micro-seismic rate. In relation to the imminence of failure, the micro-seismic rate generated in laboratory-size specimens varies with the mineral composition of the rock, the crystal size, the type of bonding between crystals, and other factors. The presence of joints, the degree of fracture, and other defects of geological origin affect the production of rock noises in megascopic bodies. Hence the micro-seismic method is not quantitative. Rather, like a stethoscope in the hands of a physician, the observations must be interpreted through experience. Usually this is not a difficult interpretation because in stable rock the micro-seismic rate is very low, whereas in rock near failure the rate may be ten to one hundred times as great.

To detect and locate small slabs of "loose" rock the geophone (described in Section 9.10) can be used with a small battery-operated amplifier and headphones. The observer aurally notes the noise rate. However, for more extensive investigations some type of recording equipment is generally employed. Because daily observations are usually made over periods of the order of hours it is not feasible to resolve the waveform of each seismic pulse. Instead, the seismic pulses are electrically integrated and recorded as a single mark (or pip) on the recording tape (Fig. 14.12.1). The amplitude of these marks is roughly proportional to the amplitude of the seismic pulse.

In practice there are two ways to delineate areas of instability. In the first, referred to as the probing method, a geophone is moved from point to point in the suspected area and the micro-seismic rate either observed aurally or recorded. As the intensity of a seismic pulse decreases with the distance from the point of origin, at some distance it will become too weak to detect. Thus the micro-seismic rate will decrease with the distance from the center of the stress area. In most areas that are stable the micro-seismic rate is virtually zero; hence any unstable body of rock that is generating noise is usually easy to locate.

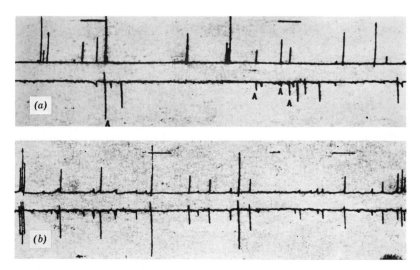

Fig. 14.12.1. (*a*) Record from two geophones 50 ft apart showing coincidences (marked *A*). (*b*) Record from two geophones in same test hole showing nearly complete coincidence.

In the second procedure, referred to as the amplitude-comparison method, a number of geophones are placed at fixed points in the proximity of the suspected area and a simultaneous recording made of the micro-seismic noise. As previously pointed out, because the amplitude of a seismic pulse picked up by the geophone nearest to the center of the generating area will have the largest amplitude, the geophone next nearest to the generating point will have the next greatest amplitude, etc. Thus, by comparing simultaneously recorded amplitudes from a given pulse, the geophone closest to the generating point can be determined. This procedure is superior to the probing method because, after the original geophone installation, access to suspected area, which may be hazardous, is not necessary.

A strong (high-intensity) pulse from a generating point will be picked up by both geophones close to, and distant from, the generating point, and the simultaneous recording of such a pulse is referred to as a "coincidence" (Fig. 14.12.1). On the other hand, if the pulse is weak, it may not be detected by the distant geophone and the record of such an event is called a "noncoincidence." An analysis of the coincident and noncoincident recordings from a suspected high-stress zone is also helpful in delineating the generating area.

<h3 style="text-align:center">14.13   SEISMIC METHODS OF DELINEATING
SUBSURFACE SUBSIDENCE</h3>

In block caving operations and other types of mining in which the overlying cover caves into the stoped area there is a need to know the limits and direction of caving as it progresses toward the surface. Although the micro-seismic method described in Section 14.12 provides a means of determining the source of seismic disturbances generated on or near the surface of underground openings, the accuracy of the determination decreases as the distance from the point of origin increases. Hence, this method is not satisfactory for following the progress of subsurface caving unless the geophones can be placed in the general proximity of the affected area, a requirement that usually necessitates placing the geophones in long holes, diamond-drilled either from the surface or from other remote points. In instances when access to the proximity of the cave has been favorable, or when existing (exploration) holes could be utilized, the micro-seismic method has been used with some success to detect and delineate subsurface subsidence.[26] Generally the drilling required for the placement of a proper network of geophones is prohibitively expensive.

Two other methods of delineating subsurface have been developed[26] referred to as the traveltime method and the traveltime difference method. The traveltime method is simple in concept. Referring to Fig. 14.13.1, a sensitive geophone (micro-seismic type) is placed on one side of a zone in which caving is contemplated and a small charge detonated on the other side of the zone. The time for the seismic pulse to travel from point of generation to the geophone is measured. This procedure is repeated periodically, and as the cave progresses upward it will intercept the direct path of the seismic pulse causing it to take a longer path. The corresponding increase in the traveltime is an indication of the limits of the cave. By placing the geophone, and detonating the charge at fixed points in the geophone and shot holes (points B_1, B_2, and A_1, A_2, A_3, Fig. 14.13.1) a network of travel paths can be established, thereby making it possible to follow the progress of the cave. Although this method also necessitates using diamond drill holes, the total length of hole is usually not as great as that required for the micro-seismic method. The charge size must be

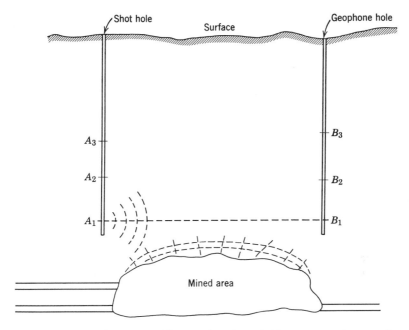

Fig. 14.13.1. Seismic setup to delineate subsurface subsidence over caving operation.

kept small—usually less than 0.1 lb of explosive, otherwise repeated shooting will damage the shot hole.

In relation to the dimensions of most mined areas, traveltimes are of the order of 100 msec; hence a relatively accurate timing device must be employed for this measurement, such as an electronic interval timer or an oscilloscope with a calibrated sweep.

Traveltime Difference Method

The micro-seismic and traveltime methods provide a means of delineating areas in which rock failure is occurring, but unless an extensive system of geophones is employed the point of failure at any specified instant cannot be ascertained. In distinction, the traveltime difference method permits, at least in concept, a determination of the actual point of failure. Suppose that at a given point a rock fails, and in doing so it creates a seismic pulse. Also, suppose that this pulse is picked up by a number of geophones placed at known positions in the surrounding rock. If the rock has the same propagation velocity in all directions, the geophone closest to the source will pick up the pulse first, the second geophone will pick up the pulse next, and so forth. If the *difference* in the traveltime of the seismic pulse to the first and second geophones, first and third geophones, and so forth, is measured, it is possible to determine the point of failure from these traveltime differences. Although the general solution of the problem of locating the source of the seismic pulse in space from traveltime differences is mathematically complicated, two special cases have been treated.[26] In the first case consider two geophones, G_0 and G_1, located on the X axis at $x = 0$ and $x = d_1$, respectively (Fig. 14.13.2) with G_0 closer to the seismic source at $P(x, y, z)$. Let p_0 be the distance from P to G_0, and t_0 be the time

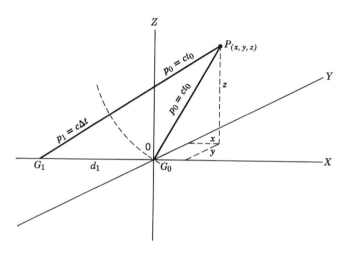

Fig. 14.13.2

for the pulse to travel from P to G_0. Then assuming a constant propagation velocity c, the distance from P to G_0 is given by

$$p_0 = ct_0 = [x^2 + y^2 + z^2]^{\frac{1}{2}} \tag{14.13.1}$$

If t_1 is the time for the pulse to travel from P to G_1, then

$$p_1 + p_0 = ct_1 = [(x - d_1)^2 + y^2 + z^2]^{\frac{1}{2}} \tag{14.13.2}$$

where $p_1 + p_0$ is the distance from P to G_1. From Eqs. 14.13.1 and 14.13.2 the distance p_1 is

$$p_1 = c\Delta t \tag{14.13.3}$$

where $\Delta t = t_1 - t_0$ is the difference in the traveltime. Eliminating p_0 and p_1 from Eqs. 14.13.1, 14.13.2, and 14.13.3 and regrouping terms gives

$$\frac{(2x - d_1)^2}{c\,\Delta t^2} - \frac{4y^2}{d_1{}^2 - \Delta t^2} - \frac{4z^2}{d_1{}^2 + c\Delta t^2} = 1 \tag{14.13.4}$$

Equation 14.13.4 represents a hyperboloid of revolution, and the origin of the seismic pulse must lie on this surface. Although this is not too informative, it illustrates the basic procedure for localizing the source of seismic pulses from traveltime differences. If three geophones are employed, two traveltime differences can be measured, and two hyperboloids of revolution can be derived, the common intersection of which will give a line in space. The source of the seismic pulse must lie on this line. Correspondingly, with

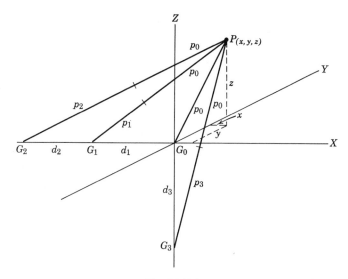

Fig. 14.13.3

four geophones, three traveltime differences can be measured, three hyperboloids of revolution derived, and the common intersection of these surfaces will give the point in space from which the pulse originates.*

In the second case four geophones, G_0, G_1, G_2, and G_3, placed in a plane, as shown in Fig. 14.13.3, can be used for the determination of three traveltime differences. Following the same mathematical steps used in the solution of the two-geophone case, three quadratic equations are developed, the simultaneous solution of which gives $P(x, y, z)$, the source of the seismic pulse. Although this result has a very practical implication, the calculation of the coordinates of $P(x, y, z)$ from the three traveltime measurements is time consuming. A simpler procedure is to employ a string model (Fig. 14.13.4.) Let the scale length of the string from G_0 to $P(x, y, z)$ be an

---

* Actually three hyperboloids of revolution have four common points of intersection. However, the geometry of the geophone placement in respect to the source is usually such that only one point is physically possible.

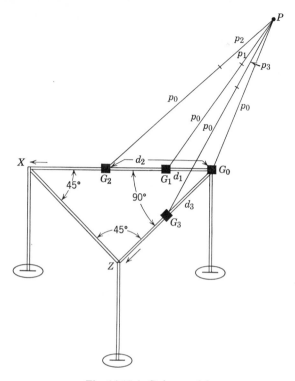

Fig. 14.13.4. String model.

unspecified length p_0; from G_1 to $P(x, y, z)$ be $p_0 + c\Delta t_{01}$; from G_2 to $P(x, y, z)$ be $p_0 + c\Delta t_{02}$; and from G_3 to $P(x, y, z)$ be $p_0 + c\Delta t_{03}$, where c is the propagation velocity of the medium and Δt_{01} is the difference between the traveltime from $P(x, y, z)$ to G_0, and from $P(x, y, z)$ to G_1, etc. If p_0 is selected such that the ends of the strings, when taut, will meet at a common point, this point is $P(x, y, z)$, the source of the pulse.

A simplified version of the traveltime difference method was employed[26] to delimit subsurface subsidence in relation to the boundary of a shaft pillar. For this investigation a hole was diamond drilled from the shaft into the suspected area (Fig. 14.13.5) and three geophones installed at specified intervals. The propagation velocity c of the rock was determined by detonating small charges at known points in the area. Electronic interval timers (Fig. 14.13.6) were used to measure traveltime differences. In Fig. 14.13.5, if P is the point of a rock fracture, geophone G_0 will receive the seismic signal first and start the three interval timers. Each timer is stopped by the arrival of the signal from its respective geophone. The differences in the measured traveltimes are Δt_{01} and Δt_{02}. It is evident that the origin

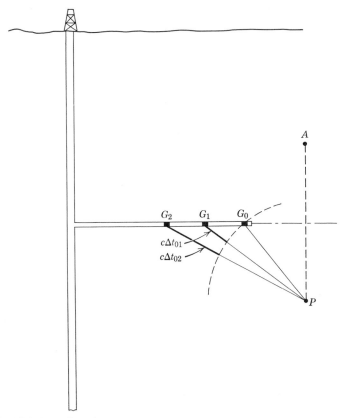

Fig. 14.13.5. Delineation of subsurface subsidence by traveltime differences.

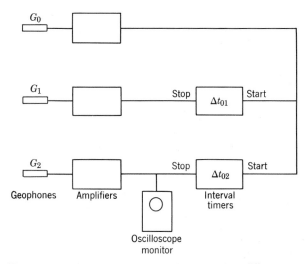

Fig. 14.13.6. Apparatus for measuring traveltime differences.

457

of the seismic signal must lie on a circle of diameter AP, lying in a plane normal to the axis of the geophone hole. A simplified string model can be used to analyze these data.

Cook[27] used the traveltime difference method to locate the epicenter of rock bursts in a deep mine in the Witswatersrand, South Africa. This method is particularly well suited for this type of problem because the seismic energy released in a rock burst is large, hence the intensity of the seismic waves is strong, a factor that simplifies the equipment used for detecting and recording weak seismic disturbances. A string model was used to locate spatially the point of the burst. The results of this investigation are discussed in Section 18.3.

REFERENCES

1. Lieurance, R. S., "Stresses in Foundation at Boulder Dam," *U.S. Bureau of Reclamation Tech. Memo.*, No. 346 (1933).
2. Olsen, O. J., "Measurement of Residual Stress by the Strain-Relief Method," *2nd An. Symp. on Rock Mech., Quarterly*, Colo. Sch. of Min., **52**, No. 3 (1957).
3. Sipprelle, E. M. and H. L. Teichmann, "Roof Studies and Mine Structure Stress Analysis, Bureau of Mines Oil-Shale Mine, Rifle, Colo.," *Trans. AIME*, **187** (1950).
4. Morgan, T. A. and L. A. Panek, "A Method for Determining Stress in Rock," *U.S. BurMines Rept. Invest.*, **6312** (1963).
5. Hast, N., "The Measurement of Rock Pressure in Mines," *Sveriges Geol. Undersokn., Årsbok*, **52**, No. 3 (1958).
6. Obert, L., "Determination of the Stress in Rock," *A State of the Art Report, ASTM*.
7. Habib, P. and R. Marchand, "Mesures des Pressions de Terrains par L'Essai de Verin Plat," *Annales de L'Institut Technique du Batiment et des Travaux Publics, serie*, Sols et Fondations, Paris, No. 58 (1952).
8. Panek, L. A. and J. A. Stock, "Development of a Rock Stress Monitoring Station Based on the Flat-Slot Method of Measuring Existing Rock Stress," *U.S. BurMines Rept. Invest.*, **6537** (1964).
9. Obert, L., "Measurement of Pressure on Rock Pillars in Underground Mines," Part I, *U.S. BurMines Rept. Invest.*, **3444** (1939); Part II, *U.S. BurMines Rept. Invest.*, **3521** (1940).
10. Tincelin, M. E., "Mesures Des Pressions De Terrains Dans Les Mines De Fer De L'Est Methode de Mesure," *Supplement Aux Annales De L'Institu Technique Du Batiment Et Des Travaux Publics* (1952).
11. Uhlmann, Manfred, "Uber die Erkundung der Spannumgsverhaltnisse in Stutzpfeilern des Kali-und Steinsalzbergbaus auf Akustischer Basis," *Freiberger Forschungshefte*, C **36** (1957).
12. Kehle, O. K., "The Determination of Tectonic Stresses through Analysis of Hydraulic Well Fracturing," *J. Geophys. Res.*, **69** (2) (1964).
13. Fairhurst, C., "Measurement of In Situ Stresses with Particular Reference to Hydraulic Fracturing," *Felsmechanik*, II (3–4) (1965).
14. Obert, L. and D. Stephenson, "Stress Conditions Under Which Core Discing Occurs," *Trans. SME, AIME*, **232** (3) (1965).
15. Obert, L., "In Situ Determination of Stress in Rock," *Mining Engr.*, **14**, No. 8 (1962).

16. Morgan, T. A., W. Fischer, and W. Sturgis, "Stress Distributions in Westvaco Mine as Determined by Borehole Stress-Relief," *U.S. BurMines Rept. Invest.*, **6675** (1965).

17. Dreyer, W., "Die Bedeutung von Modellversuchen an Salzgesteinen fur die Beurteilung gebirgsmechanischer Probleme im Kalibergbau," *Bergakademie* 16Jg., Heft 3 (1964).

18. Emery, C. L., "The Measurement of Strains in Mine Rock," *Internat. Symp. on Min. Res.*, **2**, Pergamon Press (1961).

19. Emery, C. L., "Testing Rock in Compression," *Mine and Quarry Eng.* (April and May 1960).

20. Merrill, R. H., "Roof-Span Studies in Limestone," *U.S. BurMines Rept. Invest.*, **5348** (1957).

21. Merrill, R. H., "Design of Underground Mine Openings, Oil-Shale Mine, Rifle Colo.," *U.S. BurMines Rept. Invest.*, **5089** (1954).

22. Wright, F. D. and P. B. Buckey, "Determination of Room-and-Pillar Dimensions for the Oil-Shale Mines at Rifle, Colo.," *Trans. AIME*, **181** (1949).

23. Merrill, R. H. and T. A. Morgan, "Method of Determining the Strength of a Mine Roof," *U.S. BurMines Rept. Invest.*, **5406** (1958).

24. Höfer, K. H., "Beitrag zur Frage der Standfestigkeit von Bergfesten im Kalibergbau," *Freiberger Forschungshefte*, **A-100** (1958).

25. Obert, L. and W. I. Duvall, "Micro-Seismic Method of Determining the Stability of Underground Openings," *U.S. BurMines Bull. 573* (1957).

26. Obert, L. and W. I. Duvall, "Seismic Methods of Detecting and Delineating Subsurface Subsidence," *U.S. BurMines Rept. Invest.*, **5882** (1961).

27. Cook, N. G. W., "The Seismic Location of Rockbursts," *Proc. of the Fifth Symp. on Rock Mech.*, Pergamon Press (1963).

PART THREE

DESIGN AND STABILITY
OF STRUCTURES IN ROCK

INTRODUCTION

As previously stated in the introduction to Part One, we start with the assumption when examining structural design and stability problems, that rock is an isotropic, homogeneous, linear-elastic material, and the deformational response of a rock body to an applied force can be determined from elastic theory. To provide the theoretical background for this approach the first five chapters in Part One presented the fundamentals of elastic theory, methods of solving problems in elastic theory, and the solution to a number of problems related to rock mechanics. However, because rock is not a perfectly elastic material, and therefore allowances must be made for deviations from perfect elasticity, a brief theoretical treatment of ideally inelastic materials was given in Chapter 6. Thus in Part One only the behavior of *ideal* materials is considered. In Part Two experimental methods and procedures are described for measuring the strengths, and the elastic and inelastic properties of a *real* material—rock, and for measuring the strain or deformation, or for determining the state of stress in bodies of rock.

In Part Three the theoretical and empirical results from Parts One and Two are combined to form one of the two sources of information available to the structural engineer—the other source of information is operating experience. When the knowledge from these sources is combined and reconciled, the result is a rational process that serves as one of the tools available to the civil and mining engineer for the analysis of rock structure problems.

This process may be outlined as follows:

1. A number of theories (or hypotheses) related to the behavior of rock structures have been proposed that were based only on visual observation. Some of these theories are in good agreement, and others equally inconsistent with the laws of mechanics. This rational process has been useful in

appraising these theories and in bringing some order out of controversial and sometimes conflicting ideas.

2. This knowledge has a heuristic value in that deviations in the behavior of actual structures from those anticipated from mathematical and empirical theory can be identified and investigated for cause.

3. This rationalization also makes it possible to compare the behavior of one structure with another, or to use the knowledge gained from the study of an existing structure in designing other underground openings of different dimensions or subject to a different applied stress field.

4. A theoretical and experimental knowledge of the stress distribution around underground openings is helpful in interpreting incipient or early fracture patterns sometimes observed in rock structures. Moreover, this interpretation may be helpful in anticipating when and where more general failures will occur and in prescribing remedial procedures.

5. In the absence of specific operating experience, and on the basis of only limited information, as in planning the development of a new ore body, this process provides the only rational approach to designing a system of underground openings and supporting members.

CHAPTER 15

GENERAL CONSIDERATIONS

15.1 INTRODUCTION

In the engineering evaluation of rock structure problems it is necessary to assign numerical values to at least a part of the problem variables, and in this respect structural rock mechanics is no different from any other phase of engineering. Usually the principal variables of concern are the magnitude and direction of the preexisting stresses and the strengths and other mechanical properties of the rock. Because rock strengths generally are determined in the laboratory under a state of stress unlike that in the structural problem, a criterion of failure may be required to relate laboratory-determined and in situ strengths. Also, because of uncertainties and unknowns, a safety factor should be employed that is consistent with the prevailing experience in considering problems of this kind.

As stated in the Introduction to Part One, unlike most other branches of structural engineering in which the design engineer can choose structural materials with known mechanical properties and known initial state of stress (generally assumed to be zero), an engineer in charge of the design of an underground structure usually has no latitude for selecting the site; hence, both the mechanical properties of, and initial state of stress in his structural material are unknown. Usually the first appraisal of the structural properties of subsurface rock is obtained either by visually examining exploratory drill cores or by performing laboratory tests on small specimens cut from these cores, and the preexisting state of stress generally must be assumed to be that due to the weight of the overlying rock, although this may turn out to be a poor approximation. As underground access becomes possible additional information regarding the state of stress in the rock and its mechanical structural behavior can be obtained, and the original design reevaluated. But, at best, in situ rock strengths and other mechanical properties, and the stress field and local stress concentrations can never be known precisely; hence, the engineer must exercise certain judgments in choosing numerical values for these quantities and other factors. Our purpose in this chapter is to specify criteria and guidelines that will assist the engineer in making these judgments.

15.2 STRUCTURAL CLASSIFICATION OF ROCK

The classification of rock as a structural material in which underground openings can be constructed is not a simple problem. Adjectival descriptions such as strong, weak, hard, soft, friable, etc., lack any quantitative basis and hence they have only a limited engineering value. Consideration has been given to quantitizing these terms so that, for example, a weak rock might be one with a compressive strength range from 1000 to 5000 psi; a strong rock from 5000 to 20,000 psi, etc. However, mechanical property tests are usually made on relatively small specimens of uniform rock. If the body of rock from which the specimen was taken is correspondingly uniform, classification by compressive strength, or any combination of mechanical properties, has a real value. But in most instances at the scale of an underground opening, rock contains mechanical defects such as joints, fractures, and faults, and as is well known, the in situ mechanical properties of a body of rock at this scale will depend to some indefinite degree on these defects. For example, in jointed rock, the in situ strengths will depend on the number of joints per unit volume, the attitude of the joint planes, the degree of bond across joint planes, or if the joints are unbonded, on the extent and type of decomposition product on the joint plane, and whether this product is wet (lubricated) or dry. Hence, the mechanical properties of rock as determined from laboratory tests generally do not in themselves provide a satisfactory basis for classifying rock as a structural material.

Another approach to classifying rock for structural purposes is to consider the combination of geological and mechanical rock properties that will permit the construction of a specified type of underground structure. For example, Obert, Duvall, and Merrill[1] defined competent rock as rock, which because of its mechanical and geological characteristics is capable of sustaining underground openings without the aid of any structural support except pillars and walls left during mining (stulls, light props, rock bolts, etc., are not considered structural supports). Competent rock was further divided into two classes, massive and bedded (or laminated). Massive rock was considered to be linear elastic, isotropic, and homogeneous, where isotropic and homogeneous implied not only uniformity of material but also an absence of megascopic defect (joints, partings, etc.).

This latter condition is satisfied if the spacing between joints or partings is large compared with critical dimensions of the opening, or if the strength of the bond across joints or partings is comparable with that of the rock. Bedded (or laminated) rock was considered to include any rock divided by approximately parallel planes (or surfaces) of weakness into laminae (or layers), the thicknesses of which are small compared with critical dimensions of the openings. However, the rock within any laminae was assumed to be linear elastic, isotropic, and homogeneous.

This classification is not inclusive but obviously it can be made so by defining incompetent rock as rock, which because of its mechanical and geological properties is not capable of sustaining underground openings without the aid of artificial support, such as linings, sets, or systems of props. There is no sharp demarcation between the two divisions, as all degrees of support are employed in mines. Also, time is a factor as some underground openings will remain open for limited periods without any support, but for longer periods the same opening may require substantial support.

Competent rock can be subdivided into two inclusive classes, linear elastic and inelastic. Linear elasticity is time-independent whereas inelasticity may be time-dependent, hence creep and other time-dependent phenomena may occur in this latter rock class. Since incompetent rock is generally inelastic it is not correspondingly subdivided. Megascopic isotropy and anisotropy, and homogeneity and inhomogeneity can lead to other subdivisions. However, if the divisions are made too restrictive, then identification of distinctive rock classes becomes difficult, especially for classes that have some degree of structural uniqueness.

In the following chapters the structural classification given in Table 15.2.1 is employed. These classes are defined as follows.

(1) **Competent Massive Elastic Rock.** Competent rock is any rock which because of its mechanical and geological characteristics is capable

Table 15.2.1. Structural Classification of Rock

I.	Component
	A. Massive
	1. Elastic
	2. Inelastic
	B. Laminated
	1. Elastic
	2. Inelastic
	C. Jointed
II.	Incompetent

of sustaining underground openings without the aid of any structural support except that provided by unmined rock in the form of pillars and side walls (stulls, light props, and rock bolts are not considered structural supports). Rock is massive if the spacing between joints, partings, faults, etc., is comparable to, or larger than the critical dimensions of the openings, or if the strength of the bond across partings or joints is comparable to the rock strength. Creep and other inelastic effects are evidenced by occurrences such as roof sag, floor heave, pillar shortening, or a general reduction in the dimensions of openings (in what is sometimes referred to as squeezing ground). If the rock type is such that in an underground structure these occurrences are negligible, or not requiring remedial treatment, the rock is considered to be elastic.

Typical geological types in the competent, massive, elastic class are thick-bedded sandstones and limestones, or more thinly bedded sedimentary rocks in which the partings or other planes of weakness are bonded; massive marbles, quartzites, granites, gabbros; and any jointed igneous or metamorphic rocks in which the joints are bonded.

(2) **Competent Massive Inelastic Rock.** This class is distinguished from that in (1) by the tendency of the rock to creep or flow, as evidenced by the occurrences just described. Rock types predominant in this class are the evaporite minerals: halite (salt), trona, and the potash and borate ores.

(3) **Competent Laminated Elastic Rock.** Besides meeting the specification for competent rock, this class includes all thinly laminated but relatively elastic sedimentary rocks or thinly foliated metamorphic rocks in which the laminae are separated by and/or divided into approximately parallel planes of weakness. This class is comprised mostly of the sedimentary rocks not included in (1), that is, bedded rocks in which the laminae are not cemented. In addition, some metamorphic rocks such as foliated quartzites, schists, and gneisses belong in this class.

(4) **Competent Laminated Inelastic Rock.** This class differs from (3) only in that the rock within the laminae is inelastic. Openings in this class of rock are particularly subject to floor heave or roof sag. Most of the coal measure, sedimentary rocks belong in this class; oil shale (keroginaceous marlstone) is laminated and inelastic to some degree, especially in the oil-rich members; some halite and potash deposits are laminar, the laminae being separated by clastic-filled partings across which there is virtually no bond strength.

(5) **Competent Jointed Rock.** In distinction to competent bedded rock, this class contains more than one set of virtually parallel planes of weakness that tend to divide the rock into parallelepipeds or other multisurface geometrical shapes (polyhedrons). Most rocks contain joints; it is only

when the spacing between joint planes is large, or the joints have become recemented that rock can be classified as massive. Thus the distinction between massive and jointed rock is usually one of degree.

(6) **Incompetent Rock.** As jointing becomes more closely spaced and random in direction as, for example, when there is more than one geologically distinct system of joints, rock becomes incapable of sustaining unsupported underground openings, and hence is classified as incompetent.

Fig. 15.2.1. Start of 30 ft diameter tunnel in chalk, Ft. Randall dam. The peripheral cut was made with a coal saw and the core was removed by blasting. Although the compressive strength of this rock was only 1100 psi, it would be classified as competent massive elastic rock.

The degree of incompetency is increased if the surfaces of the joint planes are altered and decomposition products are present. However, the decomposition should not result in a general disintegration of the rock since in this case the resultant product would be a soil.

The classification specified in Table 15.2.1 was formulated for the purpose of grouping the treatment of rock structure problems according to designated mechanical and geological rock characteristics. However, because the size and depth of an underground opening have a bearing on this method of classification it must be used with some judgment. For example, a low strength rock may be competent at shallow depth but incompetent at a greater depth; rock may be elastic at one depth but inelastic at another; or a bedded deposit may be massive for a small opening but laminated for a larger one. Thus chalk, which is one of the lowest strength rocks (compressive strength approximately 1100 psi) will sustain 30 ft diameter unsupported openings at a depth of 100 ft (Fig. 15.2.1) but would be completely incompetent at a depth of 1000 ft. Salt is relatively elastic at a depth of several hundred feet but becomes inelastic at a depth of several thousand feet. A typical oil shale would be considered massive in relation to a 5 × 7 ft drift but laminated in relation to a room with a 60 ft span.

15.3 IDENTIFICATION OF THE ROCK CLASS

The identification of the rock class and an evaluation of the mechanical and geological characteristics of rock as a structural material in which underground openings can be constructed is a subject which is overlooked too frequently or considered only casually. This is especially true in the exploration for mineral deposits where in many instances core drilling is limited only to the mineralized zone, and often these cores are split and partially or completely destroyed in making chemical analyses. Also drill logs may contain no more than a general identification of the rock type, an estimate of the mineral content, and remarks entered by the drilling contractor. Since the cost of a civil or mining engineering project may be strongly affected by the structural characteristics of the rock, it is generally a poor practice to economize on this phase. For example, in the exploration of a mineral deposit for only a slightly greater cost and increase in time, cores could be taken from both below and above the ore horizon, and especially in the zones that will ultimately comprise the roof and floor of stoping areas. From a careful geological examination of these cores, supplemented by laboratory mechanical property tests, the rock can be classified for the purpose of treating structrual problems. The residue from these tests can still be used for chemical analysis. Also, in the evaluation of existing underground structures the rock usually can be classified by examining cores taken from the walls and other structural parts of the openings.

One of the factors that has limited the mechanical description of rock and drill cores in particular is the lack of an adequate terminology. Terms such as joints, partings, fractures, and separations are often used synonomously or without specific definition. In the logging of cores it has been found helpful to restrict the meaning of these terms as follows:*

Joint. A break of geological origin in the continuity of a body of rock occurring either single, or more frequently in a set or system, but not attended by a displacement. Because joints are of geological origin there are usually some alteration and decomposition products on the joint surfaces which in some instances, may bond the joint.

* This terminology does not necessarily correspond to that used in structural geology.

Fracture. A fresh break in the continuity of a body of rock, not attended by a displacement and not oriented in a regular system. Fractures may be open or closed but not bonded. Fractures are often man-made, as for example, those caused by blasting.

Parting. A thin layer of deposited or altered material separating beds in sedimentary or metamorphic rocks. In sedimentary rock the depositional layer may contain carbonaceous or other organic materials. Partings generally are unbonded, but if the depositional materials have indurated, a bond strength may exist.

Separation. A relatively fresh break along a bedding plane or between beds in sedimentary or metamorphic rocks. Separations may occur on a parting plane and they are usually man-made, that is, they develop as a consequence of mining.

The spacing of joints and bedding is also a factor that has lacked any generally accepted means of specification. Deere[2] proposed the descriptive terminology given in Table 15.3.1, which is a convenience in core logging.

Table 15.3.1. Descriptive Terminology for Joint Spacing and Bedding Thickness[2]

Descriptive Term, Joint Spacing	Descriptive Term, Bedding Thickness	Spacing, in.
Very close	Very thin	Less than 2 in.*
Close	Thin	2 in.*–1 ft
Moderately close	Medium	1 ft–3 ft
Wide	Thick	3 ft–10 ft
Very wide	Very thick	Greater than 10 ft

* If the minimum joint spacing or bed thickness were 4 in., the above scale would be approximately logarithmic.

Logging the mechanical characteristics of drill cores or other rock specimens presents an even greater problem. Terms such as hard, soft, strong, weak, friable, brittle, plastic, etc., are quantitatively indefinite. At the time the cores are logged, however, the engineer must resort to this terminology to qualitatively describe the mechanical characteristics of the rock. As laboratory results become available the logs should be annotated accordingly.

Using the definitions and terminology given in the preceding paragraphs a hypothetical log of a drill cores is given in Fig. 15.3.1. Because this is primarily a log of the mechanical characteristics of the megascopic rock,

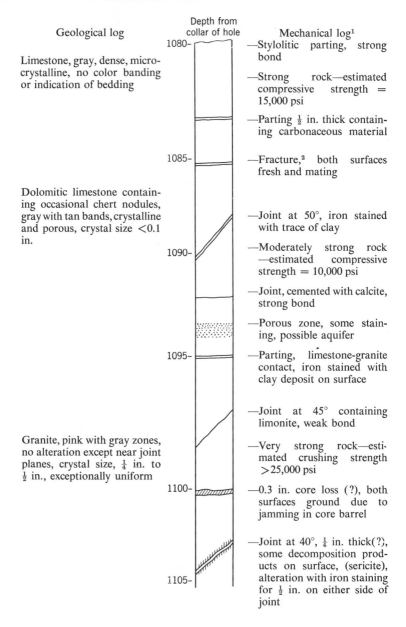

Geological log

Limestone, gray, dense, micro-crystalline, no color banding or indication of bedding

Dolomitic limestone containing occasional chert nodules, gray with tan bands, crystalline and porous, crystal size <0.1 in.

Granite, pink with gray zones, no alteration except near joint planes, crystal size, $\frac{1}{4}$ in. to $\frac{1}{2}$ in., exceptionally uniform

Depth from collar of hole

1080-

1085-

1090-

1095-

1100-

1105-

Mechanical log[1]
—Stylolitic parting, strong bond

—Strong rock—estimated compressive strength = 15,000 psi

—Parting $\frac{1}{2}$ in. thick containing carbonaceous material

—Fracture,[2] both surfaces fresh and mating

—Joint at 50°, iron stained with trace of clay

—Moderately strong rock —estimated compressive strength = 10,000 psi

—Joint, cemented with calcite, strong bond

—Porous zone, some staining, possible aquifer

—Parting, limestone-granite contact, iron stained with clay deposit on surface

—Joint at 45° containing limonite, weak bond

—Very strong rock—estimated crushing strength >25,000 psi

—0.3 in. core loss (?), both surfaces ground due to jamming in core barrel

—Joint at 40°, $\frac{1}{4}$ in. thick(?), some decomposition products on surface, (sericite), alteration with iron staining for $\frac{1}{2}$ in. on either side of joint

Fig. 15.3.1. Sample core log

[1] The double lines across the core indicate core breaks.
[2] Caused by drilling vibration or in handling core.

geological characteristics such as the geological age, fossil identification, etc., have not been included. One factor that may affect the mechanical properties of rock is the microfabric of the rock. However, like mechanical property data, this information must be obtained from laboratory study, and included as a supplement to the core log made in the field.

In competent massive rock the core recovery should be virtually 100%, and the rock should core in unbroken lengths averaging about 3 ft, with some lengths up to 10 ft or more. If the rock is jointed or bedded it should be possible to core across most joint or parting planes without the core breaking at these points.* However, massive jointed or laminated rock may contain an occasional joint or parting plane over which recementing has not been areally complete and hence a core may intercept an unbonded joint or parting. Generally, the decompositional or depositional products on these planes are indurated to the degree that no core loss occurs at these points; in fact, the ends of the core will usually mate. Thick-bedded sandstones, limestones, and the evaporite minerals are the most frequently encountered rocks in this category. Some massive igneous rocks will satisfy these requirements, although rocks in this class are more likely to be jointed, especially in the proximity of mineral deposits. Because the evaporite minerals are probably the only inelastic rocks in which the coring requirements for massive rock can be satisfied, the distinction between competent massive elastic and competent massive inelastic rock is not a difficult one.

Competent laminated rock is characterized by a single set of approximately parallel planes of weakness (partings) on most of which the core will fail as it is cored. Generally the rock between partings should not fail and the core recovery should be almost 100%. This class of rock is comprised principally of the thin bedded, sedimentary rocks. Other bedded rocks, such as quartzites and bonded cherts also belong in this group. In distinction to competent laminated rock, competent jointed rock contains more than one set of approximately parallel planes of weakness (joint planes) on most of which the core should fail as it is cored, but like bedded rock, between joint planes the rock should not fail and the core recovery should be almost 100%. Although most megascopic rock contains joints, it is only when the spacing of open or weak joints is less than the critical

---

* This specification will depend on the competency and care exercised by the driller, and on the type of core barrel employed. For example, the use of a double tube, M-type, core barrel will usually result in a better core recovery with fewer mechanical breaks due to drilling vibration or the core blocking in the core barrel. In solid rock, core fractures caused by drilling vibration can usually be identified by the fact that the fracture surfaces are fresh (no staining or indication of alteration) and will mate. Core failures due to core "freezing" in the core barrel will usually exhibit conchoidal fracture surfaces characteristic of torsional action.

dimensions of the opening under consideration that the rock should be included in this class. In general, igneous or metamorphic rocks are most likely to be in this class, and open or weak joints occur infrequently in the more plastic evaporite minerals.

Incompetent rock will core only in short length, usually less than 1 ft, or in fragments (irrespective of the direction in which the core is drilled), and core recovery may be very low. However, in this class of rock, mega-scopic failures are generally associated with the naturally occurring planes of weakness in the rock, and not necessarily with low rock strengths. In fact, some exceptionally low strength but unfractured rock will stand unsupported over large openings, whereas strong but highly fractured rock will cave freely if even comparatively small areas are unsupported. Hence, before classifying rock as incompetent it is important to establish whether core failures* were the result of joints, fractures, or other planes of weakness, or simply because the rock strength was too low to withstand the drilling action. In some highly fractured incompetent rocks, alteration and decomposition on joint planes may be extensive. If the decomposition products have not indurated, they will often be washed out by the drilling fluid, making it difficult to determine the cause of failure.

15.4   IN SITU STATE OF STRESS

From a consideration of the gravitational force acting on an element of volume in a linear-elastic sphere (for example, the earth), Phillips[3] has shown that the state of stress at a depth which is small compared to the radius of the sphere is given by

$$\sigma_v = -\gamma h \tag{15.4.1}$$

$$\sigma_h = \sigma_v \left(\frac{\nu}{1 - \nu} \right) \tag{15.4.2}$$

where σ_v is the vertical component of stress, σ_h the horizontal component of stress, and h is the vertical depth. Most stress determinations made by stress-relief and hydraulic fracturing techniques (Section 14.3) indicate that for engineering purposes Eq. 15.4.1 is reasonably valid, although significantly lower and higher vertical stresses have been reported in the proximity of megascopic defects of geological origin, for example, near faults or shear zones.

Poisson's ratio for most rocks is between 0.2 and 0.33; hence the ratio of the horizontal to vertical stress calculated from Eq. 15.4.2 should lie between 0.25 and 0.5. However, most in situ measured values of σ_h/σ_v lie

---

* Except fractures identified as "discing." See Section 14.3.

between 0.5 and 0.8 for hard rock, and between 0.8 and 1.0 for soft or inelastic rocks such as shale or salt. In a number of instances σ_h/σ_v has been found to be greater than 1.0. For example, Obert (see Section 14.3) reported that in a massive flat-bedded limestone at a depth of 2300 ft, the horizontal stress was more than twice the vertical stress. Also, horizontal stresses of the order of several thousand pounds per square inch have been measured within 50 ft of the surface in granite and granite-gneiss quarries.

In a relatively uniform subsurface area, that is, one free from faults, shear zones, etc., the stress ellipses in Fig. 15.4.1 are characteristics of possible states of stress. Stress ellipse I represents the gravity stress field

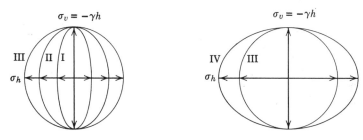

Fig. 15.4.1

given by Eqs. 15.4.1 and 15.4.2 with a Poisson's ratio of 0.25 (or $\sigma_h/\sigma_v = \frac{1}{3}$). In any subsurface area in which there is lateral constraint, this ellipse represents about the minimum lateral stress that could be present. Ellipse II represents the state of stress for (1) an elastic rock with $\nu = 0.4$ (which is about the upper limit for Poisson's ratio); or (2) an inelastic rock in which plastic or viscoelastic deformations have permitted some equalization of the vertical and horizontal stresses; or (3) an elastic or inelastic rock with $\nu < 0.4$, acted on by a horizontal compressional force of geological origin. Ellipse III (circle) represents the state of stress in an inelastic rock such as salt at great depth, in which plastic or viscoelastic effects have permitted a complete equalization of the vertical and horizontal stresses, or any elastic or inelastic rock acted on by a horizontal compressional stress of the magnitude such that $\sigma_v = \sigma_h$. In Fig. 15.4.1, ellipse IV depicts the state of stress in either an elastic or inelastic rock acted on by a horizontal compressional force. The distinction between ellipses II and III and ellipse IV is that the horizontal compressional force is necessary in the latter case.

The subsurface state of stress may be affected by the presence of large-scale mechanical anomalies. For example, let us consider a hypothetical body of rock (hatched area in Fig. 15.4.2a) which crystallized after the surrounding area had solidified, and in the process of crystallization it expanded uniformly in all directions placing the body of rock in hydro-static compression. If, in addition, this body of rock is in a gravity stress field, both the horizontal and vertical components of stress at point *A* would be greater than* those for a gravity stress field alone, as illustrated

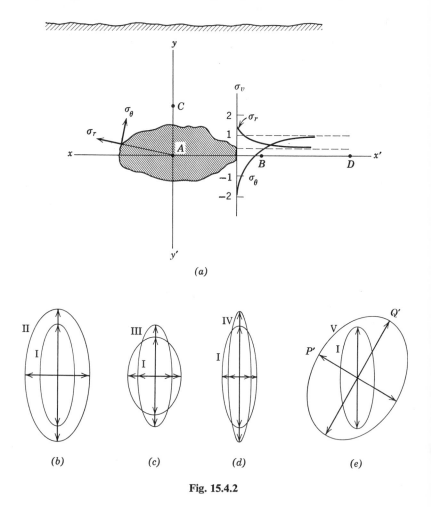

(a)

(b)   (c)   (d)   (e)

Fig. 15.4.2

in ellipse II (Fig. 15.4.2*b*) where ellipse I represents the gravity stress field. At point *B* in the stress concentration zone outside of the compressed body of rock the vertical stress would be less than, and the horizontal stress greater than the corresponding gravity components, as indicated by ellipse

* See footnote, p. 439.

III (Fig. 15.4.2c). Also, because of the equilibrium requirement the vertical stress averaged along any horizontal line through the area and adjoining stress concentration zones (as along the line x, x', Fig. 15.4.2a) would be $\bar{\sigma}_v = -\gamma h$. Similarly, at point C, Fig. 15.4.2a, the horizontal stress would be less than, and the vertical stress greater than the corresponding gravity components, as indicated in Fig. 15.4.2c by ellipse IV. If the depth of the anomaly from surface is much greater than its vertical dimension, and if the average horizontal stress outside the zone of influence (at, say, D, Fig. 15.4.2a) is $\bar{\sigma}_h$, the stress averaged along any vertical line through the anomaly and its stress concentration zones would be $\bar{\sigma}_h$.

If the stress within the anomaly is not uniform, but such that the larger secondary stress lies in some direction other than the horizontal or vertical, the stress ellipse would be like that in Fig. 15.4.2e. However, the vertical stress averaged along any horizontal line through the area would remain $\bar{\sigma}_v = -\gamma h$, and the horizontal stress averaged along any vertical line through the area would remain $\bar{\sigma}_h$.

Although this example is hypothetical it illustrates two important points. First, because the surface if free to move (is unconstrained), the vertical stress, averaged over any equidepth (h—constant) plane of sufficient area will be $-\gamma h$. The area over which the average must be taken will depend on the size of the anomaly. Geological information is insufficient for estimating this area, but for larger geological features it might be of the order of square miles. Second, tectonic forces acting in the horizontal direction will affect the horizontal stress, but the vertical stress will remain unaffected; tectonic forces acting in any direction other than vertical or horizontal will affect both the vertical and horizontal components of stress, and the axes of the stress ellipse will be in some direction other than vertical or horizontal.

These conclusions can be very helpful in analyzing in situ stress data obtained by the borehole deformation method. For example, in some rock types, and especially in jointed rock, it has been found that the magnitude of the secondary principal stresses may vary considerably from point-to-point, whereas their directions will remain almost unchanged. If the secondary principal stresses are consistently in a direction other than vertical or horizontal, it can be inferred that a tectonic force must be acting in the measurement area.

Thus for design purposes, the vertical stress can be estimated from the gravity load, using Eq. 15.4.1. The lower limit of the horizontal stress can be estimated from Eq. 15.4.2. A more realistic horizontal stress can be assumed to be between 0.6 to 0.8 of the computed vertical stress where the lower value is used for massive elastic rock and the higher value for fractured or otherwise inelastic rock. If discing is observed in exploratory cores, the horizontal stress can be computed from Eq. 14.3.26 provided that the shear strength of the rock is known. If hydraulic well-fracturing data are available, the stresses calculated from them should be used. Finally, if the measured stress is available, these values should be employed in all designed calculations.

It should be noted that joining pattern in rock or other evidences of past geological disturbance is not a satisfactory basis for estimating the present state of stress; in fact, the areas in which the greatest deviations from a gravity stress field have been measured is where the evidence of geologic disturbance has been a minimum.

15.5   IN SITU MECHANICAL PROPERTIES OF ROCK

Before underground access is possible, information about the mechanical properties of rock from a proposed underground site can be obtained only from a study of drill cores. A mechanical log of the cores can be made as described in Section 15.3, and mechanical property tests can be performed on specimens cut from the cores. The problem that remains is how to relate this information to the mechanical properties of the rock en masse.

Probably the best approach to the problem of estimating how a given rock type with known laboratory-determined mechanical properties will perform in a proposed structure is to investigate how a similar rock type performs in existing structures. This evaluation is enhanced if the shapes of the proposed and existing structures are similar, or if the existing structure has been instrumented to measure deformations or stresses, or if a structural failure, such as a pillar or roof collapse, has been either intentionally or unintentionally induced. Additional information may be gained from rock model studies, especially studies in which the model is loaded to failure.

The megascopic behavior of rock has been investigated in experimental rooms (Section 14.10) by making an analysis of stable openings, by instrumenting operating mines and other underground structures, and by examining and determining the cause of structural failures. In most of these investigations the rock stress was either measured or calculated from an assumed gravity load. Also, the rock in the walls, pillars, roof, and floor was sampled by core drilling; mechanical and geological logs were prepared, and mechanical property tests performed on specimens taken from these cores. On the basis of this information the following generalizations have been deduced equating laboratory-determined mechanical properties of rock to the performance of the same rock in situ.

Competent Massive Elastic Rock

In rock that can be classified as competent, massive elastic, the laboratory-determined, uniaxial compressive strength is a satisfactory measure of the megascopic compressive strength. Although some loss in the in situ strength may result from minor defects in the rock, this decrease in most instances is probably more than compensated by the increase in strength due to partial confinement. An exception to this generalization would be a tall pillar, that is, a pillar with a height-to-width ratio greater than 1, and in which the core of the pillar has only a minimum confinement.

The in situ shear strength of competent massive elastic rock can be taken as the laboratory-determined triaxial shear strength (Section 11.10). This value depends on the angle of internal friction and will generally lie between $\frac{1}{4}$ and $\frac{1}{2}$ of the uniaxial compressive strength of a specimen having a length-to-diameter ratio of 2. The in situ tensile strength is usually not required because in most underground openings in massive elastic rock, the structural stresses are compressive.

The stress-strain relationship for most massive elastic rock types is relatively linear. For this class of rock the laboratory-determined secant or tangent modulus reasonably approximates that of the megascopic rock. For rock that is not linear elastic (but not to the degree as to be classified as inelastic) the laboratory-measured uniaxial, or better, the biaxial or triaxial tangent modulus of elasticity determined at approximately the same stress level anticipated in service is probably a more realistic value. The laboratory-measured dynamic (sonic) elastic constants generally are not as reliable as the static elastic constants, especially for sedimentary or metamorphic rocks.

The stability of structures in this class of rock is usually not affected significantly by creep and other time-dependent phenomena, even in structures designed for a long lifetime. For example, in an underground opening in rock such as limestone ($\eta = 10^{16} - 10^{17}$ psi/sec) the creep rate would be of the order of 1 to 10 microinches/in. per year. However, local spall and minor fractures on the surface of underground openings are sometimes attributed to time-dependent relaxation of the near-surface stresses. There is evidence to indicate that this process is accompanied by micro-cracks.

Competent Massive Inelastic Rock

Whereas unconfined or partially confined competent, massive elastic rock characteristically fails as a brittle material, competent massive inelastic rock may fail either as a brittle material by fracturing, or because a limiting deformation has been reached. With the latter class of materials brittle failure generally occurs if the rock is comparatively unconfined as in a pillar. However, in short salt and potash pillars (height-to-width ratio less than 1:4) the lateral constraint is large enough to prevent brittle failure even when the axial strain is greater than 30%.

Since the coefficient of viscosity of salt and potash ores is of the order of 10^{12} to 10^{13} psi/sec, the strain rates in these materials is about 10^4 times that for limestone. In potash mines lying at a depth of 2000 to 3000 ft below surface, measured strain rates of the order of several percent per year are not unusual.

In considering problems in which the structural shape is such that brittle failure may occur, the laboratory-determined compressive strength can be used to estimate the in situ compressive strength of this class of rock. However, if the structure is of such a shape that it will deform inelastically but not fracture, the time-dependent behavior of the structure will be the problem of concern. This type of problem is not amenable to a strictly analytic treatment, especially if the geometry of the structure is complex. Model tests, such as those described in Chapter 13, are informative, and an even better approach is to utilize data obtained from existing openings. The evaporite minerals predominantly comprise this class of rock, and generally the inter- and intra-site mechanical characteristics of any one of these minerals are relatively uniform. Hence, field data from any site can be used to estimate the behavior of the rock at another site. If site-to-site differences in the material exist or if the structure is to be mined at a greater or lesser depth, that is, at a higher or lower stress level, the effect of these differences can be estimated by similitude studies.

Competent Laminated Rock

From a mechanical standpoint, laminated rock is made up of a number of approximately parallel laminae or layers, between which there may be a mechanical bond, a partial bond, or no bond. The rock layers may be of the same or different kinds or rock, and the partings between layers may contain organic materials or other minerals. Thus the partings and the layers themselves may form discontinuities that affect the mechanical properties of the megascopic rock. If the bond strength is comparable with that of the rock in the layers, the megascopic rock may be considered as massive. If the rock is unbonded or partially bonded the shear strength on the parting planes will be lower than that of the rock. How this lowered shear strength affects the megascopic rock strength will depend on the attitude of the parting planes with respect to the applied stresses. A triaxial procedure for measuring the effect of a plane or planes of weakness is described in Section 10.6. This procedure is applicable to both recemented and unbonded partings.

Before underground access is possible an estimate of some of the structural characteristics of competent bedded rock can be obtained from tests on exploratory drill cores. They are as follows:

1. If the bond is such that the rock will core across partings (excepting breaks caused by drilling vibration or impact), and if the angle of the parting in test specimens cut from the core is approximately the same as the attitude of the partings in the proposed structure, then the uniaxial compressive strength of the specimens can be used for the compressive strength of the megascopic rock.

2. For either bonded or unbonded partings, if the angle of the parting plane with respect to the axis of a drill core specimen is greater than 70°, and if the attitude of the partings in the proposed structural member is approximately the same, the uniaxial compressive strength of specimens can be used for the compressive strength of the in situ rock. This strength equivalent is especially useful in calculating the bearing strength of pillars in comparatively flat-bedded formations.

3. A thin layer of relatively weak rock lying between thicker layers of stronger rock, and in a plane roughly normal to the applied compressive stress, will not significantly lower the compressive strength of the three-layer system below that of the stronger components. This estimate of the compressive strength of the three layers is valid irrespective of the degree of bond between the layers. Thus in determining the compressive strength of layered rock in, say, a mine pillar where the layers are less than 20° from the perpendicular to the applied load, the average compressive strength of the various layers can be used as an estimate of the pillar strength.

4. For bedded rock with weak or unbonded partings lying at an angle less than 70° with respect to the direction of the applied stress, the compressive and shear strength of the collective rock becomes dependent on the shear strength and angle of internal friction of the parting or the material in the parting. For this class of rock an estimate of the joint strength can be obtained by triaxial test.

5. The tensile or flexural strength of rock measured in a direction normal to the bedding is generally lower than that of a core specimen cut parallel to the bedding. As the tensile or flexural strength is required in determining the span of a roof layer, the strength determination should be made on specimens cut from, and parallel to, the roof layer. The bedding direction in most exploratory cores is normal to the core axis. Hence, a determination of the span in layered roof usually cannot be made until underground access is possible so that properly oriented specimens can be obtained. The single-point flexural strength (Section 11.8)

has been found to approximate the in situ tensile strength of a roof layer better than the laboratory-determined uniaxial tensile strength, the latter usually being much lower. The tensile strength determined by either the Brazilian test (Section 11.6) or the two-point flexural strength is generally intermediate.

Competent Jointed Rock

Whereas in bedded rock the partings generally form a single set of weakness planes that divide the rock into layers, in jointed rock there are usually three or more sets of weakness planes that divide the rock into parallelepipeds or polyhedrons. Thus in jointed rock there is a greater probability that one of the planes of weakness will lie in an unfavorable direction with respect to the applied stresses, that is, in a direction such that failure will take place in, rather than across, the joint plane. For the purpose of designing or evaluating rock structures, the conservative assumption would be that failure in jointed rock will take place on a joint plane. To evaluate the strength of a body of rock containing such a joint or set of joints, the shear strength and coefficient of internal friction of the joint filling material as well as the direction of the joint plane with respect to the applied stresses are required (Section 10.6). Because the properties of the joint material are extremely variable and indefinitely known, the best procedure is to subject a group of specimens containing a representative joint that is oriented at an angle between 15 and 40°* to the specimen axis, to triaxial test.

Incompetent Rock

The core recovery from incompetent rock is usually so poor that it is impossible to prepare a specimen for any kind of mechanical property test. Hence no early basis for estimating the mechanical properties of this class of rock can be prescribed. The qualitative characteristics of incompetent rock are described in Chapter 18.

15.6   CRITERION OF FAILURE

A criterion of failure is a specification of the mechanical condition under which solid materials fail by fracturing or by deforming beyond some specified limit. This specification may be in terms of the stresses, strains, rate-of-change of stresses, rate-of-change of strains, or some combination of these quantities, in the materials. Because materials are seldom tested under a mechanical condition equivalent to that which they will experience in service (except in models), it is necessary for engineering

---

* Between these limits the effect of the joint angle is not critical, see Jaeger.[3]

purposes to have a valid criterion of failure. However, there is an indefiniteness regarding a criterion of failure that is valid for all classes of rock and for all mechanical conditions. Hence, before a criterion of failure is specified for use in the treatment of rock structure problems, an examination of the mechanical condition that normally exists in the rock in an around underground openings is in order.

Because of the stress concentration that exists on or near the surface of underground openings, failure is most likely to be initiated in this high stress zone. Moreover, because the component of stress normal to and at the surface is zero, the zone in which failure is most likely to initiate is either in plane (biaxial) stress or, in instances such as on the surface of a pillar, in a uniaxial state of stress. Under this limited constraint most rocks will fail as a brittle material. Another factor that effects the mode of failure is that the measured horizontal component of the stress field in underground rock is generally found to be greater than that given by $S_v[v/(1 - v)]$, where v is the laboratory determined value of Poisson's ratio and $S_v = -\gamma h$. This increased horizontal stress tends to make the stress on the boundary compressive at all points, a fact that is borne out by measurement of surface stresses (Section 14.5).

Mohr's theory of fracture stipulates that a material will fail either when the shear stress τ_n on the fracture plane has reached a limiting value dependent on the normal compressive stress σ_n acting across the plane, or when the normal tensile stress σ_n has reached a limiting value T_0 (Section 10.5), that is,

$$\tau_n = f(\sigma_n)$$
$$\sigma_n = T_0$$
(15.6.1)

A preponderance of evidence indicates that this theory is reasonably valid for brittle materials that fail by fracture, that is, by separating and forming distinct surfaces. Most rocks that are not completely confined will fail in this manner and hence this theory is valid for failure initiating on the surface of rock structures.

One of the postulates of Mohr's theory of fracture is that in a triaxially loaded body the magnitude of the intermediate stress does not affect the magnitude of the algebraically greatest and least stress at failure (although the fracture plane will pass through the direction of the intermediate stress). If, at failure, the stress normal to a rock surface is σ_1, σ_2 and σ_3 are the principal stresses parallel to the plane, and if $\sigma_1 = 0$ and $\sigma_3 < \sigma_2 < 0$ (recalling that tensile stress is positive), then this state of stress is represented by the Mohr's circle in Fig. 15.6.1. As σ_1 and σ_3 are failure stresses this circle must be tangent to the failure envelope $\tau_n = f(\sigma_n)$. Therefore the uniaxial compressive strength can be used to predict shear fracture initiating on or near the surface of a rock structure. From the second condition in Eq. 15.6.1, if on the rock surface σ_3 is tensile and reaches a value T_0, the rock will fail in tension. Thus Mohr's theory of fracture provides a satisfactory criterion of failure for use in designing or evaluating the stability of rock structures, that is, rock will fail in tension when the tensile stress exceeds the tensile strength, as determined by a standardized test. If the tensile stress is small or zero, the rock will fail in shear at a value of the compressive stress equal to the compressive

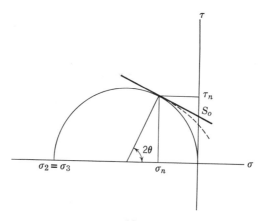

Fig. 15.6.1. Mohr's representation of the state of stress on the surface of an underground opening.

strength of the rock as determined by a standardized compressive test on a sample of the rock.

The most general expression for Mohr's fracture envelope is $\tau_n = f(\sigma_n)$ and for many rocks this envelope is curved downward in the range $\sigma_3 \leq \sigma_2 < 0$, as indicated in Fig. 15.6.1. In the same stress range Griffith's theory of brittle fracture, which is based on the propagation of microscopic cracks, predicts a parabolic envelope as indicated by the dash-line envelope in Fig. 15.6.1, whereas the Coulomb-Navier theory of failure (Section 10.5) predicts a linear envelope, $\tau_n = S_0 - \mu\sigma_n$. These theories lead to different values of the shear stress at failure, but they do not alter the basic premise that if $\sigma_1 = 0$ the uniaxial compressive strength is a measure of the maximum compressive stress that the near-surface rock will withstand. The point that is indefinite is the role of the intermediate stress σ_2. Several investigators have found that it does have some effect on the value of σ_1 and σ_3 at failure.*

* Obert and Stephenson[4] reported that failure on the interior surface of a triaxially loaded thick-wall cylinder is independent of the intermediate stress.

15.7 EFFECTS OF ANISOTROPY

The strength and elastic properties of most rocks vary with direction; for example, in sedimentary and foliated rocks, the tensile and flexural strengths and modulus of elasticity are generally lower in the direction perpendicular to the bedding or foliation than in directions parallel to the bedding or foliation. In some instances these differences are large enough so that their effects should be considered in evaluating rock structure problems. However, very little is known about the anisotropic properties of in situ rock; in fact, only meager data are available on the anisotropic properties of laboratory size specimens.

Table 15.7.1 gives the strength and modulii of elasticity of a number of rock types, measured in perpendicular directions. These data indicate that the compressive strength of most rocks is relatively independent of direction; in fact, the directional variation is generally no larger than the specimen-to-specimen variation. Hence, unless tests of a specific rock type indicate a larger variation than that given in Table 15.7.1, the effect of compressive strength variation with direction can be neglected in massive rock. On the other hand, the tensile and flexural strengths of bedded and foliated rock depend to a greater degree on direction; differences as large as tenfold in the flexural strength measured perpendicular and parallel to bedding have been measured. Thus, for example, in evaluating the stability of a laminated mine roof, the flexural strength of the roof rock should be determined from cores taken parallel to the roof span. Donath[5] measured the triaxial shear strength (Section 11.10) of a group of Martinsburg slate specimens prepared so that the cleavage plane varied in 15° increments from 0 to 90° from the direction of the larger (axial) applied compressive stress. The maximum and minimum values of the shear strength were 9400 psi and 600 psi, a difference of over fifteenfold.

The directional variation in the elastic properties of rock affect the stress distribution in the rock surrounding underground excavations and in structural members. In the most general case for anisotropic rock there are 36 elastic constants of which 21 are independent. From an engineering standpoint it would be impractical to attempt a determination of these constants; even if they could be measured, the solution of three-dimensional stress distribution problems in anisotropic rock would be intractable. However, some indication of the magnitude of the effects produced by elastic anisotropy can be gained from a consideration of a simplified case, namely, that of a circular hole in an infinite orthotropic*

* An orthotropic body has three orthogonal planes of elastic symmetry at each point, and satisfies the relationships $E_1\nu_{21} = E_2\nu_{12}$, $E_2\nu_{32} = E_3\nu_{23}$, $E_3\nu_{13} = E_1\nu_{31}$ (E_1, ν_{12}, G_{12}, etc. defined as above). An orthotropic plate is a slice from this body parallel to any one of the planes of symmetry.

Table 15.7.1. Properties of Anisotropic Rock

Rock Type	Compressive Strength, psi × 10³			Flexural Strength, psi × 10²			Tensile Strength, psi × 10²			Modulus of Elasticity, psi × 10⁶		
	$\perp^1$	$\|A^2$	$\|B$	$\perp$	$\|A$	$\|B$	$\perp$	$\|A$	$\|B$	$\perp$	$\|A$	$\|B$
Marble (Md.)	30.8	31.2	33.0	2.8	3.3		9.0			7.15	9.15	10.4
Limestone (Ind.)	10.9	9.7	10.2	1.6	1.9	1.7	5.7			4.84	5.94	5.39
Limestone (O.)	21.3	26.1	20.4	2.8	4.8	4.5	3.5	13.8	10.0	9.93	8.19	8.96
Granite (Vt.)	33.3	32.8	35.4	2.9	3.7	2.7	11.0			4.41	3.97	6.41
Granite (N.C.)	28.4	30.4		2.9	1.6						3.28	4.39
Slate (Pa.)	30.4	22.5	30.5								13.6	12.1
Granite (Colo.)	23.0	23.4		2.6	3.3		5.2	8.1		5.54	6.13	
Sandstone (O.)	10.4	8.0	7.7	0.5	0.8	0.8	11.9			0.87	0.97	1.28
Sandstone (O.)	6.1	5.2	5.1	0.4	0.7	0.8	14.7			1.03	1.54	1.63
Sandstone (U.)	8.1	4.8		0.4	0.6		4.8			1.39	1.53	
Gneiss (Ga.)	30.3			2.0						2.70	3.35	1.80
Oil Shale (Colo.)	16.6			0.4	5.2					1.80	3.10	
Oil Shale (Colo.)	15.1	25.0								3.06	4.82	
Hematite Ore (Ga.)	19.7	17.2		1.0	3.4						10.1	

[1] For bedded or foliated rock, $\perp$ indicates perpendicular to bedding or foliation; $\|A$ and $\|B$ indicate perpendicular directions in the plane of the bedding or foliation.

[2] For rock without bedding or foliation, $\perp$, $\|A$, $\|B$ indicate mutually perpendicular directions.

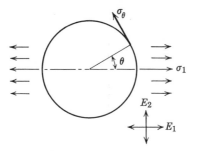

Fig. 15.7.1

plate subjected to a uniaxial applied stress σ_1 (Fig. 15.7.1). The tangential stress on the boundary of the hole σ_θ is given by*

$$\sigma_\theta = \sigma_1 \frac{E_\theta}{E_1}\left[(1 + \eta) \sin^2 \theta + \sqrt{\frac{E_1}{E_2}}\cos^2 \theta\right] \qquad (15.7.1)$$

where E_1, E_2 = the modulii of elasticity parallel and perpendicular to σ_1
$\quad\quad G_{12}$ = the modulus of rigidity in the plane of the plate
$\quad\quad \nu_{12}$ = Poisson's coefficient for the decrease in the direction perpendicular to σ_1 for a tension in the σ_1 direction

and where

$$\eta = \left[\frac{E_1}{G_{12}} - 2\nu_{12} + 2\sqrt{\frac{E_1}{E_2}}\right]^{\frac{1}{2}}$$

$$\frac{1}{E_\theta} = \frac{\sin^4 \theta}{E_1} + \left(\frac{1}{G_{12}} - \frac{2\nu_{12}}{E_1}\right) \sin^2 \theta \cos^2 \theta + \frac{\cos^4 \theta}{E_2}$$

Also, the condition $E_1\nu_{21} = E_2\nu_{12}$ must be satisfied.

If the values of the modulii of elasticity for oil shale are used, namely that $E_1 = 1.80 \times 10^6$ psi parallel to the bedding and $E_2 = 3.1 \times 10^6$ psi perpendicular to the bedding,† the condition for orthotropy is satisfied if the values of Poisson's coefficients are assumed to be $\nu_{12} = 0.18$ and $\nu_{21} = 0.31$. Also, assume a value for the modulus of rigidity of $G_{12} = 1.0 \times 10^6$ psi. For this set of elastic constants and with the bedding normal to the applied stress, the stress distribution is as shown in Fig. 15.7.2, where the radial arrows indicate the magnitude of the tangential boundary stress—compressive if directed inward, and tensile if directed outward. The corresponding stress distribution for an isotropic material

* This problem is considered in, *Theory of Elasticity of an Anisotropic Elastic Body*, by S. G. Lekhnitskii.[6]
† From row 10, Table 15.7.1; of the rock types listed in Table 15.7.1 oil shale has the greatest elastic anisotropy.

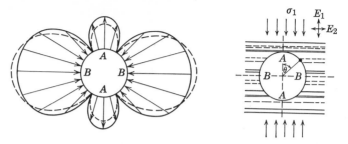

Fig. 15.7.2. Stress distribution in isotropic and anisotropic rock. Solid line = anisotropic rock, dashed line = isotropic rock.

is indicated by the dashed line. For this example the maximum effect of elastic anisotropy is to increase the tensile stress at points AA by 32% and decrease the compressive stress at points BB by 9%. If the applied stress is parallel to the bedding, the stress distribution is as shown in Fig. 15.7.3. In this case the tensile stress at points AA is decreased by 24% and the compressive stress at points BB is increased by 8%.

The effect of elastic anisotropy on the stress distribution is greatest for a uniaxial state of stress. In the example just considered, if the stress field is hydrostatic the maximum difference in the tangential boundary stress σ_θ for the isotropic and anisotropic case is 11%.

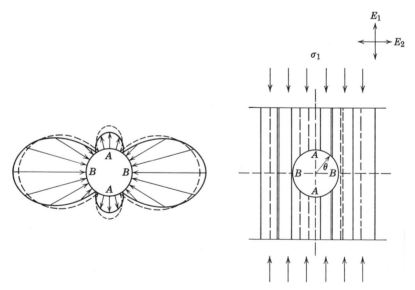

Fig. 15.7.3. Stress distribution in isotropic and anisotropic rock. Solid line = anisotropic rock, dashed line = isotropic rock.

Thus it can be inferred that for most rock the effects of elastic anisotropy are no larger than the normal variations in rock strength and hence they can be neglected. The most likely exceptions to this generalization would be strongly foliated metamorphic rocks, such as micaceous schists. If the modulii of elasticity for rock in this class differ by a factor greater than two, stress distributions around geometrically simple openings (circles and ellipsis) can be approximated from the theory for an orthotropic plate.[6]

15.8   EFFECT OF MOISTURE AND PORE PRESSURE

The laboratory-determined compressive strength of comparatively uniform specimens of marble, limestone, granite, sandstone, and slate is relatively unaffected by the moisture content in the rock. For example, in the moisture-content range between air-dry and saturated, the decrease in the compressive strengths of these rocks averaged only 12%.[7] In situ rock would probably not experience this great a variation in moisture content and hence the compressive strength of massive rock should be affected even less. The elastic properties of the same group of rocks also exhibit only a small variation in the air dry-to-saturated moisture-content range. Hence, without specific test results to the contrary it should be assumed that the properties of the more competent rocks are not significantly affected by moisture. However, some of the less competent rocks, such as shale, siltstone, and mudstone, are affected and sometimes permanently altered by moisture, and especially by a change in their moisture content (Section 10.7). These moisture changes are often accompanied by an expansion or contraction in the rock, and a general deterioration of surface or near-surface exposure. Rock in this class should always be considered suspect and an evaluation of moisture effects determined by test.

If the water in a porous rock is under pressure,* the pore pressure can affect the stability of an underground structure. Hubbert et al.[8] hypothesized that the pore fluid exerts a hydrostatic pressure on individual rock grains such that the collective effect is to change (algebraically) the stress in the rock by an amount equal to the pore pressure, a hypothesis that was confirmed by experiment. Thus for a given state of stress the addition of pore pressure acts to displace Mohr's circle toward the fracture envelope (Fig. 10.7.2) and in so doing reduces the effective rock strength.

At a depth h_w below the water table the hydraulic pressure p_w is $p_w = \gamma_w h_w$, where γ_w, the unit weight of water, is 0.036 lb/in.^{3}. At a depth h_r the vertical and horizontal components of stress in dry rock,

---

* In this treatment it will be assumed that the pore fluid is water.

$\sigma_v = \gamma_r h_r$ and $\sigma_h = \sigma_v[\nu/(1 - \nu)]$, where γ_r, the unit weight of rock, is 0.9 lb/in.[3]. If the water table is near the surface, that is, if $h_w \simeq h_r$, then $p_w \simeq 0.4\sigma_v$ and $p_w \simeq \sigma_h$. Hence, if confined, the pore pressure could appreciably affect the stability of an underground structure. However, underground drainage through joints, faults, and water courses usually lowers the water table, and in addition, a positive pressure gradient is produced between the exposed underground surface and the point where the hydraulic pressure is p_w. This pressure gradient has been measured in several comparatively deep mines and found to be of the order of 1–3 psi/ft. Thus in the zone around underground openings where the rock is most likely to fail, the effects of pore pressure should be small compared to the rock strength and therefore can be neglected. In more confined structures such as in a development drift or shaft, higher pressure gradients may occur and if suspected, they can be determined by making packer tests in boreholes.

15.9   SAFETY FACTOR

To allow for differences between the laboratory-determined and in situ strength of rock, plus the errors introduced by the various assumptions required in the computation of stresses and deformations, it is necessary to use a safety factor in the treatment of rock structure problems. The safety factor is defined as the ratio of the maximum stress in the material in service (the working load) to the strength of the material. However, as we saw in the introduction to Part one, the design specifications for an underground structure for military or industrial purposes may be vastly different from those for a mine. In the former the structure is designed for a high degree of safety over a long lifetime, whereas in a mine because of economic pressure and a desire to conserve resources there is usually a premium on extracting as much of the ore as possible. Thus the problem may become one of designing an opening that will remain open long enough to remove the mineral safely. In this sense a mine may have a safety factor in time, that is, if the required time to remove the mineral is, say, one year and, under the system of mining, the stopes will normally remain open two years, the time safety factor would be 2. Unfortunately, the time-dependent properties of rock and the mechanism of failure in inelastic rocks are not sufficiently understood so that a design can be formulated on this basis.

In practice, it has been found that a safety factor of 2 to 4 is adequate for structural members in compression such as pillars and sidewalls, whereas a safety factor of 4 to 8 is required for members in tension such as bedded roof.[1] Usually the lower values are used in the design of mines with a relatively short lifetime, and the higher values for openings with a long lifetime. These values were obtained from an analysis of a large number of existing structures. In this analysis the structural stresses were computed assuming gravity loading, and the properties of the rock were measured in the laboratory on specimens taken from the structure.

REFERENCES

1. Obert, L., W. I. Duvall, and R. H. Merrill, "Design of Underground Openings in Competent Rock," *U.S. BurMines Bull. 587* (1960).
2. Deere, Don U., "Technical Description of Rock Cores for Engineering Purposes," *Rock Mech. and Eng. Geo.*, 1-1 (1963).
3. Obert, L. and D. E. Stephenson, "Stress Conditions Under Which Core Discing Occurs," *Trans. SME AIME*, **232**(3) (1965).
4. Jaeger, J. C., "The Frictional Properties of Joints in Rocks," *Geofis. Pura Appl.*, **42-44** (1959).
5. Donath, F. A., "Experimental Study of Shear Failure in Anisotropic Rock," *Bull. Geo. Soc. Am.*, **72** (1961).
6. Lekhnitskii, S. G., *Theory of Elasticity of an Anisotropic Elastic Body*, Holden-Day Inc., San Francisco, Calif., 1963.
7. Obert, L., S. L. Windes, and W. I. Duvall, "Standardized Tests for Determining the Physical Properties of Mine Rock," *U.S. BurMines Rept. Invest.*, **3891** (1946).
8. Hubbert, M. King, and Wm. W. Rubey, "Role of Fluid Pressure in Mechanics of Overthrust Faulting," *Bull. Geol. Soc. Am.*, **70** (Feb. 1959).

CHAPTER 16

DESIGN OF A SINGLE OPENING IN COMPETENT ROCK

16.1 INTRODUCTION

If the stress distribution in the rock surrounding an underground opening is not affected by other surfaces or underground voids, it is considered to be a *single opening*. Thus a tunnel lying deep within a mountain would be a single opening, whereas a subway tunnel lying a short distance below the surface is not a single opening because its stress distribution would be affected by the surface. All underground rock is under stress because of the weight of the overlying rock and possibly because of stresses of tectonic origin. In addition, an underground opening will produce a stress concentration in the surrounding rock. If the stress in the surrounding rock exceeds its strength,* the opening will fail either by fracturing or by deforming more than some tolerable limit. Most competent rock fails by fracture, whereas incompetent types generally fail by deforming excessively and to the degree that substantial support, such as sets and linings, is required to keep the openings in service. However, openings in some competent massive inelastic and competent laminated inelastic rock, such as salt, trona, and the borate and potash ores, can be mined and kept in service for limited periods even though large displacements easily discernable to the unaided eye occur.

* Strictly speaking, because tensile stress and strength are positive quantities, and compressive stress and strength are negative quantities, failure will occur if the tensile stress is greater than the tensile strength, or if the compressive stress is less than the compressive strength. The latter statement is awkward and likely to be misinterpreted. To clarify expression statements such as "the stress is equal to, greater than, exceeds, etc., the strength" will mean that the absolute value of the compressive or tensile stress is equal to, greater than, exceeds, etc., the absolute value of the corresponding compressive or tensile strength. However, when computations are made the proper sign must be applied to all numerical values.

The classes of competent rock that generally fail by fracture are massive elastic, laminated elastic, and to a less definite degree, fractured and jointed rock. The stress distribution around openings, or in structural members in these classes of rock, can be approximated either from elastic theory (mathematical models) or from measurements on physical models made from elastic materials. The magnitude of the stress calculated or measured on models will generally be larger than the stress at corresponding points on the prototype structure. Hence, a design based on model result will be on the conservative side. The results of in situ measurements given in Chapter 14 make it possible to apply some general correction to values determined from either theory or model study to bring them more closely in line with reality. As an engineer becomes more acquainted with the measured stress pattern around openings in a given rock type, the better this type of adjustment can be made. In fact, as experience is gained an engineer will learn to evaluate various rock types in terms of the degree that they deviate from perfectly elastic materials, and to design or evaluate structures in similar rock on this basis.

The problem of designing a structure in inelastic rock is more difficult. For one thing, this class of rock may fail under one set of conditions as a brittle material by fracturing, and under another set of conditions by deforming excessively. Single openings, however, generally fail by excessive deformation although local fracture may cause surface sloughing. Another factor limiting design of an opening in inelastic rock is the lack of an adequate theory. Although theories of inelastic behavior of rock given in Chapter 6 describe in a qualitative way the behavior of inelastic materials, they do not provide quantitative answers. A part of this lack of quantitativeness results from the fact that the viscoelastic constant η, which appears in equations relating strain rate to stress is, for most inelastic rocks, even more stress dependent than the elastic constants are for the more elastic types of rock. Thus the design of structures in inelastic rock must be based more on empirical results, which may be derived either from measurement on models made from the prototype rock or from measurements made in situ.

The procedure for designing or evaluating the stability of single openings in competent rock is as follows:

1. For rock that fails as a brittle material, that is, for massive elastic, laminated elastic, or fractured and jointed rock, determine the maximum (absolute) values of the tensile and compressive stresses on the boundary of the opening or in a structural part from an analysis of a mathematical model, or from a model made from an elastic material or from the prototype rock. These maximum values are called *critical stresses*. If stress

determinations have been made in the prototype rock or in an opening in an equivalent rock that indicate that the in situ critical stresses are lower than the model values, the latter can be decreased accordingly to obtain adjusted critical stresses.

2. Measure the rock strengths, using the procedures described in Chapter 15.

3. Select an appropriate safety factor (Section 15.9).

4. For stability, the rock strength must be equal to or greater than the product of the appropriate critical stress (or adjusted critical stress) and the safety factor.

5. For rocks that fail by deforming excessively, the limiting strain rates can be determined only by similitude studies in which the more realistic result will be obtained if rock from the prototype structure is employed in the model. The limiting deformation in the prototype will depend on the method of mining.

Since the engineer responsible for designing or evaluating the stability of an underground structure seldom has an opportunity to select the rock type in which the opening is to be excavated, the problem generally reduces to one of determining the shape of opening that will minimize critical stresses or strain rates. The sections in this chapter deal primarily with the determination of critical stresses around two and three-dimensional openings in rock with assumed ideal properties, and in which the opening is subjected to various stress fields.

16.2   OPENING IN MASSIVE ROCK, VERTICAL AND HORIZONTAL AXES, TWO-DIMENSIONAL CASE

The distribution and magnitude of the stresses around a single opening in massive elastic rock can be approximated by means of elastic theory (Chapters 1 to 5), or from models made from elastic materials, as for example, photoelastic models (Chapter 12), provided that simplifying assumptions are made regarding the mechanical properties of the rock, the shape of the opening, and the stress field, that is, the state of stress in the rock before mining.

In the following treatment of this problem it will be assumed that (1) massive elastic rock is linear elastic, homogeneous, and isotropic with respect to its mechanical properties, and (2) the opening is in an infinite medium—this condition is satisfied if the distance from the opening to an adjacent boundary is greater than three times the dimension of the opening in the direction of the boundary.

To further simplify discussion, this problem will be considered first in two dimensions (a brief treatment of a single opening in three dimensions is given in Section 16.4). For this condition it is necessary that (1) the opening is long compared with its cross section, and (2) the stress distribution along the length of the opening is uniform and independent of length.

If these conditions are satisfied, the problem of determining the stress distribution around single openings reduces to one of plane strain and may be solved by considering a hole in a wide plate subject to a uniform two-directional stress field in the plane of the plate.

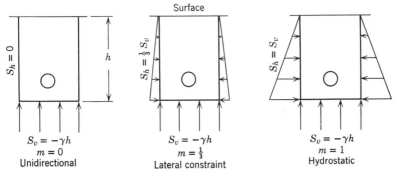

Fig. 16.2.1. Three assumed types of stress fields.

It will also be assumed that (1) the cross section of the opening can be represented by simple geometrical shapes, namely, a circle, ellipse, ovaloid, or rectangle with rounded corners; (2) the length of the opening is horizontal, and the axes of the cross section are horizontal and vertical unless otherwise specified (inclined openings are considered in Section 16.3); (3) the vertical stress is equal to the weight of the overlying rock; thus

$$S_v = -\gamma h \qquad (16.2.1)$$

where S_v is the vertical compressive stress, γ is the unit weight of the rock, and h is the vertical distance below the surface; and (4) the horizontal stress is given by

$$S_h = mS_v \qquad (16.2.2)$$

where S_h is the horizontal compressive stress and m is a constant depending on the stress field.

Three types of stress field will be considered in this chapter; these are represented graphically in Fig. 16.2.1, and they are given algebraically by Eqs. 16.2.1 and 16.2.2, where the value of m is 0, $\frac{1}{3}$, or 1. The state of stress represented by $m = 0$ might occur at shallow depths, and/or near-vertical free surfaces. The state of stress represented by $m = \frac{1}{3}$ might occur over a wide range of depths. If the body of rock is completely constrained so that there is no lateral deformation, the relation between the vertical and horizontal stress is given by (Section 15.4)

$$S_h = S_v\left(\frac{v}{1-v}\right) \tag{16.2.3}$$

where v is Poisson's ratio. When v is, say, $\frac{1}{4}$, the ratio of S_h to S_v is $\frac{1}{3}$. Thus the state of stress represented by $m = \frac{1}{3}$ corresponds to the condition of no lateral constraint in a rock having a Poisson's ratio of $\frac{1}{4}$. The state of stress corresponding to $m = 1$ might occur at great depth or in semi-viscous or plastic rocks.

Circular Opening

The exact solution for the stresses around a circular hole in an infinite plate in a biaxial stress field is (Section 4.7)

$$\sigma_r = \left(\frac{S_h + S_v}{2}\right)\left(1 - \frac{a^2}{r^2}\right) + \left(\frac{S_h - S_v}{2}\right)\left(1 - \frac{4a^2}{r^2} + \frac{3a^4}{r^4}\right)\cos 2\theta$$
$$\tag{16.2.4}$$

$$\sigma_\theta = \left(\frac{S_h + S_v}{2}\right)\left(1 + \frac{a^2}{r^2}\right) - \left(\frac{S_h - S_v}{2}\right)\left(1 + \frac{3a^4}{r^4}\right)\cos 2\theta \tag{16.2.5}$$

and

$$\tau_{r\theta} = \left(\frac{S_v - S_h}{2}\right)\left(1 + \frac{2a^2}{r^2} - \frac{3a^4}{r^4}\right)\sin 2\theta \tag{16.2.6}$$

where

S_h = horizontal applied stress*

S_v = vertical applied stress*

σ_r = radial stress

σ_θ = tangential stress

$\tau_{r\theta}$ = shear stress

a = hole radius

r = radial distance from center of hole,

θ = polar coordinate; horizontal axis represents $\theta = 0°$

Equations 16.2.4, 16.2.5, and 16.2.6, show that the stresses around the opening are independent of the elastic constants of the material and the radius of the hole. The radius a appears in these equations only in the dimensionless ratio a/r, which specifies the distance from the boundary of the hole.

---

* Compressive stress is negative and tensile stress is positive.

For convenience the magnitude of any one of the stresses near the boundary of an opening is expressed as a ratio of the stress at a point to one of the applied stresses. This ratio is referred to as the *stress concentration*. A *positive stress concentration* means that the stress at a certain point has the same sign as the applied stress. A *negative stress concentration* means that the stress at a point has the opposite sign of the

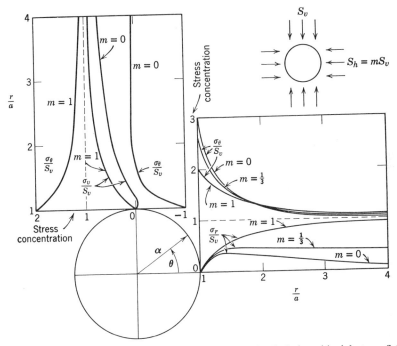

Fig. 16.2.2. Axial stress concentrations for a circular hole in a biaxial stress field.

applied stress. The maximum positive stress concentration and the minimum negative stress concentration are called *critical stress concentrations*.

Stress concentrations along the axes from a circular opening in a uniaxial ($m = 0$) and biaxial hydrostatic ($m = 1$) stress field, calculated from Eqs. 16.2.4, 16.2.5, and 16.2.6, are given in Fig. 16.2.2.* This figure shows that the tangential stress concentration is a maximum at the boundary and decreases rapidly with the distance from the boundary. Note especially that for $r < 2a$, σ_θ/S_v for $m = 1$ is less than σ_θ/S_v for $m = 0$, whereas for $r > 2a$ the opposite is true. This is necessary to satisfy the

* The stress distribution curve for $m = \frac{1}{3}$ lies between the curves for $m = 0$ and $m = 1$, but it is closer to the $m = 0$ curve.

equilibrium condition discussed in Section 14.3. As a result of this condition, for any opening of given width, the area between the horizontal axis and the σ_θ/S_v distribution curves for any value of m must be constant.

The distribution curves for σ_r/S_v show that at the boundary σ_r is zero, but this distribution changes rapidly so that at $(r/a) > 4$, σ_r is approximately equal to the applied stress in the same direction. The shear stress

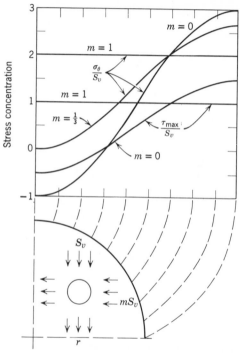

Fig. 16.2.3. Boundary-stress concentrations for a circular hole in a biaxial stress field.

in the $r\theta$-plane behaves similarly, becoming approximately equal to $S_v(1 - m)/2$ at $r/a > 4a$. Figure 16.2.3 shows the tangential and maximum shear stress distribution on the boundary of a circular opening for stress-fields corresponding to $m = 0$, $m = \frac{1}{3}$, and $m = 1$. For $m = 0$, the maximum tangential boundary stress concentration at $\theta = 0°$ and $\theta = 180°$ is 3, and at $\theta = 90°$ and $\theta = 270°$ it is -1. Thus for an applied compressive stress of $-S_v$, the maximum tangential compressive and tensile boundary stresses are $-3S_v$ and $+S_v$, respectively; for $m = \frac{1}{3}$ the maximum stress concentration is $2\frac{2}{3}$ at $\theta = 0°$ and $\theta = 180°$. At $\theta = 90°$ and $270°$, σ_θ/S_v is 0; for $m = 1$ the stress concentration is 2 at all points on the boundary of the circle.

The maximum shear stress occurs in a direction at 45° to the tangent at any point on the boundary of a circular hole. The stress concentration curve τ_{max}/S_v is also given in Fig. 16.2.3 for $m = 0$ and $m = 1$. Note that

$$\frac{\tau_{max}}{S_v} = \frac{\sigma_\theta}{2S_r}$$

Elliptical Opening

The stress distributions around elliptical openings are derived theoretically in Sections 5.2 and 5.3 for various major-to-minor axis ratios. These stress distributions, like those for circular openings, are independent of

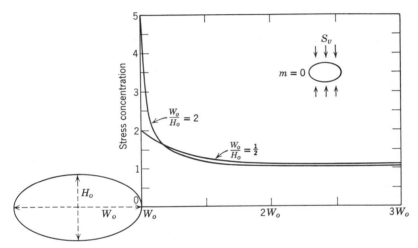

Fig. 16.2.4. Axial-stress concentrations for an elliptical hole in a uniaxial stress field.

the size of the opening and the elastic constants of the material. The magnitude of the tangential stress along the extension of the major and minor axes is given in Tables 5.3.2 and 5.3.3 for major-to-minor axis ratios of 2 and 4 (or $\frac{1}{2}$ and $\frac{1}{4}$) respectively. By letting $S_y = S_v$, $S_x = S_h$, $p = W_o/2$, and $q = H_o/2$, these tabulated results conform to the conditions assumed in this section, namely a vertical and horizontal applied stress S_v and S_h and an opening height and width of H_o and W_o.

Figure 16.2.4 shows the tangential stress distribution along the extension of the horizontal axis of ellipses having a W_o/H_o ratio of 2 and $\frac{1}{2}$, and for a uniaxial stress-field. These curves illustrate the following points:

1. The stress concentration is a maximum on the boundary of the opening and it decreases rapidly with the distance from the boundary.

2. The stress concentration varies inversely as the radius of curvature of the opening (for $W_o/H_o = \frac{1}{2}$, $\sigma_{t(max)} = 2$; for $W_o/H_o = 2$, $\sigma_{t(max)} = 5$).

3. Because the area under the stress distribution curves must be constant the $W_o/H_o = \frac{1}{2}$ curve (with $\sigma_{t(max)} = 5$) must decrease more rapidly than the $W_o/H_o = 2$ curve, hence the curves must cross.

The tangential stress distribution along the vertical and horizontal axes, for the stress fields represented by $m = 0$, and $m = 1$ are given in Fig.

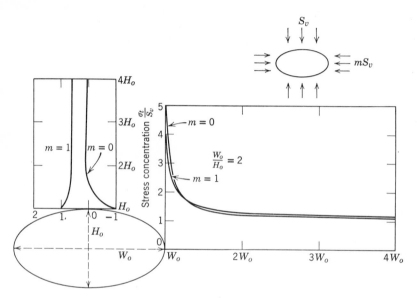

Fig. 16.2.5. Axial-stress concentrations for an elliptical hole in a biaxial stress field.

16.2.5 (the curves for $0 < m < 1$ would lie between the $m = 0$ and $m = 1$ curves). These curves show that as the applied stress field changes from uniaxial to hydrostatic the boundary stress on the horizontal axis decreases, but because the area under the distribution curve must remain constant the hydrostatic curve decreases less rapidly than the uniaxial curve.

Figure 16.2.6 shows the boundary stress distribution curves for ellipses with W_o/H_o ratios of $\frac{1}{4}$, $\frac{1}{2}$, 2, and 4, and for applied stress fields represented by $m = 0$, $m = \frac{1}{3}$, and $m = 1$. These curves show:

1. For the uniaxial stress-field $m = 0$, the maximum stress concentration at the ends of the horizontal axis increases as the W_o/H_o ratio increases, whereas the stress concentration at the top and bottom of the opening remain constant at a value of -1, signifying tension when the applied stress is compression.

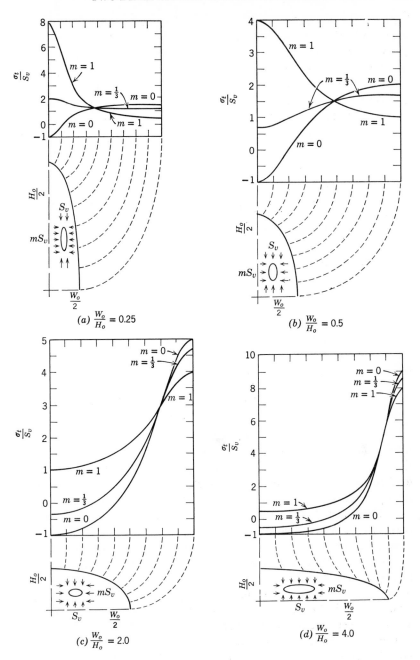

Fig. 16.2.6. Boundary-stress concentration for elliptical holes in a biaxial stress field.

2. For the biaxial stress-field $m = \frac{1}{3}$ the boundary stress concentration at the ends of the horizontal axis increases with the W_o/H_o ratio and the stress concentration at the ends of the vertical axis changes from large positive values to small negative values.

3. The hydrostatic stress-field $m = 1$ produces maximum stresses on the horizontal axis for W_o/H_o ratios greater than 1, and on the vertical axis for W_o/H_o ratios less than 1.

Table 16.2.1. Critical Stresses for Ellipses

		\multicolumn{5}{c}{W_o/H_o}			
m	θ	4	2	1	$\frac{1}{2}$	$\frac{1}{4}$
0	0°	9.0	5.0	3.0	2.0	1.5
	90°	−1.0	−1.0	−1.0	−1.0	−1.0
$\frac{1}{3}$	0°	8.33	4.33	2.66	1.66	—
	90°	−0.5	0.33	—	—	2.0
1	0°	8.0	4.0	2.0	—	—
	90°	—	—	2.0	4	8

Table 16.2.1 lists the position and magnitude of the critical stresses on the boundary. As noted in Section 5.3 if $S_h/S_v = W_o/H_o$, the boundary stress is constant and the critical stress is a minimum.

Ovaloidal Opening

The parametric mapping function (Eq. 5.2.1) will generate approximate ovaloids, and the tangential stress distribution on the boundary of these openings can be calculated from Eq. 5.2.5 for various W_o/H_o ratios, using the constants given in Table 5.2.1. This parametric procedure can also be used for determining the stress at any point outside of the boundaries of the opening. The axial tangential and radial stress distributions for ovaloids (as well as for the following treatment of rectangles with rounded corners) is characteristically similar to those for ellipses. The stress concentration occurring along any line normal to, and extending from, the boundary is a maximum or minimum at the boundary, and decreases or increases rapidly with the distance from the boundary. Also the area between the horizontal axis and the tangential stress distribution curve along the horizontal axis is dependent only on the width of the opening, and is independent of the height of the opening or the type of stress field.

The tangential boundary stress distribution for the stress fields represented by $m = 0$, $m = \frac{1}{3}$, and $m = 1$, and for W_o/H_o ratios of $\frac{1}{4}$, $\frac{1}{2}$, 2, and 4 are given in Fig. 16.2.7. Unlike elliptical openings the maximum boundary

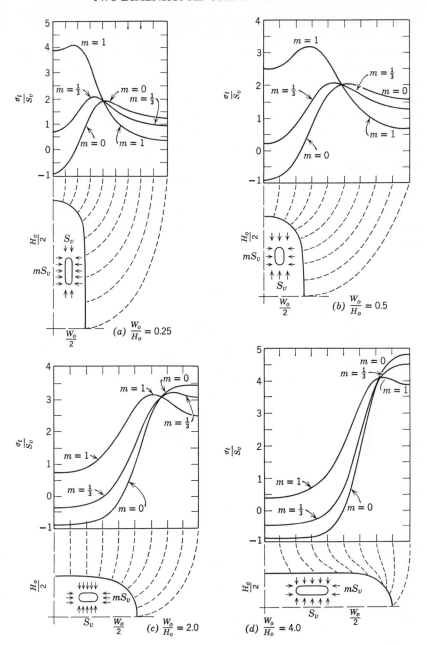

Fig. 16.2.7. Boundary-stress concentration for ovaloidal holes in a biaxial stress field.

stresses around approximate ovaloidal openings do not occur on the axes of the openings, but are shifted toward the junction of the semicircular end and the straight side. For the uniaxial stress field, $m = 0$, the maximum stress concentration along the sides of the opening increases with the width-to-height ratio. For all four width-to-height ratios, the stress concentration at the top and bottom of the opening remains approximately constant at a value of minus 1. For a biaxial stress field represented by $m = \frac{1}{3}$, the boundary stress concentration on the sides of the opening increases with the width-to-height ratio, and the stress concentration at the top and bottom of the opening changes from small positive values to small negative values. The hydrostatic stress field produces maximum stress concentrations on the sides of the opening for width-to-height ratios greater than 1 and on the top and bottom for width-to-height ratios less than 1. The position and magnitude of the critical stresses can be ascertained from these curves.

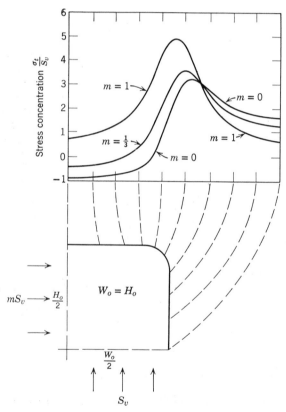

Fig. 16.2.8. Boundary-stress concentrations for a square hole in a biaxial stress field.

Rectangular Opening with Rounded Corners

Infinite stress concentrations would be produced at the right-angle corners of a conventional rectangular opening; therefore rounded or filleted corners are assumed. Normal mining conditions probably produce approximately rounded corners in a rectangular opening. The tangential boundary-stress distribution around a square opening with rounded corners is treated in Sections 5.2 and 5.3. This distribution depends on the ratio of the radius of curvature of the corner to the opening width r/W_o and the maximum stress concentration increases rapidly as this ratio becomes less than $\frac{1}{4}$ (Fig. 5.3.1). The tangential, boundary, stress distribution around a square opening with rounded corners is shown in Fig. 16.2.8 for stress fields represented by $m = 0$, $m = \frac{1}{3}$, and $m = 1$. Note that unlike a circle, the maximum compressive stress concentration occurs for a hydrostatic stress field, whereas the minimum compressive stress concentration is developed by a uniaxial stress-field. However, the latter stress field produces tension at the top and bottom of the opening.

The tangential stress concentration on the boundary of rectangular openings with rounded corners, with W_o/H_o ratios of $\frac{1}{4}$, $\frac{1}{2}$, 1, 2, and 4, and for stress fields represented by $m = \frac{1}{3}$, and $m = 0$ are shown in Fig. 16.2.9. These distributions were determined photoelastically by Duvall,[1] but they could have been calculated from the data given in Fig. 5.3.1. The position and magnitude of the critical stresses can be determined from these curves (also see Fig. 5.3.2).

Design Principles

The critical stress results given in the preceding subsections are summarized in Fig. 16.2.10. These results are valid for a two-dimensional opening with a height and width, H_o and W_o, in an elastic, isotropic, homogeneous rock (most nearly approximated by massive elastic types) and subject to applied vertical and horizontal stresses S_v and $S_h = mS_v$, with $0 < m < 1$. From these and other results given in the foregoing subsections it follows that:

1. For a uniaxial stress field ($S_h = 0$) or a biaxial stress field ($S_h = mS_v$):

   (a) The stress distribution on the boundary is dependent on the shape, but not on the size of the opening. Hence, critical stresses are dependent on the opening shape but not on the opening size.

   (b) The boundary and axial stress distribution is independent of the elastic constants of the rock. Hence, critical stresses are independent of the elastic properties of the rock.

   (c) Critical stress concentrations increase as the radius of curvature

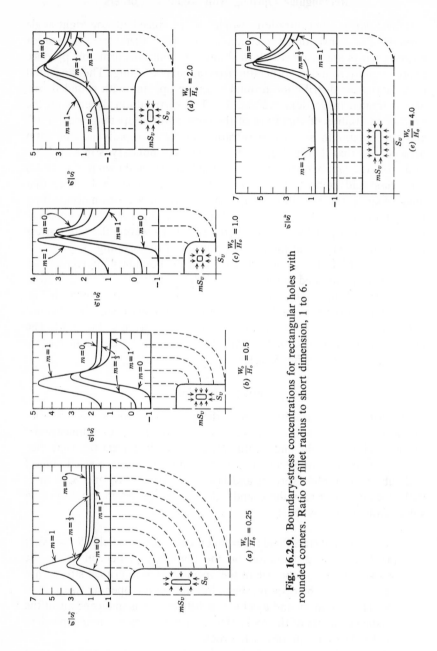

Fig. 16.2.9. Boundary-stress concentrations for rectangular holes with rounded corners. Ratio of fillet radius to short dimension, 1 to 6.

506

of the boundary decreases (except for tensile critical stress concentrations in a uniaxial stress field). Therefore, openings with sharp corners should be avoided.

(d) The tangential stress concentration on the extension of the horizontal axis through an opening of any shape is a maximum at the boundary and decreases rapidly with the distance from the boundary. Moreover, the greater the boundary-stress concentration the more rapidly the stress distribution curve will decrease with the distance from the boundary. For example, at a distance of one diameter along the horizontal axis from the boundary of a circular opening in a biaxial stress field ($m = \frac{1}{3}$) the tangential stress is less than 5% greater than the applied vertical stress. Consequently the stress distribution around an opening is not appreciably influenced by the presence of other openings or surfaces if it is separated from them by a distance equal to the length of the axis through the opening. Thus if this condition is satisfied, the opening is single.

2. For a uniaxial stress field:

(a) If $(W_o/H_o) > 1$, ovaloids or rectangles with rounded corners will give smaller critical compressive stress concentrations (at the ends of the horizontal axis) than ellipses. If $(W_o/H_o) > 1$, an ellipse is the best shape.

(b) If the vertical applied stress is parallel to the height* of any opening, a critical tensile stress concentration of -1 will develop at the ends of the vertical axis. Since the tensile strength of rock is small this tension may be significant.

3. For a biaxial stress field:

(a) An elliptical opening with its major axis parallel to the vertical applied stress will develop smaller critical compressive stress concentrations than any other opening shape. For a given stress field, if $(H_o/W_o) = (S_h/S_v)$ (see Eq. 5.2.12), then the critical stress at all points on the boundary will be compressive, constant, and a minimum. Thus an ellipse with the proper ratio of axes is the ideal opening for any stress field (other than uniaxial).

(b) If $W_o/H_o > 1$, ovaloidal or rectangular openings will induce lower critical compressive stress concentrations than elliptical openings.

(c) The tensile stress at the ends of the vertical axis of all openings in a uniaxial stress field ($m = 0$) will decrease and become compressive as m increases. For $W_o/H_o \leq 1$, this transition from tension to compression occurs for $m = \frac{1}{3}$.

* As defined in Fig. 16.3.1.

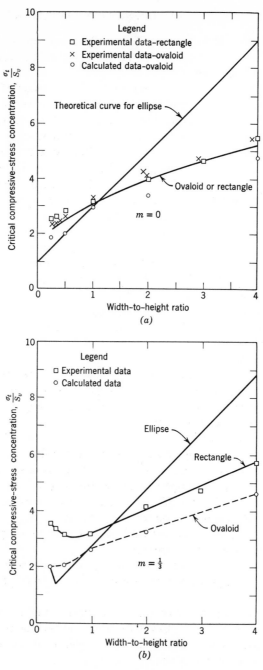

Fig. 16.2.10. Critical compressive stress concentrations for openings of various cross sections.

508

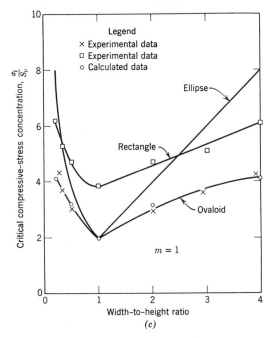

Fig. 16.2.10. (*contd.*)

4. For a hydrostatic stress field:

(a) The preferred opening is a circle. When a height-to-width ratio either smaller or greater than unity is desired, an ovaloid will induce lower critical stresses than either an ellipse or rectangle.

(b) No tensile stresses develop for any shape of opening.

The fact that the stress distribution around an opening of a given shape is independent of its size is often mistakenly interpreted as implying that the stability of the opening should be independent of its size, a conclusion that is manifestly inconsistent with mining experience. Rather, the increased instability of an opening with an increase in its dimensions results from the fact that as these dimensions increase, the probability of intercepting a mechanical defect in the rock, such as a joint or fault, increases; hence it is the strength of the rock rather than the stress that is affected by opening size. In the process of enlarging the opening size, the first manifestation of failure is usually the loosing and ultimate breaking of small pieces of rock from sidewalls and roof, sometimes referred to as slough or spall. This mode of failure is strongly affected by gravity, and in many instances it can be prevented by rock bolting. As the dimensions of the opening are increased the magnitude of these failures will most likely increase, and to such a degree that they may become the factor that limits its size. However, only in rare instances has successive sloughing lead to a general collapse of the opening. Despite the decrease in megascopic rock strength with opening size, spans in excess of 100 ft have been mined in massive elastic rock, and spans of 50 to 60 ft are commonplace.

The design equations for this class of opening in massive elastic are

$$\sigma_t \times F_s \le T_o$$
$$|\sigma_c| \times F_s \le |C_o| \tag{16.2.7}$$

where σ_t = critical tensile stress

σ_c = critical compressive stress

F_s = safety factor

T_o = tensile strength

C_o = compressive strength

Generally in underground mining the stress field is such that only compressive critical stresses are experienced. If this is the case a structural failure in massive elastic rock (excepting slough and spall) would be a rock burst. The phenomenology of rock bursts and related theory is treated in Chapter 19. However, the conditions necessary to initiate a burst are those specified in Eqs. 16.2.7.

Fig. 16.2.11. Bending in a brattice support timber caused by the plastic deformation of an adjacent stope pillar.

Fig. 16.2.12. Heavy spall from the surface of a pillar.

Because massive elastic rock is isotropic, its strengths are the same in all directions. Hence the stability of an opening will be independent of the azimuthal orientation of its long axis, provided that mS_v has the same magnitude in all directions.

Stress determinations made in massive inelastic rock such as salt or trona indicate that the stress concentrations that develop on the surface of openings at the time they are mined decrease with time due to time-dependent deformations (creep). Hence, in this class of rock, if failure is due to critical stresses exceeding rock strengths, then the surface of single openings should spall or otherwise fail at the time the opening is mined. However, failure on the surface of salt, trona, and potash and borate ores generally occur after a period of time ranging from days to years, and the factor causing failure appears to be excessive strain and/or strain rate. Figure 16.2.11 shows the sidewalls of an opening in borate ore, which is a relatively inelastic rock. In about one year the vertical convergence of this opening (calculated from the deflection of the broken timber) was more than 1.5%, which is over ten times the elastic strain this rock will tolerate.[2] Only a minimum spall occurred and the roof was intact. Normally in this class of rock, creep will continue at a constant or slowly diminishing rate[3] and as the strain increases, with time heavier spall will develop. A typical heavy spall in borate ore is shown in Fig. 16.2.12.

Thus a single opening in rock that is inelastic but sufficiently competent to mine without the aid of artificial support will generally fail by sloughing from the walls, roof, and floor rather than by sudden collapse. Because the strain rate in this class of rock depends on the applied stress and differs widely from one rock type to another, or even for a given type at different sites, the time required for a critical strain to develop, that is, a strain large enough to initiate failure, can only be determined by test.

The evaporite minerals, which predominate in the massive elastic rock class, have relatively low strengths. For example, the compressive strength of salt ranges from 4000 to 6000 psi. However, because these minerals are usually free of jointing, comparatively large openings can be mined in them. At a depth of 1600 ft in dome salt, an opening with a span of 120 ft and a length of 180 ft has stood for over 25 years, with only minor spall from the sidewalls.

16.3   OPENING IN MASSIVE ROCK, INCLINED AXES, TWO-DIMENSIONAL CASE

We will now consider those single openings having elliptical or rectangular cross sections of which the cross-sectional axes are inclined at an angle δ to the horizontal and vertical. Otherwise the assumptions required for the treatment of these openings are the same as those for single openings with horizontal and vertical cross-sectional axes (Section 16.2). Figure 16.3.1 shows cross sections of an elliptical and rectangular opening and defines the necessary variables in the problem. The x, y axes, corresponding to the direction of the applied stress field, are horizontal and vertical. The x', y' axes correspond to the major and minor axes of the cross section of the opening and are inclined at an angle δ with the x, y axes.

The problem of an inclined elliptical opening in a biaxial uniform stress field has been solved by Inglis.[4] The equation for the boundary tangential stress is

$$\sigma_t = \frac{(S_v + S_h)\sinh 2\alpha_1 + (S_v - S_h)[e^{2\alpha_1}\cos 2(\delta + \beta) - \cos 2\delta]}{\cosh 2\alpha_1 - \cos 2\beta} \quad (16.3.1)$$

where α, β = curvilinear coordinates, $x' = c \cosh \alpha \cos \beta$

$y' = c \sinh \alpha \sin \beta$; $\beta = 0$ = positive x' axis,

$\beta = 90°$ = the positive y' axis

α_1 = the constant value of α on the elliptical boundary of the opening

δ = the angle of inclination of the major axis of the ellipse (the x' axis) above the horizontal axis (x axis)

S_v, S_h = applied stress in vertical (y axis) and horizontal (x axis) directions

Panek[5] using Eq. 16.3.1 computed critical boundary stress concentrations for four W_o/H_o ratios at various angles of inclination and stress fields of three types: $m = 0$, $\frac{1}{3}$, and 1. Table 16.3.1 summarizes these

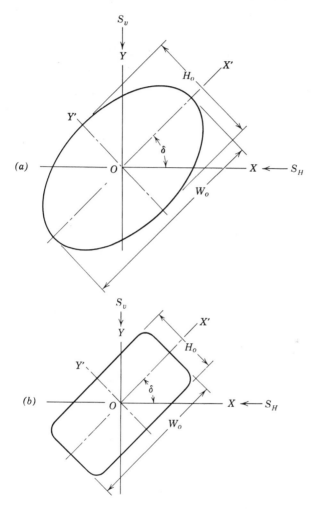

Fig. 16.3.1. Inclined openings. (*a*) Elliptical opening. (*b*) Rectangular opening.

results. The critical stress concentrations are at the ends of the major and minor axes for $\delta = 0°$ or $90°$. For other values of δ the critical stress concentrations are near the ends of the major axis. The critical compressive stress concentrations for an applied compressive stress field decrease as the major axis of the elliptical opening is rotated from the horizontal to the vertical. The maximum critical tensile stress concentrations for a uniaxial stress field, $m = 0$, occurs when $\delta = 45°$. For $m = \frac{1}{3}$ the critical stress concentrations decrease as the major axis is rotated from the horizontal to vertical. For $m = 1$, there are no critical

Table 16.3.1. Critical Stress Concentration for Elliptical Openings as Function of Angle of Inclination of Major Axis to Horizontal

$\dfrac{W_o}{H_o}$	$\delta, °$	$m = 0$ Tension	$m = 0$ Comp.[1]	$m = \frac{1}{3}$ Tension	$m = \frac{1}{3}$ Comp.[1]	$m = 1$ Comp.[1]
1	0–90	−1.0	3.0	0	2.7	2.0
2	0	−1.0	5.0	−0.33	4.7	4.0
	22.5	−1.0	4.6	−0.29	4.3	4.0
	45	−1.1	3.7	−0.13	3.5	4.0
	67.5	−1.1	2.5	(2)	2.4	4.0
	90	−1.0	2.0	(2)	1.7	4.0
3	0	−1.0	7.0	−0.44	6.7	6.0
	22.5	−1.1	6.3	−0.43	6.1	6.0
	45	−1.3	4.6	−0.29	4.7	6.0
	67.5	−1.2	2.7	(2)	3.0	6.0
	90	−1.0	1.7	(2)	1.33	6.0
4	0	−1.0	9.0	−0.50	8.7	8.0
	22.5	−1.2	8.0	−0.52	8.0	8.0
	45	−1.5	5.8	−0.42	6.1	8.0
	67.5	−1.4	3.0	(2)	3.7	8.0
	90	−1.0	1.5	(2)	2.0	8.0

[1] Compression.
[2] No critical tensile stress concentration.

tensile stress concentrations and the critical compressive stress concentrations are independent of δ.

The special case of a square opening with rounded corners, with axes inclined at $45°$, was considered briefly in Section 5.2, and Fig. 16.3.2 gives the boundary tangential stress distributions for the three stress fields, $m = 0, \frac{1}{3}$, and 1. The critical stress concentrations around other inclined rectangular openings with rounded corners was investigated photoelastically by Panek.[5] Table 16.3.2 gives the results for various angles of inclination and ratios of W_o/H_o. The ratio of the fillet radius to the height is 1 to 6.

For $m = 0$ or $\frac{1}{3}$, the critical compressive stress concentrations are maximum for $\delta = 45°$. For $m = 1$, the critical compressive stress

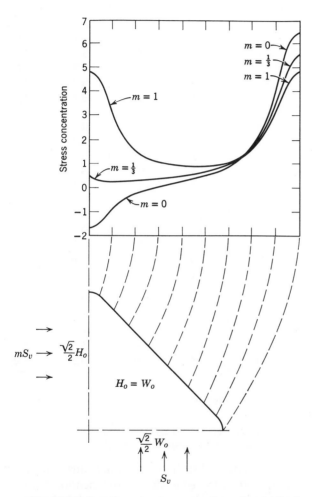

Fig. 16.3.2. Boundary-stress concentrations of an inclined square hole.

concentration is virtually independent of the angle of inclination. Critical tensile stress concentrations are large for $m = 0$ and angles of inclination of 22.5 to 67.5°. For $m = \frac{1}{3}$ critical tensile stress concentrations are small, and for $m = 1$ they are zero.

Table 16.3.2. Critical Stress Concentration for Rectangular Openings with Rounded Corners as Function of Angle of Inclination of Major Axis to Horizon

$\dfrac{W_o}{H_o}$	δ, °	$m = 0$ Tension	$m = 0$ Comp.[1]	$m = \frac{1}{3}$ Tension	$m = \frac{1}{3}$ Comp.[1]	$m = 1$ Comp.[1]
1	0	−1.0	3.1	−0.3	3.1	3.8
	22.5	−0.5	3.9	([2])	3.8	3.7
	45	−1.1	4.7	([2])	4.3	3.6
	67.5	−1.3	4.0	−0.3	3.9	3.7
	90	−1.0	3.1	−0.3	3.1	3.8
2	0	−0.8	4.0	−0.1	4.1	4.7
	22.5	−0.7	5.0	([2])	4.7	4.6
	45	−1.6	5.7	([2])	5.2	4.5
	67.5	−1.4	4.5	−0.1	4.5	4.6
	90	−1.0	2.7	−0.2	3.1	4.7
3	0	−0.8	4.6	−0.4	4.7	5.2
	22.5	−0.7	5.9	−0.1	5.5	5.2
	45	−1.8	6.5	−0.4	6.0	5.3
	67.5	−1.6	5.0	([2])	5.0	5.2
	90	−1.0	2.6	−0.1	3.3	5.2
4	0	−0.9	5.4	−0.4	5.6	6.2
	22.5	−1.0	6.5	−0.1	6.0	5.9
	45	−2.0	7.1	−0.5	6.6	5.9
	67.5	−1.9	5.5	([2])	5.5	5.9
	90	−1.0	2.5	([2])	3.5	6.2

[1] Compression.
[2] No critical tensile stress concentration.

16.4 OPENING IN MASSIVE ROCK, THREE-DIMENSIONAL CASE

Determination of the stress distribution around three-dimensional openings is more difficult than the equivalent two-dimensional problem. As a result only one class of opening has been treated theoretically, namely, ellipsoids and their degenerate forms, oblate spheroids, prolate spheroids, and spheres. To obtain a solution (see Section 5.5) the following assumptions were made:

1. If a, b, and c are the semiaxes of the ellipsoid, let c be in the z-direction (vertical) and a and b be in the horizontal directions x and y,

with $a > b > c$ (as in Fig. 5.5.3); for an oblate spheroid $a = b > c$; for a prolate spheroid $a = b > c$; and for a sphere $a = b = c$.

2. The medium is infinite, linear elastic, homogeneous, and isotropic.

3. The stress field, if uniaxial is vertical, hence S_z is parallel to the c axis; if triaxial, S_z is vertical, and S_x and S_y are parallel to the a and b axes.

For the uniaxial case the tangential stress concentration σ_z/S_z at the ends of the horizontal axes, that is, at points A and B in Fig. 5.5.3, are as shown in Fig. 5.5.4.* The results given in this figure are not sufficient to ascertain by superposition the corresponding stress concentration at A or B for any stress field other than uniaxial. However, the stress concentration around three-dimensional openings can be approximated from a consideration of similarly shaped two-dimensional openings. For example, in the ellipsoid in Fig. 5.5.3, if $a = \infty$ and $b = c$, the opening would become a horizontal tunnel with a circular cross section. In a uniaxial stress field the stress concentration σ_z/S_z at B would be 3. Now if the a axis is made finite a part of the load carried on the tunnel at B would be supported on the ends of the tunnel at A, and as the a axis is further shortened the load supported at A would increase and decrease at B. Therefore, this would lower the stress concentration at B below 3 and correspondingly decrease the area under the stress concentration curve. When $a = b = c$, that is, when the opening is a sphere the stress concentration at B would be 2 (assuming a Poisson's ratio of 0.2) as indicated in Fig. 5.5.2. Thus for this prolate spheroid with its major axis in the a-direction the stress concentration σ_z/S_z at B ranges from 3 to 2 as the a axis is shortened from ∞ to b. In a hydrostatic stress-field the corresponding stress concentration at B would range from 2 when $a = \infty$ to 1.5 when $a = b = c$ (from Eq. 5.5.2 with $a = r$), and this latter value is independent of Poisson's ratio. A similar argument could be made for a tunnel having an elliptical cross section, that is, an ellipsoid with $a = \infty$ and, say $b > c$; if the a axis is then shortened it would ultimately become an oblate spheroid with $a = b > c$. If, for the same problem, $b < c$, the opening would become a prolate spheroid where $a = b$. Thus the critical stresses on a cross section through a three-dimensional opening can be approximated by determining the critical stresses on a two-dimensional opening of similar shape provided that the applied biaxial stress field for the cross section and two-dimensional opening are equivalent. Moreover, estimates of critical stresses made on this basis will be on the conservative side.

---

* Figure 5.5.4 is most easily interpreted if it is assumed that the b-axis is of unit length; then $1 < a < \infty$ and $1 > c > 0$.

16.5 OPENING IN LAMINATED ROCK, ROOF PARALLEL TO LAMINATION

Laminated rock, as defined in Section 15.2, is composed of a succession of parallel layers whose thickness is small compared with the span of openings therein, and the layers are either unbonded or the bond strength between them is small compared with the tensile strength of the rock. Generally, if an opening is excavated in this type of rock, the roof is formed at a weakness plane* and it is relatively smooth and flat. Furthermore, because of the weak bond between laminae, the roof rock will become detached either immediately or after a time from the overlying rock and form a layer or number of layers that are loaded only by gravity. The detached layer (or layers) is called the immediate roof, and the overlying rock is called the main roof. The critical stresses or critical strain rates in the main roof, side walls, and floor should be determined on the basis of a single two- or three-dimensional opening in an elastic or inelastic rock, as considered in Sections 16.2, 16.3, and 16.4. The compressive strength of a laminated rock with the laminae planes normal to the direction of the applied stress is approximately equal to that of nonlaminated rock because the rock fails by shearing across, rather than on the planes of weakness (Section 10.6). Hence, failure on the sidewalls of a single opening in laminated rock should occur at approximately the same critical compressive stress as for an opening in massive rock. Equation 16.2.7 should be used for design purposes, where in this case C_o is the compressive strength of either a laminated or nonlaminated specimen of rock from the opening.

The limiting time-dependent deformation that can be tolerated in a gravity-loaded layer (or layers) in inelastic rock can be approximated from similitude studies, either using models made from the prototype or equivalent rock, or in large-scale in situ test rooms. The critical stresses in similar structural units in elastic rock can be determined from the theory of gravity-loaded, clamped beams or plates, provided that (1) there is uniform thickness within each layer; (2) the flexure of the beam is caused by gravity, that is, there are no end thrusts or other external forces; (3) the rock in the layers is linear-elastic, isotropic, and homogeneous; and (4) the ends of the layers are rigidly clamped by the overlying rock and sidewall.

The third assumption precludes the possibility of any fracturing in the roof layer(s). However, in a room that is long compared with its span,

---

* See Section 16.6 for opening in which the planes of weakness do not form at, or are not parallel to the roof.

vertical fractures normal to the length of the room will not increase the stress in the roof significantly. For a rock with known mechanical properties the problem of design reduces to calculating a roof span for which the critical stresses are less than the rock strength divided by an appropriate safety factor. For roofs that are long compared with their span (2:1 or more), the theory for the flexure of beams (Section 5.6) is used to calculate critical stresses; for length-to-span ratios less than 2:1 the theory for the flexure of plates (Section 5.7) is used.

Single Layer Roof

The maximum values of the deflection, shear, and tensile stress for a horizontal, gravity-loaded roof layer clamped at both ends are given by

$$\eta_{max} = \frac{\gamma L^4}{32Et^2} \qquad (16.5.1)$$

$$\tau_{max} = \frac{3\gamma L}{4} \qquad (16.5.2)$$

$$\sigma_{max} = \frac{\gamma L^2}{2t} \qquad (16.5.3)$$

where η_{max} = maximum deflection

σ_{max} = maximum stress

τ_{max} = maximum shear stress

L = span of roof layer (shorter lateral dimension of layer)

t = thickness of roof layer

E = Young's modulus

γ = unit weight of rock

The stress in a roof layer inclined $10°$ or less is only slightly lower than the stress in a horizontal layer, and hence is considered horizontal. The maximum deflection occurs at the center of the layer, and the maximum compressive, tensile, and shear stresses occur at the ends of the layer— tension on the upper surface, compression on the lower surface. Because rock is much weaker in tension than in compression, only the tensile and shear stresses in the rock layer are considered. At the center of the layer the shear stress is zero, and the tension is one-half of the maximum value. Thus failure should initiate on the top surface at the ends of the span.

The maximum shear stress varies directly as the span, whereas the maximum tensile stress varies directly as the square of the span and inversely as the thickness of the layer. The ratio of these stresses is

$$\frac{\sigma_{max}}{\tau_{max}} = \frac{2L}{3t} \qquad (16.5.4)$$

Thus if the span-to-thickness ratio is greater than 5 to 1, the tensile stress is more than three times the shear stress. Since the tensile strength of rock usually is less than the shear strength, shear stresses can be disregarded in determining the span.

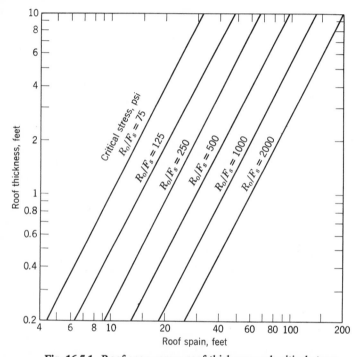

Fig. 16.5.1. Roof span versus roof thickness and critical stress.

Equation 16.5.3 may be rewritten as a design formula for the roof span by replacing σ_{max} with R_o, the modulus of rupture (outer-fiber tensile strength) of the rock, dividing R_o by F_s, a safety factor, and solving for L; thus

$$L = \sqrt{\frac{2R_o t}{\gamma F_s}} \qquad (16.5.5)$$

Figure 16.5.1 shows a graph of Eq. 16.5.5 on log-log coordinates, using a rock unit weight of 0.09 lb/in.[3] This graph relates the span to layer thickness and critical (tensile) stress.

Two-Layer Roof

For an immediate roof consisting of two layers there are two situations to consider. When the thicker layer overlies the thinner layer, each acts independently, and the stresses and deflection in each layer can be calculated by Eqs. 16.5.1, 16.5.2, and 16.5.3. When the thinner layer overlies the thicker layer, the lower layer is loaded by the upper one; the additional loading can be calculated and represented as an equivalent increase in the unit weight of the lower layer (see Section 5.6). Thus

$$\gamma_a = \frac{E_1 t_1^2(\gamma_1 t_1 + \gamma_2 t_2)}{E_1 t_1^3 + E_2 t_2^3} \qquad (16.5.6)$$

where γ_a = adjusted unit weight of lower layer

E_1 = Young's modulus of lower layer

E_2 = Young's modulus of upper layer

γ_1 = unit weight of lower layer

γ_2 = unit weight of upper layer

t_1 = thickness of lower layer

t_2 = thickness of upper layer

The maximum values of the deflection, shear, and tensile stress in the lower layer can be calculated from Eqs. 16.5.1, 16.5.2, and 16.5.3, respectively, by replacing γ_1 with γ_a. If the unit weight and Young's modulus are the same for both layers, the maximum value of γ_a is equal to $\frac{4}{3}\gamma_1$ and it occurs when the upper layer is one-half the thickness of the lower layer. Thus the maximum stress in the lower layer is $33\frac{1}{3}\%$ greater than the gravity stress in the lower layer.

Multiple-Layer Roof

When the immediate roof consists of three or more layers and the thinner layers overlie the thicker ones, the additional load on the lowest layer can be determined by calculating an adjusted unit weight for the lowest layer (Section 5.6); thus

$$\gamma_a = \frac{E_1 t_1^2(\gamma_2 t_1 + \gamma_2 t_2 + \gamma_3 t_3 + \cdots + \gamma_n t_n)}{E_1 t_1^3 + E_2 t_2^3 + E_3 t_3^3 + \cdots + E_n t_n^3}, \qquad (16.5.7)$$

where γ_a = adjusted unit weight of lowest layer

E_n = Young's modulus of nth layer

γ_n = unit weight of nth layer

t_n = thickness of nth layer

The values of the maximum deflection, shear, and tensile stress and the design equation are again obtained by substituting γ_a for γ_1 in Eqs. 16.5.1, 16.5.2, 16.5.3, and 16.5.5, respectively.

For multilayered roofs having various thicknesses and mechanical properties, the number of layers that need to be considered can be determined by using Eq. 16.5.7 stepwise; that is, the apparent unit weight is calculated for the first two layers, the first three, the first four, etc., until the adjusted unit weight shows no further increase. Only layers that produce an increase in the adjusted unit weight load the lowest layer.

Rectangular Roof

The maximum values of the deflection and tensile stress for a gravity-loaded rectangular roof layer (plate) clamped on all edges have been given in Section 5.7 (Eqs. 5.7.8)

$$\eta_{max} = \frac{\alpha \gamma a^4}{E t^2} \tag{16.5.8}$$

and

$$\sigma_{max} = \frac{6 \beta \gamma a^2}{t} \tag{16.5.9}$$

where a = shorter lateral dimension

b = longer lateral dimension

α, β = constants

Table 5.7.1 gives the values of α and β for various values of b/a and Poisson's ratio equal to 0.3.

The maximum deflection occurs at the center of the layer and the maximum tensile stress at the end of the longer dimension at the upper edge of the layer. For ratios of b to a greater than 2 the maximum stress and deflection, computed from Eqs. 16.5.1 and 16.5.3 for a beam, approximate the stress and deflection computed from Eqs. 16.5.8 and 16.5.9 for a plate. The difference in the maximum stress is less than 1%, and the difference in the maximum deflection is less than 12%.

As before, Eq. 16.5.9 can be rewritten as a design formula by replacing σ_{max} with R_o, dividing R_o by a safety factor F_s, and solving for a, the shorter lateral dimension. Thus

$$a = \sqrt{\frac{R_o t}{6 \beta \gamma F_s}} \tag{16.5.10}$$

Inclined Roof

Two types of inclined roof are considered in this section: (1) where the long axis of the roof is inclined and the short axis (span) is horizontal and (2) where the long axis is horizontal and the short axis is inclined (Fig. 16.5.2). In each type the angle measured between the roof and the horizontal is designated by θ. It is assumed that the stress in the roof

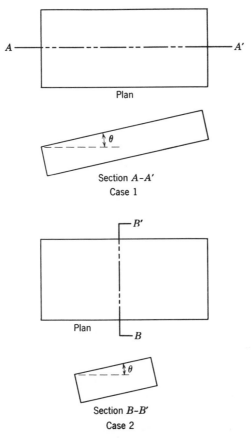

Plan

Section A-A'

Case 1

Plan

Section B-B'

Case 2

Fig. 16.5.2. Plan and section of inclined roofs.

layers is due to flexure caused by gravity. For either type the maximum stress is given by

$$\sigma_{\max} = \frac{\gamma L^2}{2t} \cos \theta \qquad (16.5.11)$$

Equation 16.5.11 can be rewritten as a design equation by the same procedure used for the horizontal roof layer. Thus

$$L = \sqrt{\frac{2R_o t}{\gamma F_s \cos \theta}} \qquad (16.5.12)$$

For large angles, if there is a lateral thrust, Eqs. 16.5.11 and 16.5.12 do not hold.

Roof Subjected to Gravity and External Forces

An added load may be imposed on a roof layer by gas or water pressure. If this pressure is considered as an added load on the layer, the maximum tensile stress and deflection become (Eq. 5.6.29)

$$\sigma_{max} = \frac{\gamma L^2}{2t} + \frac{pL^2}{2t^2} \tag{16.5.13}$$

and

$$\eta_{max} = \frac{\gamma L^4}{32Et^2} + \frac{pL^4}{32Et^3} \tag{16.5.14}$$

where p = any uniformly distributed load, for example, pressure on the layer.

Equation 16.5.13, rewritten as a design equation, becomes

$$L = \sqrt{\frac{2R_o t}{(\gamma + p/t)F_s}} \tag{16.5.15}$$

For a roof layer having a thickness of 10 in. and a unit weight of 0.1 lb/in.[3], a gas or water pressure of 1 psi will double the stress in the layer. Because a roof layer is very sensitive to hydraulic or pneumatic pressures, suspected areas should be carefully checked. Usually the pressure can be relieved by drilling holes through the layer.

A stress applied to the ends of a roof layer is additive to the flexural stress in the layer owing to gravity; hence, a compressive end stress on a roof layer decreases the maximum tensile stress and increases the maximum compressive stress in the layer. Because the ratio of the critical stress to the working stress for most thin roof layers is much lower for tensile stresses than for compressive stresses, the use of Eq. 16.5.5 for design gives an added margin of safety.

16.6 OPENINGS IN ROCK CONTAINING PLANES OF WEAKNESS NOT PARALLEL TO THE ROOF

As a general rule in mineral mining in stratified deposits the mineralization occurs in beds, and as a result the roof and floor of the openings are formed parallel to bedding (lamination) planes. However, in tunnels or openings for industrial or military purposes the attitude of openings is usually dictated by other factors, such as keeping the floor level or at some constant grade. In these instances if the opening is in laminated rock it is more likely to be intersected by a plane or planes of weakness.* These mechanical discontinuities alter both the stress distribution around the opening and the in situ mechanical properties of the rock; collectively these factors will affect the stability of the structure.

The stress distribution around an opening containing discontinuities is a problem that is not tractable to analytical treatment, but some quantitative appraisal has been obtained from photoelastic model studies. The in situ strength of stratified, foliated, or jointed rock is dependent on the attitude of the planes of weakness in the structure, and the strength and extent of the bond across them. The bond strength in tension may range from zero as, for example, in a jointed rock in which the joint planes contain decompositional products such as clay, to a value greater than the strength of unjointed rock if the joint planes are recemented with silicious products. In core drilling in jointed rock it is not uncommon to obtain cores that break across, rather than on joint planes, indicating a joint strength comparable to that of the host rock. Some estimate of the shear strength of specimens containing a natural or artificial (sawed) joint oriented at some specific angle can be determined by the procedures described in Section 10.6. Obviously this procedure will furnish data satisfactory for design purposes only if the degree of bond across, or the coefficient of internal friction of the filling material in the joint, corresponds to that in the in situ rock.

Because these factors make it difficult to ascertain either the stress around an opening or the in situ strength of rock containing planes of weakness, the treatment of stability problems in this class of rock must be made on a more empirical basis. Consider the three cases illustrated in Fig. 16.6.1, in which the z-direction is vertical and the attitude of the weakness planes are specified as follows. (1) The strike is parallel to the long axis (x axis) of the opening and the dip is at an angle α with respect to the y axis (Fig. 16.6.1a). (2) The strike is perpendicular to the long axis and the dip is at an angle β to the long axis (Fig. 16.6.1b). (3) The strike is at an angle γ to the long axis and the dip is vertical (Fig. 16.6.1c).

Case 1. If $\alpha = 0°$ this is the case considered in the preceding section; the planes of weakness form roof layers that can be considered as beams or plates; the stress concentration and the compressive strength of the rock in the sidewalls can be treated as if the rock were isotropic.

For $0° < \alpha < 20°$ the sidewall problem remains virtually the same as for $\alpha = 0°$, because the rock will fail across, rather than on the lamination

---

* In sedimentary rock, planes of weakness may result from stratification or partings; in metamorphic rock, they may result from foliation; and in all types of rock they may be due to joints or faults.

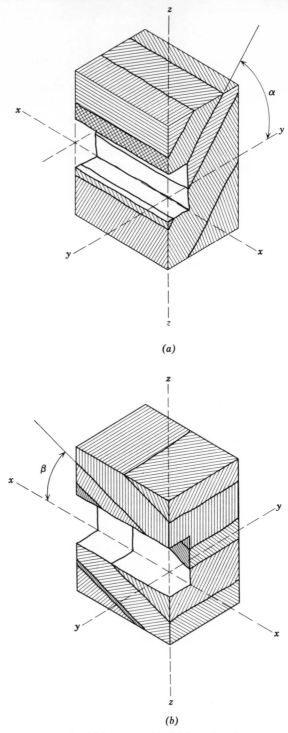

(a)

(b)

Fig. 16.6.1. Tunnel in laminated rock.

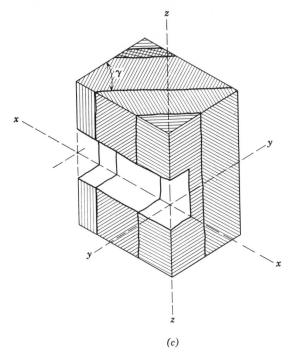

(c)

Fig. 16.6.1. (*contd.*)

planes. However, the roof problem is critical; if the roof forms along the dip (Fig. 16.6.2a), it can be considered on the basis of an inclined roof (Section 16.5). On the other hand, if the top of the opening is kept horizontal the laminae will feather into the roof at a low angle and create unsupported thin wedges of rock that are weak and often difficult to detect. Although this type of defect should not affect the general stability of the structure, it does create a roof control problem. Low angle joints running parallel to the long axis of an opening usually can be supported with rock bolts which, in this case, should be installed on a systematic basis (rather than on a spot-to-spot basis).

For $70° < \alpha < 90°$ the stability of the opening will, to a much greater extent, depend on the type of stress field. For a hydrostatic stress field, or for one in which the tangential stress in the roof is compressive, that is, for $m \geq \frac{1}{3}$ the stability of the opening can be considered on the basis of an isotropic rock, although some "loose" rock in the roof may have to be supported with rock bolts. Also wedge-shaped layers of rock may become loose or detached from the sidewalls (Fig. 16.6.2b) and rock bolting may be required. For $0 < m < \frac{1}{3}$ a tangential tensile stress will usually exist in the roof and floor, and because of the low tensile strength of laminated rock in a direction normal to the laminae, the stability of the structure will be adversely affected. In practice, this situation might arise if a tunnel is mined parallel to a cliff face or a diversion tunnel is driven parallel to the sidewall of a canyon.

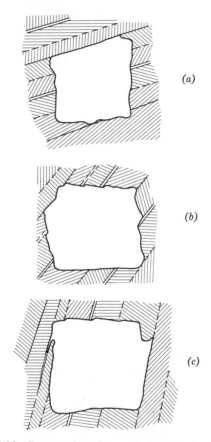

(a)

(b)

(c)

Fig. 16.6.2. Cross section of tunnel in laminated rock.

If the dip of strata or foliation is between 20 and 70°, a greater degree of shear failure in the sidewalls can be expected because for this case the rock can fail in shear along the planes of weakness (Fig. 16.6.2c). Bolting or light sets may be required to hold the loosened rock on the sidewalls and roof. In a uniaxial stress field if α increases, the tangential tensile stress in the roof may have an increasingly greater affect on the stability of the structure.

Case 2. If $\beta = 0°$ the problem is identical to $\alpha = 0°$; for $0° < \beta < 20°$ the sidewall problem is about the same as that for $0° < \alpha < 20°$, but the situation in the roof is different. The laminae will feather into the roof at a low angle (Fig. 16.6.3), but the lamina edges will strike normal to the long axis of the opening, and hence the roof cannot be formed along a plane of weakness. This relatively hazardous condition can usually be rectified by rock bolting, and systematic bolting is recommended because it is especially difficult to detect these laminae edges as the opening is being driven. Low angle jointing, stratification, or foliation in this plane should not significantly affect the general stability of the opening.

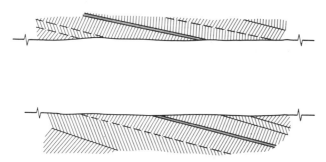

Fig. 16.6.3. Section through tunnel intersecting low angle laminae.

As β increases above 20° the problem of supporting local roof should progressively decrease, although some light support may be required to hold loose rock from lamina edges if the laminae are closely spaced. For $20° < \beta < 70°$ the in situ shear strength of the rock in the sidewalls will be lower than that of the nonlaminated rock, a factor that might have some importance in an operation subject to rock bursts. The ideal direction for mining an opening in rock containing planes of weakness is so that $\beta = 90°$, and whenever any choice is possible in selecting the orientation of an opening it should be such that this condition is satisfied.

Case 3. For $\gamma = 90°$ this case is identical to that for $\alpha = 90°$; for $\gamma = 0°$ it is identical to that for $\beta = 90°$, which is the most favorable orientation of laminae intersecting a horizontal opening. For $20° < \beta < 70°$ the roof and wall conditions are approximately the same as for $\gamma = 0°$; and for $70° < \gamma < 90°$ the roof and wall conditions approach those for $\alpha = 90°$. For the range $20° < \beta < 70°$ some light roof and wall support may be required, but in general vertical planes of weakness do not create the mining problems that arise from planes in other orientations.

REFERENCES

1. Duvall, W. I., "Stress Analysis Applied to Underground Mining Problems, Part I, Stress Analysis Applied to Single Openings," *U.S. BurMines Rept. Invest.*, **4192** (1948).
2. Obert, L. and A. Long, "Underground Borate Mining, Kern County, Calif.," *U.S. BurMines Rept. Invest.*, **6110** (1962).
3. Obert, L., "Deformational Behavior of Model Pillars Made from Salt, Trona, and Potash Ore," *Proc. Sixth Symp. on Rock Mechanics*, Univ. of Mo. (1964).
4. Inglis, C. E., "Stress in a Plate Due to Presence of Cracks and Sharp Corners," Part 1, *Trans. Inst. Naval Arch.*, London (1913).
5. Panek, L. A., *Stresses about Mine Openings in a Homogeneous Rock Body*, Edwards Bros., Ann Arbor, Mich. (1951).
6. Obert, L., W. I. Duvall, and R. Merrill, Design of Underground Openings in Competent Rock," *U.S. BurMines Bull. 587* (1960).

CHAPTER 17

DESIGN OF MULTIPLE OPENINGS IN COMPETENT ROCK

17.1 INTRODUCTION

If the stress distribution around an opening significantly affects the stress distribution around another opening and vice versa, the interacting combination is referred to as a multiple opening. Most mines are multiple openings because they require the excavation of a network of rooms, panels, or other stope configurations, although shafts and exploration drifts may be sufficiently isolated to be treated as single openings. In the preceding chapter it was determined that at a distance of two diameters from the boundary of an opening the stress field is practically unaffected by the presence of the opening. Hence, if openings are more than two diameters apart they can be considered as single, and conversely, if the openings are less than two diameters apart they are multiple.

The development of the stress concentration around single and multiple openings differs in that the maximum stress concentration around a single opening develops almost immediately as the opening is mined; for example, in the excavation of a tunnel the stress concentrations are virtually at a maximum when the tunnel face has advanced one diameter past the point of concern. In a multiple opening, however, the stress around an opening will progressively increase as the surrounding area is excavated, which in some mining operations may require months.

Because the boundaries of multiple openings are complex, the theoretical treatment of this class of problem is limited. In Chapter 5 only two multiple opening problems are considered on the basis of elastic theory, namely, two circular openings, and an infinite row of equally spaced circular openings. The stress distribution around multiple openings of other cross-sectional shapes have been determined photoelastically, but in most instances these studies have been made on two-dimensional models. In mining, the two-dimensional case corresponds to long openings separated by rib pillars, a stoping procedure sometimes practiced in the mining of relatively thin deposits such as coal or potash beds. Although this type of structure is not uncommon in mining, a more frequently encountered configuration is a three-dimensional array of openings as, for example, a system of rooms, crosscuts, and rectangular pillars. No theoretical or empirical solution has been derived describing the stress distribution in such a structure. However, structural components, such as roof laminae, or pillars, have been treated on a three-dimensional basis and the stress in other parts of the structure can be approximated from a consideration of two-dimensional models.

The procedure for designing multiple openings in massive elastic rock is essentially the same as that for single openings. The rock strength must exceed critical stresses by some specified factor of safety. However, as will be shown in some instances average stresses can be substituted for critical stresses without serious error. Also, as in the treatment of single openings, strains or strain rates that can be tolerated in structures in inelastic rock must be determined either from full-scale or model studies.

17.2 OPENINGS IN MASSIVE ROCK TWO-DIMENSIONAL CASE

The stress distribution on the boundaries and axes of two-dimensional openings in massive elastic rock can be approximated from a consideration of mathematical, photoelastic, or rock models, provided that the following simplifying assumptions are made regarding the properties of the rock, the stress field, and the shape and orientation of the openings: (1) the rock is linear-elastic, isotropic, and homogeneous; (2) the openings are in an infinite medium; (3) the long axis of each opening is horizontal; (4) the axis connecting the centers of the openings is horizontal; and (5) the biaxial stress field is S_v and S_h (Fig. 5.4.1*). These conditions are essentially the same as those assumed for the single opening problems considered in Section 16.2. After considering the elastic case the requirement for linear elasticity will be relaxed so as to consider two-dimensional pillar systems in massive inelastic rock.

Infinite Row of Circular Openings

The theoretically determined boundary and axial stress distributions for an infinite row of circular openings of width (diameter) W_o, with centers spaced at $2W_o$ (or if the width of the space between openings is

---

* In Section 5.4 the applied stresses are S_x and S_y, which correspond to S_h and S_v in this section.

W_p, then $W_o/W_p = 1$), in a vertical uniaxial stress field S_v and in a horizontal uniaxial stress field S_h are given in Figs. 5.4.2 and 5.4.3 respectively. For a vertical uniaxial stress field ($S_v \neq 0, S_h = 0$) the boundary stress concentration σ_θ/S_v for a single opening and an infinite row of openings are only slightly different (compare curve A, Fig. 5.4.2,

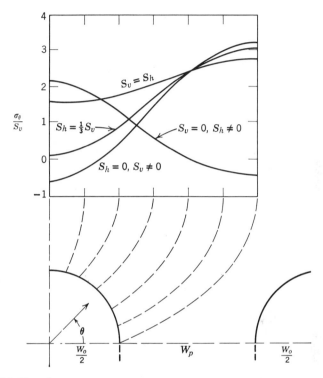

Fig. 17.2.1. Boundary stress concentration for a row of circular openings, $W_o/W_p = 1$.

with the corresponding curve for $m = 0$ in Fig. 16.2.3). The tangential stress concentration along the extensions of the horizontal axis decreases more rapidly for a single opening than for a row of openings (compare curve B, Fig. 5.4.2 with the corresponding curve for $m = 0$ in Fig. 16.2.2). However, for a horizontal uniaxial stress field ($S_h \neq 0, S_v = 0$), it is evident that one hole shields the next, especially along the axis connecting the centers of the openings, for example, for the multiple opening case at $\theta = 0°$, $\sigma_\theta/S_h = -0.39$; whereas for a single opening at $\theta = 0°$, $\sigma_\theta/S_h = -1.0$.

The boundary stress distribution for four cases; $S_v \neq 0, S_h = 0$; $S_v = 0, S_h \neq 0$; $S_h = \frac{1}{3}S_v$; and $S_v = S_h$, are given in Fig. 17.2.1. For each of these stress fields, the critical compressive stress concentration at $\theta = 0°$ is larger for multiple than for single openings, the greatest difference being for a hydrostatic stress field $S_v = S_h$. A critical tensile stress concentration occurs only at $\theta = 90°$ for uniaxial stress field $S_v \neq 0$. Assuming that for most underground sites $S_h > \frac{1}{3}S_v$, the practical implication from this comparison is that the critical stress concentrations on the boundary of any one of an infinite row of circular openings is from 15% (for $S_h = \frac{1}{3}S_v$) to 40% (for $S_h = S_v$) greater than the corresponding critical compressive stress concentration on a single circular opening, and that there is no critical tensile stress concentration.

Two-Circular Openings

The problem of two equal circular openings in a biaxial stress field specified by $S_v \neq 0$, $S_h = 0$; $S_v = S_h$; $S_v = 0$, $S_h \neq 0$ was discussed in Section 5.4 and the critical stress concentrations for different ratios of the width of opening to the width between openings W_o/W_p is given in Table 5.4.1 (here $S_h = S_x$ and $S_v = S_y$).

Considering the results for the hydrostatic stress field $S_v = S_h$, it is important to note that if W_o/W_p increases from 0 to 1 the critical compressive stress concentration, which occurs at $\theta = 0°$, increases only 20% (from 2.0 to 2.4); whereas if W_o/W_p increases from 1 to infinity, the critical stress concentration at the same point becomes infinite. Also, the stress concentrations on opposite sides of either opening (at $\theta = 0°$ and $\theta = 180°$) differ by only about 12%, the larger value being at $\theta = 0°$. In the uniaxial stress field $S_v \neq 0$, $S_h = 0$, the two openings are virtually independent of each other for W_o/W_p greater than 1; the critical stress concentration at $\theta = 0°$ increases rapidly becoming infinite for $W_o/W_p = \infty$.

Finite Row of Circular and Ovaloidal Openings

The stress distribution on the boundaries of a finite row of equally spaced two-dimensional circular and ovaloidal openings in a uniaxial stress field $S_v \neq 0$, $S_h = 0$ has been studied photoelastically by Duvall.[1] For five openings with various W_o/W_p ratios, the maximum compressive stress concentrations, which occur at points A, B, C, D, and E (Fig. 17.2.2 and 17.2.3), are given in Table 17.2.1 and 17.2.2. Note that at points D and E the stress concentrations have practically the same maximum value; hence it is concluded that for a row of five or more openings all but the two openings at each end of the row will have the same maximum value. The maximum compressive stress concentrations for a row of 2, 3, and 5 circular openings with various W_o/W_p ratios are given graphically in Fig. 17.2.4. The shapes of the curves in Fig. 17.2.4 indicates that as the number of circular openings increases the maximum compressive stress

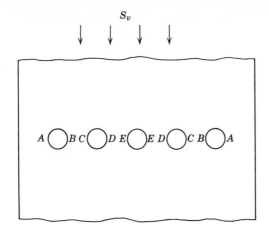

Fig. 17.2.2

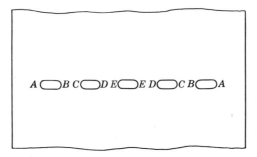

Fig. 17.2.3

Table 17.2.1. Maximum Stress Concentrations for a
Row of Five Circular Openings

| | \multicolumn{5}{c}{Stress concentrations at} |
$\dfrac{W_o}{W_p}$	A	B	C	D	E
1.07	3.29	3.29	3.29	3.29	3.29
2.21	3.63	3.72	3.89	4.03	4.03
2.96	3.53	4.08	4.22	4.39	4.39
4.35	3.96	5.12	5.22	5.28	5.28

Table 17.2.2. Maximum Stress Concentration for a
Row of Five Ovaloidal Openings, $H_o/W_o = 2.0$

$\dfrac{W_o}{W_p}$	Stress concentrations at				
	A	B	C	D	E
1.03	3.90	3.90	3.90	4.05	4.17
2.09	4.09	4.50	4.61	4.70	4.79
3.40	4.41	5.02	5.40	5.47	5.56
4.28	4.66	5.67	5.93	6.10	6.10

concentration asymptotically approaches an upper limit, which corresponds to that for an infinite row of circular openings. For example, for five circular openings spaced one diameter apart the maximum compressive stress concentration is 3.28 (at point E, Fig. 17.2.2), whereas the theoretical value for an infinite row of circular openings is 3.26. A similar inference can be made for the ovaloidal openings. A maximum compressive stress concentration curve for a row of circular openings in a hydrostatic stress field $S_v = S_h$ and with $W_o/W_p = 1$ is also given in Fig. 17.2.4. Note that as the number of openings becomes infinite the maximum stress concentration is only 15% greater for the vertical uniaxial stress field than for the hydrostatic case.

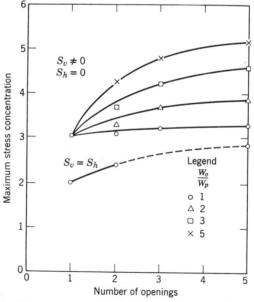

Fig. 17.2.4. Maximum stress concentration for various opening-to-pillar width ratios, and number of openings.

Design of Rib Pillars

Areas of unmined rock left between the roof and floor of an underground opening or between openings are referred to as *pillars*. These pillars together with the sidewalls of the mined area support the overlying rock; hence the stability of this support is essential to the general stability of an opening or openings. *Rib pillars* are formed between openings whose length is large compared with their cross-sectional dimensions, that is, between two-dimensional openings such as those treated in the preceding section. Therefore, the spacing between openings W_p is the pillar width (Fig. 17.2.5).

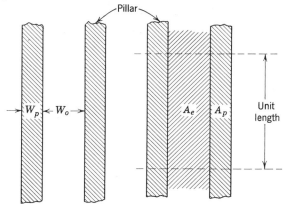

Fig. 17.2.5

For an infinite row of two-dimensional openings of width W_o separated by rib pillars of width W_p, *the average pillar stress $\bar{S}_p$ is obtained by assuming that any one rib pillar uniformly supports on its midplane the weight of the rock overlying the pillar and one-half of the opening on each side of the pillar. Thus

$$\bar{S}_p = \frac{W_p + W_o}{W_p} S_v \qquad (17.2.1)$$

The average pillar stress concentration $\bar{S}_p/S_v$ is given in Fig. 17.2.6 as a function of the opening-to-pillar width ratio. The same figure shows the maximum stress concentrations versus opening-to-pillar width ratio for a row of five circles, and a row of five ovaloids with height-to-width ratios of 0.5 and 2.0.* The following empirical equation has been fitted to these data

$$K = c + 0.09\left[\left(\frac{W_o}{W_p} + 1\right)^2 - 1\right] \qquad (17.2.2)$$

* The data for the five circular openings and five ovaloidal openings with $H_o/W_o = 0.5$ are given in the last column of Tables 17.2.1 and 17.2.2.

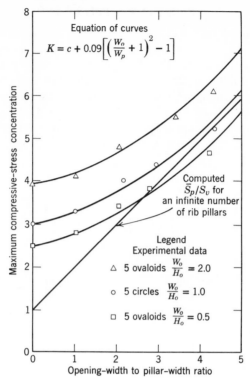

Equation of curves

$$K = c + 0.09\left[\left(\frac{W_o}{W_p} + 1\right)^2 - 1\right]$$

Computed $\bar{S}_p/S_v$ for an infinite number of rib pillars

Legend
Experimental data

△ 5 ovaloids $\frac{W_o}{H_o} = 2.0$

○ 5 circles $\frac{W_o}{H_o} = 1.0$

□ 5 ovaloids $\frac{W_o}{H_o} = 0.5$

Maximum compressive-stress concentration

Opening-width to pillar-width ratio

Fig. 17.2.6. Maximum stress concentration for various opening-to-pillar width ratios.

where K is the maximum compressive stress concentration in pillars and c is the maximum stress concentration around a single opening of the same shape in a unidirectional stress field. Thus Eq. 17.2.2 can be used to compute the maximum stress concentration in rib pillars separating a series of parallel openings. The values of c for different shaped openings can be obtained from the curves in Figs. 16.2.6, 16.2.7, and 16.2.9.

Analytical studies of the stress distribution around a row of ovaloidal openings in a two-directional stress field are inadequate for obtaining critical stress concentrations directly. The results for an infinite row of circular openings in a two-directional stress field show that confining pressures applied from the sides are very effective in reducing the tensile stresses at the top and bottom of circular openings but reduce the stress concentration in the pillars very little. If it is assumed that a row of ovaloidal openings will behave similarly, Eq. 17.2.2 can also be used to calculate the critical stress concentrations for this opening shape, provided the applied stress field corresponds to $m > \frac{1}{3}$.

Theoretical and experimental studies of the stress distribution around a row of equal size, equally spaced openings have shown that as the opening-to-pillar width ratio increases, the average pillar stress approaches the maximum stress. This effect is illustrated in Fig. 17.2.6. Also, as the average pillar stress concentration increases, the stress distribution across the midplane of the pillar becomes more uniform. For example, for an infinite row of circular openings with an opening-to-pillar width ratio of 4 the average pillar stress concentration is 5.0, and the maximum stress concentration is approximately 5.28, estimated from the data for five openings with $W_o/W_p = 5$ (Fig. 17.2.4). Hence, the stress distribution curve on the midplane would be approximately as shown in Fig. 17.2.7 (the shaded areas above and below the average pillar stress concentration line must be equal).

The fact that as the opening-to-pillar width ratio increases, the average stress approaches the maximum stress and the stress distribution becomes more uniform, is a very significant result because it implies that for design purposes, if the opening-to-pillar width ratio is large then the average pillar stress can be used as a first approximation for the critical pillar stress. Because average pillar stress will play an important role in underground design problems it seems appropriate to discuss its calculation under various conditions.

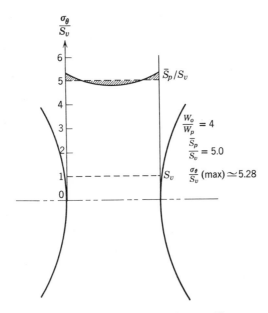

Fig. 17.2.7. Comparison of average and maximum pillar stress concentration.

The support for a system of underground openings in competent rock is the unmined rock left between and around openings. If the system of openings is mainly horizontal in extent, a horizontal plane through the openings will be composed of two parts—an excavated area, and an unexcavated or pillar area. If the assumption is made that the pillars uniformly support the entire load of the rock overlying both the pillars and excavated area, the average pillar stress is related to the applied stress by

$$\bar{S}_p A_p = S_v(A_e + A_p) \tag{17.2.3}$$

or

$$\bar{S}_p = S_v \frac{A_t}{A_p} \tag{17.2.4}$$

where $\bar{S}_p$ = average pillar stress

A_e = excavated area

A_p = pillar area

$A_t = A_e + A_p$ = total area

Thus the average pillar stress for a system of openings can be obtained from the pillar area and total area within the mineable limits. If the openings are of equal size and are equally spaced, the total area and pillar area can be calculated from the dimensions of the openings and pillars.

For a system of parallel openings separated by N rib pillars, where the width of the openings is W_o, the width of the pillars W_p, and the length of the openings and pillars is L_p, the excavated area is given by

$$A_e = L_p W_o N \tag{17.2.5}$$

and the pillar area is given by

$$A_p = L_p W_p N \tag{17.2.6}$$

Thus Eq. 17.2.4 becomes

$$\bar{S}_p = S_v \frac{W_p + W_o}{W_p} \tag{17.2.7}$$

The areal extraction ratio R_a is defined as the ratio of the excavated-to-total area,* that is,

$$R_a = \frac{A_e}{A_t} = \frac{A_t - A_p}{A_t} = 1 - \frac{A_p}{A_t} \tag{17.2.8}$$

* In mineral mining the volume extraction ratio, which is defined as the ratio of the excavated-to-total volume of a mineral deposit, usually is the only information available from mine records. The volume extraction ratio is often referred to as the extraction, extraction-ratio, recovery, or if expressed as a percentage, the percentage extraction. The areal extraction ratio is equal to the volume extraction ratio if the cross sections of the mine openings are rectangular.

Hence, Eq. 17.2.4 becomes

$$\bar{S}_p = S_v\left(\frac{1}{1 - R_a}\right) \tag{17.2.9}$$

or

$$R_a = 1 - \frac{S_v}{\bar{S}_p} \tag{17.2.10}$$

In Fig. 17.2.8 the average pillar stress concentration computed from Eq. 17.2.9 is given together with the maximum pillar stress concentrations for circular and ovaloidal openings. It is evident that for an extraction ratio greater than 0.75 the average and maximum pillar stress concentrations become practically equal, hence, for this case Eq. 17.2.10 may be rewritten as a design equation for rib pillars by replacing $\bar{S}_p/S_v$ by C_p/F_sS_v, where C_p/F_s is the safe pillar load. Thus

$$R_a = 1 - \frac{F_sS_v}{C_p} \tag{17.2.11}$$

However, if the extraction ratio is less than 0.75 the average pillar stress concentration $\bar{S}_p/S_v$ in Eq. 17.2.10 should be replaced by the maximum pillar stress concentration $\sigma_{\theta(\max)}/S_v$, and $\sigma_{\theta(\max)}/S_v$ should be replaced by C_p/F_sS_v to give Eq. 17.2.11. The value of $\sigma_{\theta(\max)}/S_v$ for any extraction ratio can be determined from the data in Fig. 17.2.8.

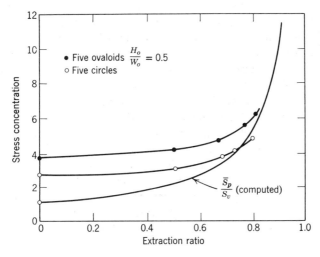

Fig. 17.2.8. Average and maximum pillar stress concentration for different extraction ratios.

Compressive Strength of Pillars

The compressive strength of a pillar C_p, as used in Eq. 17.2.11, needs some clarification. The tests described in Section 11.7 for determining the compressive strength of rock are usually made on drill core specimens having a height-to-diameter ratio of 1. However, the compressive strength of a specimen depends on the height-to-diameter (or width) ratio, and the following equation has been found to hold for specimens having a height-to-diameter ratio of 0.25 to 4.0

$$C_s = C_1\left[0.778 + 0.222\left(\frac{d}{h}\right)\right] \qquad (17.2.12)$$

where C_1 = compressive strength for specimens having $d/h = 1$

C_s = compressive strength of specimen with $d/h \neq 1$

d = diameter of specimen

h = height of specimen

The compressive strength of a rib pillar in massive elastic rock can be approximated by use of Eq. 17.2.12, provided that in this equation d and h are replaced by H_p and W_p. Thus

$$C_p = C_1\left[0.778 + 0.222\left(\frac{W_p}{H_p}\right)\right] \qquad (17.2.13)$$

Calculations made on this basis should be on the conservative side because (1) the compressive strength of a specimen of cross-sectional dimensions W and L normal to the applied load, and with $W < L$, is stronger than a specimen with a circular cross section of diameter W; (2) if the surfaces of the pillar are concave as, for example, if they are formed by circular openings, the pillar strength would be somewhat greater than that for a pillar with straight sides; (3) the end constraint on a pillar formed in a rock continuum would be greater than the end constraint normally employed in a standard compressive test, a factor that to some degree would increase the pillar strength; (4) Eq. 17.2.12 does not give as great an increase in pillar strength with W_p/H_p as is given in other formulas (see Section 13.3). For example, if $C_p = C_c(W_p/H_p)^{1/2}$, where C_c is the compressive strength of a cubical specimen, for a pillar width-to-height ratio of 4, the increase in pillar strength over that for a cubical pillar would be twofold, whereas from Eq. 17.2.12 it would increase only 1.66 times.

On the negative side, it is known (Section 13.3) that as the size of a specimen increases its compressive strength decreases, presumably because as the specimen size increases the probability of including a mechanical defect increases. Although massive elastic rock is defined as homogeneous and isotropic with respect to its mechanical properties, it does, nevertheless, include some defect. The effects of this factor are probably compensated by the positive factors listed in the preceding paragraph.

In calculating the average pillar stress (by Eq. 17.2.7) it is assumed that W_p is the minimum pillar width. In mining in relatively brittle rock it is known that blasting-induced fractures generally penetrate the surface and sidewalls of openings to a depth of about 3 ft. The development of these surface fractures is probably enhanced by the tangential stress concentration on pillar and sidewall surfaces. This fracturing weakens but does not completely destroy the bearing capacity of the rock. In making stress determinations by the overcoring method (Section 14.3), core lengths long enough (approximately 12 in.) for stress determinations usually cannot be obtained in the first 36 in. from the surface. It is inferred from this observation that the first 3 ft of rock is probably too fractured to offer much support; hence, the pillar width should be adjusted accordingly to obtain an *effective pillar width*. Obviously this surface fracture strongly affects the bearing capacity of thin pillars. For example, if the pillar width is 12 ft the effective pillar width would only be 6 ft. Thus in the design of a pillared system of openings, the pillar width should be made as large as practical. Some evaluation of the depth of blasting fractures can be obtained by observing the fractures in conventional cores taken from the surfaces of openings. The depth of surface fracture can be minimized by employing good blasting practices, for example, by firing only a minimum charge in holes breaking to pillar surfaces.

In mineral mining an effort is usually made to obtain as high an ore extraction as possible compatible with safety. A system of mining employing rib pillars may not be the best means of achieving this objective. For example, consider a flat-lying deposit 15 ft thick; because of blasting fracture the effective thickness of the pillars would only be about 9 ft. However, if the pillar height-to-width ratio is decreased to $\frac{1}{2}$ (making $W_p = 30$ ft), for an extraction ratio of 0.75 (a nominal value for hard rock mining), the room width would have to be 90 ft, a span that could be mined only in exceptionally competent rock. On the other hand, if a three-dimensional array of pillars is employed, as described in the next section, an extraction ratio of 0.75 could be achieved in the same deposit with a room and crosscut width of only 30 ft.

If high extraction is not a requisite, as is generally the practice in civil engineering projects such as the excavation of multiple-storage chambers or openings for a hydroelectric power station, a system of openings separated by rib pillars provides an exceptionally stable structure because intersections (of openings), which structurally are the vulnerable point in a three-dimensional array of openings, are kept to a minimum.

In inelastic rock the tangential stress concentrations on pillar surfaces tend to decrease with time, so that in the time normally required to mine a multiple opening the critical pillar stresses and the average pillar stress become approximately equal, even for a system of openings having an opening-to-pillar width ratio equal to unity or less. This condition should simplify the design of openings and pillars in inelastic materials such as salt or potash ore if it were not for the fact that the mechanism of pillar failure also varies with the pillar height-to-width ratio. For pillars with a height-to-width ratio equal to or greater than 1, brittle failure can occur; in fact, the sudden collapse of an extensive pillared area in a potash mine employing rib pillars with a height-to-width ratio of approximately 1 has been reported. Thus for this case the pillar design should be calculated by means of Eqs. 17.2.7, 17.2.10, and 17.2.11, using a safety factor F_s of at least 2. However, if the height-to-width ratio is less than 1, the probability of a brittle failure diminishes, and for a height-to-width ratio less than $\frac{1}{2}$ only a slow shortening of the pillars (manifest by roof-to-floor convergence) together with light to heavy spall from the pillar surfaces has been observed. In some deep ($h > 2000$ ft) European potash mines, rib pillars with a height-to-width ratio of $\frac{1}{4}$ have remained intact and stable for more than 25 years, although the ratio of the computed average pillar stress to the measured compressive strength of specimens taken from the pillars was less than 1 (indicating a safety factor less than 1). The strengthening of pillars with a small height-to-width ratio is attributed to the lateral confinement that develops over the pillar ends and interior (see Section 13.3).

The fact that in inelastic rock the stress concentrations tend to decrease with time causes a corresponding increase in the distance that the stress concentration zone extends into the rock surrounding the opening. Thus in this type of rock, openings two diameters apart will in time influence one another and hence cannot be treated as single openings. An estimate of the stress distribution in viscoelastic or plastic rock at $t = \infty$ can be inferred from the theory in Chapter 6.

The principal conclusions in this section pertinent to the design of multiple openings separated by rib pillars can be summarized as follows.

1. The maximum compressive stress concentrations that develop on sidewalls and pillar surfaces of a horizontal row of openings are dependent primarily on the vertical component of stress.

2. The maximum tensile stress concentrations that develop on the top and bottom of multiple openings in a vertical uniaxial stress field decrease

with the application of a horizontal stress, and generally become compressive if $S_h > \frac{1}{3}S_v$.

3. For a row of five or more openings in an elastic rock the maximum compressive stress concentration on all but the two pillars at each end of the row will be practically equal. The maximum stress concentration on the two end pillars and on the sidewalls will be less than that on the central pillars.

4. Thus it follows from (1), (2), and (3) that in most mining operations (in which $m > \frac{1}{3}$) the problem of designing multiple openings in massive elastic rock reduces to designing stable pillar support.

5. As the areal extraction ratio increases, the average pillar stress becomes more nearly equal to maximum pillar stress. If the areal extraction is greater than 75% the average pillar stress can be used in place of the maximum pillar stress.

6. In inelastic rock if the pillar height-to-width ratio is 0.5 or less the pillars will not fail suddenly, but will spall and deform (shorten) indefinitely.

7. In inelastic rock the maximum pillar stress will be less than that in elastic rock, and it will decrease with time to a value approaching the average pillar stress. Hence, in this class of rock for any opening-to-pillar width ratio and for any number of openings the average stress rather than the maximum stress can be used for design purposes.

17.3   THREE-DIMENSIONAL ARRAY OF PILLARS IN MASSIVE ROCK

Whereas in two-dimensional multiple opening and pillar systems the pillars are long (theoretically infinite) compared with their width, in three-dimensional systems their cross-sectional dimensions in the plane normal to the pillar axis are more nearly equal, as for example in the three-dimensional pillar patterns illustrated in Fig. 17.3.1. As noted in the Introduction to this chapter, because of mathematical complexity, problems dealing with the stress distribution around multiple three-dimensional openings have not been treated theoretically, and very little information is available from either three-dimensional rock or photoelastic model studies. However, if certain assumptions are made, the three-dimensional case can be approximated from two-dimensional experimental and theoretical results, and a design equation for the three-dimensional case specified. The assumptions are that (1) in a three-dimensional array of pillars, the pillars uniformly support the entire load of the overlying cover, and the average pillar stress can be calculated by Eq. 17.2.4. This assumption should be reasonably valid if there are more than four pillars

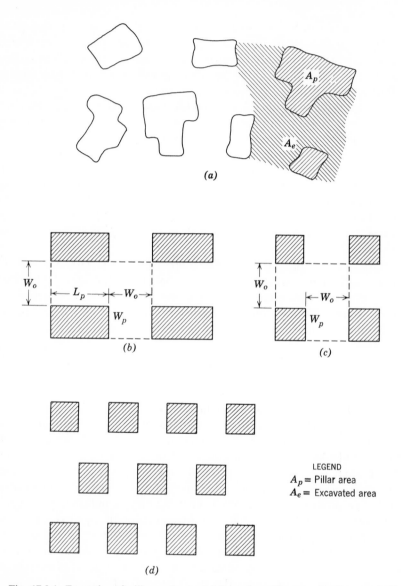

Fig. 17.3.1. Examples of pillar patterns. (*a*) Irregular pillars, randomly spaced. (*b*) Rectangular pillars, regularly spaced. (*c*) Square pillars, regularly spaced (checkerboard). (*d*) Staggered checkerboard.

in any direction across the opening, although the pillars adjoining the sidewalls will support somewhat less than this average value. (2) The extraction ratio is 0.75 or greater, a value that is consistent with the practice in many mines.* In this case the average pillar stress concentration as given by Eq. 17.2.9 can be used as a basis for design.

If these conditions are satisfied, the average pillar stress and design equations for rib pillars (Eqs. 17.2.9 and 17.2.11) are satisfactory for the design of three-dimensional pillar arrays, that is,

$$\bar{S}_p = S_r\left(\frac{1}{1 - R_a}\right) \tag{17.3.1}$$

$$R_a = 1 - \frac{F_s S_v}{C_p} \tag{17.3.2}$$

where, in this instance, C_p is the compressive strength of the pillar rock corrected by Eq. 17.2.13 for the pillar height-to-width ratio and where the pillar width is the minimum cross-sectional dimension in the plane normal to the axis of the pillar.

For a random system of irregularly shaped pillars, as illustrated in Fig. 17.3.1a, the pillar area A_p and the excavated area A_e can be evaluated by a graphical integration over the total mined area. For a regular system of rectangular pillars (Fig. 17.3.1b), the total mined area can be considered to be made of N identical elements of area $(W_o + W_p)(L_p + W_p)$, that is,

$$A_t = N(W_o + W_p)(L_p + W_p) \tag{17.3.3}$$

The total pillar area is

$$A_p = N(W_p L_p) \tag{17.3.4}$$

Hence, the total excavated area is

$$A_p = N[(W_o + W_p)(L_p + W_p) - W_p L_p] \tag{17.3.5}$$

and the extraction ratio

$$R_a = \frac{A_e}{A_t} = 1 - \frac{W_p L_p}{(W_o + W_p)(L_p + W_p)} = 1 - \frac{A_p}{A_t} \tag{17.3.6}$$

which is identical to Eq. 17.2.8.

For a checkerboard or staggered checkerboard pillar system, Fig. 17.3.1c and d, $W_p = W_o = L_p$; hence

$$R_a = 1 - \frac{W_p^2}{4W_p^2} = 0.75 \tag{17.3.7}$$

---

* If the extraction ratio is less than 0.75 the maximum pillar stress concentration calculated from the data in Fig. 17.2.8 would probably be a better value for design purposes although these data were derived from two-dimensional models.

In civil engineering projects the design of pillar systems is usually a problem of only secondary importance because in most instances there is no reason to excavate a high percentage of the rock at any site. In mineral mining the situation is just the opposite—because of the economics of mining and the desire to recover as large a percentage of the mineral as possible there is always an incentive to keep the extraction ratio high. In a number of instances, too much pillar support has been removed and as a consequence stopes, or even larger areas, have collapsed. A survey was made of open stope mines in rock that classified as massive elastic.[2]* The stope and pillar dimensions were measured and the extraction calculated (using Eq. 17.3.5); the average pillar stress was calculated from the depth of the stopes, the unit weight of the overlying rock, and the extraction (using Eq. 17.3.1, where $S_v = -\gamma h$); the compressive strength of the pillar rock was determined by measuring the compressive strength of specimens cut from drill cores taken from the pillars. From these data the pillar safety factors were calculated (using Eq. 17.3.2) which in most instances were 4 or greater and, hence, 4 is considered a nominal safety factor for pillars in this class of rock.

Although model tests indicate that the maximum pillar stress concentration is significantly greater than the average pillar stress concentration for extraction ratios less than 0.75, in situ stress determinations indicate that the tangential stress concentration on the surface of pillars is generally not much greater, and in some instances less than the computed average pillar stress. This observation supports the use of the average pillar stress rather than maximum pillar stress in designing three-dimensional pillar systems. Moreover, this approximate procedure is probably satisfactory for extraction ratios as low as 0.50. However, when these approximate methods are employed, repeated in situ pillar stress determinations are recommended.

In inelastic rock the problem of designing two- or three-dimensional pillar systems is essentially one and the same, except that creep rate data from tests of model pillars made from several inelastic rocks are available. Obert[3] found that the creep rate $\dot{\epsilon}$ ($\dot{\epsilon} = d\epsilon/dt$) in pillars made from salt, trona, and potash ore was dependent on the applied stress σ, that is,

$$\dot{\epsilon} = k\sigma^n \qquad (17.3.8)$$

where the exponent n varied from 2.4 to 3.3. In a complementary report Bradshaw, et al.[4] showed that the creep rate for one of the model pillar materials (salt) decreased as t^{-m}; hence,

$$\dot{\epsilon} = k_2\sigma^n t^{-m} \qquad (17.3.9)$$

* In each mine the excavated area had remained stable for more than five years.

where $m \approx 0.6$, and that this power law decrease was observed in the mine from which salt for the model pillar was taken. Obert[5] later reported that a similar time dependence was found for potash ore from Canada, except that $m \approx 0.9$. If a tolerable strain rate or limiting deformation is specified for a mining operation in an inelastic rock, a creep test on model pillars made from mine rock can be performed to determine k_2 and the two exponents in Eq. 17.3.9. Hence, the creep rate or, by integration, the deformation at any time t can be determined.

In either massive elastic or massive inelastic rock, pillar design is improved if the pillar width-to-height ratio is made as large as practical. However, if the extraction ratio is specified, any increase in pillar size will cause a corresponding increase in the room widths. In thinly bedded deposits the room width necessary to obtain a reasonable extraction ratio usually will not create any problem. For example, suppose the depth of a deposit, the unit weight of the overlying rock, and the compressive strength of the pillar rock are such that an extraction ratio of 0.75 can be mined safely, and a checkerboard pillar system is specified because it will provide the maximum extraction with the minimum pillar and crosscut width. Also, suppose that the thickness of the deposit is 6 ft, and that to attain the required pillar strength, a pillar width-to-height ratio of 4 is required, that is, as $H_p = 6$ ft, $W_p = 24$ ft. Thus to achieve the desired extraction, a room and crosscut width of 24 ft would be necessary, a span that is nominal in most competent massive rock. However, in a deposit, say 30 ft thick, the corresponding room width would have to be 120 ft, a span that is not generally minable in most rock types. Moreover, for this set of conditions, a decrease in room width can be achieved only by correspondingly decreasing the extraction ratio or decreasing the pillar width-to-height ratio.

17.4 OPENINGS IN ROCK CONTAINING PLANES OF WEAKNESS PARALLEL TO THE ROOF

A large part of the production of minerals, and especially coal and other nonmetallic minerals, is obtained from relatively flat-lying sedimentary deposits in which the roof and floor of mine openings are parallel to partings or other planes of weakness. The problem of pillar support in either two- or three-dimensional systems of openings in this class of rock is essentially the same as for massive rock because planes of weakness normal to the axis of pillars do not lower the pillar strength significantly. Hence, the average pillar stress can be calculated from Eq. 17.2.9, and Eq. 17.2.11 can be used for design purposes. In two-dimensional systems of openings if a layer of rock becomes detached from the main roof it can be treated as a rigidly clamped beam, as specified for single openings in Section 16.5. However, the clamping action between the main roof and the top of a pillar is not strictly a rigid clamp, for the pillar is in itself an elastic material that will deform under the load of the beam (Fig. 17.4.1). Adler[6] considered the problem of a beam supported by elastic pillars which mechanically can be thought of as a bed of closely spaced but independent springs. If w is the pillar reaction, k is the pillar modulus, and z is the pillar displacement, then

$$w = kz \qquad (17.4.1)$$

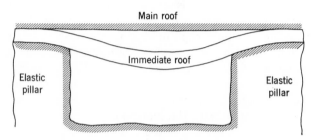

Main roof

Immediate roof

Elastic pillar   Elastic pillar

Fig. 17.4.1. Single layer roof on elastic pillars.

If the pillar is of width b and height h, the pillar modulus is related to the modulus of elasticity of the rock E_p by

$$E_p = \frac{w/b}{z/h} \qquad (17.4.2)$$

Hence

$$k = \frac{w}{z} = \frac{E_p b}{h} \qquad (17.4.3)$$

The deflection of the beam over the pillar must satisfy the differential equation (see Eq. 5.6.18)

$$(EI_y)_b \frac{d^4 z}{dx^4} = kz = \frac{E_p b}{h} \qquad (17.4.4)$$

and outside the pillar

$$(EI_y)_b \frac{d^4 z}{dx^4} = q \qquad (17.4.5)$$

where $(EI_y)_b$ is the flexural rigidity of the beam and z is the displacement. These equations must satisfy the boundary conditions which are, at the junction of the beam over and outside the pillar, the displacements and slopes must be equal.

Besides the modulus of elasticity of the beam E_b and its unit weight q, span L, and thickness t, which are required in the moments and displacements, equations for a rigidly supported beam, two additional parameters E_p and h are involved in the corresponding solutions for the elastically supported beam. These equations must be evaluated for each set of beam and pillar parameters, but they are too complex to present here completely. However, the solution shows that the maximum bending moments and hence the maximum fiber stress in an elastically supported beam (which occur near the pillar edge) are less than the maximum moments or stress

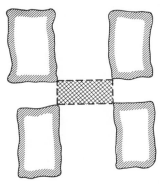

Fig. 17.4.2. Intersection area (cross-hatched) in a regular pillar system.

in a rigidly supported beam, but that the difference is small for normal values of E_p and h_p. Adler suggested that an improvement could be made by undercutting the pillar or drilling holes in the pillar surface to lower its effective modulus of elasticity and thereby decrease the maximum bending moment in the supported beam.

Thus this theoretical study leads to the conclusion that the design of a beam-type roof can be made on the basis of a rigidly clamped beam with only a small error and that a design so determined will be on the conservative side. Some lowering of the effective pillar modulus probably occurs in the surface rock of most pillars as a result of blasting fracture.

In a three-dimensional pillar system, the stability of a roof layer over the intersection of a room and crosscut (Fig. 17.4.2) is critical. However, the bending moments or stress in this layer cannot be considered on the basis of a rigidly clamped beam or plate. Woinowsky-Kreiger[7] considered the problem of an infinite slab rigidly attached to a three-dimensional array of approximately square pillars. The solution, although evaluated for what would correspond to an exceptionally high extraction ratio, indicates that the bending moment in the slab at the pillar corners can be several times that for a rigidly clamped plate.

17.5   OPENINGS IN ROCK CONTAINING PLANES OF WEAKNESS NOT PARALLEL TO ROOF

Because in two-dimensional multiple openings the immediate roof (that is, the roof layers loaded principally by gravity) in adjoining openings do not interact with one another, the roof stability problem in this case is the same as that for single openings. Generally, for optimum stability, rooms should be oriented normal to the strike of partings, joints, faults, or other planes of weakness (Fig. 16.6.1*b*). A vertical or near-vertical dip is favorable, and near-horizontal dip unfavorable. The pillar strength is only affected significantly if the strike of weakness planes is parallel to the long axis of openings. In this case the pillar strength is greatly reduced if the

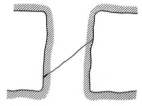

Fig. 17.5.1. Pillar with intersecting plane of weakness.

dip (angle α in Fig. 16.6.1*a*) is between 20° and 70°, and especially if the planes of weakness intersect the full cross section of the pillar, as illustrated in Fig. 17.5.1. Thus pillars with a large width-to-height ratio are favorable.

In a three-dimensional system of openings and pillars the critical part of the immediate roof is the intersections formed by rooms and crosscuts. Weakness planes dipping at any angle will create a problem if they strike through intersections. This situation can often be ameliorated by (1) expeditious placement of pillars, sometimes involving an irregular pillar placement; (2) use of rectangular pillars in which crosscuts are placed in areas of competent rock; (3) use of a staggered checkerboard pillar system in which the major weakness planes intersect alternate rows of pillars as illustrated in Fig. 17.5.2. The latter procedure may be employed as a precautionary measure to prevent roof failures that tend to progress in a given direction.

The pillar problem in three-dimensional arrays is subject to the same difficulties experienced in two-dimensional systems except that weakness planes dipping between 20 and 70° and striking in any direction are unfavorable, especially, as in the two-dimensional case, if they intersect the cross section of pillars. If this condition cannot be avoided, an effort should be made to determine an average fracture envelope for the weakness plane (Section 10.6), and to determine the pillar strength on this basis.

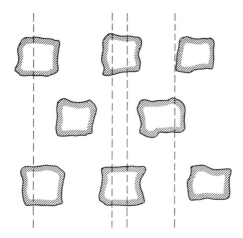

Fig. 17.5.2. Staggered pillar system oriented to intercept jointing in roof.

REFERENCES

1. Duvall, W. I., "Stress Analysis Applied to Underground Mining Problems, Part II—Stress Analysis Applied to Multiple Openings and Pillars," *U.S. BurMines Rept. Invest.*, **4387** (1948).
2. Obert, L., W. I. Duvall, and R. H. Merrill, "Design of Underground Openings in Competent Rock," *U.S. BurMines Bull. 587* (1960).
3. Obert, L., "Deformational Behavior of Model Pillars Made from Salt, Trona, and Potash Ore," *Proc., Sixth Symp. on Rock Mech.*, Univ. of Mo., Rolla, Mo. (1964).
4. Bradshaw, R. L., W. J. Boegley, and F. M. Emson, "Correlation of Convergence Measurements in Salt Mines with Laboratory Creep-Test Data," *Proc., Sixth Symp. on Rock Mech.*, Univ. of Mo., Rolla, Mo. (1964).
5. Obert, L., "Creep in Model Pillars," *U.S. BurMines Rept. Invest.* **6703** (1965).
6. Adler, L., "Rib Control of Bedded Roof Stresses," *Fourth Symp. on Rock Mech.*, Min. Industries Sta. Bull. No. 76, Univ. of Penn. (1961).
7. Woinowsky-Kreiger, S., "On Bending of a Flat Slab Supported by Square-Shaped Columns and Clamped," *J. App. Mech.*, **21** (1954).

CHAPTER 18

SUBSIDENCE AND CAVING

18.1 INTRODUCTION

Gravitational, and possibly tectonic forces, act on the rock surrounding underground excavations in such a direction as to close the opening. A part of the closure results from elastic (time-independent) deformations, a part from inelastic (time-dependent) deformations, and in many instances, a substantial part of the closure is due to the fracturing of the overlying rock which falls into the opening. Collectively these effects tend to produce a lowering of the rock overlying the opening, and this process is referred to as *subsidence*. The first manifestations of subsidence may be convergence or a succession of local failures in the rock surrounding the opening, such as repeated sloughing from the immediate roof. This phase of the process is termed *subsurface subsidence* as opposed to *surface subsidence*, which is evidenced by a depression in the overlying surface. Surface subsidence may range from a deformation that is barely measurable with precision-surveying instruments to a void having a volume comparable to that of the excavated volume of the opening.

Caving is a form of subsurface subsidence. In this treatment, however, the term is restricted to a process where either or both the rate-of-failure or the area over which the failure occurs is at least partially controlled by the mining method, as for example, in a block caving operation, and as contrasted to subsurface subsidence which to a much greater degree is an uncontrolled process. Whereas subsurface subsidence is primarily a result of the gravitational force acting on the rock surrounding or overlying an opening, caving may, in addition, be induced or encouraged by other means such as blasting or by producing local stress concentrations.

Subsidence and caving are generally associated with incompetent rock and many mining operations in which these processes occur are in rock that would be classified as such. However, there are exceptions; some evaporite minerals (salt and potash ore, for example) in room and pillar mining that normally classify as competent rock will deform without fracturing to the degree that an opening will close if the roof span is increased sufficiently. Moreover, an expression of this subsurface subsidence may be evident on surface, even through a thousand feet or more of cover. Many stratified rocks will stand unsupported over an opening having a small span, as in a room and pillar system of coal mining, but will subside if the span is increased, as in a longwall operation. In the latter case subsurface subsidence may occur by a gradual bending of the overlying strata and/or by a succession of roof failures. Usually the opening will eventually close and the deformation of the overlying strata will progress to, and cause a depression of, the surface.

Thus subsidence and caving depend on both the time-dependent and time-independent characteristics of the rock and on the stress conditions created in the rock by the geometry of the opening or the method of mining. Because of the wide variation in rock properties and in methods of mining there appear to be a number of different subsidence mechanisms, four of which are identified as *trough subsidence, subsurface caving, plug caving*, and *chimneying* (sometimes called piping or funneling). There are no completely satisfactory mathematical or empirical theories that explain any of these mechanisms, and even if theories existed the mechanical characteristics of large bodies of rock are so indefinitely known that quantitative evaluation of these processes could not be made. In succeeding sections the phenomenology of these subsidence mechanisms is described and mathematical and empirical theories are reviewed that account for some aspects of these mechanisms.

18.2 TROUGH SUBSIDENCE

The most commonly observed type of surface subsidence is that which occurs over excavations in relatively flat thin-bedded deposits overlain by stratified sedimentary rocks which, in many instances, are poorly consolidated or otherwise incompetent. The formations overlying most coal and potash mines are typical examples. Usually the first indication of subsidence in this type of excavation is a downward flexing of the immediate roof (roof sag) which may be accompanied by floor heave. As the areal extent of the excavation is enlarged, the roof sag will increase until either the roof touches the floor or the roof fractures and falls. In virtually all rock types this phase of the process is time dependent and terminates with a complete closure of the opening. Surface subsidence, manifested by vertical and lateral surface displacements, usually occurs almost immediately following the first indications of subsurface subsidence. These displacements produce lateral surface strains that in some instances increase with the increase in the dimensions of the excavation and with time until surface cracks are produced.

From an engineering standpoint trough subsidence creates two problems: support of the opening and/or control of subsurface subsidence during the period of mining, and damage to the surface or structures on the surface. Support of the opening may be effected by leaving unexcavated areas of rock in the form of pillars, or by artificial means, such as props or chocks. The treatment of artificial support is not within the scope of this book, and the support of stable openings on pillars has been covered in Chapter 17. However, subsidence can be induced by the removal of pillar support. The usual procedure is to first room-and-pillar mine a panel of from 100,000 to 500,000 ft². Then, starting at a property line or boundary next to an already caved panel, the cross-sectional area of pillars are reduced by splitting or slabbing operations until the pillars start to crush. In the terminal stage the remnant pillars are often caused to fail by blasting, a method that permits a better control of the time and place where support should be removed. If the pillar removal operation is carried out along a line, a modified longwall is created. In coal mines and in some nonmetallic mineral mines, as the span over the unsupported area is increased the immediate roof fails by a succession of sloughs and the void ultimately fills with broken rock, usually to the extent that the broken rock (often referred to as gob or self-fill) becomes compacted and partially or completely supports the overlying cover, and on the surface a subsidence trough forms over the unsupported area. However, in a potash mine overlain by a thick bed of salt, pillar support was reduced by slabbing, at such a rate that the pillars crushed slowly. The immediate roof and presumably the stratified formation thereover deformed by flexing but without fracture until the opening closed. The original room height was 12 to 14 ft; Fig. 18.2.1 shows the pillar removal area when the room height was approximately 3 ft.

Obviously the mechanics of subsurface subsidence into openings in thin-bedded deposits is so dependent on the characteristics of the rock and mining method that even empirical generalizations are not possible except under very restricted conditions. Some progress is being made in the interpretation of load cell and deformeter measurements obtained within the limits of a single property. The transfer of load from one point to another and the effect of rate of mining on pillar loads and related problems have been investigated. However, until the accumulation of data from these instrumented investigations is large enough to permit a statistical analysis, the prospect for any analytic basis for designing or controlling this type of subsurface subsidence is remote. The surface effects associated with trough subsidence appear to follow a more general pattern. The principal features of trough subsidence, based on analyses of vertical and lateral surface pin surveys (Section 9.4) have been summarized by Rellensmann.[1] Figure 18.2.2 is an idealized representation of the lateral and vertical movements characteristic of trough subsidence. Note that all lateral surface displacements are toward the point over the center of the excavation and they are

Fig. 18.2.1. Subsurface subsidence in a potash mine. (Courtesy U.S. Borax and Chemical Corp.)

zero at this point. The vertical surface displacement is a maximum over the center of the excavation, and these displacements extend well beyond the lateral limits of the excavation. The lateral surface strain is tensile at points outside the lateral limits of the excavation and compressive within the excavation limits. That is, within the limits of the subsidence zone the points of zero strain (points D in Fig. 18.2.1), which correspond to the inflection points B on the vertical surface displacement curve, are at (or near) the limits of the excavation. Rellensmann pointed out that the sum of the tensile and compressive strains must be zero. The angle of break α^* is defined as the angle between the horizontal and the line connecting the limit of excavation with the points on the surface of maximum tensile strain (points A). This line, which passes through the point of maximum tensile strain on the surface of all subsurface strata, defines the plane of

* Sometimes referred to as the angle of fracture or angle of slide.

break (although the strain may not be large enough to cause fracture). For any given rock type this line is assumed to be straight although there is some evidence to indicate that its slope increases with depth. The angle of draw* is defined as the angle between the vertical and the line connecting the limit of excavation with the point of zero vertical surface displacement (which may be indefinite because it depends on the precision with which the pin survey is made).

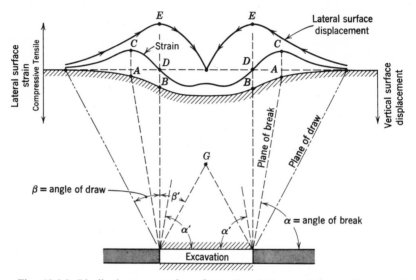

Fig. 18.2.2. Idealized representation of trough subsidence. (After Rellensmann[1]).

As the excavation is successively widened, as indicated by the extraction sections *a*, *b*, *c*, *d*, and *e*, and the corresponding surface subsidence curves (Fig. 18.2.3), the surface subsidence trough becomes progressively deeper. Rellensmann observed that the compressive strain is a maximum (curve *b*) when the width of the excavation is such that the reflections of the plane of break about the vertical (designated by the angle α' in Fig. 18.2.2) just meet at a point on the original surface. If the reflections of the plane of draw, designated by the angle β' in Fig. 18.2.2, meet at a point (*G*) below the original surface, the subsidence phase is termed subcritical; if they meet at a point above the original surface as indicated by the point *G* in Fig. 18.2.4, the subsidence phase is termed supercritical. During the subcritical phase the surface subsidence continues to increase, and it reaches a maximum when point *G* reaches the original surface, that is, at the critical point. As the excavation is further widened point *G* moves above the

* Also called the limiting angle of influence.

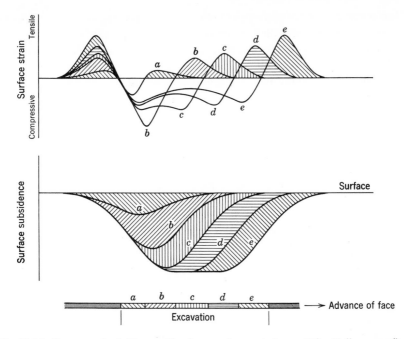

Fig. 18.2.3. Progress of subsidence with advance of excavated area. (After Rellensmann[1])

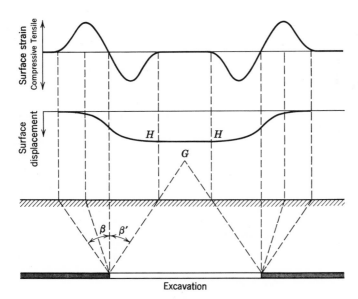

Fig. 18.2.4. Supercritical subsidence. (After Denkhaus.[2])

surface (supercritical phase) and between points HH the surface subsidence remains constant and the surface strain is zero (Fig. 18.2.4).

Some degree of success has been attained in accounting for trough subsidence on a soil mechanics basis. In Fig. 18.2.5a, assume that the cover overlying the opening is a dry granular material, whose mechanical behavior satisfies the fundamental empirically determined relationship for soils, namely

$$\tau_s = c - \sigma_s \tan \phi \qquad (18.2.1)$$

where the shearing resistance τ_s depends on the normal (compressive) stress σ_s; ϕ is the angle of friction, and c, the cohesion, is the shear resistance for zero normal stress. In Fig. 18.2.5b the line AB is a graphical

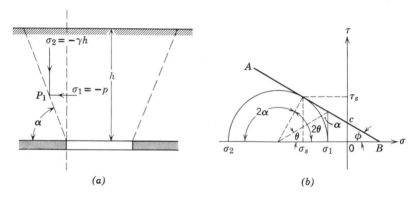

(a)	(b)

Fig. 18.2.5. Mohr representation of the state of stress on the plane of break.

representation of Mohr's failure envelope for soil, that is, of Eq. 18.2.1. In Fig. 18.2.5a let σ_1 and σ_2 be the principal (compressive) stresses at point P_1 on the plane of break. At failure, σ_1 and σ_2 will have a value such that their Mohr's circle will just touch the failure envelope. From the construction in Fig. 18.2.5b

$$2\alpha = 90° + \phi \qquad \text{or} \qquad \alpha = \frac{\phi}{2} + 45° \qquad (18.2.2)$$

and

$$\cot \phi = -\tan 2\alpha = \frac{2 \tan \alpha}{\tan^2 \alpha - 1} \qquad (18.2.3)$$

Also Eq. 18.2.1 can be written as

$$\tfrac{1}{2}(\sigma_1 - \sigma_2) \cot \alpha = c - \sigma_1 \tan \phi \qquad (18.2.4)$$

From Eqs. 18.2.2, 18.2.3, and 18.2.4

$$\sigma_2 = -2c \tan \alpha + \sigma_1 \tan^2 \alpha \qquad (18.2.5)$$

Table 18.2.1. Measured Angles of Friction and Calculated
Angles of Break for Various Rock Types (After Seldenrath[3])

Rock	Angle of Friction, Degrees	Angle of Break, Degrees
Clay	15–20	52.5–55
Sand	35–45	62.5–67.5
Moderate shale	37	63.5
Hard shale	45	67.5
Sandstone	50–70	70–80
Coal (average)	45	67.5

If, at the point P_1 (Fig. 18.2.5a), the vertical stress is $-\gamma h$ (where h is the depth and γ is the unit weight of the rock) and the lateral stress is $-p$, so that $\sigma_2 = -\gamma h$ and $\sigma_1 = -p$, then Eq. 18.2.5 becomes

$$p \tan^2 \alpha = \gamma h - 2c \tan \alpha \qquad (18.2.6)$$

This fundamental soil-mechanics equation relates the angle of the plane of break and the cohesion to the vertical and lateral stresses at a point within the soil. For cohesionless soils $c = 0$; hence Eq. 18.2.6 becomes

$$p \tan^2 \alpha = \gamma h \qquad (18.2.7)$$

As the maximum lateral tensile strain develops (at an angle α) across the plane of break, failure should occur across this plane. Observation of surface cracks generally verifies this supposition. Failure on this plane is accompanied by both downward and lateral (toward the center of the subsidence zone) displacements. Using Eq. 18.2.2, Seldenrath[3] calculated the angle of break for several rock types from their measured angles of friction. These calculated values of the angle of break, given in Table 18.2.1, compare favorably with values quoted by Rellensmann,[1] given in Table 18.2.2.

The mechanics of an elastic continuum has also been used to explain certain aspects of trough subsidence. On the basis of elastic theory Berry[4] considered the deformations caused by the closure of a thin slit in both an infinite and semi-infinite elastic, isotropic plate. This problem approximates an opening in a thin-bedded deposit, such as a coal mine. For the isotropic case the calculated surface displacements were smaller than those encountered in practice. This led Berry and Sales[5] and Berry[6] to consider the same problem in a transversely anisotropic plate, that is, a plate in which the elastic constants in directions parallel and perpendicular to the opening

Table 18.2.2. Observed Angles of Break (after Rellensmann[1])

Rock Type	Angle of Break, Degrees
Unconsolidated strata	40–60
Sand	45
Plastic strata	60–80
Clay and shales	60
Rocky strata	80–90
Hard sandstone⎫ Limestone ⎭	85

have different values. This would simulate a thin opening in a stratified rock. The resulting expressions for the surface subsidence (vertical displacement of the surface), $v_0(x)$, and the horizontal surface strain, $du_0(x)/dx$, at any point x (Fig. 18.2.6) are

$$-v_o(x) = \frac{t}{\pi(\alpha_1 - \alpha_2)} (\alpha_1 \tan^{-1} P_1 - \alpha_2 \tan^{-1} P_2) \qquad (18.2.8)$$

$$\frac{du_0(x)}{dx} = \frac{2at\alpha_1\alpha_2}{\pi(\alpha_1 - \alpha_2)} \left(\frac{Q_1}{R_1} - \frac{Q_2}{R_2}\right) \qquad (18.2.9)$$

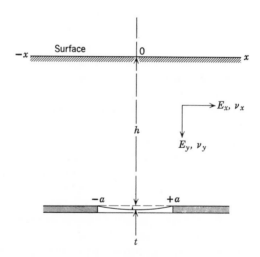

Fig. 18.2.6

where $a = $ half-width of the opening

$h = $ depth of the opening

$t = $ vertical roof convergence

$$P_j = \frac{2ah_j}{x^2 - a^2 - h_j^2}$$

$$Q_j = x^2 - a^2 - h_j^2$$

$$R_j = (Q_j + 4h_j^2 x^2)^{1/2} \qquad j = 1, 2$$

$$h_j = \frac{h}{\alpha_j}$$

α_1 and α_2 are the transverse anisotropic elastic parameters. α_1^2 and α_2^2 are the real roots in α^2 of the characteristic equation

$$\alpha^4 - 2k_2\alpha^2 + k_1^2 = 0 \qquad (18.2.10)$$

in which

$$k_1^2 = \frac{1 - v_x^2}{E_x/E_y - v_y^2}$$

$$k_2 = \frac{E_x/2G - v_y(1 + v_x)}{E_x/E_y - v_y^2}$$

Equations 18.2.8 and 18.2.9 were derived from elastic theory; hence they are time-independent, whereas it is the common experience that subsidence is largely a time-dependent phenomenon. However, Berry[5] pointed out that if the mechanical behavior of the rock is viscoelastic the elastic analysis is still valid for the final deformed state when creep has ceased (at $t = \infty$). Although rock is to a varying degree both viscoelastic and plastic, this theoretical solution gives surface subsidence and surface strain results that are in reasonable agreement with measurements made over coal deposits, as indicated in Fig. 18.2.7a and b. In this example, for the evaluation of Eqs. 18.2.8 and 18.2.9 the depth h was taken as 1973 ft, the effective convergence as 17 in. (one-third of the excavated thickness; the other two-thirds of the opening are assumed to be occupied by floor heave and broken rock), the width of the opening $2a$ as 550 ft, and the transverse elastic parameters α_1 and α_2 as 4.45 and 0.45. Despite the fact that this solution contains assumptions and approximations that introduce error and limit its applicability, it is nevertheless significant that a strictly elastic treatment can account qualitatively, and to some degree quantitatively, for the mechanism of surface subsidence. For example, the elastic solution indicates the maximum tensile strain that develops on the surface outside the projected limits of mining and it is smaller (about $\frac{1}{5}$) than the maximum

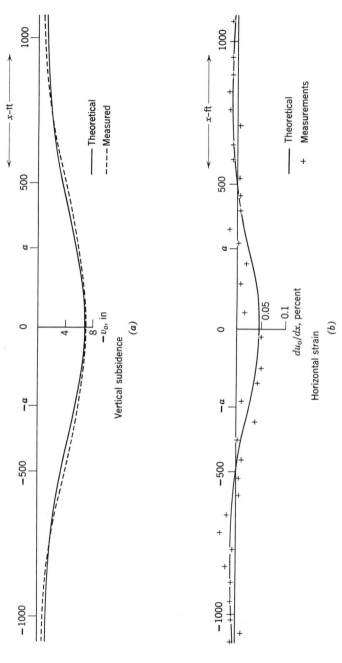

Vertical subsidence

(a)

Horizontal strain

(b)

Fig. 18.2.7. Surface subsidence over a coal mine. (After Berry and Sales[9]).

compressive strain, which occurs on the surface over the center of the excavation. Also, this solution requires that surface tensile strain extends well beyond (to infinity) the projection of the limit of mining on surface. Both these results are in general agreement with most surface subsidence surveys in which careful measurement was made beyond the projected limits of mining.

The calculated angle of break in the cited example is 73°, which is in general agreement with measured angles for the laminated rocks that overlie coal measures. Finally, and most important, this theory shows that the

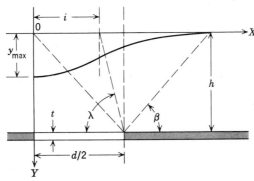

Fig. 18.2.8. (After Martos.[7])

larger part of the subsidence that occurs over thin-bed mining can be accounted for elastically. In the example given by Berry the surface subsidence was 40% of the roof convergence. Moreover, it is not necessary to assume the development of horizontal cracks or partings to explain the difference between the surface subsidence and roof convergence, but rather this difference can be accounted for by elastic strains in the vertical direction.

On the basis of both model studies and field measurements, Martos[7] found a general agreement between the measured subsidence profile and that calculated from the Gaussian subsidence profile derived by Litwitiszyn[8], namely,

$$y = y_{\max}e^{-x^2/2i^2} \tag{18.2.11}$$

where, in Fig. 18.2.8,

y = subsidence in the y-direction along x

$y_{\max}$ = the maximum (critical) subsidence at $x = 0$

i = distance from $x = 0$ to the point of inflection

Equation 18.2.11 was found to be valid for the subcritical phase if $y_{\max}$ is replaced by y', where y' is dependent on the width d, thickness t, and depth h, of an excavation. Orchard[9] has given an empirical curve relating y'/t to d/h (Fig. 18.2.9). Martos reported that the inflection point at $x = i$ generally lies over the excavated area such that the angle λ (Fig. 18.2.8) is between 75 and 85°.

Thus certain features of trough subsidence can be explained at least qualitatively by such widely dissimilar mechanisms as soil mechanics and the mechanics of an elastic continuum. Other theories have considered trough subsidence on the basis of multiple beams and plates, and on the

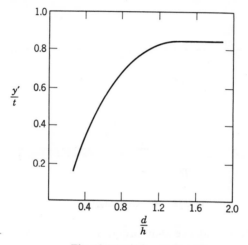

Fig. 18.2.9. (After Orchard.[9])

statistical (random) movement of particles through an aperture. However, all trough subsidence theories are inadequate because they fail to account for the time-dependent nature of the process. Also, these theories do not realistically take into account the mechanical characteristics of large rock masses.

The following three examples of trough subsidence are cited to illustrate how some of the problem variables, such as the kind of overlying rock or method of mining, affect surface subsidence. Generally, these data were obtained by measuring the vertical displacement, or the vertical and lateral displacement of a line or grid of pins installed on the surface.

Deere[10] measured the surface subsidence over a sulfur mining operation in which the sulfur was extracted by melting and pumping it from the cap rock overlying a salt dome (Frasch process). The mineralized zone was at a depth of from 1300 to 1600 ft below surface, and the overlying cover consisted of unconsolidated sediments, cemented and uncemented sandstones, gravel, and clay and clay shales. For the subsidence measurements an E–W and N–S base line over the mining area was established. Additional data were obtained by leveling to a grid of surface stations on 200-ft centers. Figure 18.2.10*a* shows the progress of the surface subsidence along the E–W base line over a 31 month interval, and Fig. 18.2.10*b* gives the corresponding surface strains. It is interesting to note that the maximum subsidence was directly above the solution zone, and that a

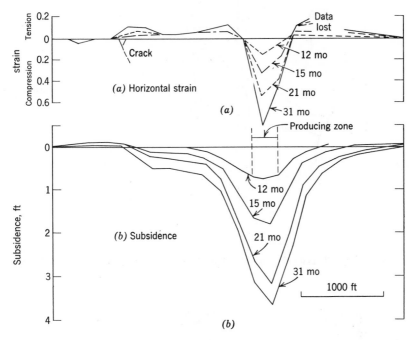

Fig. 18.2.10. Subsidence profiles and horizontal strain along E–W base line. (After Deere.[10])

small uplift (about 0.1 ft) occurred around the periphery of the subsidence area. The maximum compressive strain directly above the center of the mining area was 0.65 and the maximum tensile strains were 0.12 at the west side, and 0.2 on the east side of the E–W base line. Figure 18.2.11 is a plan showing the subsidence contours and lateral displacements (indicated by vectors) after 31 months of production. The angles of draw and break are given in Fig. 18.2.12. These angles are representative of subsidence in semiconsolidated sediments.

Pierson[11] reported the surface subsidence and strain over a New Mexico potash deposit mined by different stoping procedures. The depth of the deposit from surface was 1000 ft, and the cover consisted of

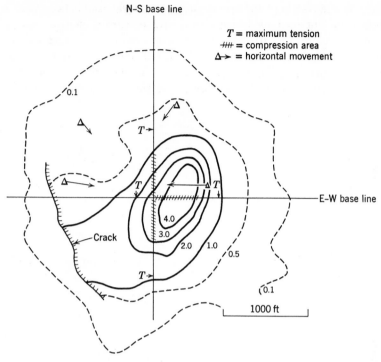

Fig. 18.2.11. Plan showing subsidence contours after 31 months of production. (After Deere.[10])

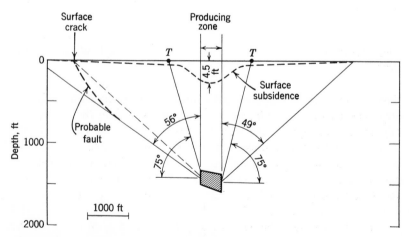

Fig. 18.2.12. Section A–A' showing surface subsidence; points of maximum tension, T; and critical angles. (After Deere.[10])

100 ft of unconsolidated sediments, 400 ft of dolomite, and 500 ft of salt immediately overlying the potash bed. The thickness of the ore varied from 7 to 15 ft. The thicker ore was originally mined by the room and pillar method (with a checkerboard pillar system), and the extraction was 62%. The pillars remained intact and only a minimum difficulty was experienced in maintaining a stable roof. After about 20 years this area was mined again by reducing (trimming) the pillars until the over-all recovery was 83%. At this extraction the pillars slowly crushed, as illustrated in Fig. 18.2.1, and the roof flexed until the opening virtually closed, but without causing the overlying formations to collapse or break.* Thirty-five days after initiating this second mining, the surface started to subside and the subsidence continued until 90 days after completion of mining. At that time the total surface subsidence was from 66 to 75% of the height of the mined area, and virtually no subsidence occurred thereafter.† Figure 18.2.13 shows the surface subsidence and surface strain over the mined area. The maximum tensile and compressive strains were 0.67 and 0.75. These maxima were approximately 400 ft apart between points A and B, and in the intervening region surface cracks and terraces occurred. Since the maximum strain was at point A over the mined area, the angle of break was greater than 90°.

Pierson also reported a second stoping method effected in a virgin area in which the height of the ore was 7.0 to 7.5 ft. First, development drifts were driven to the limits of the property. Then a retreating longwall was mined with a boring-type continuous mining machine. A regular pattern of small pillars was left behind the longwall, and their size was regulated so that the extraction averaged 92%. These pillars crushed as the longwall face retreated, allowing the opening to close and the surface to subside. A typical subsidence and strain curve is shown in Fig. 18.2.14. Note that with this stoping procedure the maximum tensile and compressive strains were 0.25 and 0.23, or only about ⅓ of that experienced in the pillar trimming operation. With the second stoping procedure the angle of break (from the limit of mining to point A) was 81°, and the angle of draw (from the limit of mining to point C was 37°).

Figure 18.2.15 shows a typical surface subsidence profile at the limit of mining over a trona mine in Wyoming.‡ The depth of the deposit is

---

* Assuming the weight of the overlying rock to be 1000 psi, the average pillar stress, calculated from Eq. 17.2.9, is

$$S_{v(\text{av})} = 1000 \left(\frac{1}{1 - 0.83} \right) = 6000 \text{ psi}$$

which is about 50% greater than the unconfined compressive strength of the ore.

† Private communication from Don U. Deere, University of Illinois.

‡ Private communication from William Fischer, FMC Corporation.

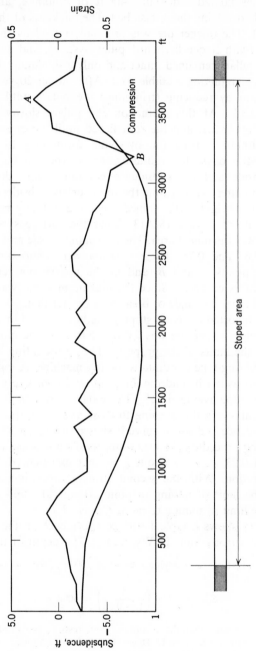

Fig. 18.2.13. Surface subsidence and strain over pillar robbing stope. (After Pierson.[11])

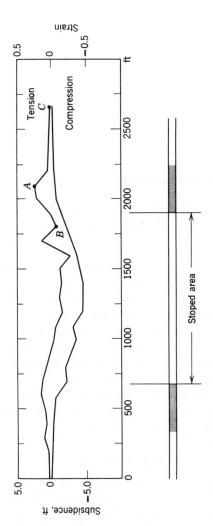

Fig. 18.2.14. Surface subsidence and strain over retreating longwall stope. (After Pierson.[11])

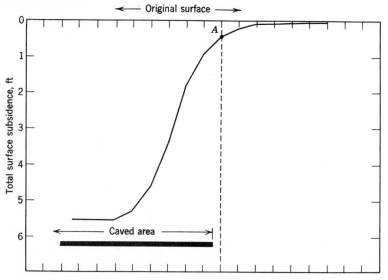

Fig. 18.2.15. Typical surface subsidence profile Westvaco mine. (Courtesy FMC Corp.)

1550 ft from surface and the cover consists of sedimentary rocks with unconsolidated sediments at surface. The mined thickness of the bed was 8 ft. The bed was mined first by the room and pillar method; the pillars were then split, and finally the remnant pillars were blasted out so that the surface subsidence could continue until the void was filled. Subsidence occurred by a gradual flexing of the surface and without the development of surface cracks. The subsidence profile was obtained two years after completion of mining, in which time the surface had come to rest. Note that the point of maximum tensile strain (corresponding to point A on the subsidence curve) was just outside the limit of mining. Surface subsidence was 70% of the thickness of the mined area.

18.3   CAVING

All subsidence processes in rock start with some manifestation of subsurface subsidence—usually an elastic or inelastic deformation in the roof of an opening that increases until the immediate roof fails. If, because of the time-independent or time-dependent characteristics of the rock, or because of the method of mining, failures in the rock overlying the opening are sustained so that the broken zone progresses toward the surface, the mechanism is referred to as caving. This process is illustrated in Figs. 18.3.1a, b, c, and d, which is an idealization of block caving. In

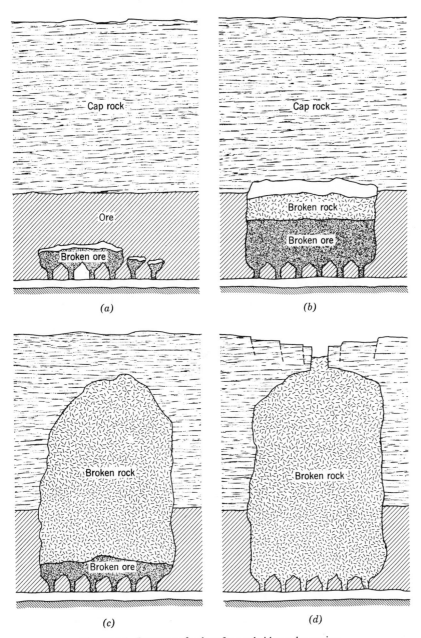

Fig. 18.3.1. Progress of subsurface subsidence by caving.

this method of mining, caving is initiated by removing support, and thereby increasing the span in the ore. At some width of span the caving will sustain itself and continue until the void is filled with broken rock. As ore is drawn from the block the overlying broken ore and rock will subside, creating a void into which additional rock can fail. Thus the cave progresses toward surface at a rate determined to a large degree by the rate that the ore is drawn. As the broken zone approaches the surface, the thickness of intact rock will decrease until it breaks through to surface.

On a megascopic scale the rocks associated with caving are incompetent, and the dominant factors that make them incompetent are jointing and low strength. Stratified sedimentary rocks, including some shales and limestones, will cave, providing their strengths are low enough. Usually in this class of rock the core recovery* from exploratory holes (from surface) is low, and it is impossible to core anything other than short lengths. Generally, casing must be used to prevent the wall of the hole from collapsing. However, other stratified rocks will cave if the strata are cut by closely spaced joints, even though a high recovery is obtained from exploratory holes and the strength of specimens cut from core is high. Highly jointed rhyolites and porphyritic rocks overlying the large disseminated copper deposits found in the western United States usually cave freely. Joint surfaces in these rocks are characteristically stained, altered, and contain decomposition products such as sericite or limonite to the degree that there is no strength across joint planes. Generally, the core recovery from this class of rock is low, often because short lengths of core from between joint planes tend to grind up in the drilling process. The compressive strengths of specimens cut from these short sections of core range from less than 1000 psi to greater than 15,000 psi. The iron formation overlying the soft hematite deposits in northern Michigan, which consists principally of thin-bedded, highly jointed jaspilites, slates, and greywackes, cave with only a minimum encouragement. In small specimens these rocks are among the stronger and harder class of rock. On the other hand, openings with large unsupported spans can be excavated in some of the weakest rocks, provided they are free from jointing. For example, in one salt mine a flat roof 120 × 180 ft has stood intact for over 25 years;[12] the compressive strength of this salt is less than 5000 psi. In an attempt to block cave a borate mineral deposit the support from an area approximately 140 × 280 ft was completely removed. Although some heavy spall resulted, a sustained cave could not be achieved. The compressive strength of this borate mineral ranges from 1100 to 6400 psi. Both the salt and borate minerals deposits were virtually free from jointing.

---

* Core recovery is the ratio of the length of core recovered to the length of the hole.

Thus the necessary and usually sufficient requirement for caving is either a low rock strength or a set or sets of closely spaced and essentially unbonded joints, or both. The rock strength should be such that in drilling exploratory holes the cores generally disintegrate so that only fragments are recovered. However, other geological factors such as dikes, sills, and faults, may strongly affect the rate of caving or the direction in which the cave progresses. Allen[13] and Boyum[14] described the premature cave that developed over the Athens mine in Northern Michigan. After three years of mining during which about three million tons of ore was removed, a cave developed and rapidly progressed to surface through over two thousand feet of cover (Fig. 18.3.2). The fact that this cave occurred so early in the life of the mine, and that it progressed so rapidly to surface are attributed to the presence of the almost vertical dikes that lie on either side of the subsidence zone. In contrast Boyum reported that in the Mather A mine in Northern Michigan, after 20 years of mining, and the removal of over 17 million tons of ore there has been virtually no indication of surface subsidence (< 0.1 ft*). However, from seismic and other observations made in drill holes from surface it has been determined that the subsurface cave had progressed to, and apparently had been arrested by a thick greenstone sill that overlies this mine (Fig. 18.3.3).

Besides the problem of ascertaining if a cave will develop in a given rock type, there are several other engineering problems that arise in connection with caving operations. First, and probably the most important, is the question of whether or not a void is being formed over the area from which ore is drawn. Usually, because of the volume expansion of broken rock, the void created by the removal of ore is filled with broken rock. However, if the overlying rock forms an arch so that the progress of the cave is arrested, a void is formed. If this void is large, there is always the danger that the intact cover will suddenly collapse, creating an air blast in the mine workings. Air blasts resulting from this cause have produced catastrophic damage.

In some mines observation drifts have been driven into the caved zone so that the progress of the cave can be examined. This is a costly procedure, especially in a deep mine where the vertical progress of the cave may be 1000 ft or more. A somewhat less costly procedure is to delimit the caved zone by probing with a diamond drill. If V_t is the volume of the caved zone, V_o the volume of the unbroken ore removed from the operation, and k is the ratio of the volumes of broken to solid rock (or ore), then the caved zone will be filled with broken rock if

$$V_t = V_o\left(\frac{k}{k-1}\right)$$

* Private communication from Bertram Boyum.

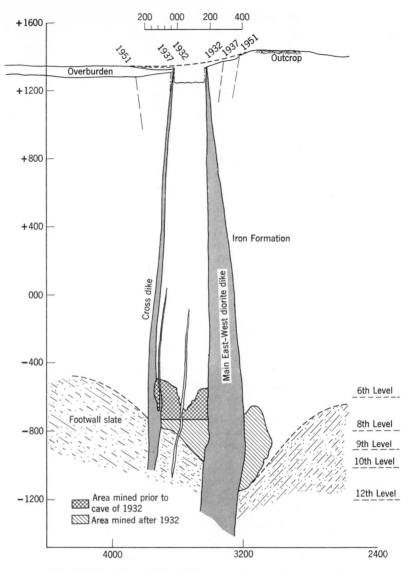

Fig. 18.3.2. Athens Mine, N–S cross section. (After Boyum.[14])

The value of k may range from 1.5 for loosely compacted pieces of hard rock to about 1.1 for highly compacted and often recemented materials, such as tailing sands. Generally in the rock types that are subject to this progressive caving mechanism, the horizontal cross section of the broken area either remains constant or decreases as the cave advances toward the surface. Hence, estimates of the volume of the caved zone would be on the high side if this volume were calculated from the height of the caved zone and an assumed constant cross-sectional area. An even less costly method is to delimit the progress of the cave by the seismic methods described in Section 14.13, although these methods may not work in some rock types.

Another problem that may require investigation is the direction of the

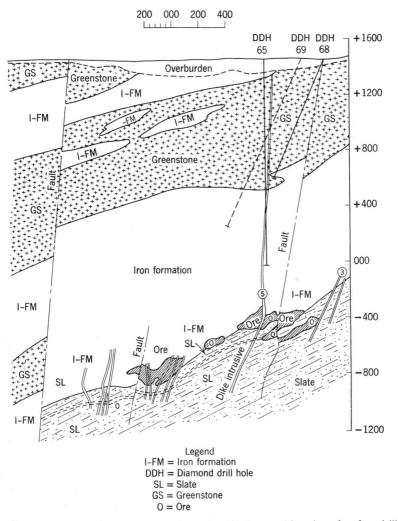

Fig. 18.3.3. Vertical section showing interpreted geology and location of surface drill holes. (After Boyum.[14])

cave as it progresses toward surface. This may be of concern in relation to other openings, especially shafts. The seismic traveltime-difference method (Section 14.13) is probably the least costly procedure for ascertaining whether or not the cave has progressed across or is approaching some limiting boundary, as for example, the boundary of a shaft pillar. By using only two geophones in a single diamond drill hole it is possible to determine if the cave is approaching or has progressed past a plane normal to the geophone hole and passing through the midpoint between the two geophones; with three geophones in a single hole it is possible to determine the distance (but not the direction) from the geophone hole to a point of failure on some plane normal to the axis of the hole.

The surface area overlying mines subject to progressive caving usually subsides to the extent that the utility of the surface ultimately is destroyed; hence the problem of determining the terminal surface void or of preserving structures on the surface is seldom of engineering interest. However, it is often important to know when manifestation of the subsurface cave will become evident on surface, or if the surface subsidence will appear directly over or at some angle to the mined area. The time, extent, and location of the initial breakthrough to surface with respect to mining, as well as the terminal areal extent and volume of the surface subsidence are dependent on the depth of the mine, the rock type, and on gross geological factors such as faulting, dikes, and sills. Collectively, these variables can affect the caving process to the degree that a quantitative evaluation of this problem is not possible, although the effects produced by the geological factors often can be appraised qualitatively. For example, the accelerating effects of the vertical dikes over the Athens mine (Fig. 18.3.2) and the retarding effect of the massive greenstone sill overlying the Mather mine (Fig. 18.3.3) are readily apparent. Because the progress of the cave over the Mather mine was carefully charted by seismic and other methods, and because of the almost negligible displacements on surface, it has been possible to utilize the surface over this mine during this 20-year period.[14]

18.4 CHIMNEYING

Chimneying is a type of cave that initiates over a relatively small area, and by a succession of failures or sloughs creates a hole or chimney that progresses rapidly to surface. Over the length of the chimney the cross-sectional area usually remains constant, and the diameter may be as small as 10 ft. Chimneys can progress through a thousand feet of cover in a matter of days, provided, of course, that the caved material is removed fast enough. Although the mechanism that causes chimneying may not be entirely different from that which produces the progressive type of caving described in Section 18.3, the fact that the affected area is so small, the rate of progress to surface so fast, and the time and place that a chimney will develop so unpredictable, makes this procedure appear to be unique.

Rice[15] described a chimney cave that resulted from opening a 14 × 28 ft roof area in a relatively incompetent graphitic slate that dipped at approximately 60°. The cave was created to provide backfill material, and as the broken slate was drawn, the cave progressed vertically through 900 ft of cover to surface in approximately one year. The cross section of the cave through the rock was about equal to the size of the initial underground opening. This report is of special interest because a number of investigators have reported that caving tends to progress in a direction normal to the dip of the beds. Chimney caves frequently develop in the cover overlying block-caving operations usually prior to any other gross indication of surface subsidence. This type of cave has occurred in such dissimilar rock as a highly jointed limestone and highly fractured porphyritic granite. In both rock types the development of the cave was first realized when small trees were found underground in the broken ore. Also in both of these cases it was inferred that the diameter of the cave was small all the way to surface, as the volume of caved material did not cause an appreciable dilution of the ore. When chimneys develop over block-caving operations, the surrounding surface area usually subsides shortly thereafter and, as this more general subsidence continues, the chimney may become obscure.

Generally the problems created by chimneying are a premature break-through to surface, and possibly some dilution of the ore which would not occur if the cave subsided uniformly. In a block-caving operation, the development of chimneys can be minimized by drawing the ore uniformly over the area of the block and, conversely, the possibility of a chimney is enhanced by the continued removal of ore from a single drawpoint.

18.5 SUBSIDENCE PLUGS

Plugs are caving processes that involve what appears to be a distinctly different subsidence mechanism. They are characterized by a sudden lowering en masse of the cover overlying an unsupported opening, such as the opening created by a block-caving operation. The suddenness of the failure is usually evidenced underground by an air blast and on surface by venting and a dust cloud. The intensity of the air blast will vary from weak to strong, depending on the volume of the void that is closed, and whether or not the excavated zone is partially filled with broken rock.

Because the body of rock subsides en masse, as opposed to the progressive type of cave, there is virtually no volume expansion of the subsiding material; hence, the volume of the surface void should be roughly equivalent to the volume of unbroken ore removed from the operation, provided, of course, that the volume created by mining was not partially filled with broken rock or ore. Also, because the subsiding mass remains intact, the angle of break (Fig. 18.2.2) must be equal to or greater than 90°.

Plug subsidence generally occurs in weakly consolidated sediments or in closely jointed but weakly bonded rocks that in most instances show some weathering or decomposition products on the joint planes. This class of rock will not core and, as a general rule, the walls of exploration holes will collapse unless cased.

In some block-caving operations it appears that the subsidence process starts with a succession of sloughs; after the cave has progressed part way to surface, the remaining cover fails as a unit. Usually preceding failure, cracks or other evidences of surface movement are not evident, at least to the unaided eye. Plug failures are more likely to occur in an area where the surface has already been broken by subsidence. Following the development of a plug, fractures, open cracks, and terraces usually develop on the surface surrounding the resulting void, and ultimately these cracks and terraces obscure the original failure.

No theory has been advanced to account for this type of failure mechanism. Obviously, soil mechanics does not provide an explanation since an angle of break of 90° should occur only in a hard rock, as indicated by the data in Table 18.2.1.

REFERENCES

1. Rellensmann, O., "Rock Mechanics in Regard to Static Loading Caused by Mining Excavation," *Second Symp. on Rock Mech., Quart.*, Colo. Sch. of Mines, **52** (1957).
2. Denkhaus, H. G., "Critical Review of Strata Movement Theories and Their Application to Practical Mining Problems," *J. So. Africa Min. and Metallurgy*, **64** (1964).
3. Seldenrath, T. R., "Can Coal Measures be Considered as Masses of Loose Structure to which the Laws of Soil Mechanics may be Applied?" *Proc. Internat. Conf. Rock Pressure*, Liege, Belgium (1951).
4. Berry, D. S., "An Elastic Treatment of Ground Movement Due to Mining; I, Isotropic Ground," *J. Mech. Phys. Solids*, **8**, Pergamon Press (1960).
5. Berry, D. S. and D. W. Sales, "An Elastic Treatment of Ground Movement Due to Mining; II, Transversely Isotropic Ground," *J. Mech. Phys. Solids*, **9**, Pergamon Press (1961).
6. Berry, D. S., "A Theoretical Elastic Model of the Complete Region Affected by Mining a Thin Seam," *Proc. Sixth Symp. Rock Mech.*, Univ. of Mo., Rolla, Mo. (1964).
7. Martos, F., "Concerning an Approximate Equation of the Subsidence Trough and Its Time Factor," *Internat. Strata Control Cong.*, Leipzig, Germany (1958).

8. Litwiniszyn, J., "Statistical Methods Applied to Problems of Rock Mechanics," *Internat. Strata Control Cong.*, Leipzig, Germany (1958).

9. Orchard, R. J., "Surface Effects of Mining—The Main Factors," *Colliery Gaurdian*, **193** (1956).

10. Deere, D. U., "Subsidence Due to Mining—A case History from the Gulf Coast Region of Texas," *Proc., Fourth Rock Mech. Symp.*, Penn. State Univ. (1961).

11. Pierson, F. L., "Application of Subsidence Observations to Development of Modified Longwall Mining System for Potash," *Trans. SME-AIME* (reprint) **65AM22** (1965).

12. Merrill, R. M., "Static Stress Determinations, Hockley (Texas) Salt Mine," *U.S. BurMines, APRL Report* (1961).

13. Allen, C. W., "Subsidence Resulting from the Athens System of Mining at Nauganee, Mich.," *Trans. AIME*, **109** (1934).

14. Boyum, B., "Subsidence Case Histories in Michigan Mines," *Fourth Rock Mech. Symp.*, Penn. State Univ. (1961).

15. Rice, G. S., "Ground Movement from Mining in Brier Hill Mine, Norway, Mich.," *AIME Tech. Pub.* No. 546 (1934).

CHAPTER 19

ROCK BURSTS, BUMPS, AND GAS OUTBURSTS

19.1 INTRODUCTION

The term, *rock burst*, has been used in a general way to describe rock failures ranging in magnitude from the expulsion of small fragments of rock from the surface of mine pillars or side walls, sometimes referred to as spitting rock, to the sudden collapse of a pillared area greater than two million square feet. The seismic effects produced by rock bursts cover as wide a range. The ground motion produced by spitting rock is detectable only with micro-seismic equipment, whereas the seismic disturbance produced by one large rock burst was recorded by a seismological station 1200 miles away. Some investigators, however, have used the term in a more restricted or qualified manner to describe rock failures associated with some phase of mining or due to some causitive factor as, for example, a pillar burst, pressure burst, strain burst, crush burst, inherent bursts, induced bursts, etc. In this treatment a rock burst is defined as any sudden and violent expulsion of rock from its surroundings, the phenomenon resulting from the static stress exceeding the static strength of the rock, and the result being of sufficient magnitude to create an engineering problem.

A *bump* is defined as a strong seismic shock resulting from a failure or a sudden displacement at some point in the rock surrounding an underground opening. The failure may be the shearing of an overlying stratum, or the displacement may occur along an existing fault. Usually the focus of the disturbance is only indefinitely known. The seismic shock may manifest itself as a thud or sharp audible report, accompanied by a ground motion strong enough to cause partially detached rock in the roof or on the walls of openings to fall. In hard rock mining because of these roof falls or other local failures, bumps are often identified as rock bursts, and in some instances when the focus of the bump is close to the opening the distinction becomes indefinite. Coal mines also experience bumps which probably originate, as in hard rock mines, with the failure of overlying strata or movement along a fault. On the other hand, in coal mining failures occurring in the seam, such as the violent failure of a pillar, are usually referred to either as bumps or coal bursts, probably because the term rock, as in rock burst, does not seem appropriate. However, at least in their physical manifestations, there is a difference between a rock burst and a bump and this distinction should be preserved whenever possible.

19.2 PHENOMENOLOGY OF ROCK BURSTS

The class of rock bursts most frequently experienced, which creates the major engineering problem, occurs in or near working areas and breaks from several hundred pounds to several thousand tons or more of rock with explosive violence. Usually smaller bursts occur in openings of limited size, such as tunnels, development drifts, or shafts, whereas larger bursts are more likely to happen in extensively mined areas. In addition, many bursts occur in worked-out and usually abandoned parts of mines. Although bursts in remote areas do not create any hazard to personnel, they may affect the general stability of the mine structure. A report prepared by the Rockburst Committee, Ontario (Canada) Mining Association[1] gives some idea of the safety problem in rock burst mines. In the four districts included in this survey, namely, Sudbury, Lakeshore, Timmins, and Little Long Lac, in a seven-year period a total of 1167 bursts were reported, of which 115 were rated as heavy or strong. These bursts caused 21 fatalities and 64 nonfatal compensation cases; of the total fatalities from all causes, 7.5% were due to rock bursts. In the Witwatersrand (Africa) gold mines in a seven-year period, 9% of the total fatalities resulted from rock bursts and in the Kolar (India) gold-field rock bursts accounted for over 50% of the total fatalities in a three-year period.[2] Besides the hazard to personnel, the loss of production resulting from the clean-up and retimbering period following bursts, together with the added cost of extra ground support, makes rock bursts one of the principal operational problems in deep mining.

That both the severity and the frequency of rock bursts increase with depth is a well-known fact, and the cause of this increase is usually attributed to the increasing weight of the overlying rock and, correspondingly, the increasing stress in the rock with depth. However, the weight of the overlying rock is not the only force that can cause or contribute to the cause of rock bursts. Minor rock bursts have been reported in surface quarries, and bursts of greater magnitude have been reported in comparatively shallow mining, that is, at depths less than 1000 ft.

Undoubtedly these near-surface bursts are primarily the result of tectonic forces acting in a direction approximately parallel to the surface. This explanation appears plausible in the light of results from an investigation in which compressive stresses of the order of several thousand psi were measured parallel to and within 50 ft of the surface of a granite gneiss quarry.[3] In general these near-surface bursts occur infrequently and are not of sufficient intensity to create any major operational problem. In deep mines that are rock-burst prone, the depth at which bursts are first experienced is usually below 2000 ft and, in most instances, they do not become a serious problem until depths greater than 3000 ft are reached. However, a number of mines have operated at depths greater than 5000 ft without any bursts, which indicates that factors other than depth affect the production of bursts.

Apart from stress, the next important causitive factor is probably the mechanical characteristics of the rock. The class of rock generally associated with rock bursts is qualitatively described as hard, strong, and brittle. In terms of mechanical properties this class of rock may have an unconfined compressive strength range from 15,000 to 60,000 psi, and a modulus of elasticity range from 6×10^6 to 14×10^6 psi. Other mechanical properties, such as shear strength and hardness, probably relate in some degree to bursting although both shear strength and hardness data are meager. Moreover, neither shear strength tests nor hardness tests are sufficiently standardized to permit a satisfactory comparison among the various laboratories engaged in rock-burst research.

On the other hand, rock types in the same strength and elastic property range are known to be minable at depth without bursting. The strain energy that can be stored in rock has also been considered as some measure of the tendency to burst, that is, the higher the maximum strain energy that can be stored in a given type of rock, the more likely the rock will be a type subject to bursting. As the strain energy per unit volume in a uniaxially loaded body of rock is $\sigma^2/2E$, the maximum strain energy per unit volume would be $C_0/2E$, where C_0 is the uniaxial compressive strength. Hence, on a strain energy basis for rock types in the same stress environment with all mechanical characteristics identical except compressive strength, the type with the lowest compressive strength should be least likely to burst, a conclusion that is consistent with experience in some mines. For example, in the deep copper mines in northern Michigan, the low strength amygdaloidal basalt ($C_0 = 5000$–$15,000$ psi) are less likely to burst than the more glassy basalts with a compressive strength range from 15,000 to 30,000 psi.

The mechanical characteristic that seems to distinguish bursting from nonbursting rock is that the latter tends to plastically and/or visco-elastically deform under stress to the degree that failure, at least on a megascopic scale, takes place slowly. Such a characteristic has not been reduced to any quantitative scale, although some attempt has been made in this direction. Denkhaus et al.[4] and Grobbelaar[5] have described a test for measuring what is termed the "relative violence of rupture." For this

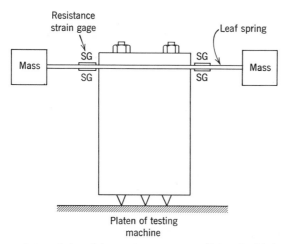

Fig. 19.2.1. Relative violence of rupture gage. (After Grobbelaar.[5])

test a device was developed for measuring the impulsive rebound of the bed of a testing machine, produced when a compressive test specimen fails (Fig. 19.2.1). For specimens that fail more violently, it is presumed that the rebound will be greater. This test is dependent on the mechanical characteristics of the testing machine as well as that of the specimen; hence the measured results are in arbitrary units. Table 19.2.1 gives the mechanical properties of a number of rock types from the South African gold fields.

Brittleness and other characteristics related to bursting probably can be more closely identified with petrology rather than with other mechanical properties of the rock. Igneous and metamorphic rocks are generally more rock-burst prone than the sedimentary rocks. In terms of mineral composition, the more silicious rocks and those containing other hard minerals belong in the rock-burst class, whereas the carbonates and other soft minerals are associated with the nonbursting class. Crystal size also appears to be a factor, because in the progression from coarse to fine grain to microcrystalline to glassy or amorphous, rock tends to become more disposed to bursting.

Table 19.2.1. Properties of Some Witwatersrand Rock*

Type of Rock	Mine	Density, lb/ft³	Uniaxial Compressive Strength, lb/in.²	Modulus of Elasticity, 10⁶ lb/in.²	Poisson's Ratio	Relative Violence of Fracture
Shale†	Modderfontein East (East Rand)	171.0	32,900	10.2	0.26	1.9
Shale (normal to the bedding planes)†	East Rand proprietary mines	172.2	24,700	11.0	0.25	0.8
Shale (parallel to bedding planes)†	Central Rand	173.3	29,700	15.0	0.20	1.4
Kimberley shale†	Western deep levels (West Rand)	178.1	24,300	11.3	0.47	0.7
Jeppestown shale†	(West Rand)	177.2	17,800	12.1	0.38	—
Hanging wall quartzite	East Rand	168.1	41,000	11.9	0.15	4.1
Reef bands	Proprietary mines	172.8	43,000	12.9	0.17	5.6
Footwall quartzite	(Central Rand)	169.9	33,700	12.1	0.20	3.0
Main Bird quartzite†	Western deep levels (West Rand)	169.4	35,400	11.8	0.26	3.4
Kimberley quartzite	Western deep levels	164.8	57,500	11.6	0.18	8.1
Dolomite†	Western deep levels (West Rand)	178.7	58,500	14.1	0.36	8.4
Chert	East Rand	164.2	69,000	12.2	0.26	18.5
Diabase (fresh)	Proprietary mines	182.9	64,600	15.3	0.27	15.3
Diabase (slightly decomposed)	Central Rand	173.1	35,400	10.0	0.23	2.7
Porphyrite lava†	Western deep levels	176.6	60,700	13.2	0.29	9.7
Amygdaloidal lava†	West Rand	172.6	37,400	12.2	0.38	3.1

* After Hill and Denkhaus.[6]

† Preliminary results only.

Major geological features also play a role in the production of bursts. Faults and dykes may cause weaknesses in the mine structure, and a related increase in burst production in their proximity. The occurrence of rock bursts in the neighborhood of dykes and faults was investigated by Hill and Denkhaus[6] with the typical result shown in Fig. 19.2.2. These

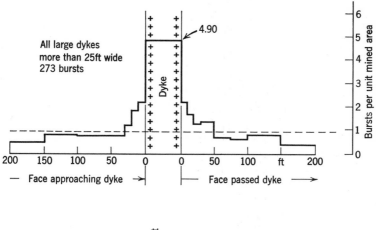

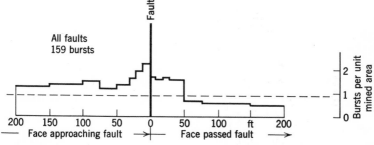

Fig. 19.2.2. Influence of dykes and faults on burst production. (After Hill and Denkhaus.[6])

data were obtained from the Witwatersrand district in mines operating at depths between 6000 to 9000 ft.

In mining, rock properties and geological features are factors over which the engineer has no control. Also, because the stress increases with depth, it is inevitable that rock-burst problems will be encountered in deep mining. To some degree the frequency or severity of bursts can be mitigated through the application of the mining techniques discussed in Section 19.6.

19.3   DETECTION AND LOCALIZATION OF ROCK BURSTS

An accurate record of rock bursts should be kept by the engineering staff of every affected mine. The record should indicate if possible the time; place of occurrence (spatial coordinates, level, stope number, etc.); magnitude in the mine, and in terms of the ground motion on surface; audible report; quantity of rock broken or size of area affected; violence, as evidenced by the displacement or throw of broken rock; and the relationship of the event to the mining procedure, such as the time or method of blasting, or to the configuration of the stope or details not usually maintained in mine records or maps. Such a record is especially useful in reconstructing the rock-burst history of a mine or in evaluating changes occurring over long periods, such as those resulting from the deepening of the mine or modifications in the mining method or procedure.

The rock-burst record is usually the responsibility of the mine super-intendent, who collects information from underground foremen, who in turn get information from miners or in some instances from an individual or team assigned to this task. This procedure has limitations. For one thing, the information should be collected on a 24-hour per day basis; even a three-shift basis is not sufficient if there is a period between shifts when men are not in stoping areas (especially if shot firing occurs at the end of the shift, as it does in most mines). Furthermore, it is difficult to ascertain the place of occurrence or the magnitude of the burst in worked-out or otherwise inaccessible areas.

To further the collection of data some mines have installed seismographs either on the surface, or underground, or both. From the seismograph records, the number and diurnal or longer period distribution of bursts can be determined accurately, and the magnitude and/or violence can be estimated from the measured amplitude of the ground motion. Usually the number of recorded bursts exceeds the reported number, sometimes by a factor of ten or more. However, a part of the bursts identified on the records may not produce a sufficient underground effect to be classified by miners as rock bursts. In addition, many of the unreported bursts probably occur in remote areas or possibly are the result of sudden displacements in unmined rock along faults or at the contact with dykes, sills, or faults.

A better degree of localization can be achieved using micro-seismic type equipment, Section 9.10 (operated at comparatively low sensitivity). A net of geophones can be installed in a rock-burst prone area, and by amplitude triangulation (Section 14.12), the proximity of each burst can be established. Micro-seismic type equipment has an advantage over seismographs in that the micro-seismic geophone can transmit signals over long cables (up to several thousand feet). Hence, geophones (which operate unattended) can be placed in rock-burst prone areas that, with the progress of mining, may become hazardous or inaccessible. Because they are relatively inexpensive some geophones may be sacrificed in the interest of obtaining data from remote areas.

The traveltime triangulation method provides a more sophisticated approach to localizing bursts. A modified micro-seismic equipment can be used for this purpose, or any seismograph in which the paper speed is fast enough to permit the resolution of relatively small time differences (of the order of milliseconds). Cook[7] reported an investigation made in one of the deep gold mines in South Africa in which eight seismometers were placed in the proximity of a rock-burst prone area. Using a string model (Section 14.13) and the measured traveltime difference, the foci of bursts could be determined to within ± 16 ft (90% confidence limit).

19.4 SEISMIC ENERGY FROM ROCK BURSTS

That seismic energy is generated and radiated outward in all directions when a rock burst occurs is well known. However, it was only recently that much consideration was given to the source of this energy and the mechanisms by which it was generated. A review of the early literature on rock bursts shows that a number of investigators calculated the energy released by rock bursts, but none of these investigators gave the basis for the calculations or a reference to the derivation of any formulas. The general impression obtained from these early papers is that the main source of the seismic energy released by a burst is the strain energy stored in the rock before it is fractured. In 1945 Weiss[8] suggested that the source of the seismic energy might be in the solid rock surrounding the opening.

Black and Starfield[9] gave a theoretical solution for the strain energy in a stressed plate with and without a circular hole. Also, Press and Archambeau[10] gave the theoretical solution for the seismic energy released when a spherical cavity is created in rock subjected to a general three-dimensional stress field. Both these papers start with the condition of no opening and calculate the strain energy before and after·the opening is suddenly formed.

Duvall and Stephenson[11] calculated the work done by the applied stress field when a cylindrical or a spherical cavity is suddenly enlarged. These two theoretical problems seem to be most appropriate, as the rock burst problem is concerned with the sudden loss of support of a section of solid rock near an opening which results in the enlargement of the opening. Because the work of Duvall and Stephenson is basic to an understanding of the source of the seismic energy released by rock bursts, a part of their derivation is reproduced here.

The stresses around a spherical cavity of radius a in an infinite elastic rock mass subjected to uniform stress $-S$ in the radial direction at some large distance from the origin are given by Eqs. 4.7.8 as

$$\sigma_r = -S\left(1 - \frac{a^3}{r^3}\right)$$

$$\sigma_\theta = \sigma_\Phi = -S\left(1 + \frac{a^3}{2r^3}\right) \tag{19.4.1}$$

$$\tau_{r\theta} = \tau_{\theta\Phi} = \tau_{\Phi r} = 0$$

where r, θ, and Φ are the spherical coordinates.

Because of spherical symmetry all displacements are radial. From Hooke's law equations and the strain-displacement relationship for spherical coordinates (Eqs. 3.10.5), we obtain

$$\frac{\partial u}{\partial r} = \frac{1}{E}\left[\sigma_r - \nu(\sigma_\theta + \sigma_\Phi)\right] \tag{19.4.2}$$

where u is the radial displacement. Substitution of Eqs. 19.4.1 into 19.4.2 and integration gives

$$u = -\frac{S}{E}\left[(1 - 2\nu)r + (1 + \nu)\frac{a^3}{2r^2}\right] \tag{19.4.3}$$

The strain energy per unit volume W_a in the rock surrounding the cavity is from Eqs. 3.9.2 and 19.4.1

$$W_a = \frac{3S^2}{2E}\left[(1 - 2\nu) + (1 + \nu)\frac{a^6}{2r^6}\right] \tag{19.4.4}$$

If the cavity radius is changed from a to c (where $c > a$), the strain energy per unit volume at any distance r also changes and the new value is given by

$$W_c = \frac{3S^2}{2E}\left[(1 - 2\nu) + (1 + \nu)\frac{c^6}{2r^6}\right] \tag{19.4.5}$$

Thus as the cavity radius is increased from a to c, the increase in the strain energy per unit volume at any distance $r \geq c$ is given by

$$W_c - W_a = \frac{3}{4}\frac{S^2}{E}(1 - \nu)\left(\frac{c^6 - a^6}{r^6}\right) \tag{19.4.6}$$

The increase in the strain energy in a spherical shell of thickness dr is

$$4\pi r^2(W_c - W_a)\, dr$$

The total increase in the strain energy ΔW_{cR} in the medium from radius c to some large radius R is

$$\Delta W_{cR} = \int_c^R 4\pi r^2(W_c - W_a)\, dr \tag{19.4.7}$$

Integration of Eq. 19.4.7 after substitution of Eq. 19.4.6 gives

$$\Delta W_{cR} = \frac{\pi S^2(1 + \nu)(c^6 - a^6)}{E}\left(\frac{1}{c^3} - \frac{1}{R^3}\right) \tag{19.4.8}$$

If $R \gg c > a$, Eq. 19.4.8 becomes

$$\Delta W_{cR} = \frac{3}{4}\frac{S^2(1 + \nu)}{E}\left(1 + \frac{a^3}{c^3}\right)V_{ac} \tag{19.4.9}$$

where V_{ac} is the volume of material removed when the radius of the cavity is changed from a to c.

As a result of removing a volume of material V_{ac}, the strain energy in the medium surrounding the cavity has increased by an amount ΔW_{cR}. Therefore, the medium surrounding the cavity has had work done on it by the applied stress field. The amount of this work can be calculated for any spherical surface of radius R. The area of a spherical surface is $4\pi R^2$, and the total radial force acting on this surface is $4\pi R^2 \sigma_r$ where σ_r is given by Eq. 19.4.1 when $r = R$. Therefore, the work done by this force is given by

$$\Delta W_{ac} = \int r\pi R^2 \sigma_r\, du \tag{19.4.10}$$

where du is the displacement that takes place when the radius of the cavity changes from a to c. The quantity du can be obtained by differentiating Eq. 19.4.3 with respect to a, thus

$$du = \frac{-3S(1 + \nu)a^2}{2ER^2}\, da \tag{19.4.11}$$

Substitution of Eqs. 19.4.1 and 19.4.11 into Eq. 19.4.10 gives

$$\Delta W_{ac} = \frac{6\pi S^2(1 + \nu)}{E}\int_a^c \left(a^2 - \frac{a^5}{R^3}\right) da \tag{19.4.12}$$

Integration of Eq. 19.4.12 gives

$$\Delta W_{ac} = \frac{6\pi S^2(1 + \nu)}{E}\left(\frac{c^3 - a^3}{3} - \frac{c^6 - a^6}{6R^3}\right) \tag{19.4.13}$$

If $R \gg c$ Eq. 19.4.13 becomes

$$\Delta W_{ac} = \frac{3}{2} \frac{S^2(1 + \nu)}{E} V_{ac} \qquad (19.4.14)$$

where V_{ac} is the volume of material removed as the radius of the cavity is changed from a to c.

Equation 19.4.14 gives the work done by the applied stress field at a radius of R when the cavity size is changed from $r = a$ to $r = c$, and Eq. 19.4.9 gives the increase in the total strain energy in the rock from $r = c$ to $r = R$. The difference between these two values is the amount of seismic energy W_s that is released. Thus

$$W_s = \Delta W_{ac} - \Delta W_{cR} = \frac{3}{4} \frac{S^2(1 + \nu)}{E} V_{ac}\left(1 - \frac{a^3}{c^3}\right) \qquad (19.4.15)$$

If $c \gg a$, or if $a = 0$, Eq. 19.4.15 becomes

$$W_s = \frac{3}{4} \frac{S^2(1 + \nu)V_{ac}}{E} = \frac{3}{8} \frac{S^2 V_{ac}}{G} \qquad (19.4.16)$$

The expression $1 - a^3/c^3$ in Eq. 19.4.14 is a small correction factor that results because a finite spherical cavity was assumed as a starting point. This correction factor approaches 1 very rapidly as c becomes greater than a.

Duvall and Stephenson also calculated the seismic energy released when a circular tunnel in a hydrostatic stress field is suddenly enlarged in diameter. The method of derivation is similar to that given here and their final equation for W_s is

$$W_s = \frac{1}{2} \frac{S^2}{G} V_{ac} \qquad (19.4.17)$$

Equations 19.4.17 and 19.4.16 are similar, differing only in the numerical factors of $\frac{1}{2}$ and $\frac{3}{8}$. Thus it can be assumed that for a single opening the exact shape of the outer boundary of the fractured rock has little effect upon the seismic energy released during a rock burst.

Equation 19.4.16 is shown graphically on log-log coordinates in Fig. 19.4.1. The first six lines from the bottom of the graph give the relation between seismic energy W_s in ft-lb and the volume of broken rock V_{ac} in ft^3 for various depths encountered in. mining. The value of G has been assumed as 4×10^6 psi and the hydrostatic pressure $-S$ has been assumed equal numerically to the depth in feet. The next three lines represent the earthquake region for shallow, intermediate, and deep-focus earthquakes. In constructing these curves the assumed value of G was 10^7 psi.

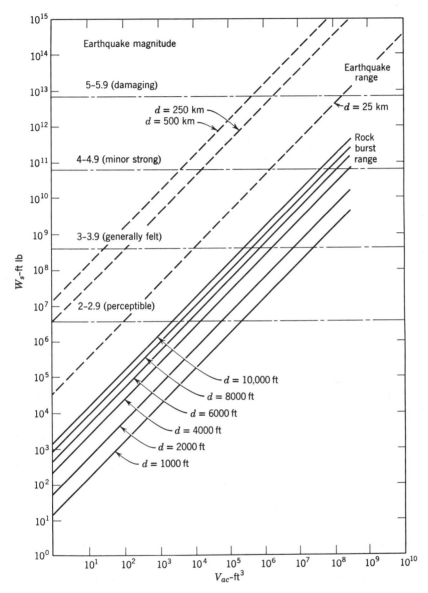

Fig. 19.4.1. Available seismic energy as a function of depth and volume of broken rock.

The horizontal dashed lines in Fig. 19.4.1 represent the estimated energies released by earthquakes of various magnitudes. These energies were calculated from the Gutenberg-Richter formula[10]

$$\log W_s = 9.4 + 2.14M \qquad (19.4.18)$$

where M is the magnitude of the earthquake and W_s is the released seismic energy in ergs. From this figure we can see that rock bursts which relieve the stress on a volume of rock equal to 10^5 to 10^6 ft^3 correspond to perceptible earthquakes and that rock bursts involving 10^7 to 10^8 ft^3 of rock correspond to commonly felt earthquakes.

19.5   FACTORS AFFECTING THE PRODUCTION OF ROCK BURSTS

In deep mines with a rock burst problem bursts usually are experienced first at depths around 2000 to 3000 ft. Because the condition necessary for a burst requires that the stress in the rock exceed its strength, it might be presumed that once a critical depth had been reached where the stress is sufficient to cause a burst, any further increase in depth and corresponding increase in stress would cause the rock-burst rate to increase rapidly to the point of limiting the minable depth. This presumption, of course, has not proved to be the case. Although the rock-burst rate and severity do increase with depth, there are mines operating at depths greater than 11,000 ft, and it is planned to deepen some to 13,000 ft or more. The fact that a critical depth is not reached, as might be expected, poses the question: What are the factors affecting the production of rock bursts, and how do they vary with depth? To examine this question consider first the relevant factors that preexist mining, the geology, mechanical properties of the rock, and the stress field. Geological factors such as joint spacing, bond strength across joints, and the number of faults, dykes, and sills vary widely from mine to mine or even within the same mine. At least some of these mechanical defects in the rock are known to affect markedly the production of bursts (Section 19.2). However, there is no evidence to indicate that these geological factors systematically change with depth in such a manner that they would lessen the probability of bursts. Also, the mechanical properties of the rock in rock-burst prone mines often vary widely from point-to-point, but generally these properties do not change systematically with depth so as to increase the rock-burst rate.

The stress field may have a strong bearing on burst production, and especially at shallow depths. Small bursts have been reported at depths less than 100 ft in granite quarries. The only rational explanation for these bursts is that they result from large horizontal forces of tectonic origin and, in fact, large horizontal stresses have been measured at shallow depth in granite quarries. If horizontal forces of this magnitude are present in the proximity of underground openings, they could account for the occurrence of bursts at relatively shallow depths. Moreover, because the horizontal stress in a body of rock loaded only by its own weight would increase with the depth at a rate greater than in a body of rock acted on by large horizontal tectonic forces, the incidence of bursts would increase less rapidly in the latter case. Also, if the stress field tends to become more nearly hydrostatic with depth, both shear stresses and the stress concentrations on underground surfaces would be minimized. This factor could lessen the probability of bursts at depth.

Following the excavation of an underground opening in an inelastic rock, the stress concentration that immediately develops in the rock surrounding the opening is redistributed progressively over larger areas, thereby reducing the maximum stress concentration. Moreover, because the inelastic deformations are greatest at the surface, the maximum stress concentration which initially is at the surface, moves away from this boundary (Fig. 19.5.1). In rocks that are relatively viscoelastic, such as

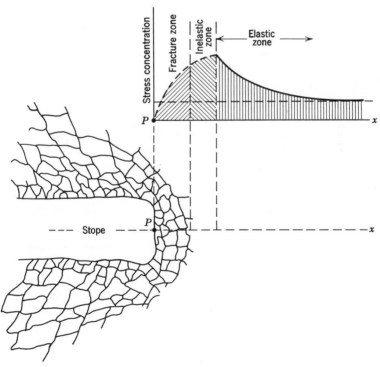

Fig. 19.5.1. Stress concentration near the working face.

halite, this readjustment occurs almost immediately after the opening is excavated and, if the opening is at moderate depth (1000 to 2000 ft), to the degree that the maximum tangential stress is only slightly greater than the vertical component of stress.[12] In the strong, relatively brittle rocks that are characteristic of rock-burst prone mines, this readjustment probably takes place more slowly. As a consequence, in areas where the stress concentration is large there is always considerable fracturing in the near-surface rock and occasionally rock bursts. However, no other factor plays as important a role in keeping the number of rock bursts within tolerable limits as the readjustment of stress resulting from inelastic deformations in the stress concentration zone near the boundary of openings.

Although the fractures in wall rock caused by blasting may penetrate for several feet into the wall of a drift or stope, the fracture zone that develops around an opening in a deep mine may extend many feet into the rock from the surface of the opening. Hodgson[13] determined that in some instances seismic contact with solid rock was not established for distances as great as 30 ft from the wall of a stope in a rock-burst prone mine. Similar determinations have been made by examining the fractures in diamond drill cores taken from the walls of openings in deep mines. Hill and Denkhaus[6] examined the fracture pattern in diamond drill holes with a petroscope* and they concluded that in stopes lying at a depth between 8000 and 9000 ft the fracture zone extended 15 to 20 ft ahead of the face. On the basis of observed fractures in foot and hanging wall drifts and raises driven above and below the center of a wide stoping area lying at a depth of 8000 ft, Leeman[14] reported the fracture zone extended at least 100 ft above and below the stope. However, in the same stope the depth of the fracture zone ahead of the working face was estimated to be 6 to 10 ft. Thus the fracture zone and the stress distribution near the working face is as shown in Fig. 19.5.1.

The fractured zone overlying a stope, sometimes referred to as the intra-dosal zone, is probably acted on more by gravity than by forces caused by the weight of the overlying rock. Some investigators consider that this zone of broken rock forms a voussoir arch, that is, an arch of unbonded blocks held in place by the resistance offered by the abutments in a manner similar to a masonry arch. Because relatively large deformations occur in the fractured rock ahead of the face, the compressive and shear stresses in this zone are reduced, usually below the level required to produce a burst. Rather, bursts are more likely to occur in the solid rock ahead of the fracture zone, presumably at or near the point of maximum stress. The

---

* An optical instrument similar to the stratascope described in Section 9.9.

broken rock ahead of the working face may lessen the effects of bursts by acting as a cushion or barrier, thereby restricting the violent expulsion of rock from the bursting zone.

19.6   PREVENTION OF ROCK BURSTS

As noted in the last section, both the tendency of the stress field to become hydrostatic with depth and the increase in the creep rate with depth act to inhibit the production of rock bursts and, to some extent, account for the fact that a critical depth below which an opening cannot be sustained has not been reached. These factors are inherent to the site of the opening and hence are variables over which the engineer has no control. However, it does not appear that these are the only factors that have made it possible for the engineer to extend the depth of mining from, say, 2000 to 3000 ft, where rock bursts are first encountered, to present depths of over 11,000 ft. From all indications, changes and modifications in the method of mining based on an accumulation of operating experience have had an equally if not more important part in making mining possible at great depth. It is both interesting and informative to analyze the reasoning behind some of these changes and developments, and to note in retrospect how both theory (such as the doming theory) and practice have evolved so as to conform with our present knowledge about the stress condition in, and the mechanical behavior of, rock structures. Although it is not the purpose of this chapter to review the history of these developments, a summary of the more important conclusions is given together with whatever rationalizations can be made.

Rock bursts have been experienced in a variety of underground operations including coal and nonmetallic mineral mines, but it is in the deep metallic mineral mines that bursts have become a critical operating problem, and it is only in this class of mine that any substantial amount of operating data have been obtained. Generally, the mineral deposit is of the vein type varying from a few feet to possibly 20 ft in thickness, extending several thousand feet along the strike, and dipping from 30° to approximately vertical. The rock type is either metamorphic or igneous, hard, brittle, and with a uniaxial compressive strength ranging from 10,000 to 50,000 psi or more. Examples are the gold mines in Ontario, Canada; Witwatersrand district, South Africa; Mysore district, India; the copper mines in Northern Michigan; and the zinc mines in the Coeur d'Alene district, Idaho. Usually the grade of ore makes it economically desirable to extract a high percentage of the deposit; hence, a sheet of ground several thousand feet along the strike, as great as 11,000 feet down the dip, and from a few feet to a few tens of feet in thickness is removed. Support of this void may be by several means: by remnant pillars usually spaced at random and in areas of low grade ore; by uniformly spaced pillars usually of minimum size; by self-fill, that is, backfill created by the progressive sloughing or more general failure of the roof; and by placed-fill or packs. The latter may be hydraulically placed mill sand or sands from other sources. In the Mysore district, India, quarried granite block packs are employed, a practice that would not be economically feasible in most mining districts. Obviously wood or steel props would be completely ineffective in supporting anything other than the immediate roof. For example, at a depth of 4500 ft, the weight of the overlying cover is approximately 5000 psi or 360 tons/ft². Even the strongest steel props could not be spaced close enough together to support this weight.

Backfill can provide effective support although its function is sometimes misunderstood. To develop a bearing capacity backfilling material must be compressed. Hence, as fill is placed in a stope it offers virtually no support—it is not until the opening starts to close and thereby compresses the filling material that it offers resistance. Because subsidence or closure is a time-dependent process, this resistance must similarly be time-dependent. Thus the stress condition that develops in and around a filled stope is as follows. Consider a typical longwall operation in which the excavated area is backfilled, and that at some phase of the operation the weight of the overlying rock has reached an equilibrium condition, that is, a part being carried on the unexcavated rock and a part on the backfill, and that the fracture zone and stress concentration ahead of the face and in the backfill are as pictured in Fig. 19.6.1a. If the face is now advanced from C to C', the fracture zone ahead of the face is reduced to approximately that caused by blasting and, because support has been removed, the stress concentration ahead of the face is increased to its maximum value (Fig. 19.6.1b). Next, the excavated area from B to B' is filled. As the roof settles on the fill, and while the fracture zone advances, the stress concentration ahead of the face and in the fill (Fig. 19.6.1c) becomes identical with that in Fig. 19.6.1a. At the time the face is advanced, the stress concentration in the solid rock is a maximum and hence the probability of a rock burst is a maximum. As the imposed load redistributes the stress, the concentration decreases and correspondingly the probability of a burst diminishes. This probability can only increase thereafter if weight is transferred to this face from mining at some other point.

The bearing capacity of backfill is related to the included voids in the material. Thus the particle-size distribution as well as the degree of compaction are important factors. Most mill sands make good backfill materials, and they will usually develop a satisfactory bearing capacity

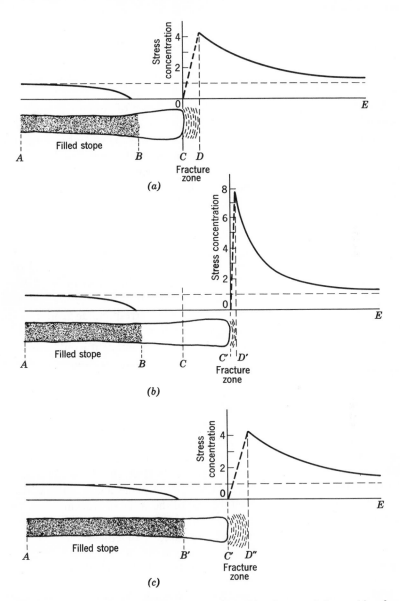

Fig. 19.6.1. Change in the stress concentration with advance of the working face.

with a compaction of the order of 10%.* Assuming that a compaction of 10% is necessary to develop the required bearing capacity, it is evident that the stope closure (by volume) must be of this magnitude. Since the creep rate in the hard rocks associated with rock-burst production is very low, it is important to use a filling material that will develop its required bearing strength with a minimum compaction so that the effectiveness of the fill is achieved in a minimum time.

Besides affecting stope closure and the degree of support provided by backfill, time also is a factor in developing the fracture zone surrounding an opening. As previously noted, this zone lessens the violent expulsion of rock from a burst, and thereby lessens the hazard to personnel. The self-generated seismic disturbances created by rock fracture (Section 14.12) indicate that following a stoping round, there is a continuing fracture in the rock that diminishes with time. In a shallow mine in strong rock this fracture period may be relatively short—approximately 15 minutes—following which fracturing virtually ceases. On the other hand, analysis of micro-seismic data from rock-burst prone mines shows a high fracture rate following all rounds, which decreases slowly to some background rate. This background rate, which to some degree is related to the rock-burst rate in the mine, indicates that the development of the fracture system is continuing and probably accompanied by a corresponding readjustment in stress. Thus time may affect the production or severity of rock bursts, and if so there should be an optimum rate of mining that would minimize burst production. This supposition has had only meager and inconclusive investigation.

On the basis of years of operating experience, stoping procedures have evolved that tend to minimize the production or severity of rock bursts. Recalling that in most of the rock-burst prone mines a vein of relatively uniform thickness is being mined, it is evident there is virtually no latitude for altering the stope geometry in the plane normal to the dip, and that in this section the stope appears as a narrow slit (Fig. 19.6.1). Because of the relatively large stope length-to-thickness in this plane, a large stress concentration should exist at (or near) the end of the stope or working face. This concentration is further increased if in the plane of the dip small areas of rock are left unmined and relatively intact (sometimes referred to as remnant pillars, islands, peninsulas, etc.), or if obtuse, or even worse, if acute angles are formed on the stope boundaries as illustrated in Fig.

---

* Mill tailings generally are a combination of coarser particles (sands) and finer particles which in some backfilling operations are removed by washing. Weyment[15] found that a better bearing capacity for a given degree of compaction could be obtained by allowing some of the finer material to remain in the product, probably because a better particle-size distribution is attained.

19.6.2. These areas of increased stress concentration are known to be subject to bursting and should be avoided. Because a longwall advancing toward the boundaries of the property or the limits of the ore provides for a high percentage extraction and at the same time a working face with a minimum of angle cuts, this procedure is favored in the Witwatersrand District, South Africa.[6] Of the two methods for advancing longwall faces, by underhand or overhand mining, the underhand method is considered better because the angle created by the working face with the level or overlying mined area is obtuse, whereas with the overhand method it is acute. Note that if the longwall retreats from the property or ore boundary toward the center of the deposit, a remnant area would be created as they converge, a condition that is not recommended.

In the deep copper mines in northern Michigan, a modified longwall stoping method was developed that at vertical depths up to 5000 ft has been found to minimize burst production and at the same time optimize the transportation of ore, men, or materials. A retreating overhand longwall is created between levels, but across a group of four levels a staggered retreating longwall is produced (Fig. 19.6.3). Behind the longwall face between levels, pillars are left so small in cross-sectional area that they ultimately crush, allowing the roof first to sag and then fail. This roof failure is extensive and creates a self-fill which ultimately, partially or completely supports the weight of the overlying rock. This system has the advantage that the transportation of ore, men, and materials is through unmined rock rather than through mined areas. The system has the disadvantage that a number of angle cuts are formed on the retreating longwall and that, as the face advances towards the level above, a decreasing unmined area is created which is burst-prone.

The effectiveness of longwall mining in reducing the number and severity of bursts results from the fact that along the length of the face the superincumbent load is distributed more or less uniformly. By avoiding sharp corners and other configurations that cause or enhance stress concentrations, and by properly placing a high density fill in mined areas, burst production is further minimized. However, these methods do not altogether eliminate bursts along the longwall face, and it is these bursts that create the greatest hazard to personnel. In 1954 Roux and Denkhaus[16] proposed a relatively simple procedure, referred to as de-stressing, for ameliorating burst conditions in a working face. Mining engineers had observed that bursts are more likely to occur in unfractured rock than in fractured rock. Hence, if additional fracture could be induced in the rock ahead of the face or if the depth of the fracture zone could be extended, the probability of a burst might be diminished. From a stress distribution standpoint, the probable change in the stress concentration zone caused by

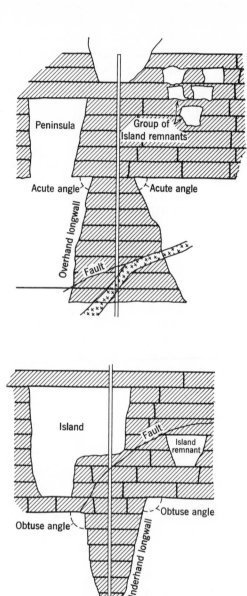

Fig. 19.6.2. Overhand and underhand longwall mining. (After Hill and Denkhaus.[6])

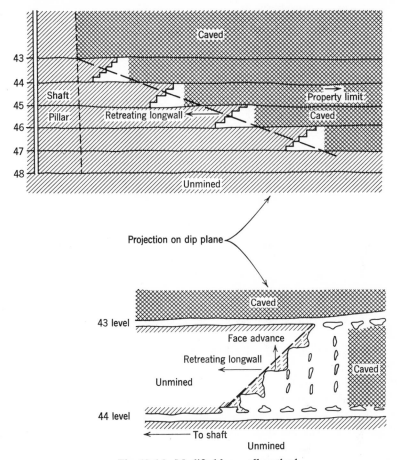

Fig. 19.6.3. Modified longwall method.

de-stressing should be as illustrated in Fig. 19.6.4. Moreover, if a burst did occur it would most likely take place in the solid rock and hence the effects would be mitigated by the additional cushion of broken rock ahead of the face.

To increase the fracture density or extend the fracture zone Roux and Denkhaus suggested that a fraction of the shot holes in each face should be extended past the depth that breakage occurs, that is, past the normal burden, and these holes should be loaded and fired together with the round. A part of the energy released by the explosive in the back part of the de-stressing holes would be expended in creating additional fracture. In a later report Roux and Denkhaus stated that this technique had been put into effect in the Witwatersrand District as a two-part procedure. First,

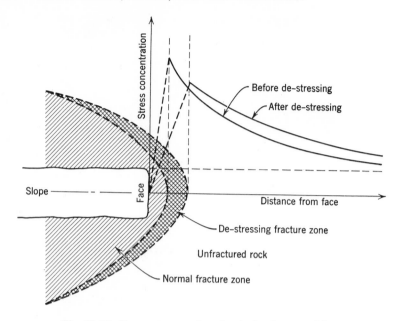

Fig. 19.6.4. Stress concentration ahead of a de-stressed face.

the de-stressing holes were drilled to a depth of 10 ft, the back 5 ft of each hole was loaded, stemmed, and detonated. Then the normal shot holes were drilled to a depth of about 5 ft, loaded, and detonated. On the basis of an initial investigation, these researchers reported that de-stressing reduced the total number of bursts along the face, and more important, shifted a large fraction of the bursts that normally occurred on-shift to off-shift periods (Table 19.6.1).

Hill and Plewman[17] continued this investigation and reported the results obtained during the stoping of over 100,000 ft². These data, given in Table 19.6.2, indicate the effectiveness of de-stressing, as practiced in the Witwatersrand district, and confirm that this procedure reduces the total

Table 19.6.1

Time of Burst	Before De-stressing		After De-stressing	
	Number	Percent	Number	Percent
During day shift	13	27	1	3
Off-shift periods	35	73	31	97
Total	48	100	32	100

Table 19.6.2. Effect of De-stressing*

	Before De-stressing	After De-stressing	Percentage Improvement
Number of stopes	17	17	—
Square feet stoped	112,800	115,800	2.7
Incidence:			
Number of bursts	48	32	33
Incidence of bursts per 10,000 ft² stoped	1.53	1.00	35
Severity:			
Number of severe bursts	26	6	77
Percentage of severe bursts to total bursts	54	19	—
Incidence of severe bursts per 10,000 ft² stoped	0.83	0.19	77
Time:			
Number of bursts during day shift	13	1	92
Percentage of bursts during day shift to total bursts	27	3	—
Number of severe bursts during day shift	6	0	100
Percentage severe bursts during day shift to total bursts	12.5	0	—
Casualties:			
Number killed	6	—	
Number injured	38	1	
Total casualties	44	1	

* After Hill and Plewman.[17]

number of bursts and shifts a large fraction of the bursts from on-shift to off-shift periods. The reduction in casualties is especially impressive.

Another procedure that shows some promise of reducing the number of rock bursts that occur in the proximity of the face during working periods is to blast the shot holes in each stope in as short an interval as possible, or even better to fire all stopes simultaneously. Rounds fired with milli-second delay detonators probably take the minimum time consistent with good breakage and fragmentation. This procedure has been found to significantly increase the number of bursts that occurs in the face area immediately after blasting, and to decrease the number that occurs during the working shift. Thus, if the rounds are fired at the end of the shift and if there is an idle period of about an hour more after the shift, the hazard to personnel may be reduced. Mines working a two-shift operation can employ an 8-hour shift followed by a 4-hour idle period to advantage.

Two explanations have been advanced to account for the increase in bursts following the rapid firing of shot holes in a stope. One explanation assumes that the intensity of the seismic shock is enhanced by these rapid detonations, that is, that there is a cumulative effect and that this increased vibration will trigger bursts in an already high stress area.

However, Duvall and Devine[18] reported that the vibrations produced by conventional millisecond delay rounds generally show no significant cumulative effect. Moreover, the seismic vibration produced by blasting persists only a fraction of a second and this effect should trigger bursts only during this period. Although bursts sometimes occur almost concurrent with shot firing, the interval of increased burst production may persist for an hour or more after blasting. A more plausible explanation is that the load carried by the rock that was blasted from the face is transferred to the intact rock behind the face. Hence, the affected area is more subject to burst.

19.7 PREDICTION OF ROCK BURSTS

In a mine subject to rock burst the principal effort should be corrective, that is, it should be directed toward reducing the number or severity of these failures by modifying the mining method or by some remedial procedures such as de-stressing or simultaneous shot firing. However, another approach that might be considered is the development of a procedure for predicting rock bursts so that precautionary measures could be taken to protect personnel and mining machinery. For example, the procedure might involve the repeated measurement of the stress at points within a suspected area; if the stress approaches the rock strength, a burst would be predicted. Probably because of the difficulties in making repeated in situ stress and rock strength determinations, this procedure has not been attempted.

However, the use of the micro-seismic method to predict rock bursts has been investigated by Obert and Duvall.[19] This method has the advantage that the measured quantity, namely, the micro-seismic rate, is in itself at least qualitatively related to the imminence of failure (Section 14.12). In a study made in a deep copper mine in Northern Michigan over a 40-day period, it was found, most significantly, that the micro-seismic rate fluctuates between wide extremes, indicating that the rock stress undergoes almost continual readjustment, with maxima occurring every few days (Fig. 19.7.1). The magnitude of the micro-seismic rate was not in itself a satisfactory basis for predicting rock bursts, probably because this rate depends not only on the intensity of the stress but also on the distance from the focus of the stress area to the geophones. Thus the further inference was made that within the 165 × 200 ft stoping area under investigation high stress zones develop sporadically and then diminish, either through gradual readjustment or bursting, thus transferring the load to another point. However, a large increase in micro-seismic rate was found to precede most bursts; hence, a criterion was developed, namely, that if

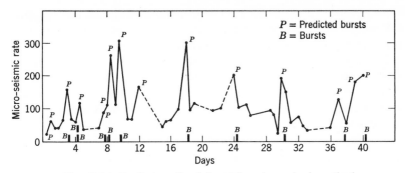

Fig. 19.7.1. Prediction of rock bursts by micro-seismic method.

the micro-seismic rate increased in any interval not exceeding 24 hours by a factor of two or more, a burst is imminent. Moreover, if after such an increase the micro-seismic rate continued to increase, the possibility of a burst is assumed to persist. On the basis of this posteriori criterion, 14 burst predictions would have been made during the 40-day period; 9 would have been followed by rock bursts within a 12-hour period; 5 would not have been followed by bursts; and 2 bursts would not have been predicted (both of which occurred on the same day).

At first it might seem that any method that could predict 9 out of 11 bursts would be useful. Moreover, as this investigation was a first, it is probable that with a greater accumulation of data and with the development of new or improved equipment, the ratio of the predicted-to-total bursts could be made even higher. The value of such a procedure, however, depends not only on being able to predict a high percentage of the total bursts, but it is even more dependent on being able to keep the ratio of the predicted bursts to total predictions high. From all indications because of the variability of the in situ rock strength, and because the stress in a potential burst area may be relieved by readjustment rather than by a burst, predictions will be made that are not followed by bursts. In the cited investigation 5 out of 14 predictions were not followed by bursts, and a generally similar result was experienced in a deep gold mine in Canada.[20] If a mining operation is to benefit by some method of prediction, men and equipment should be removed from a suspected area as soon as a burst is forecast, and stoping in the area should not be resumed until either the burst occurs or a stress readjustment takes place, a process that might require one to several shifts. The success of such an undertaking will be strongly affected by the number of times predictions are made that are not followed by bursts since miners develop an adverse psychological reaction to returning to an area in which a predicted burst did not materialize. Another factor that should be considered is, if mining is suspended because of burst.predictions, the rate of mining will be affected. In most rock burst mines the rate of mining has been established on the basis of operating experience so as to minimize the occurrence of bursts; hence work delays resulting from burst predictions might cause a generally worsened condition. In any event, a mining company should consider the effects of these factors before engaging in a research program whose objective is burst prediction. The same effort directed toward a reduction in the number or severity of bursts might be more productive.

19.8 PHENOMENOLOGY OF BUMPS

As defined in Section 19.1, rock (or coal) bursts and bumps are separate and distinct phenomena, the former being characterized by a sudden and violent failure occurring on or near a rock surface, whereas the latter is identified as a failure or movement within a body of rock (as along a fault). Bumps generate locally strong seismic shocks that may cause rock falls in mined areas. In addition, a bump produced, say, by the shearing of an overlying stratum may transfer to pillars or other parts of underground structure a sudden load of sufficient magnitude to cause a burst. Also, a burst may remove support from the surrounding rock, and in so doing induce a bump. Thus it is often difficult to ascertain from either seismic effects or audible reports which failure mechanism is associated with an event; except that in the case of bursts there is always evidence of a violent failure, whereas in bumps there may be no evidence of a violent failure. As previously stated, where a seismograph is used to record seismic events only a small fraction of the recorded events is known to be bursts. Probably a part of the unidentified events is due to bumps and the remainder to bursts occurring in inaccessible areas.

Because rock bursts that occur in working areas are usually identified correctly, an abundance of data has become available for statistical analysis and the determination of cause and effect relationships. However, the preventive measures that have been developed to reduce the incidence or intensity of bursts appear to be ineffective in reducing bumps, probably because the failure mechanisms are so unrelated. On the other hand, because the documentation of bumps has been meager and often inaccurate, an understanding of this phenomena has not been developed. In hard rock the cause of bumps is generally presumed to result from movements along faults, although virtually no evidence has been obtained to verify this supposition. In coal mines both movements on faults and failure (shearing) of overlying strata are suspected causes. Dunrud and Osterwald[21] presented evidence to show that some bumps are associated with movement on faults. The tests were performed in the coal mines in

Central Utah (Sunnyside District). The epicenter of the burst was determined from the compressional and shear wave propagation of seismic waves as determined by seismograph stations placed on the surface overlying these mines. The time resolution was such that it was not possible to determine with this seismic method the true focus of the bump. Also, it is possible that the part of the evidence identified as bumps was actually bursts since Dunrud and Osterwald did not make any distinction between the two phenomena.

19.9 GAS OUTBURSTS

Gas outbursts occur when, in the course of mining, gas under pressure is liberated in substantial quantity. The event is often accompanied by a rock (or coal) failure which in many instances is violent. Gas outbursts are most commonly experienced in mining coal, and the liberated gas is generally a mixture of methane, carbon dioxide, and nitrogen in explosive proportion. The liberation of the gas creates a hazard which usually is rectified by adequate ventilation. The rock failure also creates some local hazard, but this type of failure usually is not of sufficient proportion to cause a structural problem.

Outbursts in coal have been investigated extensively, and many events have been documented. The containment of the gas in coal and the mechanism of failure accompanying an outburst are not fully understood, although some explanations have been offered as, for example, that by Patching.[22] It appears that the gas is adsorbed primarily in the pore structure of coal, but migrates only through fractures, partings, joints, cleats, etc. Usually these migration paths are restricted to the degree that coal en masse is relatively impermeable and, as a result, attempts to drain (bleed off) the gas through boreholes have been relatively ineffective except in small areas.

Gas outbursts are also experienced in salt, potash, and other evaporite mineral deposits. In these minerals the gas may occur in partings, especially those containing some organic material, or more frequently the gas may be contained (or adsorbed) interstitially between or possibly within crystals. Hand-size specimens of rock salt from an outburst area have been found to retain interstitial gas over long periods and under sufficient pressure to cause individual crystals to be expelled with a sharp audible report when the specimen is slightly strained by hand. The gas in evaporate deposits includes carbon dioxide, nitrogen, and methane. In some instances, the methane content is high enough for the product to be explosive.

If an area is mined in a salt or potash ore deposit containing interstitial gas, a moderately violent to violent expulsion of the mineral from the floor, roof, or wall may occur. Gas outbursts of this kind, sometimes referred to as runs, have created openings as large as 30 ft in diameter and over 100 ft long; often large quantities of gas (of the order of 10^5 ft³) are liberated. No generally acceptable procedure or remedial action has been prescribed for preventing or mitigating this type of outburst. As a consequence, areas in salt and potash mines that are outburst-prone are usually abandoned.

If gas occurs in a separation or parting overlying or underlying an opening, a load is imposed on the roof or floor member (Section 14.10, Eq. 14.10.4) which may cause a roof or floor failure. For example, a gas pressure of 14 psi would impose a load of one ton per square foot. Gas pressure as high as 90 psi has been measured in partings in a limestone roof, and this pressure has caused roof members several feet thick to fail. Roof and floor outbursts in both coal and potash mines have also been attributed to this cause. To eliminate this hazard regularly spaced bleeder holes should be drilled in the roof or floor to relieve the pressure. The proper spacing of holes drilled for this purpose will depend on the ability of the gas to migrate from point-to-point. The relief of pressure can be determined by placing a packer and pressure gage in an isolated bleeder hole and allowing the gas pressure to reach a constant value. Then, starting at some distant point, a succession of bleeder holes should be drilled at progressively shorter distances from the pressure measuring point, until a distance is reached that relieves the pressure below some minimum acceptable value.

REFERENCES

1. Parker, R. D., *Report of the Rock Burst Committee of the Ontario Mining Assn.*, 1941.
2. Morrison, R. G. K., *Rockburst Situation in Ontario Mines* (a report to the Ontario Mining Assn.), 1940.
3. Hooker, V. E., H. R. Nicholls, and W. I. Duvall, "In Situ Stress Determinations in a Lithonia Gneiss Outcrop," *Earthquake Notes* (abstract) *Eastern Section*, **35**, Nos. 3–4, Seismological Soc. of Am. (1964).
4. Denkhaus, H. G., A. J. A. Roux, and C. Gobbelaar, "A Study into the Mechanical Properties of Rock with Special Reference to their Bearing on the Occurrence of Rock Bursts," *S. African J. Mech. Engrs.* (1955).
5. Grobbelaar, C., "Some Properties of Rock from a Deep Level Mine of the Central Witwatersrand," *Assoc. Mine Mgrs. S. Africa* (1958).
6. Hill, F. G. and H. G. Denkhaus, "Rock Mechanics Research in S. Africa, with Special Reference to Rock Bursts and Strata Movement in Deep Level Gold Mines,' *Trans. Seventh Commonwealth Min. and Met. Cong. S. Africa*, **2** (1961).
7. Cook, N. G. W., "The Seismic Location of Rock Bursts; Rock Mechanics," *Proc. of the Fifth Symp.*, edited by C. Fairhurst, Pergamon Press, 1963.
8. Weiss, O., "Rockburst, A Symposium," *AIME Trans.*, **163** (1945).
9. Black, R. A. L. and A. M. Starfield, "A Dynamic or Energy Approach to Strata Control Theory and Practice," *Fourth Intern. Conf. on Strata Cont. and Rock Mech.* Columbia University, 1964.

10. Press, F. and C. Archembeau, "Release of Tectonic Strain by Underground Nuclear Explosions," *J. Geophys. Res.*, **67**, No. 1 (1962).

11. Duvall, W. I. and D. Stephenson, "Seismic Energy Available from Rock Bursts and Underground Explosions," *Trans. SME, AIME* (1965).

12. Merrill, R. H., "Static Stress Determinations in Salt, Site Cowboy," *Appl. Phys. Res. Lab., Rept. APRL 38-3-1* (1960).

13. Hodgson, E. A., "Recent Developments in Rock Burst Research at Lake Shore Mines," *Can. Inst. of Mining Met., Trans.* 46 (1943).

14. Leeman, E. R., "Some Underground Observations Relating to the Extent of the Fracture Zone Around Excavations in Some Central Rand Mines," *Assoc. Mine Mgrs. S. Africa* (1958).

15. Wayment, W. R. and D. E. Nicholson, "Improving Effectiveness of Backfill, *Mining Congr. J.* (Aug. 1965).

16. Roux, A. J. A. and H. G. Denkhaus, "An Investigation into the Problem of Rock Bursts, An Operational Research Project," "Part II, An Analysis of the Problem of Rock Bursts In Deep Level Mining," *J. Chem. Met. and Min. Soc. S. Africa*, **55**, No. 5 (1954).

17. Hill, F. G. and R. P. Plewman, "Implementing De-Stressing, with a Discussion on the Results so far Obtained," *Assoc. Mine Mgrs. S. Africa* (1957).

18. Duvall, W. I. and J. Devine, "Vibrations from Instantaneous and Millisecond-Delayed Quarry Blasts," *U.S. BurMines Rept. Invest.*, **6151** (1963).

19. Obert, L. and W. I. Duvall, "Micro-Seismic Method of Determining the Stability of Underground Openings," *U.S. BurMines Bull. 573* (1957).

20. Hodgson, E. A. and Z. E. Gibbs, "Seismic Research Program, Rock Burst Problem, Lake Shore Mines," *Dep. Mines Resources Surveys, Engr. Br., Rept.*, No. 14 (1944–1945).

21. Dunrud, C. R. and F. W. Osterwald, "Seismic Study of Coal Mine Bumps, Carbon and Emery Countries," *Trans. SME, AIME, No. 2* (1965).

22. Patching, T. H., "Investigation Related to the Sudden Outburst of Coal Gas," *Proc. Rock Mech. Symp.*, McGill University, 1962.

CHAPTER 20

ROCK BOLTING

20.1 INTRODUCTION

As noted in the Introduction to Part One, this volume is concerned with rock structures that are self-supporting, that is, structures in which the load due to the weight of the overlying rock and/or tectonic forces is supported by the sidewalls, pillars, and other unexcavated areas of rock. In general, openings in which a part of the applied loads are carried on artificial supports (in distinction to the natural support offered by the rock), such as chocks, sets, packs, linings, arches, etc., have not been considered. Also, the design and function of this class of support has not been treated. Although rock bolting is in a sense an artificial support, an exception is made to include this subject because this means of reinforcing or stabilizing the rock on the surface of underground openings* does not contribute appreciably to the support of the rock structure. Rather, the principal function of rock bolting is to reinforce and support partially detached ("loose"), thinly laminated, or otherwise incompetent rock that would be subject to failure under the action of gravity. Moreover, as opposed to other methods of reinforcing or supporting local areas of rock, rock bolting is superior in a number of ways, and to the extent that, in some instances, underground structures, that would not be serviceable if the surface rock were not stabilized and that would become too costly if supported with linings or sets, are made serviceable within economic limits through the use of rock bolting.

Some of the advantages of rock bolting are the following. (1) In coal mines or other flat-bedded mineral deposits, the cost of supporting mine

---

* Rock bolting is also used to support rock walls and slopes around surface excavations, such as road cuts and spillways. Most of the information in this chapter pertaining to the installation and servicing of rock bolts is equally applicable to this use of rock bolts.

roof with rock bolts is comparable with the cost of timber supports. However, rock bolting is more permanent, hence maintenance costs are reduced. (2) Because rock bolts are less subject to damage from blasting or other mining operations than metal or timber props, bolting can be installed close to the working face. Whereas timber support usually interfers with underground haulage and the movement of machinery, rock bolts do not. (3) In larger mine openings or in industrial or military installations, timber support is usually impractical and the cost of linings or steel sets or arches can become prohibitively costly. In this type of opening rock bolts may provide an effective means of reinforcing surface rock in both laminated and jointed, or fractured formations. The fact that rock bolting is relatively permanent and requires a minimum of maintenance makes this procedure especially suitable for all installations designed for a long lifetime.

20.2 REINFORCEMENT OF LAMINATED ROOF BY SUSPENSION

By far the most common use for rock bolts is to reinforce laminated roof in relatively flat-bedded coal, metallic, and nonmetallic mineral mines. The function of the bolts in producing this stabilizing action is due to a suspension effect, a friction effect, and, more generally, a combined suspension and friction effect. The suspension effect is illustrated by the following examples. Consider, first, a mine roof composed of an unsupported lamina underlying a very thick body of rock (Fig. 20.2.1a), and that this lamina is suspended by a two-dimensional array of uniformly spaced rock bolts. Although the probability of encountering such a roof structure in an underground opening is remote, this model serves to illustrate the suspension function of a roof bolt. For a given bolt-spacing and lamina thickness, this model sets an upper limit on the load carried by the bolts.

For example, consider a lamina of length L, width B, thickness t, and unit weight γ. If there are n_1 rows of bolts containing n_2 bolts per row the bolt spacing would be $L/(n_1 + 1)$ and $B/(n_2 + 1)$. If the lamina is completely suspended by the bolts, the load per bolt W_b is given by

$$W_b = \frac{\gamma t B L}{(n_1 + 1)(n_2 + 1)} \qquad (20.2.1)$$

Thus, if $L = 40$ ft, $B = 24$ ft, $t = 4$ ft, $\gamma = 156$ lb/ft^3 and the bolts are placed on 4×4 ft centers, that is, $n_1 = 9$ and $n_2 = 5$, the load per bolt would be approximately 10,000 lb, which is about 84% of the yield strength of a $\frac{3}{4}$ in. diameter mild steel bolt.

If the same lamina is held at its edges so that it acts as a clamped beam (Fig. 20.2.1b) and a sufficient number of bolts are installed and tightened so that over its entire surface the lamina is just brought in contact with the overlying body of rock, this case is identical to that of two clamped beams

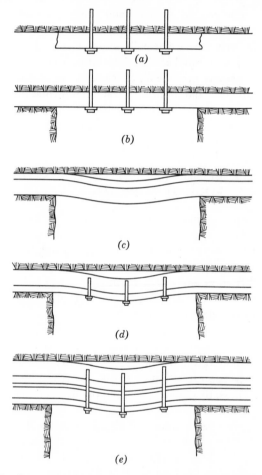

Fig. 20.2.1. Roof supported by bolting. (a) Supported. (b) Supported with fixed ends. (c) Supported from thick lamina. (d) Supported including thick lamina.

treated in Section 5.6, except in this instance the load transfer from the supported lamina to the thick member is effected through the rock bolts. Let q_1, E_1, and I_1, and q_2, E_2, and I_2 be the loads per unit length, moduli of elasticity, and moments of inertia of the two beams. If $q_1/E_1I_1 > q_2/E_2I_2$ and if Δq is the load per unit length transferred from one beam to the other,

Eq. 5.6.47 is applicable

$$\frac{q_1 + \Delta q}{E_1 I_1} = \frac{q_2 - \Delta q}{E_2 I_2} \tag{20.2.2}$$

Since $I_1 \propto t_1{}^3$, when t_1 becomes large, I_1 approaches ∞ and the left-hand side of Eq. 20.2.2 approaches zero; hence

$$\Delta q = q_2 \tag{20.2.3}$$

Since $q_2 = \gamma_2 t_2 B$, the weight of the lamina is

$$W = L q_2 = \gamma_2 t_2 B L$$

If this lamina is bolted on 4×4 ft centers, the load per bolt W_b will be

$$W_b = \frac{\gamma_2 t_2 B L}{(n_1 + 1)(n_2 + 1)} \tag{20.2.4}$$

That is, the bolt load in this case is just the same as for the case when the ends of the beam were unsupported (Eq. 20.2.1).

Next, consider a two-member roof model (Fig. 20.2.1c) in which the laminae act as clamped beams, and the contact between them is frictionless. If the order of the beams is such that the ratio of the load per unit length to the flexural rigidity of the lower member is greater than that for the upper member, the latter will rest on and be partially suspended by the lower member. The maximum deflection $\eta_{(\max)}$ and the maximum stress $\sigma_{(\max)}$ in the two-member unit can be determined by substituting $\bar{\gamma}$ and $\overline{Et^2}$, given by Eqs. 5.6.54, for γ and Et^2 in Eqs. 5.6.29. Thus

$$\eta_{(\max)} = \frac{L^4(\gamma_1 t_1 + \gamma_2 t_2)}{32(E_1 t_1{}^3 + E_2 t_2{}^3)}$$

$$\sigma_{\max(1 \text{ or } 2)} = \frac{L^2}{t_{(1 \text{ or } 2)}} \left(\frac{\gamma_1 t_1 + \gamma_2 t_2}{t_1 + t_2} \right) \tag{20.2.5}$$

where the subscripts 1 and 2 apply to the thicker and thinner laminae, and where $\sigma_{(\max)(1)}$ is the maximum stress in the lamina whose thickness is t_1, etc. The load transfer is given by Eq. 5.6.48,

$$\Delta q = \frac{q_2 E_1 I_1 - q_1 E_2 I_2}{E_1 I_1 + E_2 I_2} \tag{20.2.6}$$

If both the unit weight and modulus of elasticity of the laminae are approximately equal, as is often the case in successive laminae, the load transfer is from the thinner to thicker member and, correspondingly, the strain (and stress) will be decreased in the thinner member and increased in the thicker member. Obviously, for this case rock bolting cannot affect the load transfer; hence the suspension effect due to bolting is zero.

If the order of the two-members roof is reversed so that the ratio of the load per unit length to the flexural rigidity is greater for the upper member (Fig. 20.2.1d), without rock bolting the two members will flex independently and a separation between them will result. However, if rock bolts are installed and tightened so that the lower member and upper members just touch (in frictionless contact) and the bolt spacing is close enough together so that the distributed load is uniformly transferred along both the length and width of the beam, this case is identical to the two-member case given above, except that the load transfer is made through the bolts. Thus the suspension effect is given by Eq. 20.2.6 and the load per bolt (n bolts per row) is

$$W_b = \frac{\Delta q L}{n} \tag{20.2.7}$$

This example can be generalized for k beams for which

$$\frac{q_1}{E_1 I_1} > \frac{q_2}{E_2 I_2} > \cdots \frac{q_k}{E_k I_k}$$

by substituting

$$\overline{Et^2} = \frac{\left(\sum_i^k E_i t_i^3\right)}{\sum_i^k t_i}$$

and

$$\bar{\gamma} = \frac{\left(\sum_i^k \gamma_i t_i\right)}{\sum_i^k t_i} \tag{20.2.8}$$

for Et^2 and γ in Eq. 5.6.29 to derive the maximum deflection of the bolted assembly and the maximum stress in each member.

Generally a laminated roof is made up of a number of beams of different thicknesses that may occur in any possible arrangement, as illustrated in Fig. 20.2.1e. In this case without bolting some beams or combinations of beams may rest on other beams or combinations of beams, and between other beams or combinations of beams separations may occur. A general treatment of the suspension effect for this problem cannot be written, but by considering the ratios of the load per unit length to the flexural rigidity of individual members and subgroups to determine which will load (rest on) other members or groups and which will separate, the problem can be analyzed by the procedure given above.

The examples in Fig. 20.2.1 illustrate the suspension effect due to bolting. The two principal assumptions made were that the bolted beams were in frictionless contact, and that the load transfer through the bolts was equivalent to the transfer of distributed load when one beam rests on another. From an analysis of the bending strains in centrifugally loaded models of bolted roof laminae, Panek[1] found that bolts cannot be tightened so that the beams just touch without any frictional effect. Rather the efficiency of the suspension effect is lowered because of this cause. Panek also found that as the number of bolts per row was increased the load transfer rapidly approached that corresponding to a uniform distribution of transferred load. In fact, in bolted roof models in which the number of bolts per row was three or more (so that the spacing between bolts was $\frac{1}{4}L$ or less) the maximum bending strain and the maximum deflection very nearly corresponded to that for a uniform distribution of load transfer. Other results that indicated agreement between experiment and theory are the following. (1) The bending strain in a bolted unit increased proportionally to the applied centrifugal load, (2) For models in which

$$\frac{q_1}{E_1 I_1} > \frac{q_2}{E_2 I_2} \cdots \text{etc.}$$

and in which without bolting a separation of the laminae would normally occur, the bending strain was greatest in the lamina with the least ratio of distributed load to flexural rigidity, and vice versa. That is, for laminae of the same material the bending strain is greatest in the thinnest lamina and least in the thickest lamina. (3) For the same model when bolted the greatest bending strain develops in the thickest lamina and is least in the thinnest lamina. Thus in a bolted roof, the thickest lamina is usually the critical one, and failure of this member will cause a failure of the complete unit.

20.3   REINFORCEMENT OF LAMINATED ROOF BY FRICTION EFFECT

The reinforcement of a laminated horizontal roof by the friction effect results from the clamping action of tensioned rock bolts, which creates a frictional resistance to slip on the interface between laminae, thereby reducing the flexure and, correspondingly, the stress and strain in the laminae. For example, consider two clamped beams of the same material, the first composed of a single member of thickness t and the second composed of four laminae of thickness $t/4$ (Fig. 20.3.1). Also assume that there is not frictional resistance to slip on the interface between laminae. Equation 5.6.29 shows that for the single-member beam the maximum stress or strain is only $\frac{1}{4}$, and the maximum deflection $\frac{1}{16}$ that for the four-member beam. However, if the frictional resistance produced by tensioned bolting can completely prevent slip on the interface between laminae, the maximum stress, strain, and deflection in the two units would be equal.

The friction effect was investigated by Panek[2,3,4] by centrifugally loading laminated beams clamped at each end (Fig. 13.1.1). Laminae of the same material and thickness were used so that for all loading conditions each deflected equally without load transfer. Under this condition there can be

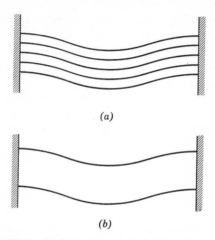

(a)

(b)

Fig. 20.3.1. Multiple (a), and single (b) lamina roof.

no suspension effect; hence only the tension applied to the bolts is effective in producing a reinforcing action. The various factors that affect the strain in a centrifugally loaded model are K, the centrifugal loading factor (Section 13.6); γ, the unit weight of the model material; L, the span; t, the lamina thickness; b, the spacing between rows of bolts; N, the number of bolts per row; F_b, the bolt tension; h, the bolt length; and E, the modulus of elasticity of the model material. The model strain, ϵ_x can be expressed as a function of dimensionless products of these variables (Section 13.2), that is,

$$\epsilon_x = f_1\left(\frac{K\gamma L}{E}, \frac{L}{t}, \frac{L}{b}, N, \frac{F_b}{EL^2}, \frac{h}{t}\right) \qquad (20.3.1)$$

The maximum bending strain in a clamped beam occurs at the clamped ends and is given by (Eq. 5.6.29)

$$\epsilon_{max} = \frac{\gamma L^2}{2Et} \qquad (20.3.2)$$

The maximum bending strain in the unbolted model specified above is also given by Eq. 20.3.2 but is designated by ϵ_{nfs} (where the subscript *nfs* designates no friction or suspension). If ϵ_f is the maximum strain in the bolted model, then the decrease in strain due to bolting $\Delta\epsilon_f$ is (where the subscript *f* indicates with friction)

$$\Delta\epsilon_f = \epsilon_f - \epsilon_{nfs} \qquad (20.3.3)$$

From a regressional analysis of the data from model tests, Panek[5] determined ϵ_{nfs} and ϵ_f as a function of the dimensionless products in Eq. 20.3.1. This relationship, expressed as the ratio of $\Delta\epsilon_f/\epsilon_{nfs}$, is

$$\frac{\Delta\epsilon_f}{\epsilon_{nfs}} = -0.265(bL)^{-\frac{1}{2}}\left[NF_b\frac{\left(\frac{h}{t}-1\right)}{\gamma}\right]^{\frac{1}{3}} \qquad (20.3.4)$$

The reinforcement factor *RF* due to the frictional effect is defined as

$$RF_t = \frac{1}{1 + \left(\frac{\Delta\epsilon}{\epsilon_{nfs}}\right)} \qquad (20.3.5)$$

Equations 20.3.3 and 20.3.4 serve as design equations for determining the degree of reinforcement produced by tension bolting equal-thickness laminae in either a model or the prototype structure. Because these equations do not contain any quantity related to the strength of the laminae material, they cannot be used as such to determine the strength of either a model or prototype roof. If $\Delta\epsilon_{max}/\epsilon_{nfs}$ in Eq. 20.3.4 is substituted in Eq. 20.3.5, the reinforcement factor can be expressed in terms of the parameters of the model or prototype. This expression is presented graphically in Fig. 20.3.2 in which it is assumed that all laminae have the same unit weight. Thus, if

Lamina thickness	$t = 3$ in.
Bolt length	$h = 4$ ft
Bolt tension	$F_b = 10,000$ psi
Number of bolts per row	$N = 3*$
Spacing of rows	$b = 4$ ft
Roof span	$L = 16$ ft*

the reinforcement factor *RF* is 1.9 as indicated by following the chart along the path *abcdefg*.

---

* For $L = 16$ ft, $N = 3$, and $b = 4$ ft, the bolt spacing is on 4-ft centers (with no bolts at supporting ends).

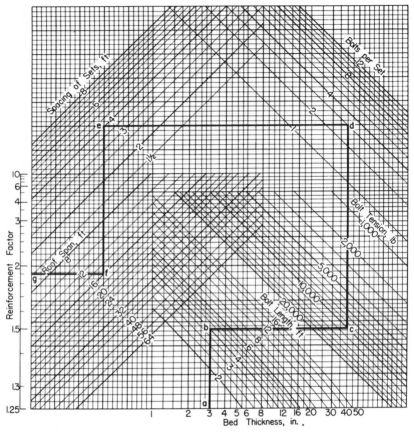

Fig. 20.3.2. Roof-bolting design chart for friction effect.

20.4 REINFORCEMENT OF LAMINATED ROOF BY COMBINED FRICTION AND SUSPENSION

Using a combined theoretical and experimental (model) approach, Panek[5] investigated the degree of reinforcement produced in laminated roof rock by the combined action of friction and suspension. In the most general case the thickness, unit weight, and modulus of elasticity of each lamina in a bolted unit will be different. Also, the friction between lamina will vary and the laminae may occur in any order. To write a design equation that can treat this general case would be difficult and not too practical because of the problems involved in determining the exact parameters of the prototype roof. However, by making several simplifying assumptions that were found to have only a negligible effect, a modified method was developed.

Table 20.4.1. Values of α for Various N

Bolts per Set, N	1	2	3	4	5	6	7
α	0.750	0.889	0.938	0.960	0.972	0.980	0.984

Considering the friction effect first, a regressional analysis of the data from centrifugal tests on models made of laminae of different thickness and unit weight showed that the average thickness t_{av} and the average unit weight γ_{av} can replace t and γ in Eq. 20.3.4 without serious error. Since $\Delta\epsilon_f/\epsilon_{nfs} = \Delta\sigma_f/\sigma_{nfs}$ (because the laminae are assumed to be perfectly elastic), Eq. 20.3.4 can be written*

$$\frac{\Delta\sigma_f}{\sigma_{nfs}} = -0.265(bL)^{-\frac{1}{2}}\left[NF_b\frac{\left(\dfrac{h}{t_{av}}-1\right)}{\gamma_{av}}\right]^{\frac{1}{3}} \qquad (20.4.1)$$

where $\Delta\sigma_f = \sigma_f - \sigma_{nfs}$.

The model tests also disclosed that the combined effects of friction and suspension are multiplicative, and can be expressed in the form

$$\sigma_{fs} = \sigma_{nfs}\left(1 + \frac{\Delta\sigma_f}{\sigma_{nfs}}\right)\left(1 + \frac{\Delta\sigma_s}{\sigma_{nfs}}\right) \qquad (20.4.2)$$

where $\Delta\sigma_s = \sigma_s - \sigma_{nfs}$, and the subscripts s, fs, f, nfs denote the effects due to suspension, friction and suspension, friction, and no friction or suspension, respectively. Panek also determined that

$$\frac{\Delta\sigma_s}{\sigma_{nfs}} = \alpha C u_i \qquad (20.4.3)$$

where α is a constant depending on the bolt spacing, that is, the number of bolts per row N, and C is a constant depending on the number of laminae in a bolted unit. The value of α and C are given in Tables 20.4.1 and 20.4.2. The quantity $1 + u_i$ is the ratio of the flexural rigidity of the

Table 20.4.2. Values of C for Various Number of Strata

Number of Strata	3	4	5	6	8	10	12
C	0.953	0.900	0.865	0.838	0.800	0.772	0.751

* Panek[6] pointed out that the numerical constant multiplying the right-hand side of Eq. 20.4.1 was determined on the basis of a coefficient of friction between laminae of $F_r = 0.7$. If the coefficient of friction is known to be otherwise, the multiplier should be $0.375F_r$.

ith lamina to the average flexural rigidity of all laminae, that is, for a bolted unit of i laminae

$$1 + u_i = \frac{(\gamma_1 t_1/\gamma_i t_i) + (\gamma_2 t_2/\gamma_i t_i) + \cdots + (\gamma_j t_j/\gamma_i t_i)}{(E_1 t_1^3/E_i t_i) + (E_2 t_2^3/E_i t_i) + \cdots + (E_j t_j^3/E_i t_i)} \quad (20.4.4)$$

If all laminae have equal E and γ (which is a reasonable approximation for bedded sedimentary rocks such as those commonly found over coal deposits Eq. 20.4.4 reduces to

$$1 + u_i = t_i^2 \frac{\sum t_j}{\sum t_j^3} \quad (20.4.5)$$

Thus, if Eqs. 20.4.1, 20.4.3, and 20.4.4, or 20.4.5 are evaluated in terms of the parameters of a prototype roof, the reinforcement can then be determined from Eq. 20.4.2.

As an example,* consider a twelve-lamina prototype roof (similar to the seven-laminae model, Fig. 13.1.1) in which the laminae are made from the same rock (that is equal E's and γ's), and the coefficient of friction between laminae is 0.7. The roof dimensions and other parameters are: roof span, $L = 240$ in.; number of bolts per row, $N = 4$; bolt tension, $F_b = 5000$ psi; spacing between rows of bolts, $b = 48$ in.; the total thickness of bolted strata, $h = 48$ in.; the average lamina thickness, $t_m = h/12 = 4$ in.; and the unit weight, $\gamma = 0.09$ lb/in.[3]. The values of u_i and the bending stresses for the laminae are given in Table 20.4.3.

Comparison of columns four and five (Table 20.4.3) shows that the bending stress in thin laminae is reduced by the transfer of a part of their body load to thicker laminae and, conversely, the bending stress in thick laminae is increased by the transfer of body load from thin laminae. If the flexural strength of the roof rock is assumed to be 800 psi, which is a representative value for sedimentary rocks, the lower eight laminae would fail without bolting. With bolts tightened just enough to prevent laminae separation, the suspension effect reduces the bending stresses in all laminae below the point of failure, and if the bolts are tensioned, a further reduction of approximately 35% is achieved, as indicated in column 6. The safety factor for this tension-bolted unit, which is given by the ratio of the rock strength to the maximum working stress in the unit, is $800/338 = 2.4$.

Although it may be difficult to determine the various parameters of an in situ roof structure so that this procedure can be applied to ascertain the margin of safety of a bolted roof, these combined theoretical and experimental analyses serve to illustrate how reinforcement in a bolted

* From Panek.[6]

Table 20.4.3

Lamina No.	t_i, in.	u_i*	$\Delta\sigma_s/\sigma_{nfs}$†	σ_{nfs},‡ psi	σ_s,§ psi	σ_{nfs},‖ psi
12	7	0.013	0.009	370	373	241
11	3	−0.814	−0.587	864	357	231
10	3	−0.814	−0.587	864	357	231
9	12	1.977	1.425	216	524	338
8	3	−0.814	−0.587	864	357	231
7	3	−0.814	−0.587	864	357	231
6	3	−0.814	−0.587	864	357	231
5	3	−0.814	−0.587	864	357	231
4	3	−0.814	−0.587	864	357	231
3	3	−0.814	−0.587	864	357	231
2	3	−0.814	−0.587	864	357	231
1	2	−0.917	−0.661	1296	439	284

* Calculated from Eq. 20.4.5.
† Calculated from Eq. 20.4.3.
‡ Calculated from beam theory, $\sigma_{nfs} = L^2/2t_i$.
§ Calculated from $\sigma_s = \sigma_{nfs}(1 + \Delta\sigma_s/\sigma_{nfs})$; evaluation of Eq. 20.4.2 for suspension effect only.
‖ Calculated either from Eq. 20.4.2 with $\Delta\sigma_f/\sigma_{nfs}$ calculated from Eq. 20.4.1.

roof is achieved by suspension and friction effect. They also bring to attention two important points: first, that a bolted roof composed of laminae of approximately equal thickness will be reinforced only by the friction effect, whereas bolted units containing one or more thick members will be reinforced by both suspension and friction; and second, rock bolts must be tensioned to develop a friction effect. The bolt tension must be sufficient to close or prevent strata separation and, in addition, to produce a frictional resistance to bedding plane slip.

20.5 REINFORCEMENT OF FRACTURED AND JOINTED ROCK

The treatment of flat-laminated roof (Sections 20.2, 20.3, and 20.4) was considered on the basis of beam theory. In this treatment it was assumed that the roof laminae were not intersected by joints, cutters, fractures, or other planes of weakness. If this type of roof is intersected by occasional joints running parallel to the span, beam theory is still valid; but if the planes of weakness run perpendicular to the span, the bending moments of the roof laminae will be strongly affected and beam theory is not applicable, although the latter type of roof can be stabilized by suspension, provided that the rock bolts can be anchored in a thick competent lamina. For this case the bolting design should be based on dead weight loading (Eq. 20.2.1).

Igneous and many metamorphic rocks are not laminated but generally contain one or more sets of joints, which may be completely bonded, partially bonded, or competely unbonded. The spacing and spatial orientation of these sets of joints will vary from point-to-point. Also, these rocks may contain other planes of weakness such as shear zones and faults. When an underground excavation is created by conventional means, that is, by drilling and blasting, the joints and faults in the proximity of the excavation are loosened by blasting vibration and, in addition, randomly oriented fractures are created. Core drilling, stress determinations, and seismic tests made in the walls of underground openings indicate that these fractures and loosened joints are present to a depth ranging from 3 to 6 ft or more from the opening surface, thus creating a zone of relatively incompetent or "loose" rock surrounding the opening as illustrated in Fig. 20.5.1.

If an excavation such as that shown in Fig. 20.5.1 is in a gravity stress field or if the stress field is known from measurement, the distribution along any line normal to the surface can be determined (or reasonably approximated) from theory, provided that (among other things) the rock is assumed to be homogeneous.

However, stress measurements made in the wall rock of underground openings show that the tangential stress in the plane normal to the axis of the opening σ_θ does not correspond to that predicted by theory; but for the case of a gravity stress field, the distribution is more like that indicated in Fig. 20.5.1. Thus on the horizontal line normal to the axis of the opening, the maximum tangential stress is not at the surface but rather at a point approximately corresponding to the limit of the loosened and fractured zone. The relaxation of the near-surface stress along this line could result from creep in the rock, but for the more elastic rocks, such as granite, it is more likely that it occurs as a consequence of the near-surface fracture and loosening of the joints. Also, stress measurements have indicated that the tangential stress around the cross-sectional periphery of an opening is more uniformly compressive than predicted by theory, which is equivalent to assuming that the stress field is more nearly hydrostatic than a strictly gravitational stress field. As might be expected in this loosened and fractured rock, the measured stress often varies erratically from point-to-point; for example, see Fig. 14.5.1.

The complex state of stress and the lack of homogeneity in fractured and jointed rock are such that an analytic treatment of the reinforcement furnished by rock bolting is not feasible. Lang[7] furnished some insight into

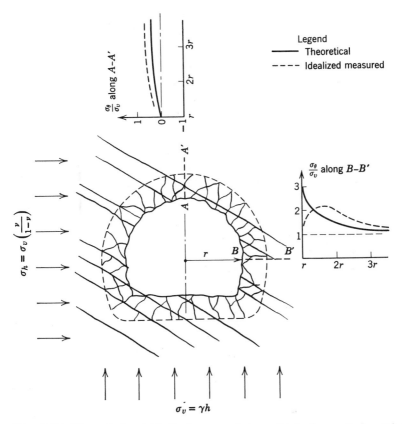

Fig. 20.5.1. Theoretical and idealized measured stress distribution around opening in gravity stress field ($v = 0.25$).

the manner in which this reinforcement is achieved by considering the forces acting across joint planes. In Fig. 20.5.2a, if F_b is the bolt load normal to a unit area of the fracture plane, F_p is the force parallel to the surface acting on the same unit area, $\tan \phi$ is the coefficient of friction of the joint plane, and α is the angle the normal to the joint plane makes with the surface, then the force acting parallel and normal to the joint plane is $F_p \sin \alpha$ and $F_b + F_p \cos \alpha$. The condition necessary for stability is

$$\frac{F_p \sin \alpha}{F_b + F_p \cos \alpha} < \tan \phi \qquad (20.5.1)$$

which can be written as

$$\frac{F_b}{F_p} > \sin \alpha (\cot \phi - \cot \alpha) \qquad (20.5.2)$$

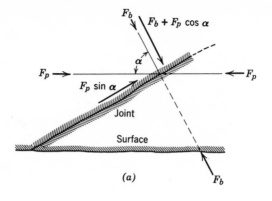

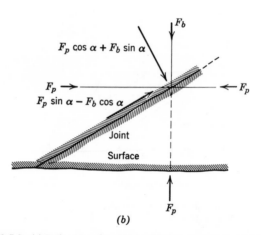

Fig. 20.5.2. (*a*) Bolt normal to joint. (*b*) Bolt normal to surface.

If $\alpha < \phi$ no bolt is necessary, as can be seen by letting $F_b = 0$ in Eq. 20.5.1. On the other hand, F_p must be very small if bolting is to be effective in stabilizing the joint. For example, for a coefficient of friction of 0.7, $\tan^{-1} \phi = 35°$. If $\alpha \geq 40°$, $F_b/F_p \leq 0.15$. For bolts installed on 4-ft centers and tensioned to 14,000 lb, the force per unit area on the joint plane F_b would be only about 6 psi, and hence $F_p \leq 40$ psi.

If the bolt is installed perpendicular to the surface (Fig. 20.5.2*b*), the condition necessary for stability is

$$\frac{F_p \sin \alpha - F_b \cos \alpha}{F_p \cos \alpha + F_b \sin \alpha} < \tan \phi \qquad (20.5.3)$$

but, as in the previous example, bolting is not effective unless F_p is small.

Whereas the friction effect in fractured and jointed rock produces only minor reinforcement, suspension can be very effective in stabilizing this type of rock. A 29/32 in. D mild steel bolt (yield load = 20,000 lb) placed on 4-ft centers and anchored to withstand the full yield strength of the bolt will support an 8.5-ft thick layer of completely detached rock. If a safety factor of 2 is allowed for loss of anchorage, this system of bolting would still suspend over 4 ft of completely detached rock, which is about the average depth from the surface that most rocks in this category are loosened and fractured by mining operations. Since this thickness of rock is never completely detached, at least at the time of mining, and as some degree of reinforcement is afforded by friction, a bolting design based on suspension, that is, dead-weight loading, would be on the conservative side. To produce the friction effect the bolts must be tensioned, preferably at the time of installation and as close to the working face as possible.

20.6 BOLTING MATERIALS

Rock bolts are produced by a number of manufacturers in a variety of forms. The principal features of representative types are described below.

The slotted-type rock bolt assembly (Fig. 20.6.1) consists of a steel bolt threaded on one end and with a swaged or flame-cut slot in the other end, a wedge, a bearing plate, and a nut and washer. The specifications for two typical bolts are given in Table 20.6.1. Typical wedges are $5\frac{1}{2}$ in. long and taper from $\frac{3}{4}$ to $\frac{1}{16}$ in.

The bolt assemblies are installed by first drilling a $1\frac{1}{4}$-in. diameter hole to a depth 2 in. less than the bolt length. The wedge is placed in the slot and the wedge-slot end of the bolt rammed into the bolt hole by hand, thereby securing an initial anchorage. The bolt is then driven over the wedge by applying an impact hammer or stopper to the threaded end of the bolt. The driving time should be from 20 to 30 sec. The bearing plate, washer,

Table 20.6.1. Slotted-Type Bolt Specifications

Bolt Type	A	B
Diameter, in.*	29/32	1
Length, ft*	2–10	2–10
Slot type	Swaged	Flame-cut
Slot length, in.	6	6
Thread type	Rolled	Cut
Thread length, in.	4	4
Yield load, lb	20,000	20,000

* Other diameters and lengths are available.

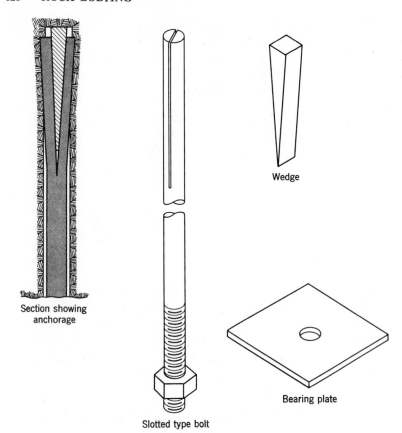

Section showing
anchorage

Wedge

Slotted type bolt

Bearing plate

Fig. 20.6.1. Slotted-type rock bolt.

and nut are placed on the threaded end and the nut tightened with an impact wrench to a torque of approximately 260 ft lb. This produces a normal bolt tension of 10,000 lb (provided that a satisfactory anchorage can be obtained, see Section 20.7).

Except for drilling the bolt hole to the proper depth, the procedure for installing slotted-type bolts is not critical. However, the bolt will drive over the wedge easier and optimum anchorage will be obtained if the hole diameter is not less than $1\frac{1}{4}$ in. or greater than $1\frac{5}{16}$ in.

Slotted-type bolts are stronger and subject to less slip (Section 20.7) than expansion shell-type bolts. They are considered the best general-purpose bolt for installation* in permanent underground openings such as storage chambers, hydroelectric stations, and hardened military bases.

---

* However, in areas requiring special treatment, such as faults and shear zones, grouted-type bolts may be more effective.

Expansion-Type Bolts

The expansion-type bolt assembly consists of a headed bolt (Fig. 20.6.2*a*) which is threaded onto an internally tapered one- or two-piece shell containing a mating tapered plug (nut) (Fig. 20.6.2*b*). Tightening the bolt causes the shell to expand and effect an anchor. Most expansion-type bolts

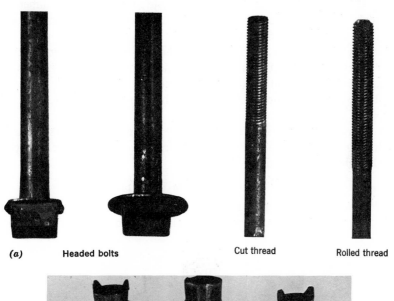

(a)　　　**Headed bolts**　　　　　　Cut thread　　　　Rolled thread

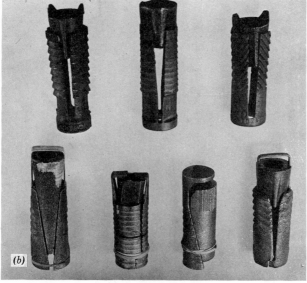

(b)

Fig. 20.6.2. (*a*) Expansion-type rock bolts. (*b*) Expansion shells and plugs.

are made to fit into holes ranging in diameter from $1\frac{3}{8}$ in. to $2\frac{1}{2}$ in. The length of the hole is not critical provided it is longer than the bolt. Two-bolt diameters are in common use; a $\frac{3}{4}$ in. D mild steel bolt, which with rolled threads has a yield load of 14,000 lb, and a $\frac{5}{8}$ in. D alloy steel bolt which also has a yield load of 14,000 lb.

In the installation of expansion-type bolts, the bolt hole diameter is critical and should not exceed the manufacturer's recommended value (usually the ideal diameter is one in which the shell will just pass). The assembly is placed in the hole and tightened with an impact wrench. The maximum torque that should be applied to a $\frac{3}{4}$-in. D bolt is 200 ft lb, which produces a bolt tension of approximately 8000 lb. For a $\frac{5}{8}$ in. D bolt, the maximum torque is 175 ft lb, which produces a bolt tension of 7000 lb.

Expansion-shell bolts are especially well adapted to routine (systematic) bolting in mines, although they are not as strong and are more subject to anchor slip than slotted-type bolts.

Williams Groutable Rock Bolt

The Williams groutable rock bolt consists of a hollow-cored high-strength alloy steel reinforcing bar, threaded on each end. An expansion

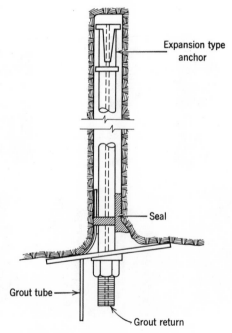

Fig. 20.6.3. Williams bolt installation.

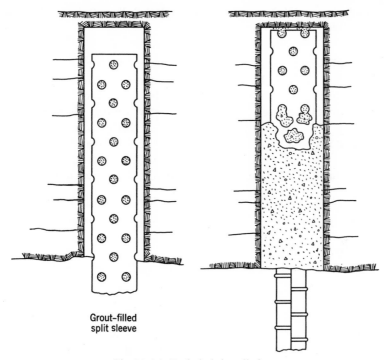

Grout-filled
split sleeve

Fig. 20.6.4. Perfo bolt installation.

shell anchor is attached to one end, and a nut and bearing plate to the other end (Fig. 20.6.3). To install, the bolt and anchor are first inserted in the bolt hole and torqued to approximately 50 ft-lb to set the anchor. The bolt is then removed (leaving the anchor in the hole), a plastic grout tube is taped to the bearing plate end of the bolt, and the assembly rethreaded into the anchor. Next, the hole is sealed at the collar with a quick setting (2 min) cement (Fig. 20.6.4). A plastic grout is then injected into the grout tube until the hole is filled and the grout returns from the end of the hollow bolt. Note that the hollow bar prevents air from being trapped in the end of the bolt hole. The bolt is then tensioned by torquing to approximately 250 ft lb per square inch of cross section, and the grout allowed to harden.

With this system a tensioned and fully grouted bolt installation is achieved, a procedure that is desirable for long-life installations. In addition, grouting improves the anchorage. A 1-in. D Williams bolt when fully grouted in a $1\frac{5}{8}$ in. bolt hole will develop an ultimate strength of 40,000 lb and carry a safe working load of 30,000 lb. Standard bolt lengths are 8, 10, and 15 ft; couplings can be employed to increase the bolt length to multiples or combinations of these lengths. Bolts with larger diameters are available.

Perfo Rock Bolt

The Perfo system consists of perforated steel half-sleeves that are filled with mortar, tied together, and inserted in the drill hole. A steel bar which may be a reinforcing bar, a smooth bar, or a threaded bar is then pushed through the sleeve, forcing the mortar through the perforations, completely filling the entire drill hole. The Perfo system provides a distributed anchorage along the entire length of the hole, thereby making it possible to obtain anchorage in soft rocks. This procedure will provide a nontensioned installation; if a tensioned bolt is required, a threaded bar

Table 20.6.2. Perfo Sleeve, Reinforcing Bar, and Drill Hole Sizes

Bar diameter, in.	Drill hole, in.	Perfo sleeve, in.
$\frac{3}{4}$ (No. 6 bar)	$1\frac{1}{4}$	$1\frac{1}{16}$
$\frac{7}{8}$ (No. 6 bar)	$1\frac{3}{8}$	$1\frac{1}{16}$
1 (No. 7 bar)	$1\frac{1}{2}$	$1\frac{1}{4}$

can be pushed into a short length of Perfo sleeve placed in the back end of the drill hole. The bolt is tensioned after the mortar has hardened. Perfo sleeves come in two sizes, $1\frac{1}{4}$ and $1\frac{1}{16}$ in. ID, and in lengths up to 20 ft.* Table 20.6.2 gives the sleeve, bar, and hole-size combinations.

The Perfo system is satisfactory for long-life installations especially in soft rocks that do not provide a normal anchorage for conventional bolts or in harder but jointed or fractured rocks, where anchorage over the length of the hole is desirable.

20.7   ANCHORAGE TESTING

In the theoretical treatment of the reinforcement of laminated and fractured and jointed rock in Sections 20.6.2, 20.6.3, and 20.6.4, it was assumed that rock bolts would develop and maintain a required load-bearing capacity. This assumption in turn requires that the bolt can be satisfactorily anchored in the bolt hole. A satisfactory anchorage is one that will provide a maximum bearing capacity load with a minimum displacement of the bolt head with respect to the initial point of anchorage as the bolt load is increased. In practice, the bolt-head displacement is the result of the following factors: an elastic elongation of the bolt (for a given bolt length, which can only be reduced by using bolts with larger cross-sectional areas); a displacement (slip) of the anchor in the hole as the

* Up to the full strength of the reinforcing bar develops as the mortar curves.

bolt is loaded; and, in expansion-type bolts, a displacement of the plug within the shell as the bolt is loaded.

If, after installing and tensioning a bolt in a laminated roof, head displacement occurs as a consequence of an additional load being imposed on the bolt,* a displacement of the roof laminae will result which will cause increased bending stresses in the laminae and possibly a further rupture of the partial bonds between laminae. In a fractured or jointed rock bolt-head displacements can also cause a relative movement among joints and fractures and further loosening of the near-surface rock.

To check effectiveness of a bolt installation it is necessary only to measure the bolt-head displacement as a function of the applied bolt load. This measurement can be made with the equipment shown in Fig. 20.7.1.

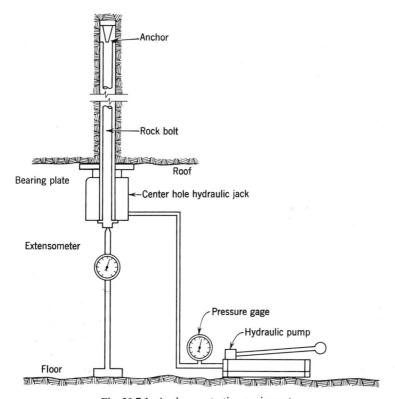

Fig. 20.7.1. Anchorage testing equipment.

* Bolt-head displacement may also result from a time-dependent anchor slip (creep). This effect has not been investigated to the extent that preventive measures can be recommended. Hence, it will be assumed that an anchorage that minimizes the bolt-head displacement as the bolt is loaded will also produce the least time-dependent slip.

The procedure for slotted and expansion-type bolts differs in only one respect. With the slotted-type bolt, the bolt is first anchored in the usual manner by driving it over the wedge. Then a coupling and extension is threaded onto the bolt so that it will extend through the centerhole hydraulic jack. A nut on the free end of the extension holds the jack against the bearing plate. With the expansion type, the assembly is first anchored and tensioned in the prescribed manner. The bolt is then removed and replaced with a longer (but otherwise identical) bolt so that the

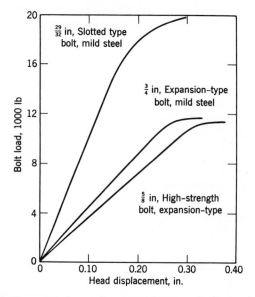

Fig. 20.7.2. Representative anchorage performance in firm rock (after Panek.[8])

jack can be included. *Neither the slotted nor expansion bolts should be pretensioned before applying the load with the jack.* The test consists of loading the bolt with the jack and measuring the bolt-head displacement with the extensometer. Typical results for properly installed bolts are given in Fig. 20.7.2.[8] As previously noted, the 1 in. slotted type bolt is superior to either the $\frac{3}{4}$ in. or $\frac{5}{8}$ in. high-strength bolt because it will sustain a greater bolt load, and because bolt-head displacement per unit load is smaller.

20.8 ROCK BOLT INSTALLATION

Although a satisfactory anchorage can be developed in most rock types, some care must be exercised in the installation of rock bolts to assure an optimum anchorage and bearing capacity. The first recommended step is to make a stratoscope (Section 9.9) survey of the immediate roof or area to be bolted. This is especially desirable in virgin areas in which no previous knowledge is available regarding the structural composition of the rock. In bedded and foliated rock, anchorage should be made in a relatively thick lamina and, if possible, near the top of the lamina. Also, if one or more thick lamina can be included within the bolted unit, the strength of the bolted unit may be significantly greater. For the same bolting cost (the bolting cost is roughly proportional to length of bolt per unit area), a 6 ft-bolt length on 5 ft centers may give a stronger bolted unit than a 4 ft bolt on 4 ft centers if the additional 2 ft of bolt length permits anchorage in or the inclusion of a thick lamina. In jointed rock a stratoscope examination may assist in determining the altitude of the joint planes or possibly the weakest set of joints so that the bolts may be installed in a direction approximately perpendicular to this plane. Also, weathered zones or areas of incompetent rock may be detected.

In the installation of expansion shell bolts, the length of the bolt hole is not critical, provided, of course, that it is longer than the bolt. However, for slotted-type bolts the hole length must be such that when the bolt is properly driven over the wedge enough of the threaded end of the bolt should extend from the hole to allow for the bearing plate and nut (usually 2 to 4 in.).

In the installation of either expansion shell and slotted-type bolts, the diameter of the bolt hole is critical. With slotted-type bolts if the hole is too large, the bolt may be driven over the complete length of the wedge without expanding the prongs enough to produce a sufficient contact pressure (and gouge) on the wall of the hole. If the hole is undersize, a proper gouge will not develop and a loss of anchorage capacity will result. Expansion-shell bolts will not anchor properly in an oversize hole because the wedge may be drawn completely within the shell without producing a sufficient shell-to-hole contact pressure. However, for this bolt type the minimum hole size is not critical; hence, the best practice is to keep the hole size just large enough to permit insertion of the bolt Because the hole diameter is critical in the routine installation of bolts, the hole diameters should be checked periodically with a hole gage to insure proper size.

To prevent or reduce the development of bending stress in laminated roof, the shearing of partial bonds between lamina, and time-dependent displacements, rock bolts should be installed as close to the working face as practical and immediately tensioned. Although the suspension effect is not dependent on pretensioning bolts, it has been the common experience that a better reinforcement is obtained in all rock bolt installations if the bolts are pretensioned. In routine installations the bolts are tightened with a pneumatic impact wrench. These wrenches can be preset to tighten the bolt to a specified torque. The recommended upper limits on the torque that should be applied to mild steel bolts are for a $\frac{5}{8}$ in. D bolt, 175 ft lb; for a $\frac{3}{4}$ in. D bolt, 200 ft lb; and for a $\frac{29}{32}$ in. D bolt, 300 ft lb. Since 1 ft lb torque will produce a bolt tension of approximately 40 lb, these recommended torques correspond to bolt tensions of: for a $\frac{5}{8}$ in. D bolt, 6000 lb; for a $\frac{3}{4}$ in. D bolt 8000 lb; and for a $\frac{29}{32}$ in. bolt, 12,000 lb. The bolt load versus head displacement curves for $\frac{29}{32}$ in. D and $\frac{3}{4}$ in. D mild steel bolts (Fig. 20.7.2) show these tensions to be well within the yield load of the corresponding size bolt. Torquing above these specified values has been found to cause the bolts to yield in shear with an elongation of the bolt and a loss in tension.

For the few rock types in which a satisfactory anchorage cannot be developed (due to anchor slip) in the conventional manner, an improved anchorage can sometimes be achieved by grouting the anchor end of a slotted-type bolt (including the wedge) into the hole. After the grout has

Fig. 20.8.1. Wire mesh secured with rock bolts. (Courtesy U.S. Air Force.)

Fig. 20.8.2. Light channel secured with rock bolts. (Courtesy U.S. Air Force.)

set, the bolt is tensioned in the usual way. Quick setting grouts (lumnite) will expedite this process. Also, epoxy cements have been found to produce a superior quick setting bolt-to-rock bond of high strength.[9]

In jointed or thin-bedded formations, there is a tendency for the surface rock between bolts to loosen and slough in the course of time. Generally this slough will not affect the over-all stability of the bolted area, although in some instances erosion of the surface will continue until bolt tension is lost. In any event, loose rock falling from the surface is a hazard and special surface treatment is required. One such treatment is to place a wire mesh under the bearing plates as is illustrated in Fig. 20.8.1. A 2 in. number 8 wire mesh bolted on 4 ft centers will hold at least 12 in. of broken rock, and in some instances the constraint offered by this broken rock will arrest further erosion. Another procedure is to bolt light channel iron to the roof as shown in Fig. 20.8.2. Also lagging secured by heavier channel may be bolted to the roof although the bolts cannot be properly tensioned if high compliance materials (wood lagging) is employed.

REFERENCES

1. Panek, L. A., "The Effects of Suspension in Bolting Bedded Mine Roof," *U.S. BurMines Rept. Invest.*, **6138** (1962).
2. Panek, L. A., "Theory of Model Testing as Applied to Roof Bolting," *U.S. BurMines Rept. Invest.*, **5154** (1956).
3. Panek, L. A., "Design of Bolting Systems to Reinforce Bedded Mine Roof," *U.S. BurMines Rept. Invest.*, **5155** (1956).
4. Panek, L. A., "Principles of Reinforcing Bedded Mine Roof with Bolts," *U.S. BurMines Rept. Invest.*, **5156** (1956).
5. Panek, L. A., "The Combined Effects of Friction and Suspension in Bolting Bedded Mine Roof," *U.S. BurMines Rept. Invest.*, **6139** (1962).
6. Panek, L. A., "Design for Bolting Stratified Roof," *Trans. SME-AIME*, **229** (1964).
7. Lang, T. A., "Theory and Practice of Rock Bolting," *Trans. SME-AIME*, **220** (1961).
8. Panek, L. A., "Anchorage Characteristics of Roof Bolts," *Mining Cong. J.* (Nov. 1957).
9. McLean, D. C., "Use of Resins in Mine Roof Support," *Mining Eng.* (Jan. 1964).

AUTHOR INDEX

SUBJECT INDEX